Advances in Refining Catalysis

Advances in Refining Catalysis

Edited by
Deniz Uner

Professor, Chemical Engineering
Middle East Technical University, Ankara, Turkey
&
Visiting Professor, Chemical and Biomolecular Engineering
University of California, Berkeley, CA

CRC Press is an imprint of the
Taylor & Francis Group, an **informa** business

CRC Press
Taylor & Francis Group
6000 Broken Sound Parkway NW, Suite 300
Boca Raton, FL 33487-2742

First issued in paperback 2019

ISBN-13: 978-1-4987-2997-0 (hbk)
ISBN-13: 978-0-367-87360-8 (pbk)

Library of Congress Cataloging-in-Publication Data

Names: Üner, Deniz, editor.
Title: Advances in refining catalysis / edited by Deniz Üner.
Description: New York : Routledge, [2017] | Includes bibliographical references and index.
Identifiers: LCCN 2016029763 | ISBN 9781498729970 (hardback)
Subjects: LCSH: Catalysis. | Catalytic reforming. | Chemical engineering.
Classification: LCC TP156.C35 A35 2017 | DDC 660/.2995--dc23
LC record available at https://lccn.loc.gov/2016029763

Visit the Taylor & Francis Web site at
http://www.taylorandfrancis.com

and the CRC Press Web site at
http://www.crcpress.com

Contents

Preface

Oil refining is not a trivial technology. The raw material is complex in composition and prone to price fluctuations, driven by global and local politics. The technology developments in this field are rather slow to take shape and take effect due to the large scales of economies involved. The investments are very large, and it takes several years to pay back. Therefore, technology changes are not as quickly reflected as in other industries, such as automotive and electronics. The economics of the refinery operations depend mainly on the efficiency of the design, skills of the engineers in the field, and last but not the least, the quality of the catalysts used. This volume intends to bring together some of the developments in the refining catalysis taking place when the oil price fluctuations were at their peak. While the evolving and emerging trends in selected refinery operations are emphasized, two different perspectives on the concept of biorefinery are also presented. In Chapter 10, a thorough review of the recent literature on catalysis for biomass conversions is presented. In Chapter 11, a perspective for integrating the biorefineries into existing value chain from the point of view of catalysis can be found.

Deniz Uner
Berkeley, California

Acknowledgments

This book has its origin in the Refinery Engineer Training Program (RAYEP) and the related project that was conducted under its auspices between the Middle East Technical University Chemical Engineering Department and TUPRAS—Turkish Petroleum Refineries Co. I would like to express my thanks to general directors Yavuz Erkut, who passed the torch (or the flare) at the end of 2015 to Ibrahim Yelmenoglu, and on their behalf to all who contributed to make this volume possible. My appreciation extends to all members of the METU RAYEP team, particularly Prof. Erdogan Alper, from whom I learned immensely about refining and refinery engineering.

This volume came to shape during my sabbatical at the Chemical and Biomolecular Engineering Department of the University of California Berkeley. First and foremost, I am grateful to my host and the Department Chair Prof. Jeffrey A. Reimer for his invitation and making my sabbatical a truly enriching experience. I thank Prof. Enrique Iglesia for stimulating discussions in and outside the classrooms of the Chemical Reaction Engineering and Catalysis courses during the 2015–2016 academic year, Prof. Alexander Katz for his contagious enthusiasm in catalysis, Prof. Alexis T. Bell for his refined wisdom, and last but not the least Prof. John M. Prausnitz, who is a great mentor in his gentle presence. I also would like to thank Prof. Halil Kalipcilar, the chair of the Chemical Engineering Department at the Middle East Technical University, for making my sabbatical at UC Berkeley a pleasant journey.

It is my hope that the positive changes brought about in the twentieth century by the internal combustion engine technology and its driving force—oil—will be surpassed by the advances-in-the-making of the twenty-first century. Our efforts are for the children of our children.

Contributors

Hayim Abrevaya
UOP Research
Des Plaines, Illinois

Deniz Onay Atmaca
R&D Center
Turkish Petroleum Refineries Co (TÜPRAŞ)
Kocaeli, Turkey

Volkan Balci
Department of Chemical and Biological Engineering
Koç University
and
Koç University TÜPRAŞ Energy Center (KUTEM)
Istanbul, Turkey

and

R&D Center
Turkish Petroleum Refineries Co (TÜPRAŞ)
Kocaeli, Turkey

Ayşegül Bayat
R&D Center
Turkish Petroleum Refineries Co (TÜPRAŞ)
Kocaeli, Turkey

Jeffery C. Bricker
UOP Research
Des Plaines, Illinois

Maureen Bricker
UOP Development
Des Plaines, Illinois

Peter Broadhurst
Johnson Matthey
Billingham, UK

Basar Caglar
Middle East Technical University
Chemical Engineering Department
Ankara, Turkey

Gabriele Centi
University of Messina
and
ERIC aisbl
and
CASPE/INSTM
Department MIFT
Section Industrial Chemistry V.le F. Stagno D'Alcontras 31
Messina, Italy

Elif Ispir Gurbuz
Rennovia Inc.
Santa Clara, California

Arzu Kanca
Ataturk University
Chemical Engineering Department
Erzurum, Turkey

Paola Lanzafame
University of Messina
and
ERIC aisbl
and
CASPE/INSTM
Department MIFT
Section Industrial Chemistry V.le F. Stagno D'Alcontras 31
Messina, Italy

Raimon Perea Marin
Johnson Matthey
Billingham, UK

Siglinda Perathoner
University of Messina
and
ERIC aisbl
and
CASPE/INSTM
Department ChiBioFarAm
Section Industrial Chemistry V.le F. Stagno D'Alcontras 31
Messina, Italy

James E. Rekoske
UOP CTO
Des Plaines, Illinois

İbrahim Şahin
Department of Chemical and Biological Engineering
Koç University
and
Koç University TÜPRAŞ Energy Center (KUTEM)
Istanbul, Turkey

Nazife Işık Semerci
Energy and Materials Engineering
Ankara University
Ankara, Turkey

Jumal Shah
Johnson Matthey
Billingham, UK

Melek Bardakcı Türkmen
R&D Center
Turkish Petroleum Refineries Co (TÜPRAŞ)
Kocaeli, Turkey

Alper Uzun
Department of Chemical and Biological Engineering
Koç University
and
Koç University TÜPRAŞ Energy Center (KUTEM)
Istanbul, Turkey

Burcu Yüzüak
R&D Center
Turkish Petroleum Refineries Co (TÜPRAŞ)
Kocaeli, Turkey

Xin Zhu
UOP Engineering
Des Plaines, Illinois

Section I

Refining Technologies Outlook

1 Advances in Refining Technologies

James E. Rekoske, Hayim Abrevaya, Jeffery C. Bricker, Xin Zhu, and Maureen Bricker

CONTENTS

1.1 INTRODUCTION

Although oil price fluctuations have recently slowed capital investments, the fundamental long-term market drivers for growth of fuels and petrochemicals are still very strong, for example, the growth of middle class. By 2030, 3 billion people are expected to enter the middle class, mostly in emerging markets, and create demand for quality fuels and products derived from petrochemicals. This chapter discusses key refining advances as well as petrochemical integration including shale gas, high-quality fuels, renewables, energy efficiency, and refinery–petrochemical integration.

With worldwide petroleum refining capacity expected to grow steadily for several decades, breakthrough new technologies will offer a competitive advantage for early adopters, allowing improved high-quality fuel production for transportation as well as potential petrochemical integration. Clean, high-quality fuels produced in high yields will remain a primary focus, driven by environmental legislation requirements and better economics. These technology advancements can be accomplished by development of new process technology often made possible by breakthroughs in new catalytic materials as discussed in this chapter.

More rapid petrochemical demand growth rates relative to demand for fuels create new opportunities for refiners to improve value generation by adding petrochemical feedstock production to their product slate. We expect that the multinational climate agreement from Paris negotiations in 2015 should lead to a longer-term expansion of renewable content into transportation fuels. As discussed in this chapter, technology for advanced fuels from renewable hydrocarbon sources has been commercialized successfully.

Although the market fluctuations may modify techno economic choices by refiners, the overall strategy to implement the most economic, efficient, and environmentally sustainable technologies creates the best opportunities for growth and profitability in the refining industry.

1.2 FUEL QUALITY

This section will focus on the fundamentals of important catalyst design considerations.

Presently, various environmental regulations stipulate a low level of sulfur and in some cases a low level of aromatic composition in both gasoline and diesel products, which has spurred a growth in a variety of hydroprocessing complexes around the world. As the sulfur level required in gasoline and diesel reached low levels approaching zero, a large increase in both hydrotreating and hydrocracking capacity grew in the world, and is still projected to grow over the next decade.[1] Even when a refiner in a certain area of the world does not need the stringent specification for their region, the process is designed for the stiffer requirement so that the refiner is able to export their products to any area of the world.

Hydrocracking and hydrotreating are the two major technologies that allow refiners to meet fuel specifications. There have been excellent recent reviews of the overall technology.[2]

1.2.1 HYDROCRACKING

Hydrocracking is a flexible catalytic refining process that is commonly applied to upgrade the heavier fractions obtained from the distillation of crude oils, including residue. The process adds hydrogen, which improves the hydrogen-to-carbon ratio of the net reactor effluent, removes impurities like sulfur to produce a product that meets the environmental specifications, and converts the heavy feed to a desired boiling range. The chemistry involves the conversion of heavy molecular weight compounds to lower molecular weight compounds through carbon–carbon bond breaking and hydrogen addition. Hydrocracking feeds can range from heavy vacuum gas oils (VGOs) and coker gas oils to atmospheric gas oils to blends of the combination of VGO, light cycle oils (LCOs), coker gas oils, and tar sands products. The demand for hydrocracking is predicted to be a stable growth market at 6%.

1.2.2 FLOW SCHEMES

Hydrocracking units can be configured in a number of ways. The unit can consist of one or two reactors with either one or multiple catalysts. The process can use one or two stages and be operated in once-through or recycle mode. The choice of the configuration depends on the feed properties and the specific product slate desired by the refiner. The five main types of operating units are mild, moderate/medium pressure, conventional, partial conversion, and resid hydrocracking. These are shown in Table 1.1.

As described in Section 1.2.4, the reaction section fulfills two functions: pretreating/hydrotreating and hydrocracking. This is shown in Figure 1.1 as separate reactors, though both functions can be achieved in a single reactor when using some types

TABLE 1.1
Types of Hydrocracking Processes

Unit Type	Typical Conversion	Total Pressure (bar/psig)	Hydrogen Partial Pressure (bar/psig)	Reactor Temperature (°C/°F)
Mild (MHC)	20–40	60–100/870–1450	20–55/290–840	350–440/662–824
Moderate/medium pressure	40–70	100–110/1450–1600	50–95/725–1380	340–435/644–815
Conventional	50–100	110–200/1600–2900	95–140/1390–2030	340–435/662–842
Resid Hydrocracking (LC-fining)	65–100	97–340/1400–3500	73–255/1050–2625	385–450/725–914

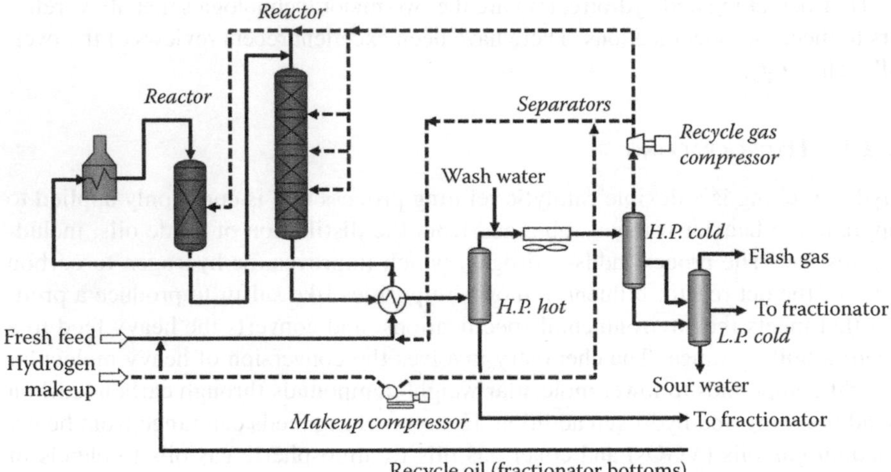

FIGURE 1.1 Typical flow diagram of reactor section of single-stage hydrocracking unit.

of catalysts, e.g., an amorphous catalyst. In most modern configurations, different catalysts are used for the hydrotreating and the hydrocracking sections. However, both types of catalyst can be loaded in the same reactor; separate vessels are not needed. When using both pretreatment and cracking configurations, the first catalyst (a hydrotreating catalyst) converts organic sulfur, oxygenates, and nitrogen from hetero compounds in the feedstock to hydrogen sulfide, water, and ammonia, respectively. The deleterious effect of gas phase H_2S and NH_3 on hydrocracking catalysts is considerably less than that of the corresponding organic hetero compounds in the liquid phase.[3,4] The hydrotreating catalyst also facilitates the hydrogenation of aromatics. In the single-stage configuration, the products from the hydrotreating reaction zone are passed over a hydrocracking catalyst where most of the hydrocracking takes place. The conversion occurs in the presence of NH_3, H_2S, and small amounts of unconverted hetero compounds. The hydrotreating catalyst is designed to convert the hetero compounds in the feedstock. Typically, such catalysts comprise sulfided molybdenum and nickel on an alumina support. The reactor operates at temperatures varying from 570°F to 800°F (300–425°C) and hydrogen pressures between 1250 and 2500 psig (85–170 bar). Under these conditions, heteroatom elimination, significant hydrogenation, and some cracking also take place. The cracking catalyst operates at the same hydrogen pressures but at temperatures varying from 570°F to as high as 840°F (300–450°C) for amorphous hydrocracking catalysts and up to 825°F (440°C) for zeolite-containing catalysts.

1.2.3 Two-Stage Recycle Hydrocracking

The two-stage hydrocracking process configuration is also widely used, especially for large capacity units. In two-stage units, the hydrotreating and some cracking take place in the first stage. The effluent from the first stage is separated and fractionated, with the unconverted oil passing to the second stage for further reaction. The reactor effluent from the second stage reaction section goes back to the common fractionator. A simplified schematic of a two-stage hydrocracker is shown in Figure 1.2. The key point is that the catalyst in the second stage may be operating in the near absence of ammonia and hydrogen sulfide. Thus, the catalysts used in the second stage need to be tailored for that reaction environment to maximize desired product selectivity.

1.2.4 Chemistry

The chemistry of hydrotreating and hydrocracking is commonly taken together and termed hydroprocessing and is similar for both sections of the hydroprocessing unit. Hydrotreating converts heteroatoms, and hydrocracking converts higher carbon number feed molecules to lower molecular weight products by cracking the side chains and by saturating the aromatics and olefins. Hydrocracking catalysts will also

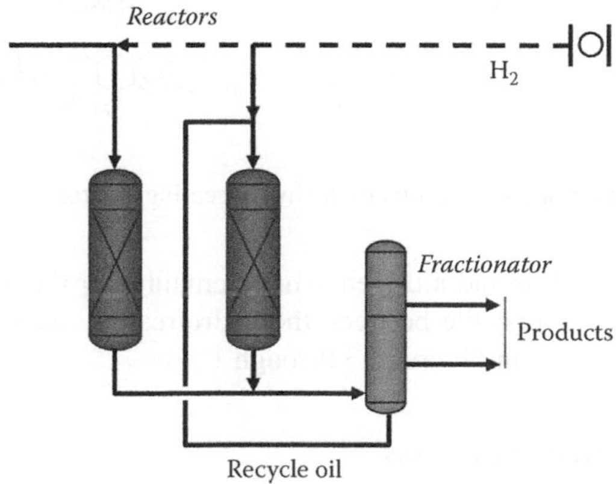

FIGURE 1.2 Two-stage hydrocracking.

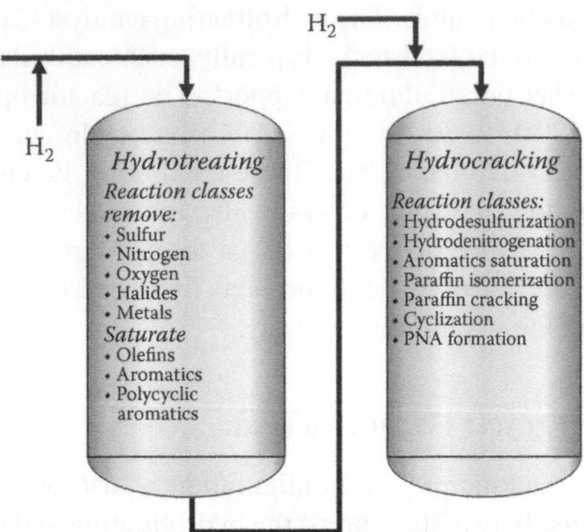

FIGURE 1.3 Summary of hydroprocessing reaction classes in the sections of the hydrotreating and hydrocracking reactors.

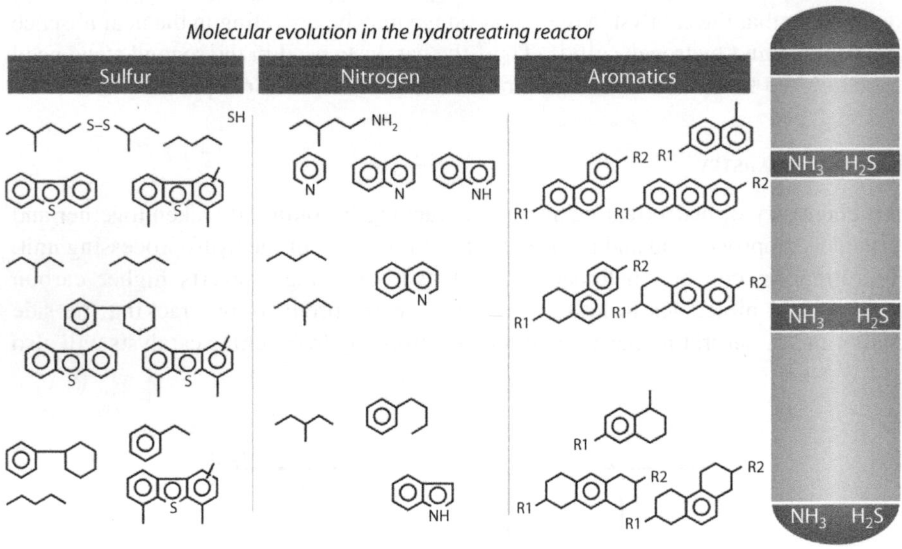

FIGURE 1.4 Evolution of the reactions in the hydrotreating reactor.

remove any residual sulfur and nitrogen, which remain after the hydrotreating. The evolution of the reaction profile between the hydrotreating reactor and the hydrocracking reactor is shown in Figures 1.3 through 1.5.

1.2.5 Hydrotreating Reactions

The hydrotreating reactions proceed in the following order: metal removal, olefin saturation, sulfur removal, nitrogen removal, oxygen removal, and halide removal and aromatic saturation. Figure 1.4 shows the evolution of the reactions in the hydrotreating reactor. Hydrogen is consumed in all of the treating reactions.

1.2.6 Hydrocracking Reactions

Hydrocracking reactions proceed through a bifunctional mechanism.[5–9] Two distinct types of catalytic sites are required to catalyze the separate steps in the reaction sequence. The first function is the acid function, which provides for cracking and isomerization, and the second function is the metal function, which provides for olefin formation and hydrogenation. The cracking reaction requires heat, while the hydrogenation reaction generates heat. Figure 1.5 shows the evolution of the reactions in the hydrocracking reactor. Overall, there is a heat release during hydrocracking reactions; the heat release is a function of the hydrogen consumption, where higher hydrogen consumption will generate a larger temperature increase. Generally, the hydrogen consumption in hydrocracking (including the pretreating section) is 1200–2400 SCFB (200–420 Nm^3/m^3) resulting in a typical heat release of 50–100 Btu/SCF H_2 (2.1–4.2 kcal/m^3 H_2), which translates into a temperature increase of about 0.1°F/SCF H_2 consumed. This includes the heat release generated in the hydrotreating section.

In general, the hydrocracking reaction starts with the generation of an olefin or a cyclo-olefin on a metal site on the catalyst. Next, an acid site adds a proton to the olefin or cyclo-olefin to produce a carbenium ion. The carbenium ion cracks to a smaller carbenium ion and a smaller olefin. These products are the primary hydrocracking products. These primary products can react further to produce still smaller

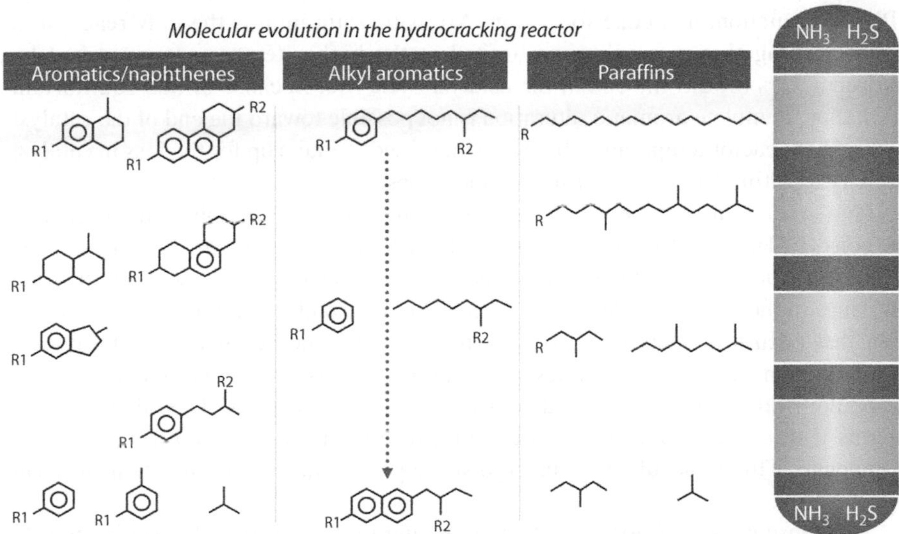

FIGURE 1.5 Evolution of the reactions in the hydrocracking reactor.

secondary hydrocracking products. The reaction sequence can be terminated at primary products by abstracting a proton from the carbenium ion to form an olefin at an acid site and by saturating the olefin at a metal site. Figure 1.6 illustrates the specific steps involved in the hydrocracking of paraffins. The reaction begins with the generation of an olefin and the conversion of the olefin to a carbenium ion. The carbenium ion typically isomerizes to form a more stable tertiary carbenium ion. Next, the cracking reaction occurs at a bond that is beta to the carbenium ion charge. The beta position is the second bond from the ionic charge. Carbenium ions can react with olefins to transfer charge from one fragment to the other. In this way, charge can be transferred from a smaller hydrocarbon fragment to a larger fragment that can better accommodate the charge. Finally, olefin hydrogenation completes the mechanism. This selectivity is due in part to a more favorable equilibrium for the formation of higher carbon number olefins. In addition, large paraffins adsorb more strongly. The carbenium ion intermediate causes extensive isomerization of the products, especially to a-methyl isomers, because tertiary carbenium ions are more stable. Finally, the production of C_1 to C_3 is low because the production of these light gases involves the unfavorable formation of primary and secondary carbenium ions. Other molecular species such as alkyl naphthenes, alkyl aromatics, and so on react via similar mechanisms, e.g., via the carbenium ion mechanism.

The hydrocracking reactions tend to favor conversion of large molecules because the equilibrium for olefin formation is more favorable for large molecules, and because the relative strength of adsorption is greater for large molecules. In hydrocracking, the products are highly isomerized, C_1 and C_3 formation is low, and single rings are relatively stable.

In addition to treating and hydrocracking, several other important reactions take place in hydrocrackers. These are aromatic saturation, polynuclear aromatics (PNAs) formation, and coke formation. Aromatic saturation is the only reaction in hydrocracking that is equilibrium-limited at the higher temperatures reached by hydrocrackers toward the end of the catalyst cycle life. Because of this equilibrium limitation, complete aromatic saturation is not possible toward the end of the catalyst cycle when reactor temperature has to be increased to make up for the loss in catalyst activity resulting from coke formation and deposition.

PNAs, sometimes called polycyclic aromatics (PCAs), or polyaromatic hydrocarbons (PAHs), are compounds containing at least two benzene (BZ) rings in the molecule. Normally, the feed to a hydrocracker can contain PNAs with up to seven BZ rings in the molecule. PNA formation is an important, though undesirable, reaction that occurs in hydrocrackers. Figure 1.7 shows the competing pathways for conversion of multiring aromatics. One pathway starts with metal-catalyzed ring saturation and continues with acid-catalyzed cracking reactions. The other pathway begins with an acid-catalyzed condensation reaction to form a multiring aromatic compound. This molecule may undergo subsequent condensation reactions to form a large PNA.

The consequence of operating hydrocracking units with recycled oil is the creation of large PNA molecules that can contain more than seven aromatic rings in the molecule. These are called heavy PNAs (HPNAs) whose formation is due to condensation reactions of smaller PNAs on the catalyst surface. The HPNAs produced on

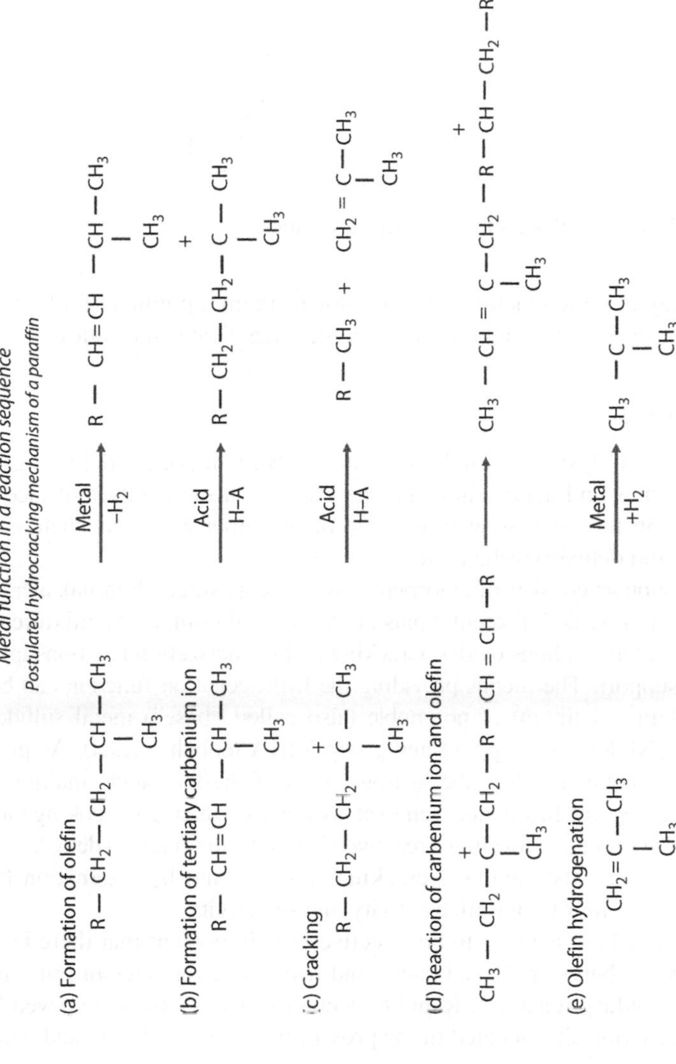

FIGURE 1.6 Steps involved in the hydrocracking of paraffin molecules.

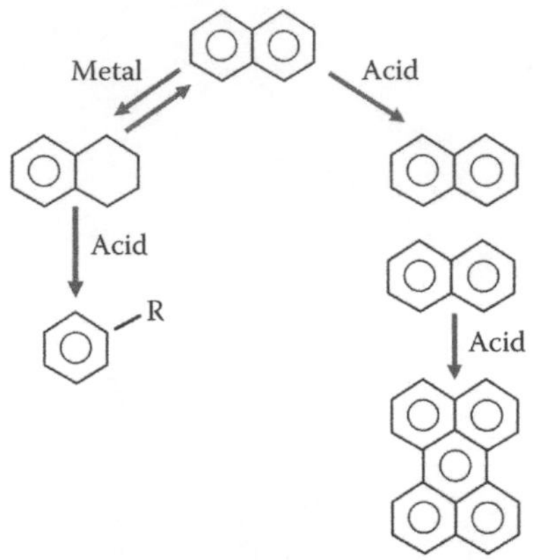

FIGURE 1.7 Possible pathways for multi-ring aromatics.

the catalyst may exit the reactor and cause downstream equipment fouling, or they may deposit on the catalyst and form coke, which deactivates the catalyst.

1.2.7 Catalysts

Hydrocracking catalysts are dual functional.[10] Both metallic dehydrogenation/hydrogenation sites and acidic sites must be present on the catalyst surface for the cracking reaction to occur as well as some of the other reactions such as hydroisomerization and dehydrocyclization.

The acidic support consists of amorphous oxides (e.g., silica-alumina), a crystalline zeolite such as a (modified Y zeolite) plus binder (e.g., alumina), or a mixture of crystalline zeolite and amorphous oxides. Cracking and isomerization reactions take place on the acidic support. The metals providing the hydrogenation function can be noble metals (palladium, platinum) or non-noble (also called "base") metal sulfides from group VIA (molybdenum, tungsten) and group VIIIA (cobalt, nickel). As previously discussed, these metals catalyze the hydrogenation of the feedstock, making it more reactive for cracking and heteroatom removal, as well as reducing the coking rate. They also initiate the cracking by forming a reactive olefin intermediate via dehydrogenation.

The ratio between the catalyst's cracking function and hydrogenation function can be adjusted in order to optimize activity and selectivity.

For a hydrocracking catalyst to be effective, it is important that there be a rapid molecular transfer between the acid sites and hydrogenation sites in order to avoid undesirable secondary reactions. Rapid molecular transfer can be achieved by having the hydrogenation sites located in the proximity of the cracking (acid) sites.

1.2.8 Acid Function of the Catalyst

A solid oxide support material supplies the acid function of the hydrocracking catalyst. Amorphous silica-alumina (ASA) provides the cracking function of amorphous catalysts and serves as support for the hydrogenation metals. Sometimes, ASA catalysts or a combination of ASA and zeolite can be used to produce high-yield distillate hydrocracking catalysts. ASA also plays a catalytic role in low-zeolite catalysts. Zeolites are commonly used in high activity distillate-selective catalysts and in hydrocracking catalysts for the production of naphtha.

Amorphous mixed metal oxide supports are acidic because of the difference in charge between adjacent cations in the oxide structure. The advantages of ASA for hydrocracking are pores, which permit access of bulky feedstock molecules to the acidic sites, and moderate activity, which makes the metal–acid balance needed for distillate selectivity easier to obtain.

Zeolites are used in hydrocracking catalysts because they provided high activity due to their higher acidity compared to the ASA materials. Zeolites are three-dimensional, microporous, crystalline solids with well-defined structures that contain aluminum, silicon, and oxygen in their regular framework; cations and water are located in the pores. The silicon and aluminum atoms are tetrahedrally coordinated with each other through shared oxygen atoms. When the cations in zeolites are exchanged with ammonium ion and calcined, an acidic material can be formed.

One zeolite used in hydrocracking, Y zeolite, has a structure nearly identical to the naturally found zeolite Faujasite.[9] The Y zeolite has both a relatively large free aperture, which controls access of reactants to acid sites, and a three-dimensional pore structure, which allows diffusion of the reactants in and products out with minimal interference. Both Bronsted and Lewis acids are possible in zeolites. The number and the strength of the acid sites can be varied in various synthesis steps. These sites are highly uniform, but each zeolite may have more than one type of site. The following factors influence the number and strength of acid sites in zeolites:

- The types of cations occupying the ion exchange sites
- Thermal treatments of the zeolite
- The framework silica-to-alumina ratio in the zeolite

For example, Y zeolite can be treated to modify the Si/Al ratio; common methods to accomplish this are either a thermal or a hydrothermal treatment. Figure 1.8 shows an image after hydrothermal treatment of stabilized Y zeolite. When aluminum is removed, the total number of acid sites is decreased because each proton is associated with framework aluminum. As can be seen in Figure 1.8, there is also a generation of mesoporosity in the zeolite. However, the reduction of the alumina sites increases the strength of the acid sites of the remaining alumina to a certain point. As a result, the total acidity of the zeolite, which is a product of the number of sites and strength per site, peaks at an intermediate extent of dealumination. The crystallinity of the zeolite can also be modified depending on the treatment history. The acid site concentration and strength of the zeolite will affect the final hydrocracking catalyst properties.

1.2.9 Catalyst Manufacturing

Hydroprocessing catalysts can be manufactured by a variety of methods. The method chosen usually represents a balance between manufacturing cost and the degree to

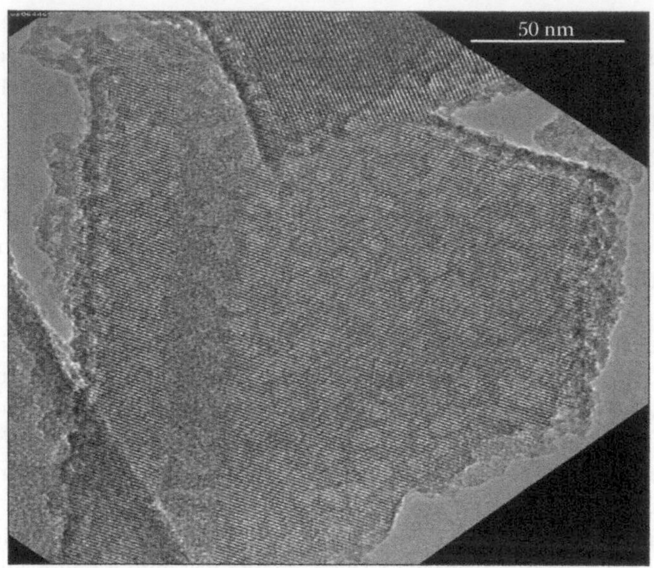

FIGURE 1.8 Stabilized Y after thermal treatments.

which the desired chemical and physical properties are achieved. While the chemical composition of the catalyst plays a decisive role in its performance, the physical and mechanical properties also play a major role. The preparation of hydroprocessing catalysts involves several steps, the details of which are usually trade secrets of catalyst producers, but include precipitation, filtration (decantation, centrifugation), washing, drying, forming, calcination, and metal impregnation.

1.2.10 Catalyst Activation

Base metal hydrocracking catalysts have to be prepared in the final state through a sulfiding procedure in order to create the active species, the metal sulfides. Several names are used for this treatment, such as sulfiding or pre-sulfiding. The metals on the greatest majority of catalysts are in an oxide form at the completion of the manufacturing process.

Noble metal catalysts are activated by hydrogen reduction of the finished catalyst, in which the metal is also in an oxide form.

The active phase is similar for the CoMo or NiMo supported on Al_2O_3 or SiO_2 and by analogy Ni/W on the more complex hydrocracking catalyst.[10] These consist of Co or Ni atoms on the edges of MoS_2 slabs and are referred to as CoMoS and NiMoS phases, and by analogy NiWS phase. W and Mo exhibit different behavior with respect to sulfidation. Ni/W is more difficult to sulfide than Ni/Mo, i.e., requires higher temperature. The presence of the Ni enhances the sulfidation of either Ni or Mo. The slabs of Mo or S have a specific length and appear as strands with different levels of stacking. An image[11] of Ni/W in a hydrocracking catalyst is shown in Figure 1.9. In this figure, the individual atoms of W can be seen.

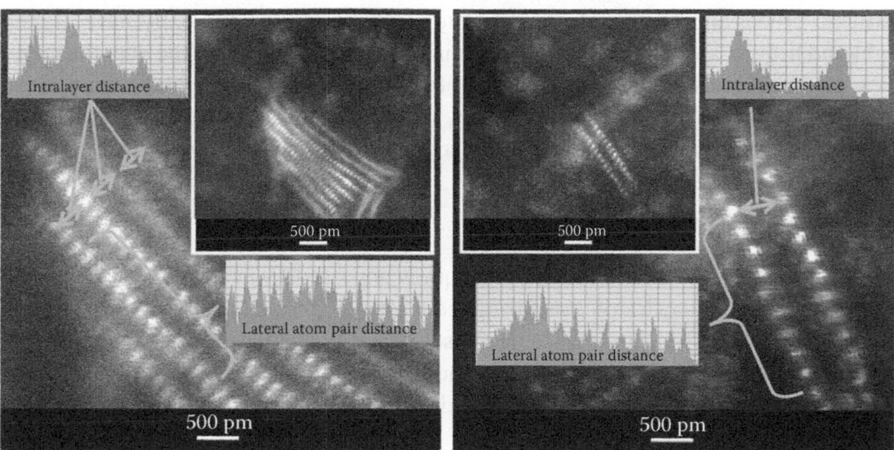

FIGURE 1.9 Image of NiWS active phase using UOP aberration-corrected Titan 80-300 Super X electron microscope.

Sulfiding is accomplished mainly in situ, though some refiners have started to do the activation outside the unit (ex situ). More and more refiners will opt to receive the catalyst at the refinery site in pre-sulfided state to accelerate the start-up of the unit. In situ sulfiding can be accomplished either in vapor or liquid phase. In vapor phase sulfiding, the activation of the catalyst is accomplished by injecting a chemical that easily decomposes to H_2S, such as dimethyl-disulfide (DMDS) or di-methyl-sulfide (DMS).

1.2.11 Summary

The growth in clean fuels has led to a strong growth in hydroprocessing technologies. The core of hydroprocessing technologies are the catalysts, which can be designed for ultrahigh activity for heteroatom removal with selective cracking to distillate, jet, or naphtha depending on the refiner needs. The catalyst design and reactor stacking takes into account the cascading reactions, which must all be catalyzed efficiently for the optimum reactor yields, run length, and product quality. UOP has invested heavily in hydroprocessing tools including pilot plants, advanced material characterization, new materials, and catalysis to be the leading provider of hydroprocessing technology in the world today.

1.3 IMPACT OF SHALE OIL ON FUTURE REFINING

The US shale crude boom of 2010–2014 has been tempered in 2015 by lower crude oil prices. Nonetheless, it is anticipated that, worldwide, light shale crudes will play an important role as traditional crude supply/demand comes back into balance. This section attempts to explore some of the opportunities and challenges associated with shale crudes, aka tight shale oil.

Tight shale crude is typically characterized as light and sweet with a very different yield pattern for refined products, as well as contaminant levels, impacting the refiners' processing units and product slate. Tight oil crudes typically have higher light and heavy naphtha yields, presenting increasing challenges to the naphtha complex, typically consisting of naphtha hydrotreating (NHT), light naphtha isomerization, and UOP CCR™ Platforming units. Advances, discussed herein, in each of these technologies are important for refiners to capture the value of shale oil.

1.3.1 Market Overview

The US supply of domestic natural gas liquids and crude was drastically impacted by improvements in hydraulic fracturing technology.[12] The supply of natural gas was first impacted, resulting in a price decrease, which presented refiners with a lower cost fuel and feedstock for hydrogen. The second major application of hydraulic fracturing technology was to recover liquid hydrocarbons from existing shale plays, which were previously uneconomical.

Production of tight shale oil from these fields had increased rapidly such that North Dakota is now the second largest producer of crude oil in the United States (US EIA data). The recently issued BP World Energy Outlook estimated that US tight shale oil production could add as much as 5 MMBPD by 2030.[13] Worldwide, abundant shale oil deposits have been discovered as shown in Figure 1.10 and will prove to be an important resource in the future.

1.3.2 Tight Oil Crude Yields Shift toward Lighter Hydrocarbons

The crude assay reflects the yield pattern and is key information for determining the refinery products. Shown in Figure 1.10 are yields of liquified petroleum gas (LPG), naphtha, kerosene, diesel, VGO, and vacuum residue for several crudes with varying American Petroleum Institute (API) gravities. Table 1.2 provides key crude properties for the tight shale and conventional crudes. The tight oil crudes, Bakken, Eagle Ford, and Utica, are typically lighter crudes, have higher API, and have predominantly higher LPG and naphtha with less heavy VGO and vacuum residue material. The general trends for tight oils (high API) are as follows:

- Lower sulfur, nitrogen contaminants
  - Lower H_2 demand for hydrotreating and hydrocracking units
- Lower vacuum residue yield
  - Lower VGO, coker, and fluid catalytic cracking (FCC) rates
  - Lower FCC rates, lower $C_3^=/C_4^=$ to alkylation unit, less alkylate
  - Less FCC naphtha, alkylate in gasoline pool
- Higher paraffin concentration
  - Diesel cut: higher cetane number with poorer cold-flow properties
  - Naphtha cut: leaner reformer feed, lower C_{5+} and H_2 yields
- Higher light and heavy naphtha yields
  - Larger increase in light naphtha; isomerization feed rates increase
  - Increase in feed rate to platforming unit (fixed bed or CCR)
  - Higher quantity of isomerate and reformate in gasoline pool

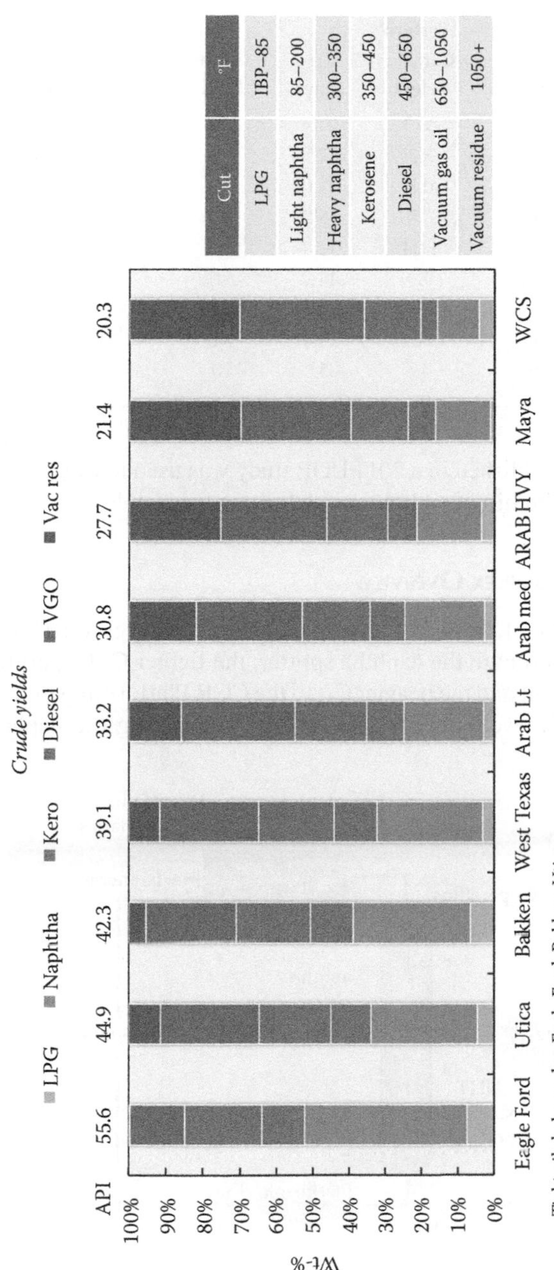

Cut	°F
LPG	IBP–85
Light naphtha	85–200
Heavy naphtha	300–350
Kerosene	350–450
Diesel	450–650
Vacuum gas oil	650–1050
Vacuum residue	1050+

FIGURE 1.10 Crude yields (wt%).

TABLE 1.2
Crude Properties

	Units	Western Canada Select	Maya	Arab Med	WTI	Bakken	Eagle Ford
API GRAVITY	API	20.3	21.4	30.8	39.1	42.3	46.5
Sulfur	wt%	3.3	3.5	2.6	0.3	0.1	0.1
Neut ot TAN No.	mgKOH/g	0.8	0.1	0.3		0.0	0.1
Nitrogen	ppm	2770	3573	1210	1000	500	41
Hydrogen	wt%	11.4	12.1	12.7	13.3	13.8	14.3
Rams bottom carbon	wt%	9.0	11.2	5.6		0.8	0.0
Iron	ppm	8	5	4		2	1
Vanadium	ppm	118	286	34		0	0
Nickel	ppm	49	53	10		1	0

A linear program published in a 2014 UOP study was used to evaluate the refinery-wide impact when switching to a lighter tight oil feed slate.[14,15]

1.3.3 Naphtha Complex Overview

A simplified schematic of the naphtha complex is shown in Figure 1.11. Hydrotreated feed from the NHT is sent to the naphtha splitter; the lighter C_5/C_6 paraffins are sent to the isomerization unit and the heavier C_{7+} to the CCR Platforming unit. Controlling the gasoline pool BZ level is a key specification, impacting the naphtha complex

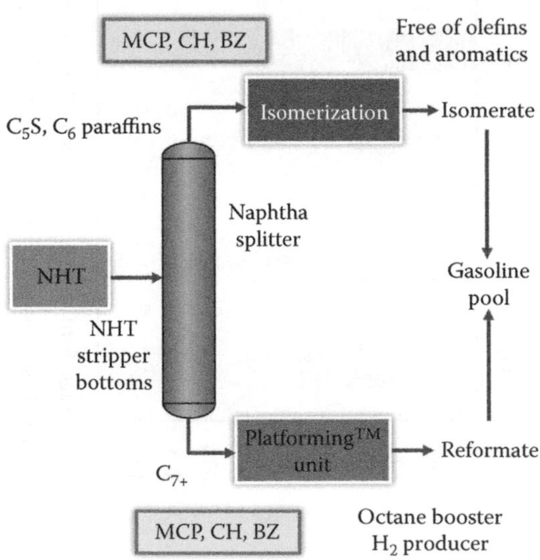

FIGURE 1.11 Naphtha complex.

design. The naphtha splitter is designed for BZ control, splitting the BZ precursors (methycyclopentane [MCP], cyclohexane [CH], and BZ) either to the isomerization or CCR Platforming unit. When BZ production is targeted, BZ precursors are sent to the reformer unit. If minimum BZ is targeted, then the BZ precursors would be sent to the isomerization unit, which is excellent for BZ saturation and conversion to high-octane C_6 paraffins.

Removing the C_6 components from the Platforming unit feed has many advantages. First, a higher C_{7+} feed rate can be achieved to the CCR Platforming unit. Second, since C_6 components are among the hardest to reform, the CCR Platforming unit severity is reduced. Third, the BZ concentration of the reformate product is greatly reduced.

Processing the C_5 and C_6 components in an isomerization unit allows the refiner to recover lost octane due to BZ reduction.

1.3.3.1 Optimizing the NHT Unit—Catalyst Improvements

Catalyst improvements have allowed refiners to increase feed rate and/or reduce utilities. UOP has introduced two new catalysts, HYT-1118 for straight-run, low nitrogen feeds and HYT-1119 for cracked stocks and high nitrogen feeds.

HYT-1118 catalyst is UOP's latest-generation cobalt molybdenum hydrotreating catalyst, building upon the proven performance of UOP's S-120 catalyst. HYT-1118 catalyst provides the benefits of higher hydrodesulfurization (HDS) and hydrodenitrification (HDN) activity while being more than 30% less dense than the previous-generation S-120 catalyst.

UOP's HYT-1119 is a robust catalyst designed for use in NHT applications to remove sulfur and nitrogen compounds with cracked stocks and high nitrogen feeds. HYT-1119 has high stability due to high pore diameter and pore volume, resulting in long catalyst life. In addition, standard regeneration methods can be used to minimize catalyst life-cycle costs.

Shown below are the features and benefits of HYT-1119:

- Improved metal distribution
- 20% higher surface area for higher silicon pickup
- No phosphorus for increased stability and higher silicon pickup
- Lower aromatic saturation
- Reduced loaded density for lower fill cost
- Higher HDS and HDN activity

1.3.4 Isomerization

Light naphtha, C_5–C_6, from the naphtha splitter overhead is sent to the isomerization unit. UOP offers several isomerization designs in which the product research octane number clear (RONC) can range from ~82 to ~92.

- UOP Par-Isom™ Process design: ~82–85 RONC
- UOP Penex™ Process design: HOT Penex—Hydrogen Once Through ~82–85 RONC
- UOP Penex™ design with recycle: ~87–92 RONC

1.3.4.1 Par-Isom Unit with PI-242 or PI-244 Catalyst

The Par-Isom unit uses PI-242 or PI-244 (lower Pt) catalyst, while the Penex unit uses a chlorided alumina-type catalyst. The PI-242/244 catalyst is robust to water and sulfur upsets and does not require a high chloride concentration; consequently, driers and caustic scrubbing for chloride removal are not required, reducing capital expenditures (CAPEX). Both processes can produce 82 to 85 RONC without a liquid recycle.

UOP PI-242 and UOP PI-244 Par-Isom catalyst features and benefits:

- UOP's PI-242 or PI-244 offers an alternative to chlorided alumina catalyst with nearly equivalent selectivity and activity along with tolerance to contaminants such as sulfur and water.

Fully regenerable and long life.

- Use of UOP Par-Isom catalyst can reduce the required catalyst volume from 10% to up to 50% compared to competitive products, thus reducing the fill cost. For new units, UOP Par-Isom catalyst can reduce catalyst and Pt fill cost by 58% and 63% respectively, along with reduction in CAPEX and operating expenditures (OPEX).

1.3.4.2 Penex Process with I-122, I-82 I-84 Chlorided Alumina Catalysts

A benefit of the Penex low coking tendency is that the unit could be designed for "Hydrogen Once Through" (HOT). In the HOT Penex process, recycle gas is not required, thus eliminating the need for a product condenser, product separator, stabilizer feed/bottoms exchanger, and recycle gas compressor. The lower equipment and utility costs significantly lower CAPEX and OPEX, and is the current standard design for Penex units. The product RONC is 83–85 RONC for a single hydrocarbon pass.

1.3.4.3 UOP I-82 and UOP I-84 Penex Catalysts

I-82 catalyst is the highest activity, light paraffin isomerization catalyst commercially available. The catalyst is an amorphous, chloride alumina catalyst containing platinum and is a robust product for maximizing isomerate octane-barrels. I-82 is optimized for the UOP Penex process and is particularly suited for feedstocks that contain a high concentration of BZ and C_6+ cyclic hydrocarbons. The catalyst selectively converts normal butane, pentane, and hexane to higher octane branched hydrocarbons. In addition, I-82 saturates BZ and is designed to operate over a wide range of reaction conditions and feedstocks.

I-84 catalyst (0.18% Pt) is an extension of I-82 catalyst (0.24 wt% Pt), but with 25% lower platinum concentration and is suited for feedstocks that have moderate levels of BZ and C_{6+} cyclic hydrocarbons.

The yield and RONC/motor octane number clear (MONC) benefits are dependent upon the X factor of the feed as shown in Figure 1.12. The performance benefit can be significant for I-82, for all X factors, which improved the product value from $6 million/year (base case) to $12 million/year (case 2) for the highest X factor feeds.[16]

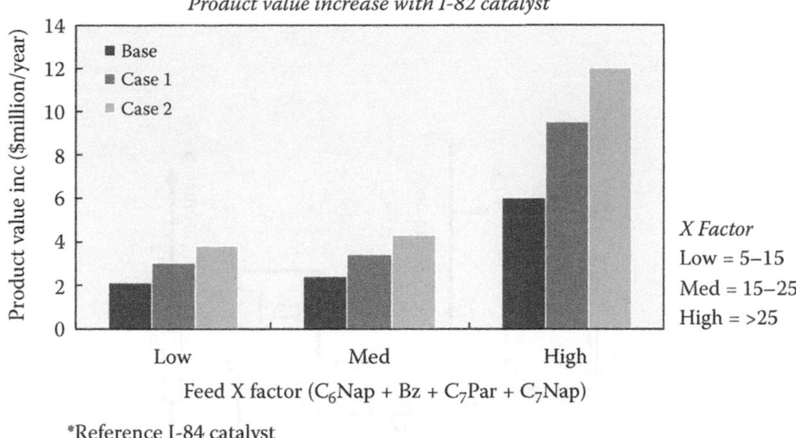

FIGURE 1.12 Product value increase with I-82 with varying X factor feeds.

The higher isomerate RONC-barrels (BBLs) allowed the CCR Platforming unit to decrease severity and increased C_{5+} yields.

1.3.5 CCR Platforming Unit

Heavy naphtha, C_{7+}–C_{11}, from the naphtha splitter bottoms is sent to the CCR Reformer unit. The heavy naphtha is reformed to convert paraffins and naphthenes to an aromatics-rich reformate and hydrogen. Shown in Figure 1.13 is the process flow for the CCR Platforming unit.

1.3.5.1 Process Flow

Hydrotreated straight-run naphtha is mixed with recycle hydrogen, then preheated by exchange with reactor effluent in the combined feed exchanger. The combined feed is raised to the reaction temperature in the charge heater and sent to the reactor section. The predominant reactions are endothermic; consequently, an interheater is used between each reactor to reheat to the reaction temperature.

Catalyst flows vertically by gravity down a reactor stack of three to four reactors. Over time, coke builds up on the catalyst and requires regeneration. Coked catalyst is continually withdrawn from the bottom of the last reactor in the stack and transferred to UOP's CycleMax™ CCR™ regenerator for catalyst regeneration, consisting of four steps: coke burning, oxychlorination, drying, and reduction. The first three steps of coke burning, oxychlorination, and drying occur in the regeneration tower, while the fourth step, reduction, occurs in the reduction zone on top of the first reactor in the reactor stack.

1.3.5.2 CCR Platforming Catalyst Options

UOP continues to develop new CCR Platforming catalysts to address customers' demands and deliver higher performance. UOP has an extensive catalyst portfolio, addressing various customers' needs for motor fuel and/or aromatics production.

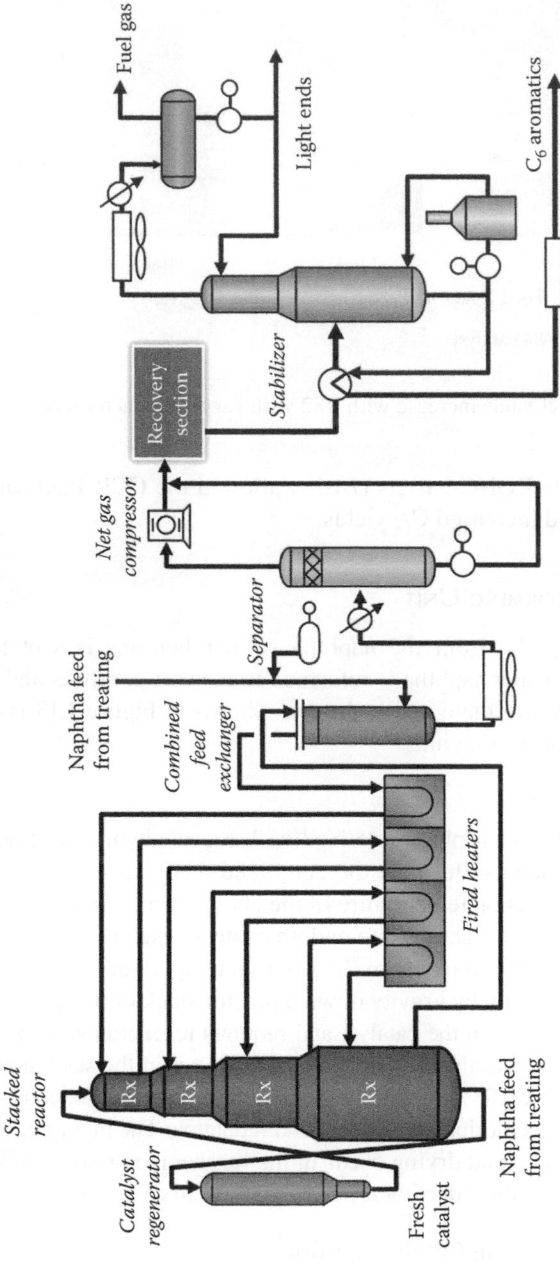

FIGURE 1.13 CCR Platforming unit process flow.

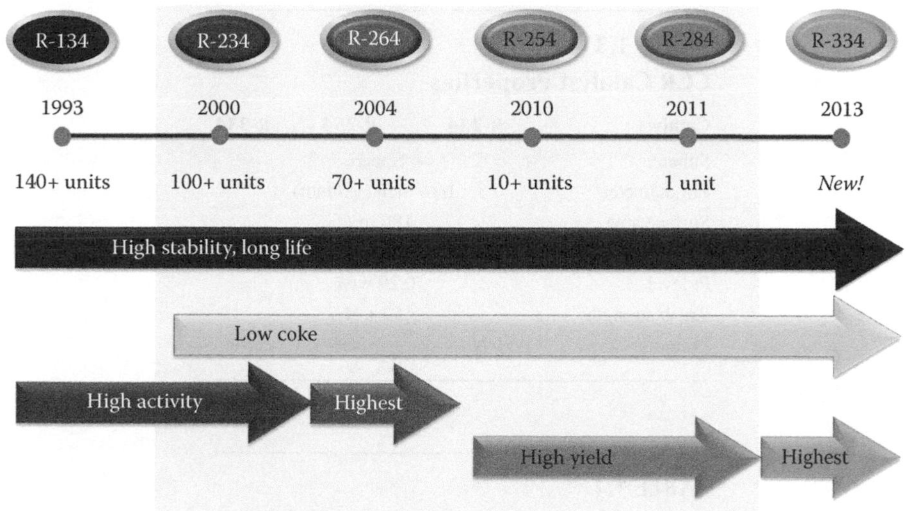

FIGURE 1.14 UOP CCR Platforming catalyst innovations.

Figure 1.14 shows the evolution of UOP's CCR Reforming catalyst: R-134™, R-234™, R-264™, R-254™, R-284™, and the new R-334™ catalyst.

The R-130 series catalyst was first commercialized in 1993. Features of the R-130 series catalyst are high activity (lower reactor inlet temperatures), good surface area stability, and chloride retention.

Due to changes within the market, a low coke R-234 catalyst, was developed, providing an initial 20–25% lower coke make and higher C_{5+} yield versus R-134 catalyst. R-234 catalyst also maintained the excellent surface area stability and chloride retention as the R-130 series catalyst.

R-264 catalyst was developed specifically for customers who targeted to increase throughput to their unit. The R-264 is a high-yield, higher-density catalyst, providing an increased reactor pinning margin. For BZ/toluene/xylene (BTX) operation, the unit could be operated at a higher feed rate, and coupled with the higher activity, maximum BTX is produced.

UOP also developed promoted catalyst, R-254 and R-284 catalysts, both providing significantly higher yields. There are over 10 units operating with R-254 catalyst and others with R-284 catalyst.

UOP's latest catalyst offering is R-334 catalyst, designating the first 300 series product. The R-334 catalyst is UOP's highest yield catalyst and also features long life, low coke. Table 1.3 shows the catalyst properties for R-234, R-254, and R-334 catalysts. It is interesting to note that the high-yield R-334 catalyst does not include a promoter, obtaining the optimal performance by proprietary base and manufacturing technique.

1.3.6 Maximizing Profitability with High-Yield R-334 Catalyst

The high-yield R-334 catalyst reduces cracking, decreasing fuel gas, LPG, and C_5's. Table 1.4 quantifies the benefits of higher yields with R-334 catalyst for 75% Eagle

TABLE 1.3
CCR Catalyst Properties

Catalyst	R-234	R-254	R-334
Sphere		Sphere	
Pill diameter		1/16 inch (1.6 mm)	
Surface area		180 m^2/g	
Density (ABD)		36 lb/ft^2 (560 kg/m^3)	
Pt level		0.29 wt%	
Crush strength		50 + N	
Promoter	No	Yes	No

TABLE 1.4
R-334 Yield Benefits

		R-234	R-334
Case		2	
Crude		Eagle Ford/WCS	
Feed rate	BPD	39,317	39,317
Reformate	RONC	101.3	101.3
P	Lv-%	75	75
N	Lv-%	16	16
A	Lv-%	9	9
Product yields			Delta
H_2	SCFB	1775	90
Fuel gas	wt%	2.57	−0.52
C_3	Lv%	2.33	−0.53
C_4	Lv%	2.8	−0.6
C_5	Lv%	2.1	−0.47
C_{5+}	Lv%	80.9	1
C_{5+} RVP	psi	2.1	−0.2
RONC—bbls		Base	39,825

Ford/25% WCS. The feed composition and feed rate were kept constant, but the catalyst was changed from R-234 catalyst to R-334 catalyst. The benefits of R-334 catalyst include higher yield, but in addition, the reduced cracking reduces C_5's, decreasing the reformate reid vapor pressure (RVP) by 0.2 psig (1.4 kPa). The reduction in RVP allows additional 110 BPD of C_4's to be blended into the gasoline pool while maintaining the 9 psig (62 kPa) RVP target.

The key to these advanced CCR catalysts are the control of olefin cyclization versus cracking to light ends. An extensive body of mechanistic work on catalytic reforming has been published[16] with an overall mechanism shown in Figure 1.15.

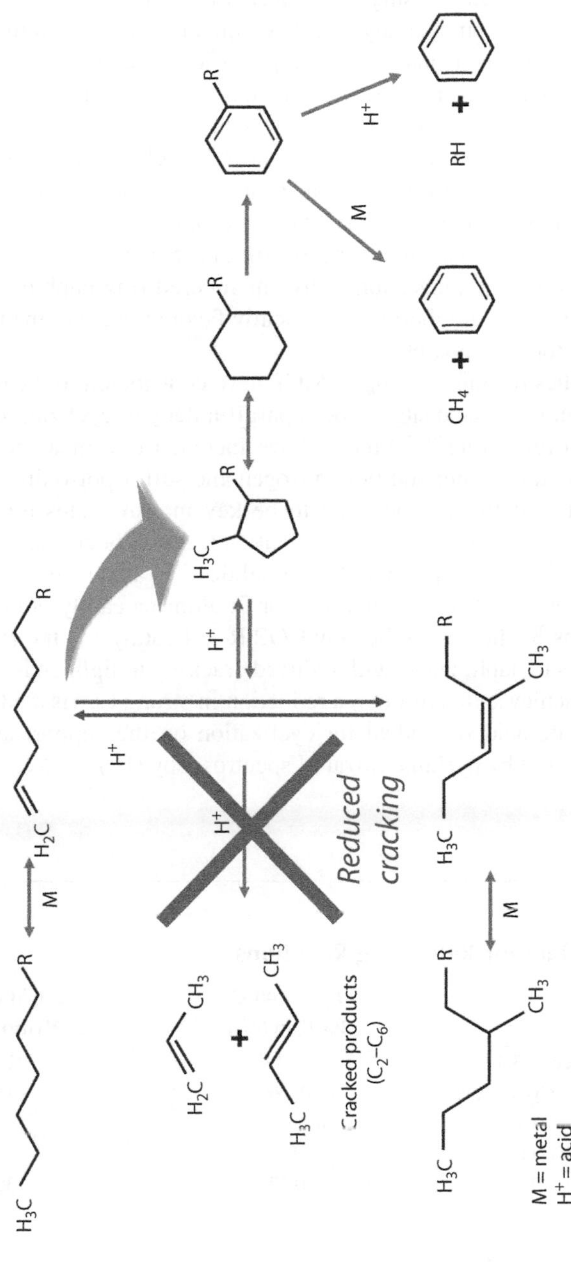

FIGURE 1.15 Schematic of a mechanism for reforming over a bifunctional Pt–alumina catalyst.

Early studies used pure components to probe the reaction chemistry.[17] High liquid hourly space velocity (LHSV) testing allowed the identification of primary cyclic olefin products in conversion of CH and MCP.[18] The bifunctional nature of catalytic reforming has been investigated using the conversion of MCP to BZ. This reaction has been studied as a function of catalyst acid and metal content, with the result that the acid-catalyzed naphthene isomerization step was identified as rate-controlling.[19]

Dehydrogenation reactions that form olefins and aromatics are highly endothermic, and the large heat of reaction for reforming is an important consideration in the design of reactor systems. Paraffin isomerization is relatively neutral, while cracking reactions are exothermic. Table 1.5 shows the heats of reaction for the main reforming reactions for carbon number 6.[20] Formation of aromatics is thermodynamically favored and provides a driving force for paraffin and naphthene conversion. This is particularly important for conversion of five-membered ring naphthenes, as five-membered ring naphthenes are thermodynamically favored over six-membered ring naphthenes under reforming conditions.

Mechanistic studies of ring opening of MCP have contributed to the understanding of the role of metal and acid sites in both paraffin dehydrocyclization and naphthene conversion in reforming.[21,22] Studies have included rate measurements with hydrogen partial pressure, water addition, nitrogen and sulfur poisoning, and mixed catalyst experiments. Olefins are believed to be key intermediates for isomerization, cracking, and aromatization. Olefin cyclization has also been specifically studied over alumina and supports promoted with halides.[23,24] The results indicate that diolefins are not intermediates in cyclization for Pt-alumina catalysts under typical reforming conditions.[25] The key to the new UOP R-334 catalyst is the selectivity to cyclization of olefins to naphthenes, with reduced cracking to light ends as shown in Figure 1.15. R-334 achieves this through a reduction in strong Lewis acid sites while maintaining moderate acid sites ideal for cyclization on the support as shown in Figure 1.16 as measured by pyridine infrared spectroscopy (IR).

TABLE 1.5
Thermodynamic Data for Reforming Reactions

Reaction	K_P^a at 500°C for P_i in MPa	ΔH_r kJ mol^{-1} of Hydrocarbon
Cyclohexane $\rightleftarrows$ Benzene + $3H_2$	6×10^8	221
Methylcyclopentane $\rightleftarrows$ Cyclohexane	0.086	−16
n-Hexane $\rightleftarrows$ Benzene + $4H_2$	8.4×10^8	266
n-Hexane $\rightleftarrows$ 2-Methylpentane	1.1	−5.9
n-Hexane $\rightleftarrows$ 1-Hexane + H_2	0.37	130

[a] For the reaction $(HC)_1 \rightleftarrows (HC)_2 + nH_2$ the equilibrium constant is defined as: $K_P = \frac{^P(HC)_2^{P^n} H_2}{^P(HC)_1}$

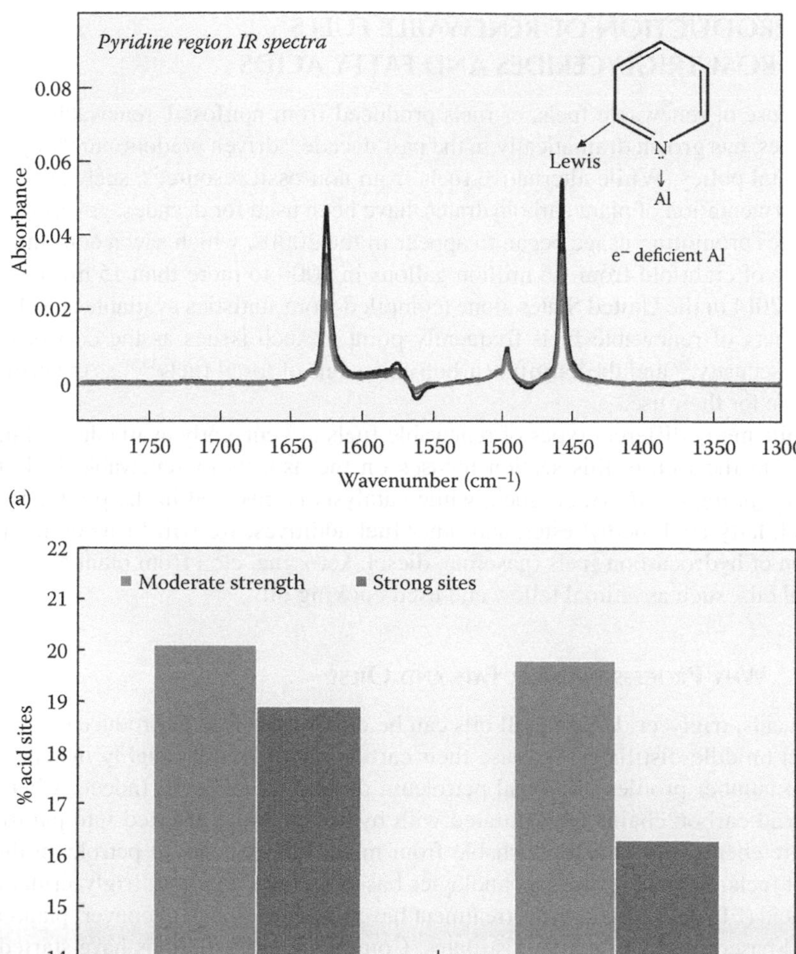

FIGURE 1.16 R-334 optimizes acid site strength. (a) Pyridine IR spectra showing the lewis acid sites, (b) is the relative strong and moderate acid sites for the two catalysts.

1.3.7 Summary

Processing lower-cost tight oil shale crudes such as Bakken, Utica, and Eagle Ford can be accommodated and provides significant margin opportunity; however, it presents new challenges. With higher light naphtha yields, optimizing the naphtha complex is a key component to profitability. Profitability can be increased further when employing the latest high-yield catalysts, such as UOP HYT-1119 hydrotreating catalyst, *UOP* R-334 Platforming catalyst, and UOP I-82 catalyst for paraffin isomerization.

1.4 PRODUCTION OF RENEWABLE FUELS FROM TRIGLYCERIDES AND FATTY ACIDS

Global use of renewable fuels, or fuels produced from nonfossil, renewable carbon resources, has grown dramatically in the past decade[26] driven predominantly by governmental policy. While alternative fuels from nonfossil resources, such as ethanol from fermentation of plant carbohydrates, have been used for decades, governmental mandates promoting usage began to appear in the 2000s, which saw a consumption increase of eightfold from 1.6 million gallons in 2000 to more than 15 million gallons in 2014 in the United States alone (compiled from statistics available at Ref. 27). Supporters of renewable fuels frequently point to such issues as the cost of oil,[28] energy security,[29] and the harmful carbon emissions of fossil fuels[30,31] as the primary rationale for their use.

While many different types of renewable fuels are currently available, and more will be in the future, this section focuses on the aspects of renewable fuels that impact *refining catalysis*. As such, while catalysis is involved in the production of ethanol, fatty acid methyl ester, and other fuel additives, we will focus on the production of hydrocarbon fuels (gasoline, diesel, kerosene, etc.) from plant and waste natural oils, such as animal tallow and used cooking oil.

1.4.1 Why Process Natural Fats and Oils?

Fatty acids, triglycerides, and tall oils can be effectively used to produce diesel and jet fuel (middle distillates) because their carbon chain length roughly matches the carbon number profiles of typical petroleum diesel and jet fuels. Indeed, when the fatty acid carbon chains are saturated with hydrogen and converted into paraffins, they are chemically indistinguishable from major components of petroleum diesel and jet fuels. Several process technologies based on the concept of triglyceride, free fatty acid (FFA), or tall oil hydrotreatment have been developed to convert plant- and animal-based oils into middle distillates. Commercial installations have started up over the past few years that produce renewable diesel from triglyceride, FFAs, or tall oil feed sources, as shown in Table 1.6.[32–40]

1.4.2 Triglycerides and Fatty Acids

Triglycerides are molecules composed of three fatty acids bound to a glycerol backbone by an ester linkage.[41] They are the fundamental unit of energy storage in living organisms. When the fatty acids are no longer linked to the glycerol backbone, they are called FFAs and are composed of a nonpolar hydrocarbon chain with a polar carboxylic acid functional group. Tall oil is primarily composed of FFAs and rosin acids, and is a commercial by-product of wood pulping. General structures for triglycerides and FFAs are shown in Figure 1.17. The shorthand notation used to describe fatty acids is carbon number followed by the number of carbon–carbon double bonds, also known as the degree of unsaturation, separated by a colon (carbon number:degree of unsaturation). The fatty acid carbon chain length and the degree

TABLE 1.6
Commercial Triglyceride, FFA, and Tall Oil Hydroprocessing Installations

Company Name	Capacity (BPD)	Technology	Location	Primary Product	Start-Up Date
Neste	3900	NEXBTL™	Porvoo, Finland (1)	Diesel	2007
	3900	NEXBTL™	Porvoo, Finland (1)	Diesel	2009
	16,600	NEXBTL™	Singapore	Diesel	2010
	16,600	NEXBTL™	Rotterdam, Netherlands	Diesel	2011
Dynamic Fuels	5000[a]	Bio-Synfining™	Geismar, LA, USA	Diesel	2010
Preem	10,000 (30% RTD[b] Co-feed)	HydroFlex™	Gothenburg, Sweden	Diesel	2010
Diamond Green Diesel	10000	Ecofining™ [c]	Norco, LA, USA	Diesel	2013
ENI	7200[a]	Ecofining™ [c]	Venice, Italy	Diesel	2014
AltAir Fuels	2500	Ecofining™ [c]	Paramount, CA, USA	Jet Fuel	2016

[a] Denotes production capacity rather than feed capacity.
[b] Raw tall diesel is a derivative of tall oil, in which the free fatty acids of tall oil are esterified to make fatty acid methyl esters. The rosin acids of tall oil are largely inert to esterification.
[c] Ecofining™ refers to the UOP/ENI Ecofining process.

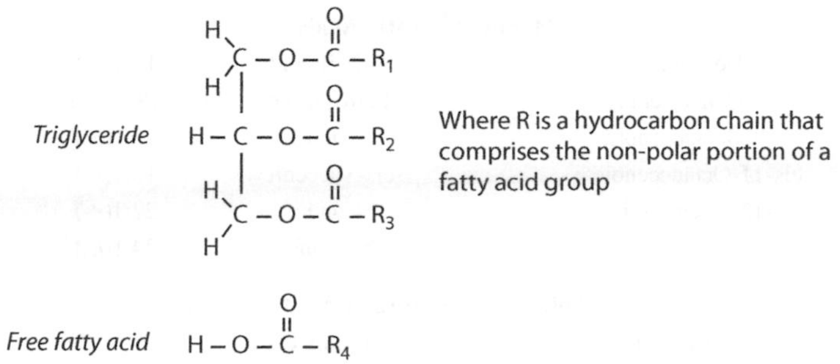

Triglyceride

$$\mathrm{H_2C{-}O{-}C({=}O){-}R_1}$$
$$\mathrm{H{-}C{-}O{-}C({=}O){-}R_2}$$
$$\mathrm{H_2C{-}O{-}C({=}O){-}R_3}$$

Where R is a hydrocarbon chain that comprises the non-polar portion of a fatty acid group

Free fatty acid

$$\mathrm{H{-}O{-}C({=}O){-}R_4}$$

FIGURE 1.17 General structure of triglyceride and free fatty acid molecules.

of hydrogen saturation can vary significantly, usually depending on the source of the triglyceride. For example, coconut triglycerides are rich in lauric (12:0) and myristic (14:0) fatty acids, whereas soybean triglycerides are rich in oleic (18:1), linoleic (18:2), and linolenic (18:3) fatty acids. As is apparent in these two examples, fatty acids are nearly exclusively found in nature with even-numbered carbon chains of approximately 8 to 22 carbon atoms. The combination of fatty acids that can make up a triglyceride varies widely among all organisms, and can even vary, to a lesser extent, within the same organism, depending on the season of the year, region of the world, or strain of organism. In general, most animal-derived triglycerides are more saturated than plant triglycerides. Table 1.7 shows nomenclature, carbon chain length, and degrees of unsaturation for several common fatty acids.[41] Table 1.8 shows common triglyceride oil compositions in terms of the concentration of fatty acids.[42] The method used determines the fatty acid composition by capillary column gas-liquid chromatography and provides compositions in relative (area %) values.

TABLE 1.7
Common Fatty Acids Found in Plants and Animals

Systematic Name	Trivial Name	Shorthand
	Saturated Fatty Acids	
Ethanoic	Acetic	2:0
Butanoic	Butyric	4:0
Hexanoic	Caproic	6:0
Octanoic	Caprylic	8:0
Decanoic	Capric	10:0
Dodecanoic	Lauric	12:0
Tetradecanoic	Myristic	14:0
Hexadecanoic	Palmitic	16:0
Octadecanoic	Stearic	18:0
Eicosanoic	Arachidic	20:0
Docosanoic	Behenic	22:0
	Monoenoic Fatty Acids	
cis-9-Hexadecenoic	Palmitoleic	$16{:}1(n_7)$
cis-6-Octadecenoic	Petroselinic	$18{:}1(n_{12})$
cis-9-Octadecenoic	Oleic	$18{:}1(n_9)$
cis-11-Octadecenoic	cis-Vaccenic	$18{:}1(n_7)$
cis-13-Docosenoic	Erucic	$22{:}1(n_9)$
cis-15-Tetracosenoic	Nervonic	$24{:}1(n_9)$
	Polyunsaturated Fatty Acids[a]	
9,12-Octadecadienoic	Linoleic	18:2(*n*-6)
6,9,12-Octadecatrienoic	γ-Linolenic	18:3(*n*-6)
9,12,15-Octadecatrienoic	α-Linolenic	18:3(*n*-3)
5,8,11,14-Eicosatetraenoic	Arachidonic	20:4(*n*-6)
5,8,11,14,17-Eicosapentaenoic	EPA	20:5(*n*-3)
4,7,10,13,16,19 Docosahexaenoic	DHA	22:6(*n*-3)

Source: Christie, W. What is a lipid? The common fatty acids of animal and plant origin. *The AOCS Lipid Library*, July 25, 2013. Web. Accessed on May 7, 2014, http://lipidlibrary.aocs.org/Lipids/whatlip/index.htm.

[a] All the double bonds are of the cis configuration.

TABLE 1.8
Fatty Acid Composition of Common Triglyceride Oils

Feedstock	C_{8}:0	C_{10}:0	C_{12}:0	C_{14}:0	C_{15}:0	C_{16}:0	C_{16}:1	C_{17}:0	C_{17}:1	C_{18}:0	C_{18}:1	C_{18}:1*n*-9 (OH)	C_{18}:2	C_{18}:3	C_{18}:39c, 11t, 13t	C_{20}:0	C_{20}:1	C_{20}:1*n*-n,11t, (OH)	C_{20}:2	C_{20}:5	C_{22}:0	C_{22}:1	C_{24}:0	C_{24}:1	Unknowns
Algae 1				0.6		6.9	0.2			3.0	75.2		12.4	1.2		0.4					0.1				
Babassu	0.5	3.8	48.8	17.2		9.7				4.0	14.2		1.8												
Beef tallow			0.2	2.9	0.6	24.3	2.1	1.2	0.4	22.8	40.2		3.3	0.7		0.2	0.6								0.5
Borage						9.3				3.8	17.1		38.7	26.1							0.2	2.5		1.5	0.8
Camelina oil						5.0				2.2	17.1		18.0	37.9		1.4	9.8		1.6		0.4	4.5	0.3	0.2	1.0
Canola oil						3.8	0.3			1.9	63.9		19.0	9.7		0.6					0.4		0.2	0.2	
Castor[a]						0.9				1.1	3.1	90.3	4.0	0.6											
Choice white grease				1.3		21.6	2.8	0.2	0.3	9.0	50.4		12.2	1.0		0.2	0.5					0.3			0.2
Coconut	6.3	6.0	49.2	18.5		9.1				2.7	6.5		1.7												
Coffee						11.0	0.5			3.4	70.0		12.7	0.8		0.6	0.1				0.2		0.1		0.6
Corn						12.1	0.1		0.1	1.8	27.2		56.2	1.3		0.4					0.2				0.6
Cuphea viscosissima			0.2	4.7		18.2				3.5	46.9		22.8	2.3		0.6					0.4		0.6		
Evening primrose						6.0				1.8	6.6		76.3	9.0		0.3									
Fish			0.2	7.7		18.8	9.3	0.3	0.3	3.9	15.0		4.6	0.3		0.2	1.4			25.1	0.7	1.3		0.4	10.5
Hemp						5.2				2.4	13.1		57.1	20.0		0.7					0.5				0.7
Hepar, high IV			0.2	1.0		20.7	2.7	0.3	0.3	8.9	46.7		15.6	0.5		0.2	0.8		1.3		0.2	0.4		0.1	0.1
Hepar, low IV		0.1	0.1	1.5		28.0	1.9	0.3	0.2	20.2	36.1		9.7	0.3		0.2	0.7		0.4			0.3			
Jathropa						12.7	0.7			5.5	39.1		41.6	0.2		0.2									

(Continued)

TABLE 1.8 (CONTINUED)
Fatty Acid Composition of Common Triglyceride Oils

Feedstock	C_8:0	C_{10}:0	C_{12}:0	C_{14}:0	C_{15}:0	C_{16}:0	C_{16}:1	C_{17}:0	C_{17}:1	C_{18}:0	C_{18}:1	C_{18}:1*n*-9 (OH)	C_{18}:2	C_{18}:3	C_{18}:39c, 11t, 13t	C_{20}:0	C_{20}:1	C_{20}:1*n*-n,11t, (OH)	C_{20}:2	C_{20}:5	C_{22}:0	C_{22}:1	C_{24}:0	C_{24}:1	Unknowns
Lesquerella fendlen[a]				0.1		0.9	0.3			1.7	13.0		5.8	10.6		0.7		66.5						0.4	
Linseed						4.4				3.8	20.7		15.9	54.6		0.2					0.3		0.1		
Moringa oleifera						5.5	1.2			5.8	76.3		0.7			3.1	2.0				4.2		0.4		0.8
Mustard						2.6	0.2			1.2	20.6		20.6	13.3		0.9	10.7		1.0		0.5	25.6	0.2	1.5	1.1
Neem						14.9	0.1			20.6	43.9		17.9	0.4		1.6					0.3		0.3		
Palm			0.2	0.5		43.4	0.1			4.6	41.9		8.6	0.3		0.3					0.1				
Perilla seed						5.3	0.1			2.2	16.6		13.7	62.1											
Poultry fat			0.1	1.0		16.6	3.2	0.3	0.2	7.5	36.8		28.4	2.0		0.1			0.1		0.3	0.4			
Rice bran				0.3		12.5				2.1	47.5		35.4	1.1		0.6					0.3		0.2		
Soybean						9.4				4.1	22.0		55.3	8.9							0.3				
Stillingia			0.4	0.1		7.5				2.3	16.7		31.5	41.5											
Sunflower						4.2				3.3	63.6		27.6	0.2							0.7		0.4		
Tung						1.8				2.1	5.3		6.8	0.7		0.2			0.1				10.4		0.4
Used cooking oil			0.1	0.1		11.8	0.4	0.1	0.1	4.4	25.3		49.5	7.1		0.3					0.4	0.3	0.1		
Yellow grease			0.1	0.05		14.3	1.1	0.3	0.2	8.0	35.6		35.0	4.0		0.3					0.3	0.2			

Source: Christie, W. What is a lipid? The common fatty acids of animal and plant origin. *The AOCS Lipid Library*, July 25, 2013 Web. Accessed on May 7, 2014, http://lipidlibrary.aocs.org/Lipids/whatlip/index.htm.

[a] In the GC/FID chromatogram, the hydroxy ester peaks were missing. The quantity of the hydroxy peaks was estimated from the hydroxyl value with the assumption that all the hydroxyl value was the primary hydroxy acid in the sample.

1.4.3 A Brief History of Processing Natural Oils for Fuels

It was identified in as early as 1939,[43] and perhaps even earlier, that fats, greases, mineral oils, and other biological compounds containing heteroatoms like oxygen could be converted via high-pressure hydrogenation processes into hydrocarbons for reuse in various applications. Over the intervening years, researchers experimented with new materials (e.g., zeolites) and process combinations (e.g., pyrolysis followed by hydrogenation) to improve the applicability and economics of the use of natural oils. One of these such studies was conducted in Brazil[44] in 1983. Alencar et al. showed that pyrolysis followed by hydrogenation in a Parr reactor using a Pt-based catalyst resulted in the formation of a mixture of hydrocarbons from palm, babassu, and piqui oils. Yields were reported to be quite high—over 95% for palm oil.[44]

In the early 1990s, the Saskatchewan Research Council led a Canadian consortium that investigated the prospects of coprocessing vegetable oils with petroleum feedstocks using typical hydroprocessing catalysts.[45] Stumborg et al. reported the production of super cetane, a high-cetane diesel fuel additive from the conversion of canola or other vegetable oils on hydroprocessing catalysts.[35] Tall oil, a by-product of the krafting process (particularly from pine trees), was also used as a feedstock. Conversion over typical Ni–Mo hydroprocessing catalysts produced a liquid hydrocarbon product with a cetane index ranging from 55 to 90, but the cold-flow properties were such that the additive could not be used in winter. In addition, the authors reported that there were remaining challenges associated with catalyst deactivation.[46]

The conversion of natural fats and oils via processes similar to the hydroprocessing of petroleum feedstocks has been the subject of considerable research over the past 20 years. While the subject has been a portion of several comprehensive reviews on the broader subject of bioresources for fuels and chemicals,[44–48] perhaps the best summary of the hydroprocessing of natural fats and oils to date has been provided by Furimsky[49] and Kordulis et al.[50] The reader is referred to these two critical reviews for an understanding of the state of the art.

1.4.4 The Chemistry of Triglyceride and Fatty Acid Conversion to Fuels

1.4.4.1 Deoxygenation and Saturation

There are substantial differences in the chemical composition of triglycerides and petroleum. Unlike triglycerides, middle distillate fuels derived from petroleum do not contain oxygen or olefin species. With three atoms of oxygen present per triglyceride molecule, removal of oxygen, or deoxygenation, and saturation of double bonds are important chemical reactions in the conversion of triglycerides to hydrocarbon biofuels.

Deoxygenation and saturation are typically done at elevated temperature and pressure over base metal catalyst in the presence of excess hydrogen.[51] The triglyceride oil feed is mixed with hydrogen, heated to reaction temperature, contacted with the catalyst in the reactor, and the reactor effluent is cooled and separated. Saturation reactions occur most rapidly, where hydrogen deficiencies along the fatty chains are hydrogenated. Deoxygenation can occur via two major pathways: (1) hydrodeoxygenation, where the oxygen atoms of the triglyceride combine with available hydrogen to form water (H_2O), or (2) decarbonylation/decarboxylation, where the triglyceride backbone retains a linkage to the first carbon of the fatty chain forming carbon dioxide (CO_2) or carbon monoxide (CO).

The saturation and deoxygenation reactions both consume hydrogen and evolve significant heat. Commercial-scale production facilities must be designed to manage reactor exotherm to keep operating temperatures within equipment constraints and minimize thermal fouling that could result in buildup of deposits that lead to reactor plugging. With CO, CO_2, H_2O, and H_2 simultaneously present, the water–gas shift reaction and its reverse will interconvert the molecules, depending on the catalyst selection and reaction conditions.[52] Once deoxygenated and saturated, the fatty acid chains of the original triglyceride have been converted to long linear paraffin chains (normal alkanes) that are fully hydrocarbon. **Reactions 1** and **2** show the reaction scheme for hydrodeoxygenation and decarboxylation on a model triglyceride molecule composed of myristic (14:0), palmitic (16:0), and oleic (18:1) acids. **Reaction 3** shows the reverse water–gas shift reaction.

Reaction 1

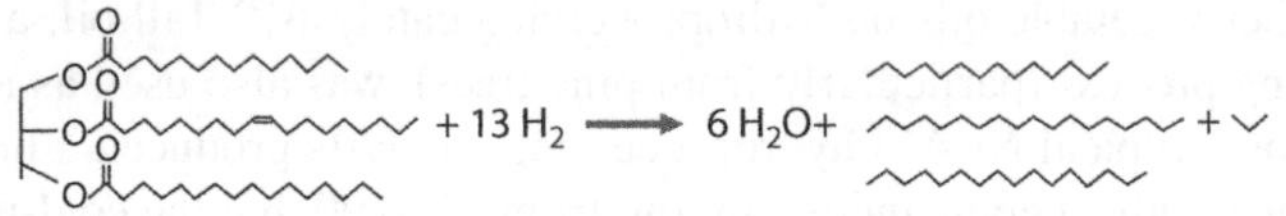

$$\text{Triglyceride} + 13\,H_2 \longrightarrow 6\,H_2O + \text{n-paraffins} + \text{propane}$$

Reaction 2

$$\text{Triglyceride} + 4\,H_2 \longrightarrow 3\,CO_2 + \text{n-paraffins} + \text{propane}$$

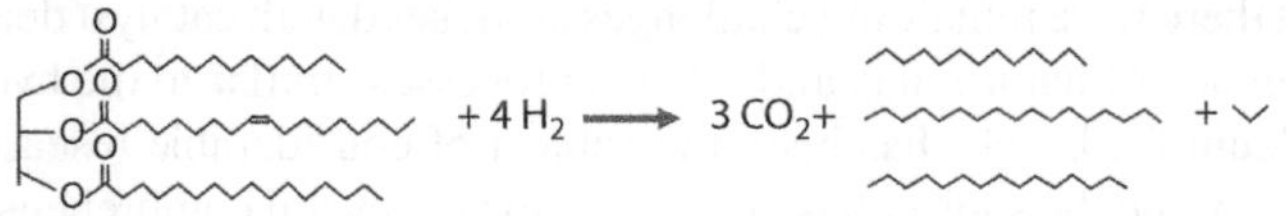

Reaction 3

$$CO_2 + H_2 \rightleftharpoons H_2O + CO$$

Table 1.9 shows the theoretical range of products and by-products that would be expected at various levels of hydrodeoxygenation and decarboxylation of camelina oil. While hydrodeoxygenation prevails relative to decarboxylation/decarbonylation,

TABLE 1.9
Theoretical Deoxygenation Product Profile of Camelina Oil at Various Levels of Hydrodeoxygenation (HDO) and Decarboxylation/Decarbonylation ($DeCO_x$)

Fraction HDO	mole frac.	0.1	0.5	0.9
Fraction $DeCO_x$	mole frac.	0.9	0.5	0.1
Consumed H_2	mass%	2.2	3.0	3.8
H_2O	mass%	1.2	6.0	10.9
CO + CO_2	mass%	13.3	7.4	1.5
Propane backbone	mass%	4.9	4.9	4.9
Naphtha + jet + diesel *n*-paraffins	mass%	82.7	84.6	86.5

the yield of paraffinic product increases. This is because hydrodeoxygenation retains the entire fatty acid carbon chain intact and converts it to an even-number *n*-paraffin, whereas decarboxylation/decarbonylation shortens the fatty acid chain by one carbon in making CO_2 or CO and converts it to an odd-number *n*-paraffin. Because of this, the quantity of paraffins containing an odd number of carbons can be used as a measure of the decarbonylation/decarboxylation reaction. The yield advantage of hydrodeoxygenation is counteracted by the greater consumption of hydrogen relative to decarboxylation/decarbonylation. The deoxygenation/saturation operation must be managed to complete the deoxygenation, saturate all carbon–carbon double bonds, and reduce trace contaminants that could poison the isomerization section catalyst or lead to off-specification products.

1.4.4.2 Isomerization and Cracking

The *n*-paraffins that result from deoxygenation/saturation of animal and plant triglyceride oils are most often of carbon number 15 (nC_{15}, *n*-pentadecane) through 18 (nC_{18}, *n*-octadecane). In terms of boiling points, the *n*-paraffins bracketed by nC_{15} and nC_{18}, with normal boiling points of 270°C and 316°C, respectively, fall firmly in the typical diesel boiling range (140–370°C) and nearby the typical jet boiling range (140–300°C). However, the waxy nature of *n*-paraffins at normal ambient temperatures precludes them from serving as drop-in diesel or jet fuels in most situations. Therefore, some amount of dewaxing via isomerization and cracking is required to convert the *n*-paraffins into usable diesel and/or jet fuels composed of normal and *iso*-paraffins.

Isomerization and cracking reactions can be carried out over various types of catalysts, such as bifunctional noble metal–acid catalysts and base metal–acid sulfided catalysts. Isomerization and cracking are done at elevated temperature and pressure in the presence of hydrogen. As the *n*-paraffin rearranges to form branched isomers (*iso*-paraffins), the more-branched isomers have increased susceptibility to cracking. Therefore, as the reaction conditions are adjusted to increase isomerization, inevitably the amount of cracking also increases, making the balance of process conditions and product properties a delicate one. Figure 1.18 shows the general reaction pathway for isomerization and associated cracking on a bifunctional catalyst.[51,53–56]

For diesel production, it is most economical to operate the isomerization–cracking section such that enough isomerization occurs to meet the target cold-flow properties, such as cloud point, without allowing any more isomerization or cracking than necessary. Cracking converts paraffins that are already appropriate carbon number

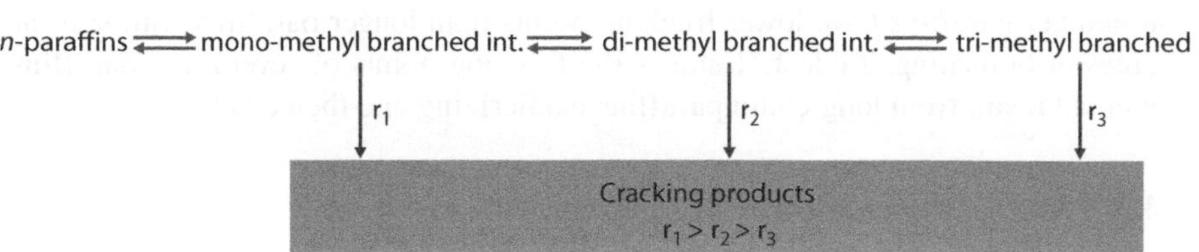

FIGURE 1.18 Isomerization and cracking reaction pathway via branched intermediates (int.) on bifunctional catalyst.

TABLE 1.10
Freezing Points of Isomers of Nonane

Species	n/i	Freezing Point (°C)
n-Nonane	normal	−53
2-Methyloctane	iso	−81
2,2-Dimethylheptane	iso	−113
2,2,4-Trimethylhexane	iso	−120
2,2,3,4-Tetramethylpentane	iso	−121

TABLE 1.11
Freezing Points of *Iso*-Paraffin Cracking Products

Species	Carbon No.	Freezing Point (°C)
iso-Butane	4	−159
Isopentane	5	−160
2-Methylpentane	6	−154
2-Methylhexane	7	−118
2-Methylheptane	8	−109
2-Methyloctane	9	−81
2-Methylnonane	10	−74

for diesel into paraffins that are only suitable for lighter products, such as jet, naphtha, or lighter by-products. Increasing the severity of the process conditions will produce more cracking and therefore more jet fuel; however, there is a point where cracking reactions begin to dominate, and the benefit of increasing reaction severity is diminished due to jet and diesel yield losses.

The combination of isomerization and cracking of triglyceride-derived normal paraffins can have a dramatic effect on the cold temperature performance of the biofuel products. As a paraffin molecule of a given carbon number is rearranged to have branches, the freezing point generally decreases relative to the normal paraffin of the same carbon number. Table 1.10 shows the freezing points of several pure paraffin isomers of nonane (C_9H_{20}). Conversion of long-chain paraffins into cracked products also helps improve the cold temperature performance of a fuel because lower carbon number paraffins have lower freezing points than longer paraffins with similar degrees of branching. Table 1.11 shows the freezing points of several *iso*-paraffins that could result from long-chain paraffins isomerizing and then cracking.[57]

1.4.5 Metals Removal from Biofuel Feedstocks

It is quite common to need to remove nickel (Ni) and vanadium (V) from petroleum-derived feeds to hydroprocessing units as they are known to be poisons to the catalysts. Renewable triglyceride oils more often contain trace levels of different metals: iron (Fe), calcium (Ca), magnesium (Mg), potassium (K), and sodium (Na). Additionally, phosphorous (P), typically from residual phosphatides, is a common elemental contaminant. It has been shown that these "bio-metals" can be deoxygenation catalyst poisons.[50] Methods to deal with such contaminants may include more rigorous triglyceride oil refining, especially in the bleaching step, or use of demetalization catalysts similar to those used in petroleum hydrotreating to remove metals upstream of the deoxygenation/saturation catalyst. If feed metals and phosphorous are allowed to break through the entire process, they can cause the final products to be off-specification. This can particularly be a problem for jet fuels derived from triglyceride oils. For example, the ASTM D7566 specification[58] for renewable jet fuels has a tight limit of 0.1 mass-ppm on all individual metals and phosphorous.

1.4.6 Recent Insights and Remaining Questions

Academic interest has developed rather slowly in hydrotreating of natural fats and oils, and only particularly in the past 10 years since the commercial launch of related process technology. Research has been focused on three basic areas:

- Coprocessing of natural fats and oils with petroleum fractions in existing infrastructure
- Evaluation of the interactions that occur in simultaneous removal of different heteroatoms (e.g., O, S, and N removal)
- Understanding the impact of oil type and source on the yields and performance of different catalysts

Although the focus was on a much broader set of biosourced feedstocks, the 2013 review of hydroprocessing of biofeeds by Furimsky[49] offers a good review of the subject. Very recently, Kordulis et al.[50] published a comprehensive review of nickel-based catalysts. Based on our own experiences, and factoring in the work conducted since Furimsky's review was published, we would offer the following observations about the state of the art in hydroprocessing of natural fats and oils.

To achieve the objective of producing fuels that are chemically indistinguishable from their petroleum-derived counterparts, total elimination of oxygen from the liquid product must be accomplished. As can be seen by nearly every example presented in the literature (for example, see Ref. 59) and noted by Furimsky,[49] this is very difficult to achieve even at the start of operation with fresh catalyst. There are several possible reasons for this, ranging from the presence of phospholipids and other contaminants in the feedstock[39] to the simplistic and inefficient laboratory reactors employed in most studies. It is fair to suggest, however, that catalysts with optimized activity and selectivity for the conversion of natural fats and oils should provide an advantage over catalysts designed for S and N removal from petroleum compounds. In addition to start of run activity, the activity stability of catalysts reported in the literature[59] is also an area of concern for industrial process development.

It is clear that the issues of fit-for-purpose catalysts designed for natural fats and oil feedstocks with suitable stability can be resolved. Both Neste[32,33,35] and UOP[39,40,60] have commercial units in long operation with run lengths matching the observed run lengths for petroleum-derived hydroprocessing units. However, despite the commercial success of these operating units, there are several issues that remain as challenges to be resolved through enhancements in the science and engineering of these units.

Kubička and Horáček[61] investigated the performance of sulfide $CoMo/Al_2O_3$ catalysts for the hydrotreatment of rapeseed oil containing different impurities. The oils were obtained at different points in a standard edible oil processing facility and ranged from the most pure grade (edible oil) to degummed oil to crude rapeseed oil. As a result, the feeds had different levels of FFA, phospholipids, water, and other organic impurities. The authors showed that, under identical processing conditions, the deoxygenation rate stability was strongly dependent on the presence of impurities in the feed—the higher the levels of impurities, the faster the CoMo catalyst deactivated. The authors postulated that the presence of P and alkali metals in the feed was the primary cause of catalyst deactivation.[61] This suggests that better pretreatment/demetallation catalysts could be of interest.

Similarly, it has been postulated that the support material needs to be optimized to handle the presence of higher metals, P, and other "biometals," and the waxy feedstocks. Tiwari et al. from the Indian Institute of Petroleum in Dehradun have investigated the performance of Ni–W on mesoporous silica-alumina and Ni–Mo on mesoporous alumina[62] for conversion of soy oil coprocessed with refinery gas oil. The authors concluded that the Ni–W hydrocracking catalyst had a higher propensity for the decarbonylation/decarboxylation reaction pathway, while the Ni–Mo hydrotreating catalyst had a slightly higher propensity to prefer the hydrodeoxygenation pathway. The authors indicate that coprocessing did not inhibit the ability of the catalysts to remove sulfur from the gas oil, indicating that there is sufficient activity on the catalyst to convert both S and O compounds.[62] However, no information regarding catalyst stability is reported.

In addition to these features, other aspects of the process and catalyst are less understood and either will become important or are currently important factors in application of the technology. For example, catalysts that provide a tunable balance between hydroxygenation and decarbonylation/decarboxylation pathways could be important as the balance between the value of feedstock and H_2 has been quite volatile in recent years. These and other challenges make the conversion of triglyceride feedstocks into fuels a fertile ground for continued research well into the future.

1.5 REFINERY–PETROCHEMICAL INTEGRATION

1.5.1 Refinery–Petrochemical Integration Flow Schemes

More rapid petrochemical demand growth relative to demand for fuels, coupled with increase in naphtha content of crudes from the influx of shale-based production in the United States, is resulting in a widening gap between gasoline and naphtha-derived petrochemical margins, creating new opportunities for refiners to improve value by adding petrochemicals to their product slate. While uncertainty in the oil markets can lead to delays in CAPEX, the value increase from petrochemicals is stable despite fluctuations in crude oil prices.

While product streams, such as naphtha or LPG, have provided a common interface between refinery and petrochemical businesses, investing in "on purpose" integration is a relatively recent industry trend. In most developed regions, refinery configurations are similar to that shown in Figure 1.19, with conversion units such as residue upgrading, fluid catalytic cracking for gasoline production, catalytic reforming for gasoline upgrading, and distillate hydrotreating units to meet fuel quality specifications.[63]

1.5.2 Olefins and Aromatics Production

In many emerging regions, it is now common practice to take an integrated approach that offers the benefits of improved refinery margins from production of higher-valued petrochemical products by upgrading internal refinery streams. An example of full refinery–petrochemical integration is shown in Figure 1.20.[63] Such refiners contribute both to olefins and aromatics chain value. Here, propylene and ethylene generated under high-severity FCC conditions, at the expense of olefins in the gasoline range, are recovered in a steam cracker, which enables additional light olefin production from straight-run and FCC-generated light naphtha (LCN) and LPG. Typical-size FCC units, though, are large enough to support world-scale polypropylene facilities; hence integration to steam cracker is not necessary to justify propylene recovery. Heavy naphtha from high-severity FCC (HCN) is rich in aromatics, hence may not need to be sent to the reformer and can be blended, prior to the hydrotreating step, with pyrolysis gasoline from the steam cracker, which is also rich in aromatics. Selective HDS of aromatics-rich streams has been reported in the literature.[64] The hydrotreated HCN/pyrolysis gasoline can be mixed with the reformer effluent, ready to be sent to the aromatics complex, after having diverted sufficient portion of the stream for blending to the gasoline pool. However, due its high aromatic content, blending this stream with alkylate may be necessary to meet the overall gasoline fuel specifications. In this case, additional mixed butylenes produced under high-severity FCC become a key advantage, as these butylenes can be used to make a high-octane alkylate that is both aromatic- and sulfur-free.[65] New catalysts and higher severity can also be used to increase the aromatics production in the reforming unit. Additional aromatics can also be produced in the LCO Upgrading unit.

To meet EPA MSAT I and MSAT II regulations of 0.62v% BZ in gasoline, BZ may need to be extracted from the reformate/pyrolysis gasoline/HCN pool though, as shown in Figure 1.21. This can be accomplished, for example, through the UOP Sulfolane™ process. Other aromatics extraction technologies are also available.[66] Short of a full integration, this latter option will involve a fractionation unit that will generate a valuable mixed xylene stream, in addition to petrochemical-grade BZ and other aromatics. Toluene, in turn, can be transalkylated with A_9/A_{10} stream from the fractionation to make more mixed xylenes, for example, in a UOP Tatoray™ unit.[67] In case the aromatic pool methyl:phenyl ratio is less than 2.0, some of the toluene can be alkylated with methanol to make more xylenes.[68] The paraffinic raffinate from the aromatics extraction step and the A_{11+} can then be blended for the gasoline pool.[67]

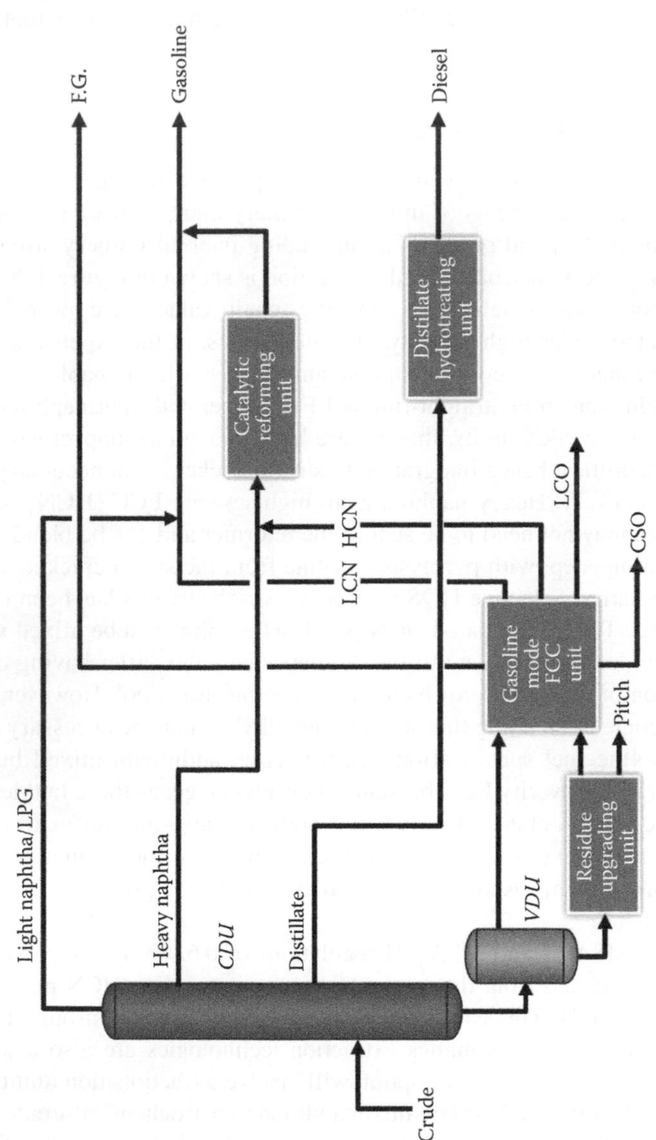

FIGURE 1.19 Traditional fuel refinery configuration.

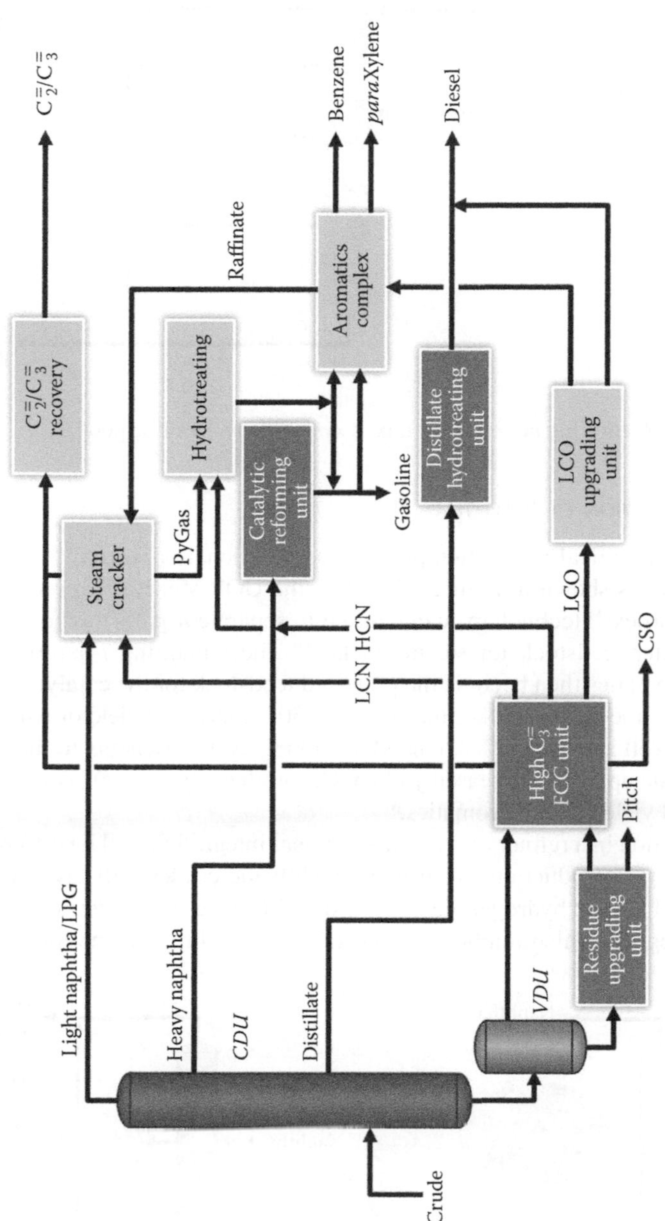

FIGURE 1.20 Fully integrated refinery–petrochemical complex.

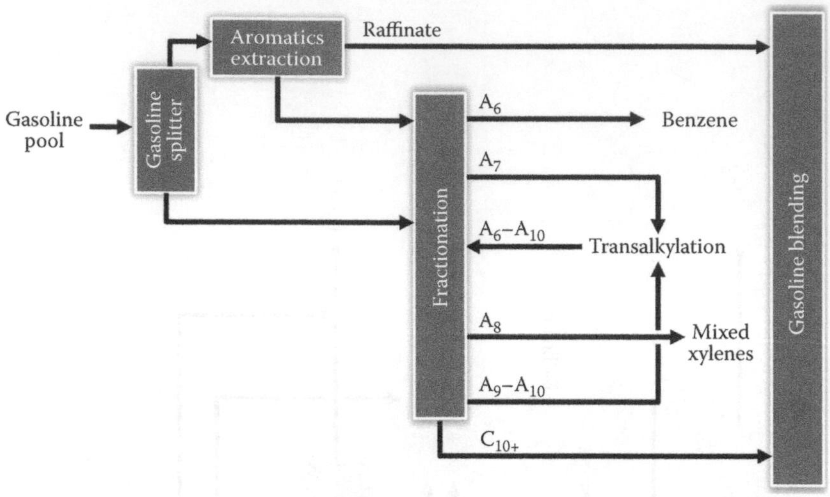

FIGURE 1.21 Recovering benzene and mixed xylenes from gasoline pool.

1.5.3 Segregation of Naphtha

Refinery–petrochemical integration provides an additional opportunity by segregation of naphtha, as shown in Figure 1.22. Here the UOP MaxEne™ process, which uses UOP's Sorbex™ technology, can be used to separate *n*-paraffins from naphtha to make a better feedstock for steam cracker.[69] The remaining *i*-paraffins, naphthenes, and aromatics then become the preferred feedstock for the catalytic reformer unit. Benefits to the steam cracker include up to 30% increased yield of ethylene and propylene, as well as up to 50% extended de-coking cycles. Benefits to the catalytic reformer include up to 7% increased yield of C_{5+} products (gasoline), as well as up to 12% increased yield of total aromatics.[67]

Crude selection in a refinery with petrochemical integration will be different than for traditional fuel production, since more naphthenic crudes will favor aromatics production and higher hydrogen content feeds will make more light olefins in the FCC unit. Integration also implies that the site will need to become more intimate

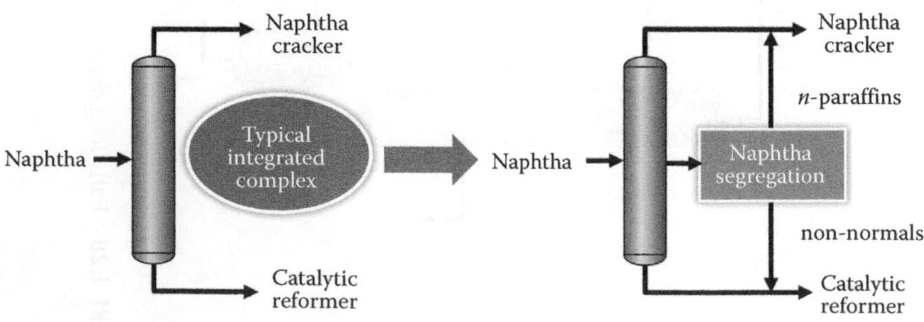

FIGURE 1.22 Segregation of naphtha feed for steam cracker and reformer.

with the characterization, modeling, and manipulation of high-value petrochemicals, such as aromatic ring potential, through a reformer or olefin selectivity in FCC.

1.5.4 Requirements and Characteristics for Key Catalysts in a Refinery–Petrochemical Complex

FCC process is the key enabler for refinery–petrochemical integration. Recent developments in FCC catalysis, including description of key catalytic components, structure–function relations, approaches to increase propylene selectivity, approaches to improve structure stability, as well as the use of new zeolite framework types, have been summarized in a recent review article.[70] The key cracking component of the FCC catalyst is USY zeolite, and an indicator of its content in the catalyst is the micropore surface area. USY has a low unit cell size and is rare earth–exchanged to provide an adequate level of hydrogen transfer activity. A second component of the catalyst is an active alumina, which is part of the matrix and helps in the cracking of the larger feed molecules. The mesopore surface area of the catalyst gives an indication of the active alumina content. In addition, a shape-selective zeolite additive with a pentasil structure, i.e., ZSM-5, is used to enhance propylene production. The most common mode of operation of the FCC unit is aimed at the maximum production of gasoline. Higher severity operation with additive results in production of additional light olefins and a higher octane gasoline.[65] Increasing the ZSM-5 additive from 3–5 to ~30 wt%, increasing the temperature from 490–540°C to >550°C, and introducing as much as 10–30 wt% steam lead to propylene yield increase from 4–7% to 12–22% and to butylene yield increase from 4–8% to 8–14% at the expense of gasoline, which is lowered from 47–53% to 30–40%. To increase propylene yield by an additional 2–3 wt%, it is useful to have a recycle naphtha cracking reactor, coupled to the VGO-fed riser.[65] Such naphtha cracking reactors operate at even higher temperatures, i.e., much higher than 560°C, and higher cat/oil ratio, i.e., ~30. These reactors benefit from minimizing back-mixing, and a down-flow reactor is one way of achieving such plug flow behavior. A catalytic naphtha cracking reactor in a refinery–petrochemical complex can be stand-alone and does not need to be coupled to an FCC riser, which processes VGO and can process feed streams like straight-run, coker, and visbreaker naphthas, pyrolysis gasoline, condensate, and C_4 to C_6 olefinic feeds.[65]

Energy balancing the endothermic cracking reaction in the presence of low delta coke requires high-temperature regeneration at 750–800°C; hence hydrothermal stabilities of USY and especially of ZSM-5 are critical. ZSM-5 dealumination from the framework can be monitored by solid-state Al NMR or by XRD unit cell volume measurements and will occur unless P-type modifiers are used to stabilize the aluminum. The role of postsynthetic P-modification of ZSM-5 for improving hydrothermal stability under FCC conditions has been the subject of a recent review article.[71] Phosphorous–aluminum interaction is the key driving force behind physicochemical changes in P-modified ZSM-5. P interacts with framework Al to form hydrothermally stable silicon aluminum phosphate (SAPO) interfaces, and P interaction with extra-framework Al results in an amorphous $AlPO_4$ species. These interactions lead to changes that impact not only structure stability of ZSM-5 but also its acidity and shape selectivity. The hydrothermal stability of such P-modified catalysts is often verified in lab-scale prescreening measurements that involve treatment under 100% steam at ~800°C.[72] Presence of Ni and V in the feed is another source of concern, as these metals accelerate catalyst deactivation by blocking pores and by disrupting distribution of acid sites through reaction with the high surface area matrix element. Metal poisoning in FCC catalysts is managed by optimizing matrix porosity and designing metal trap.[72,73]

While steam cracking is a thermal process operating at 750–800°C based on a free radical mechanism, under high-severity FCC, feed molecules are activated at ~550°C by forming carbenium and carbonium ions. Carbenium ions are formed over catalytic acid sites by protonation of the light olefins that are generated by the small amount of thermal cracking that takes place under FCC conditions. Chain propagation mostly occurs inside the large cages of Y zeolite by hydride transfer from large feed molecules to small carbenium ions, which then form the more stable large carbenium ions. The latter can, in turn, crack into multiple small carbenium ions in successive beta-scissions steps, which give a smaller olefin and a smaller carbenium ion each time. In parallel with this bimolecular mechanism, a monomolecular mechanism that involves direct protonation of the paraffinic compounds occurs. The carbonium ions can then split into a smaller paraffin, i.e., methane, ethane, propane, and a smaller carbenium ion, which can, in turn, deprotonate to give the olefin. It is critical to keep the hydride transfer activity of the Y zeolite just at the right level to ensure facilitation of bimolecular activation without causing hydride transfer to valuable propylene. To maximize light olefins, hydride transfer activity of ZSM-5 is also minimized by carefully regulating the Si/Al ratio and utilizing appropriate modifiers. Under FCC conditions, large molecules in the VGO range are first cracked inside Y zeolite to LCO and gasoline range.[74] This is called primary cracking. Secondary cracking of gasoline range olefins and production of light olefins occur mostly inside ZSM-5.[73] Large mesopores in USY enable access of large feed molecules to acid sites in micropores, as well as the rapid removal of valuable products. Recent innovations in the field include catalysts for which mesoporosity is provided by zeolite alone, rather than by a matrix element.[75] Gasoline-range paraffins are not reactive under FCC conditions unless temperatures much higher than 560°C are used. The monomolecular mechanism becomes increasingly favorable over strong acid sites, in medium pore size zeolites, at high temperatures, and at low hydrocarbon partial pressures.[76]

UOP's PetroFCC™ can produce a range of propylene yields from 7% to 19% by utilizing a combination of specific processing conditions, selected mechanical hardware, and a high concentration of ZSM-5 additive in a single riser.[77] Even with high-severity operation, due to equilibrium constraints, the C_4 and C_5 olefins will always exist in significant quantities. In case there is no need to use mixed butylenes to make an alkylate, the C_4/C_5 olefin stream can be recycled to the riser and recracked to make additional propylene, along with some ethylene. However, the single-pass conversion of C_4/C_5 olefins will be relatively low under FCC conditions. One way to overcome this becomes possible by integration of FCC with Catolene™, the UOP technology for dimerizing and trimerizing C_4/C_5 olefins. With C_4/C_5 oligomer recycle, the propylene yield can be increased to 22–24%.[77]

The key characteristics and status of various other FCC-based technologies targeting olefin production have been summarized in the multiclient 2013 study by

The Catalyst Group Resources and in various other publications.[78] Sinopec's Deep Catalytic Cracking (DCC) is the only commercial FCC process with double risers and is licensed only in China. Sinopec's Catalytic Pyrolysis Process (CPP) is licensed outside China by Stone & Webster. Both DCC and CPP process VGO-type heavy feeds at 550–595°C in the presence of 6–30 wt% H_2O and can produce 20% or higher propylene.

HS-FCC is developed by an alliance of Saudi Aramco, JX Nippon Oil & Energy, and King Fahd University of Petroleum and Minerals and is licensed by Axens/Technip/Stone & Webster. HS-FCC provides a propylene yield of 15–19% from a variety of VGO feeds in a down-flow reactor at 550–650°C and 20–40 cat/oil ratio.[79] Advanced Catalytic Olefins (ACO™) is developed by SK Global Chemical and licensed by Kellogg Brown & Root (KBR), and is used to selectively crack paraffinic naphtha and light distillate streams into light olefins at ~650°C. Following recovery of aromatics, non-aromatics are recycled back to the riser to achieve a total yield of 65% propylene + ethylene $\left(1C_3^= : 1C_2^=\right)$.

1.6 IMPROVING ENERGY EFFICIENCY WITH SIMPLIFIED PROCESS DESIGN

1.6.1 Challenges for Increased Process Energy Efficiency

In the nineteenth century, oil refineries processed crude oil primarily to recover kerosene for lanterns, and the refinery design was very simple. The invention of the automobile shifted the demand to gasoline and diesel, which remain the primary refined products today. Transportation fuel production for automobiles required more conversion and hence increased plant complexity. In the 1990s, stringent environmental standards for low sulfur drove refineries to include more hydroprocessing, leading to greater refinery complexity. In recent times, higher energy efficiency and lower greenhouse emission requirements have resulted in designs incorporating more heat recovery with additional equipment and thus have further elevated the process complexity. This trend is illustrated by Figure 1.23.

As a leading technology company in the refining and petrochemical industry, UOP is at the forefront of efforts to tackle this challenge. In the last few years, UOP has demonstrated good vision and has invested in developing process technology that is first class in energy-efficient design while maintaining low capital costs through application of process integration methodology.[80] As a result, major improvements have been and continue to be made to UOP key process technologies including FCC, MTO, Oleflex, hydrocracking, Uniflex, and diesel hydrotreating processes, as well as to overall aromatics complexes, phenol complexes, and naphtha complexes.

1.6.2 Systematic Process Integration Methodology

Traditional energy efficiency improvements have a narrow focus on energy recovery alone with little consideration of interactions with process flowsheeting, equipment design, and process conditions. As a result, energy-saving projects usually have limited economic benefits and thus have difficulty in competing with capacity

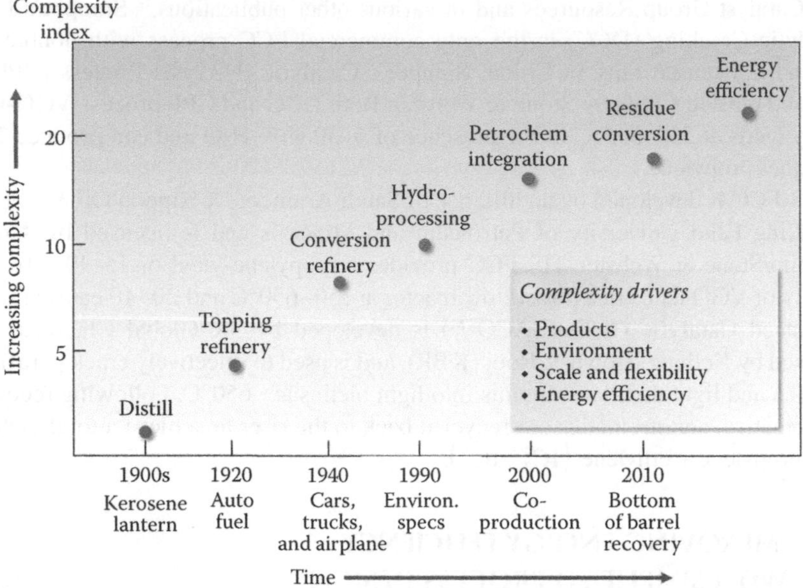

FIGURE 1.23 Trend of increased refinery complexity over time.

and yield-related projects. In contrast, the process integration methodology[80] as explained in Figure 1.24 takes a different approach in that energy optimization is closely integrated with changes to both process flowsheeting and conditions, as well as equipment design.

This methodology consists of four core components, namely process simulation development, equipment rating analysis, process integration analysis, and opportunity interaction optimization. This methodology has been applied to numerous design projects with common features such as

- Clearly defined needs, objectives, scope, and basis
- Reduced capital investment and operating cost to achieve the objectives
- Simplified process design and enhanced equipment performance
- Increased possibility of getting customer satisfaction

The purpose of process simulation development is to represent current plant design for the base case, defined in terms of key operating parameters and their interactions. Thus, the simulation can provide the specifications for equipment rating assessment.

The key role of equipment rating analysis is to assess equipment performance and identify equipment spare capacity and limitations. Utilization of spare capacity can enable expansion up to 10–20% in general and accommodate improvement projects with low capital cost investment. When equipment reaches hard limitations—for example, a fractionation tower reaches its jet flood limit, a compressor reaches its flow rate limit, or a furnace reaches its heat flux limit—it could

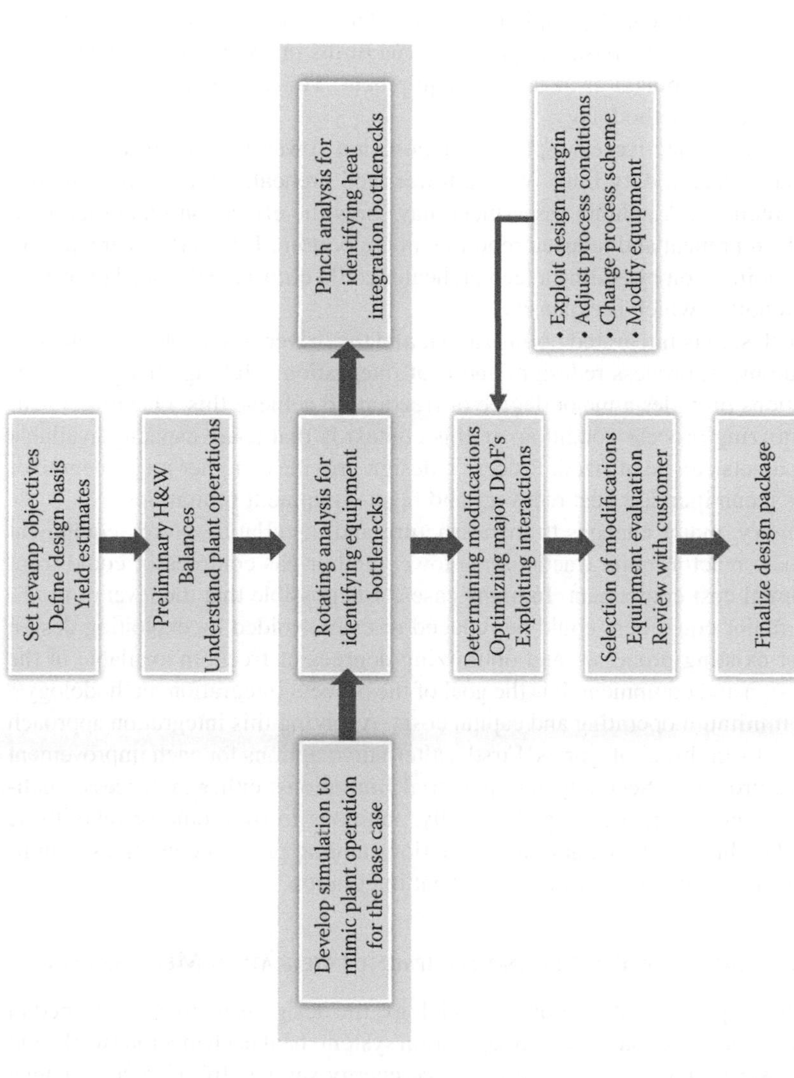

FIGURE 1.24 Process integration methodology. (From Zhu, X.X. *Energy and Process Optimization for the Process Industries*, Wiley/AICHE, New Jersey, 2014.)

be expensive to replace or install new equipment. The important part of a feasibility study is to find ways to overcome these constraints, which is accomplished in the next two steps.

The third step is to apply the process integration methods to exploit interactions and identify changes to process conditions, equipment, process redesign, and utility systems, with the purpose of shifting plant bottlenecks from more expensive to less expensive equipment. By capitalizing on interactions, it is possible to utilize equipment spare capacity, push equipment to true limits in order to avoid the need to replace existing equipment or install new equipment. This is a major feature of this process integration methodology.

A simple and effective example is fractionation tower feed preheat. A tower reboiler could reach a duty limit. With a tower feed preheater, the required reboiling duty is reduced. A column assessment may show the effects on separation with increased feed preheat and reduced reboiling at the bottom. If the effects are acceptable, this modification by adding a feed preheater could eliminate the need of installing a new reboiler, which is expensive.

The fourth step is integrated optimization, and the driver is to exploit interactions among equipment, process redesign, and heat integration. Making changes to process conditions provides a major degree of freedom to achieve this. One direct benefit of optimizing process conditions in this context is that spare capacity available in existing assets can be utilized. Process redesign provides another major degree of freedom as it can increase heat recovery and relax equipment limitations.

In summary, major changes to infrastructure and installation of key equipment such as a new reactor, main fractionation tower, and/or gas compressor could form a major capital cost component. In many cases, it is possible that the level of modification to major equipment could be reduced or even avoided by exploiting design margins for existing processes and optimizing degrees of freedom available in the existing design and equipment. It is the goal of the process integration methodology[80] to achieve minimum operating and capital costs. Applying this integration approach can give results in three categories. Firstly, alternative options for each improvement idea will be provided. Secondly, any potential limitations, either in process conditions or equipment, will be flagged. Thirdly, solutions to overcome or relax these limits will be obtained by exploiting interactions among process conditions, equipment performance, process redesign, and heat integration.

1.6.3 Application of Process and Equipment Integration Methodology

By applying the process integration methodology, the design is no longer confined in a subsystem, from reaction system to separation system, heat exchanger network, and heat and power systems. In many studies, the energy savings from process change analysis far outweigh those from heat recovery projects. The following examples are used to demonstrate the effectiveness of the process integration methodology. The first example involves improving energy efficiency for a single process design, while the second deals with optimizing an overall complex consisting of multiple process units.

Example 1.1: Improving Hydrocracking Process Efficiency

Consider improvements to a hydrocracking process unit. A traditional hydrocracking process design features one common stripper receiving two feeds containing very different compositions, with one feed being from the cold flash drum and the other being from the hot flash drum; a typical design is shown in Figure 1.25. The processing objective of the stripper is to remove H_2S from the feeds. These two feeds originate from the reaction effluent, which first goes to the hot separator. The overhead vapor containing light products of the hot separator goes to the cold flash drum, while the bottoms containing relatively heavy products of the hot separator go to the hot flash drum. Eventually, the liquids of both hot and cold drums are fed to the same stripper. The stripper bottoms then become the feed for the main fractionator. The shortcoming of this process sequence can be summarized as follows: separation and then mixing followed by separation again.

Thus, the inefficiency of this single stripper design is rooted in the mixing of the hot flash drum and cold flash drum liquids, which undo the separations upstream. To avoid this inefficient mixing, it is proposed to use two strippers, namely a hot stripper that receives the hot flash drum liquid as the feed and a cold stripper that is used for the cold flash drum liquid.[81] Furthermore, the cold stripper bottoms do not pass through the main fractionator feed heater but go directly to the main fractionator. Only the hot stripper bottoms go to the main fractionator feed heater. This proposed design is shown in Figure 1.26. Implementing this design in simulation indicates that a reduction of 110 MMBtu/h is achieved by the proposed two-stripper design, i.e., 42% reduction of the fractionator heater duty or 23% total energy reduction for the overall hydrocracking unit. This is a very significant improvement, but the process design becomes more complex as a new stripper column must be added together with associated equipment. The next question is, can we simplify the design and reduce the capital cost while improving energy efficiency?

To simplify the overall process design, a stacked column design is adopted as shown in Figure 1.27 in which the cold stripper is stacked over the hot stripper to form a single column with a common overhead system.[82] The vapor from the hot flash stripper, which is below the cold flash stripper, is drawn from the side of the column and is sent to the upper section of the combined stripper. The stripped cold flash liquid leaves the side of the column just above the hot flash stripper top tray. An internal head separates the two columns. This simplified design reduces equipment count, simplifies column overhead operation, and reduces capital cost from the design in Figure 1.26. A more detailed cost/benefit analysis indicates operating cost savings of $2.5 million per year as well as equipment cost savings of $3.0 million.

Example 1.2: Energy Optimization for Overall Refinery Site

This example is about how energy efficiency improvements can support capacity expansion and yield improvement. A refinery wishes to increase hydrocracking capacity by 15% in order to meet new diesel demand. However, a screening study based on simulation and equipment rating indicated several major bottlenecks, which could require significant capital investment and make the expansion too

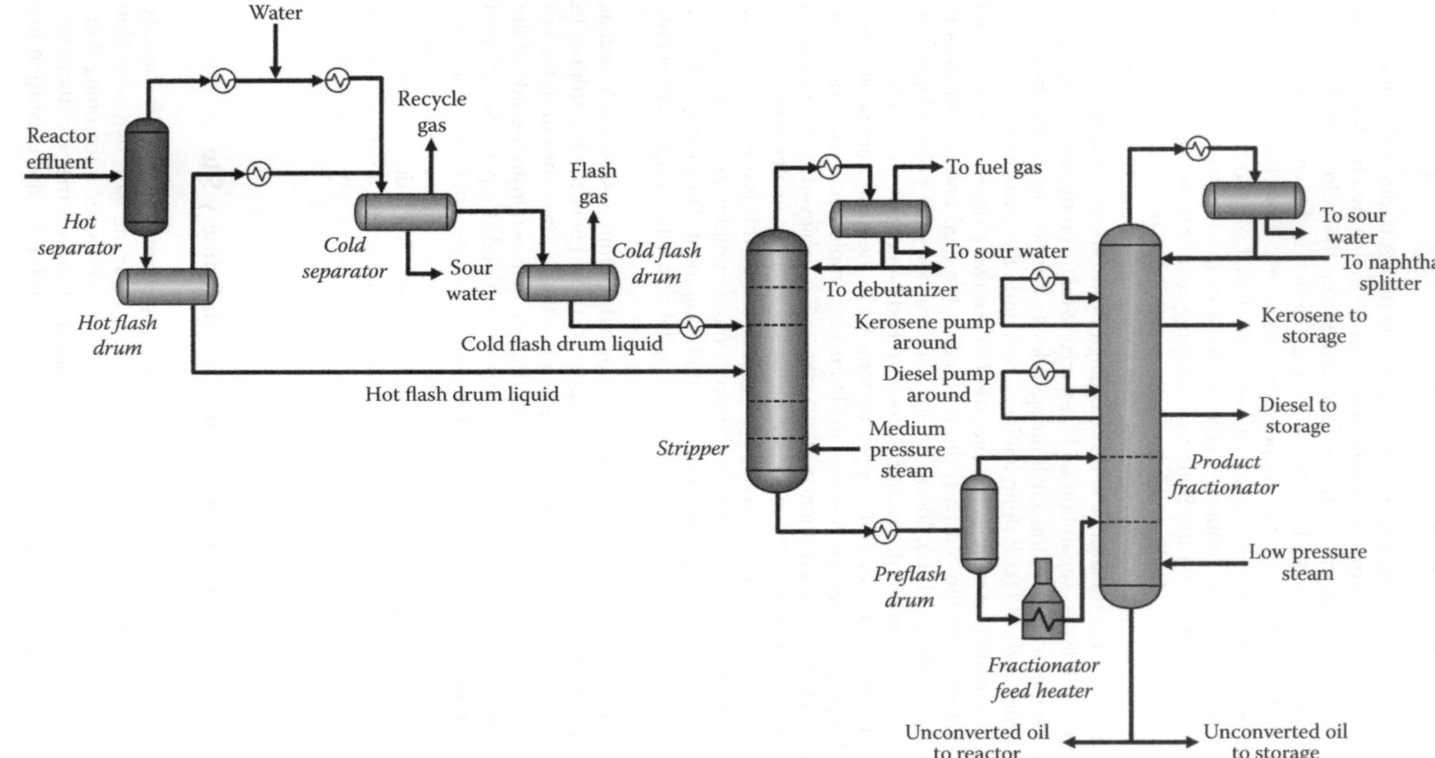

FIGURE 1.25 Single stripper fractionation scheme.

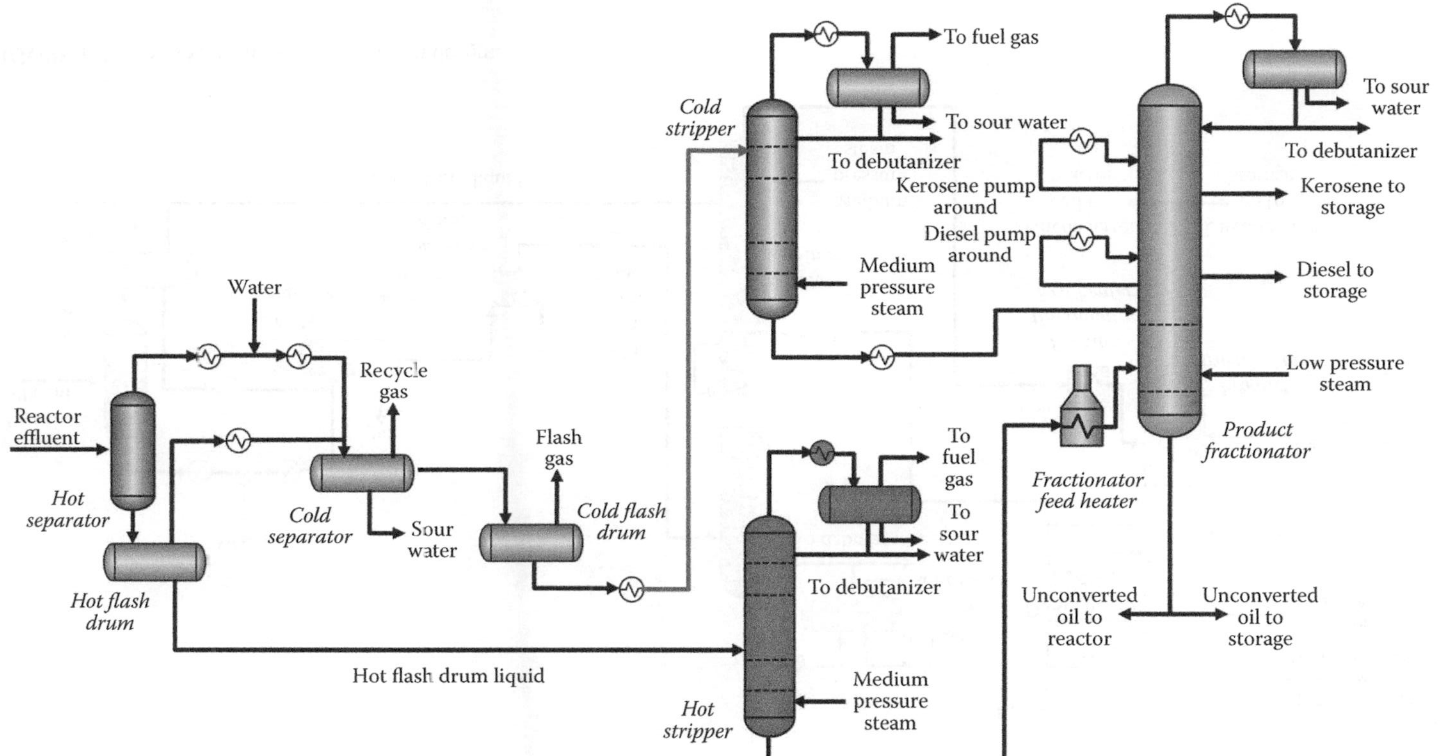

FIGURE 1.26 Proposed two-stripper column design.

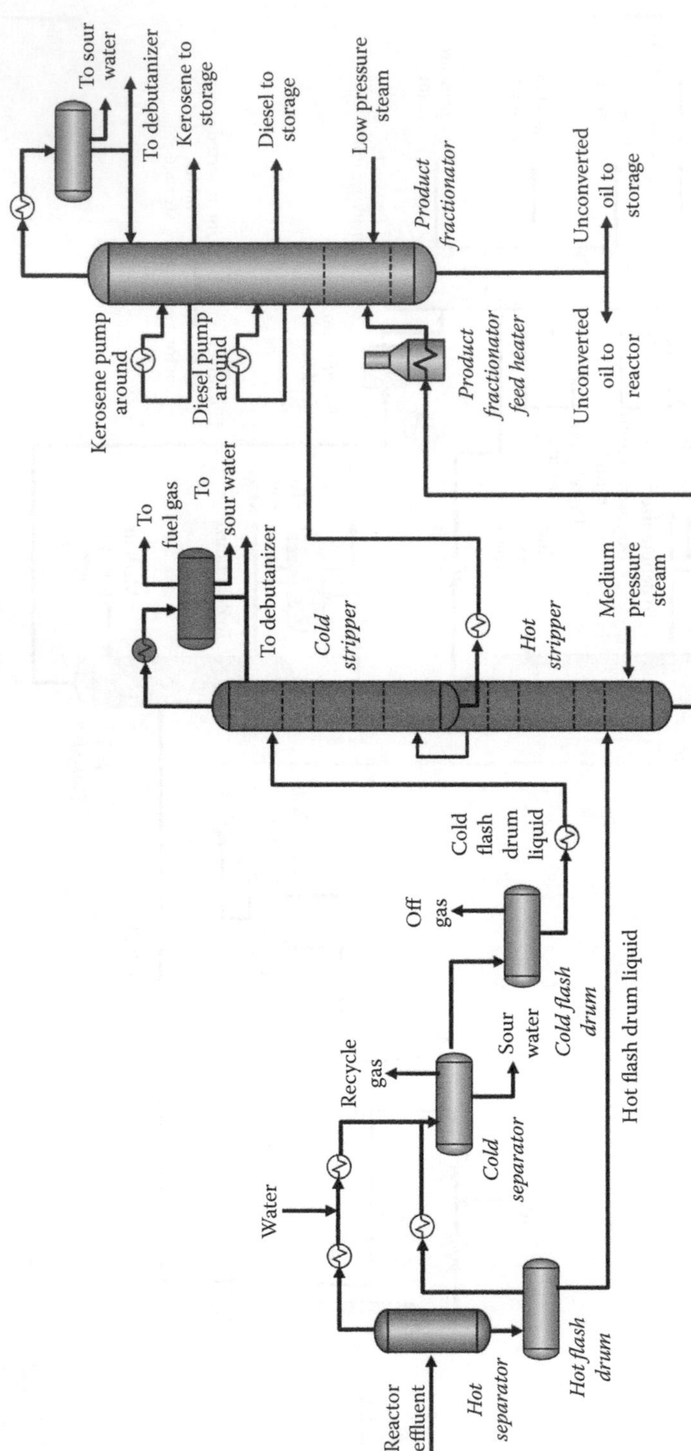

FIGURE 1.27 Stacked two-stripper column design.

expensive. In the screening study, it was found that the hydrocracker was unable to handle a 15% expansion in feed rate because

- The heaters for the reaction and fractionation sections would be too small.
- The space velocity would be too high for the existing reactors.
- The fractionation tower and debutanizer tower would have severe flooding.

By applying process and energy integration methodology, the bottlenecks are overcome with acceptable capital costs. The results are summarized as follows.

Removing the heater bottlenecks: To do this, the network pinch method[83] was applied to the existing heat exchanger network. The method indicated that installation of four heat exchangers was required to use the reactor effluent heat to preheat the reactor feed and the fractionation feed. These new exchangers could reduce total heater duty by 20% or 100 MMBtu/h with a payback of less than 2 years based on energy savings alone. Thus, the need to revamp the reactor charge heaters and the fractionator heater was avoided.

Removing the reactor bottlenecks: The hydrocracking unit was designed in the 1970s. It was not surprising to find that the existing reactor internals experienced poor distribution. By installing better mixing and distribution devices, gas and liquid distribution was improved, and the existing reactors were able to handle the higher space velocity due to the increased feed rate. A new catalyst was also used to deal with diesel cold-flow property issues.

Removing the fractionation bottlenecks: In evaluating both the fractionation tower and debutanizer tower, the simulation model predicted excessive liquid loading and thus downcomer backup flooding. To avoid this, UOP ECMD trays were considered for retrofitting because the enhanced capacity multiple downcomer (ECMD) trays feature multiple downcomers with equalized downcomer loadings. This feature could mitigate downcomer backup flooding and thus allow towers to accommodate more than a 15% increase in feed rate.

During evaluation, a new opportunity was furthermore identified. In the current operation, a 58°C overlap exists between the recycle oil and the gas oil product, which corresponds to a 340°C temperature boiling point (TBP) cut point. Newly designed units will typically have a TBP cut point of 380–410°C and a gap of 10–30°C. The poor separation in the fractionation column results in approximately 13 wt% of the recycle oil being in the gas oil boiling range in order to meet the specifications for the gas oil product. Due to the large amount of gas oil range material that is being recycled, the distillate yield is reduced from overconversion of the gas oil, and excess hydrogen is consumed. Additionally, a higher reactor severity is needed, leading to a higher activity catalyst (or shorter catalyst life) and higher temperatures when compared to a unit with better separation. Also, more HPNA components are made at the higher reactor severity. Recovery of gas oil from the recycle oil (fractionation column bottoms) could be worth $20 million per year for the refinery margin. A possible solution was to add several trays in the column bottom section, and the good news was that there was enough space available to accommodate these trays.

In summary, the modifications for the process unit include the following:

- Four new heat exchangers to reduce duty of both the reaction and fractionation heaters by more than 20%. Thus, the heater bottlenecks were resolved.
- Reactor internals with better mixing and distribution devices to avoid addition of a new reactor or major revamp of the existing two reactors.
- New catalyst to deal with diesel cold-flow property issues.
- Replacement of tower internals in the fractionator and debutanizer columns to mitigate liquid loading issues and avoid severe downcomer flooding.
- Addition of several trays in the bottom section of the fractionation column to reduce the gas oil in the recycle oil from 13 to less than 4 wt%.
- A power recovery turbine was not selected in this revamp package as it was not required for feed expansion but could be considered as a major feature in a future energy-saving project.

The overall study identified energy savings and allowed the desired throughput increase to be attained with minimal capital costs with less than one year payback because expensive modifications were avoided. The synergy between energy savings and process technology know-how was critical to achieving these results.

1.6.4 Summary of Potential Energy Efficiency Improvements

Table 1.12 combines all of the potential saving opportunities to provide a perspective on the level of benefits that could be achieved for a typical refinery of 100,000 barrels per day (BPSD) by adopting the comprehensive process integration methodology. Recent studies[80] indicate that typical benefits of 15% energy reduction are expected and may rise to 30% for refinery plants operating in the fourth quartile in energy efficiency. The resultant energy cost saving opportunity is in the range of $11–23 million per year. The carbon credits associated with the reduction in greenhouse gas emissions range from 180,000 to over 360,000 metric tons per year. For US oil refining

TABLE 1.12
Typical Energy Saving from Different Categories of Opportunities

Saving Opportunities	Energy Improvement (%)	Energy Saving (MM$/Year)	CO_2 Reduction (kMT/Year)
Group 1: Gat basics right	3–5	2.2–3.7	36–60
Group 2: Improved operation and control	2–5	1.5–3.7	24–60
Group 3: Improved heat recovery	5–10	3.7–7.5	60–120
Group 4: Advanced process technology	3–7	2.2–5.2	36–84
Group 5: Utilities optimization	2–3	1.5–2.2	24–36
Total	15–30	11–23	180–360

Note: Basis: 100,000 BPSD refinery; energy price: $6/MMBtu; 330 operating days/year.

capacity at 9 million barrels per day, the energy cost saving opportunity could be in the range of $1–2 billion per year, and greenhouse gas emissions reductions could be in the range of 16–32 million metric tons per year.

1.7 SUMMARY

Changes in the landscape of hydrocarbon feedstock cost and availability have occurred continuously over the >100-year history of the refining industry, providing new opportunities for refiners to capture value through new technology implementation. As this chapter illustrates, breakthrough technologies are now available, which take advantage of hydrocarbon feedstocks as diverse as shale gas liquids, renewable vegetable oils, and heavy gas oils. It is now possible to maximize refinery profitability through molecule management, producing a slate of refinery and petrochemical products with the highest margins while exceeding all environmental and energy efficiency targets. The hydroprocessing process schemes now allow maximum flexibility to produce distillate products meeting the most stringent fuel specifications. Refinery–petrochemical integration will increase refinery complexity, providing additional opportunities for projects addressing higher energy efficiency, which should, in turn, help meet the more stringent environmental regulations of the future. Advanced technologies for shale gas liquid conversion allow refiners to maximize octane of the gasoline pool while producing more high-value petrochemical products from heavier naphtha fractions. The golden age of refining and petrochemicals is not over; it is just beginning.

REFERENCES

1. OPEC, http://www.opec.org/opec_web/static_files_project/media/downloads/publications/WOO_2013.pdf
2. Bricker, M.L., Thakkar, V., Petri, J. *Hydrocracking in Petroleum Processing*, Springer International Publishing, Cham, Switzerland, 2014.
3. Baral, W.J. and Huffman, H.C. Eighth World Petroleum Congress, Moscow, 1971, 4, 119–127.
4. Bolton, A.P., Zeolite chemistry and catalysis, A.C.S. Monograph Series No. 117, Washington, D.C., 1976, p. 714.
5. Mills, G.A., Heinemann, H., Milliken, T.H., Oblad, A.G. *Ind. Eng. Chem.* 1953, 45, 134.
6. Weisz, D.B. *Adv. Catal.* 1962, 13, 137.
7. Sinfeld, J.H. *Adv. Chem. Eng.* 1964, 5, 37.
8. Sinfeld, J.H. *Bimetallic Catalysts: Discoveries, Concepts and Applications*, Wiley, New York, 1983.
9. Baerlocher Ch. and McCusker, L.B. Database of zeolite structures, http://www.iza-structure.org/databases/
10. Topsoe, H.I., Clausen, BS., Massoth, F.E. *Hydrotreating Catalysis*, Springer-Verlag, Berlin, 1996.
11. Bricker, M., Schaal, M.T., Bradley, S., Sanchez, S. 244th ACS National Meeting & Exposition Energy and Fuels Division, 8th International Symposium on Hydrotreating/Hydrocracking Technologies, August 19–23, 2012.
12. US Department of Energy Office of Fossil Energy and National Energy Technology Laboratory, "Modern Shale Gas Development in the United States: A Primer," April 2009. Prepared by Ground Water Protection Council, Oklahoma City, OK.

13. BP Energy Outlook 2030, http://www.bp.com/content/dam/bp/pdf/energy-economics/energy-outlook-2015/bp-energy-outlook-booklet_2013.pdf
14. Wier, M.J., Sioui, D., Metro, S., Sabitov, A., Lapinski, M. "Optimizing Naphtha Complexes in the Tight Oil Boom," AFPM Annual Meeting AM-14-35, March 2014.
15. Huovie, C., Wier, M.J., Rossi, R., Sioui, D. "Solutions for FCC Refiners in the Shale Oil Era," (AFPM) Annual meeting, Paper AM-13-06, March 2013.
16. Paal, Z. *Catalytic Naphtha Reforming*, 2nd edn., G.J. Antos, A.M. Aitani (eds.), *Chemical Industries*, Vol. 100, Marcel Dekker, New York, 2004, p. 35.
17. Pollitzer, E.L., Hayes, J.C., Haensel, V. *Adv. Chem. Ser.* 1970, 97, 20.
18. Haensel, V., Donaldson, G.R., Riedl, F.J. *Proceedings of the 3rd International Congress on Catalysis*, Vol. 1, G.C.A. Schuit, P. Zweitering, W.M.H. Sachtler (eds.), North-Holland, Amsterdam, 1965, p. 294.
19. Haensel, V. and Addison, G.E. *Proceedings of the 7th World Petroleum Congress*, Vol. 4, J.W. Scott, N.J. Patterson (eds.), Elsevier Publ. Co. Ltd., Barking, England, 1968, p. 113.
20. Smith, R.L., Naro, P.A., Silvestri, A.J. *J. Catal.* 1971, 20, 359.
21. Donnis, B.B. *Ind. Eng. Chem. Prod. Res. Dev.* 1976, 15, 254.
22. Brandenberger, S.G., Callender, W.L., Meerbott, W.K. *J. Catal.* 1976, 42, 282.
23. Hoeffkes, H., Baumgarten, E., Hollenberg, D. *J. Catal.* 1982, 77, 257.
24. Callender, W.L., Brandenberger, S.G., Meerbott, W.K. *Proc. Int. Congr. Catal.*, 5th, Vol. 2, J.W. Hightower (ed.), North-Holland, Amsterdam, 1973, p. 1265.
25. Sinfelt, J.H. Bifunctional catalysts, T.B. Drew, J.W. Hoopes, Jr., T. Vermeulen, G.R. Cokelet (eds.), *Advances in Chemical Engineering*, Vol. 5, Academic Press, New York, 1964, p. 37.
26. International Energy Agency, *Medium-Term Renewable Energy Market Report 2015.*
27. *Ethanol Industry Outlook: 2008–2015 Reports*, Renewable Fuels Association; http://www.afdc.energy.gov/fuels/ethanol_fuel_basics.html, Alternative Fuels Data Center, Energy Efficiency & Renewable Energy, US Department of Energy, and http://www.eia.gov/, US Energy Information Administration (EIA).
28. Schnepf, R. and Yacobucci, B.D. *Renewable Fuel Standard (RFS): Overview and Issues*, Congressional Research Services, March 2013.
29. US Department of Defense, *Operational Energy Strategy: Implementation Plan*, March 2012.
30. Demirbas, A. *Appl. Energ.* 2009, 86, S108–S117.
31. Fatih, M. and Demirbas, A. *Appl. Energ.* 2009, 86, S151–S161.
32. Neste Oil, "Cleaner solutions," *Neste Oil*, n.d. Web. Accessed on May 15, 2014, http://www.nesteoil.com/default.asp?path=1,41,11991,22708,22709
33. Neste Oil, "Production capacity," *Neste Oil*, n.d. Web. Accessed on May 15, 2014, http://www.nesteoil.com/default.asp?path=1,41,11991,22708,22720
34. Lane, J. "Albemarle: *Biofuels Digest*'s 5-minute guide," *Biofuels Digest*, January 29, 2013. Web. Accessed on May 15, 2014, http://www.biofuelsdigest.com/bdigest/2013/01/29/albemarle-biofuels-digests-5-minute-guide/
35. Lane, J. "Neste Oil: *Biofuels Digest*'s 5-minute guide," *Biofuels Digest*, November 6, 2012. Web. Accessed on May 15, 2014, http://www.biofuelsdigest.com/bdigest/2012/11/06/neste-oil-biofuels-digests-5-minute-guide/
36. Schill, S. "Hydroprocessing goes small scale," *Biodiesel Magazine*, August 19, 2009. Web. Accessed on May 15, 2014, http://www.biodieselmagazine.com/articles/3666/hydroprocessing-goes-small-scale
37. Dynamic Fuels. "About," *Dynamic Fuels*, n.d. Web. Accessed on May 15, 2014, http://www.dynamicfuelsllc.com/about.aspx
38. Egeberg, R., Knudsen, K., Nyström, S., Grennfelt, E.L., Efraimsson, E. "Industrial-scale production of renewable diesel," PTQ Q3 2011, *Haldor Topsoe*, n.d. Web. Accessed on May 15, 2014, http://www.topsoe.com/business_areas/refining/~/media/PDF%20files/Refining/paper_industrial_scale_prod_of_renewable_diesel.ashx

39. Diamond Green Diesel. "Home," *Diamond Green Diesel*, n. d. Web. Accessed on May 15, 2014, http://www.diamondgreendiesel.com/Pages/default.aspx
40. Honeywell. "Italy's largest refiner to use Honeywell's UOP/Eni Ecofining™ Process Technology at Venice refinery," *Honeywell*, November 12, 2012. Web. Accessed on May 15, 2014, http://honeywell.com/News/Pages/Italy%E2%80%99s-Largest-Refiner-To-Use-Honeywell%E2%80%99s-UOP-Eni-Ecofining-Process-Tech-At-Venice-Refinery.aspx
41. Christie, W. What is a lipid? The common fatty acids of animal and plant origin. *The AOCS Lipid Library*, July 25, 2013 Web. Accessed on May 7, 2014, http://lipidlibrary.aocs.org/Lipids/whatlip/index.htm
42. Sanford, S.D., White, J.M., Shah, P.S., Wee, C., Valverde, M.A., Meier, G.R. "Feedstock and biodiesel characteristics report," Renewable Energy Group, Inc., http://www.regfuel.com, 2009.
43. Schrauth, W. US Patent Number 2,163,563, June 20, 1939.
44. Alencar, J.W., Alves, P.B., Craveiro, A.A. *J. Agric. Food Chem.* 1983, 31, 1268–1270.
45. Stumborg, M., Wong, A., Hogan, E. *Biores. Technol.* 1996, 56, 13–18.
46. Huber, G.W., Iborra, S., Corma, A. *Chem. Rev.* 2006, 106, 4044–4098.
47. Naik, S.N., Rout, P.K., Dalai, A.K. *Renew. Sust. Energy Rev.* 2010, 14, 578.
48. Demirbas, A. *Appl. Energ.* 2011, 88, 17.
49. Furimsky, E. *Catal. Today.* 2013, 217, 13–56.
50. Kordulis, C., Bourikas, K., Gousi, M., Kordouli, E., Lycourghiotis, A. *Appl. Catal. B: Environ.* 2016, 181, 156–196.
51. Egeberg, R., Michaelsen, N.H., Skyum, L. "Novel Hydrotreating Technology for Production of Green Diesel," Research Technology Catalysts, *Haldor Topsoe*, n.d. Web. Accessed on May 15, 2014, http://www.topsoe.com/business_areas/refining/~/media/PDF%20files/Refining/novel_hydrotreating_technology_for_production_of_green_diesel.ashx
52. Roberge, T.M. Novel blends of sulfur-tolerant water–gas shift catalysts for biofuel applications, University of South Florida, M.S. Thesis, 2012.
53. Weitkamp, J. and Ernst, S. In *Guidelines for Mastering the Properties of Molecular Sieves*, D. Barthomeuf, E.G. Derouane, W. Holderich (eds.), Plenum Press, New York, 1990, p. 343.
54. Girigis, M.J. and Tsao, Y.P. *Ind. Eng. Chem. Res.* 1996, 35, 386.
55. Beecher, R. and Voorhies Jr., A. *Ind. Eng. Chem. Prod. Res. Dev.* 1969, 8(4), 366.
56. Calemma, V., Pertatello, S., Perego, C. "Hydroisomerization and hydrocracking of long chain n-alkanes on Pt/amorphous SiO_2–Al_2O_3 catalyst," *App. Catal. A: Gen.* 2000, 190, 207–218.
57. Petrochemical and Supply Division Chemicals Group. "Reference Data for Hydrocarbons and Petro-Sulfur Compounds," Petrochemical and Supply Division Chemicals Group, Phillips Petroleum Company, OK, 1974.
58. ASTM. *ASTM D7566-12a Standard Specification for Aviation Turbine Fuel Containing Synthesized Hydrocarbons*, ASTM International, West Conshohocken, PA. Table A2.1.
59. Bezergianni, S. and Kalogianni, A. *Biores. Tech.* 2009, 100, 3927–3932.
60. Biorrefineria. "AltAir retrofitted the existing refinery through process technology developed by Honeywell UOP. Using its proprietary technology, it can convert a variety of sustainable feedstocks into Honeywell Green Jet Fuel™ and Honeywell Green Diesel™. AltAir Fuels is a renewable fuels company headquartered in Paramount, CA." Web. Accessed on Mar 12, 2016, https://biorrefineria.blogspot.com/2016/03/AltAir-Fuels-Honeywell-UOP-biorefinery-California-HVO.html
61. Kubička D. and Horáček, J. *Appl. Catal. A: Gen.* 2011, 394, 9–17.
62. Tiwari, R., Rana, B.S., Kumar, R., Verma, D., Kumar, R., Joshi, R.K., Garg, M.O., Sinha, A.K. *Catal. Comm.* 2011, 12, 559–562.

63. Lippmann, M. The Integration paradigm: How fuel producers can overcome barriers to entry into petrochemical markets, AM-15-64 in American Fuels & Petrochemicals Manufacturers 2015 Annual Meeting, San Antonio, TX, 2015.
64. Bruno, K. An innovative approach: Catapults refiners to the next level in FCC catalyst selection, AM-15-29 in American Fuels & Petrochemicals Manufacturers 2015 Annual Meeting, San Antonio, TX.
65. Letzsch, W. and Dean, C. How to make anything in a catalytic cracker, AM-14-62 in American Fuels & Petrochemicals Manufacturers 2014 Annual Meeting, Orlando, FL, 2014.
66. Jap, C., Chiu, J., Wu, K., Yoo, S., Lin, T., Hwang, J. Advanced aromatics recovery technology with AED-BTX process, In AIChE Spring National 2014 meeting, New Orleans, LA, 2015.
67. Piotrowski, P. Refinery & petrochemicals integration: Managing molecules to maximize value, In AIChE Spring National 2014 meeting, New Orleans, LA, 2014.
68. Chou, C. Maximizing paraxylene production by optimization of the methyl group, In AIChE Spring National 2014 meeting, New Orleans, LA, 2014.
69. Funk, G., Boehm, E., Sohn, S., MaxEne™ Process, in *Handbook of Petroleum Refining Processes*, 3rd edn., Meyers, R.A. (ed.). To be published in 2016.
70. Vogt, E.T.C. and Weckhuysen, B.M. *Chem. Soc. Rev.* 2015, 44, 7342–7370.
71. Van der Bij, H. and Weckhuysen, B.M. *Chem. Soc. Rev.* 2015, 44, 7406–7428.
72. Singh, U., Knoll, J., Ziebarth, M., Fougret, C., Cheng, W., Brandt, S., Nicolich, J. Advances in $C_3^=$ maximization from the FCC unit, AM-14-25 in American Fuels & Petrochemicals Manufacturers 2014 annual meeting, Orlando, FL, 2014.
73. Turner, K., Loewnthal, E., Jones, A., Stein, L. Generating value with FCC catalyst innovation at Placid refining, AM-15-64 in American Fuels & Petrochemicals Manufacturers 2015 annual meeting, San Antonio, TX, 2015.
74. Fletcher, R., Sexton, J., Skurka, M. Best practices for achieving maximum ZSM-5 value in FCC, AM-13-01 in American Fuels & Petrochemicals Manufacturers 2013 annual meeting, San Antonio, TX, 2013.
75. Humphries, A., Copper, C., Seidal, J. Increasing butylenes production from the FCC Unit through Rive's Molecular Highway™ technology, AM-15-32 in American Fuels & Petrochemicals Manufacturers 2015 annual meeting, San Antonio, TX, 2015.
76. Abrevaya, H. Unique aspects of mechanisms and requirements for zeolite catalysis in refining and petrochemicals. In *Zeolites in Industrial Separation and Catalysis*, Kulpratipanja, S. (ed.), Wiley-VCH, 2010.
77. Hemler, C., Smith, L., Davydov, L., Fei, Z. UOP Fluid Catalytic Cracking Process. In *Handbook of Petroleum Refining Processes*, 3rd edn., Meyers, R.A. (ed.). McGraw-Hill, New York, 2016.
78. TCGR, The Catalyst Group Resources. Unconventional catalytic olefins production: Commercial vision and breakout? In a multi-client study, Spring House, PA, 2013.
79. Lambert, N., Ogasawara, I., Abba, I., Redhwi, H., Santner, C. HS-FCC for propylene: Concept to commercial operation, in PTQ Q1, 2014.
80. Zhu, X.X. *Energy and Process Optimization for the Process Industries*, Wiley/AICHE, New Jersey, 2014.
81. Hoehn, R.K., Bowman, D.M., Zhu, X.X. Process for recovering hydroprocessed hydrocarbons with two strippers, US Patent 2013/0046125 A1, 2013.
82. Hoehn, R.K., Bowman, D.M., Zhu, X.X. Apparatus for recovering hydroprocessed hydrocarbons with two strippers in one vessel, US Patent 8,715,596.
83. Zhu, X.X. and Asante, N.D.K. Diagnosis and optimization approach for heat exchanger network retrofit, *AIChE J.* 1999, 45(7), 1488–1503.

2 Recent Developments in Refining Catalysts—Patent Survey Statistics for a Corporate and Geographical Outlook

Deniz Onay Atmaca

CONTENTS

2.1 WORLD REFINING CATALYSIS OUTLOOK

In this chapter, patents issued between 2008 and 2014 on hydrocracking (HYC), hydrotreatment (HT), H_2 production, and fluid catalytic cracking (FCC) catalysts as well as FCC additives were reviewed. The analysis approach was both statistical and technical. Each of the topics indicated above will be given special emphasis in their respective sections. In this section, an overall analysis is presented.

Here the statistical analysis was based on the following criteria:

- Number of patents for each process (HYC, HT, FCC catalysts and additives, H_2 production)
- Number of patents by years for the period 2008–2014

- Number of patents by the applicant category (refinery, catalyst manufacturer, universities/research institutes/R&D companies and collaboration with each other)
- Number of patents by the continental origin (Europe, America, Asia)

Detailed separate surveys for the investigated refinery catalyst (HYC catalysts, HT catalysts, FCC catalysts and additives, and H_2 production catalysts) have been presented in relevant sections, respectively. Each section is structured as follows.

The relevant technology was reviewed to give the reader a historical perspective in terms of the technology development including the process chemistry and catalysts. A statistical analysis of the issued patents on the corresponding technology catalysts in terms of number, year, continental origin, and applicant category, as well as a statistical analysis for the main technical topics are defined specifically according to each catalyst type separately. The breakdown of the main technical topics to the subtopics was specifically defined as particular targets. The technical developments within the particular targets were discussed. Finally, future prospects in the corresponding technology combining the historical background with the projection of the investigated patents were discussed.

Main technical topics and target topics in each chapter are the result of a subjective analysis based on the general items stated within the investigated patents. Therefore, several classifications for each of the topics can be both interrelated and equivalent in terms of ultimate target that is to be fulfilled. Moreover, during the analysis of the technical topics, in some of the cases, not fully competent classifications can be present for the patents including more than one of the topics and targets. However, a technical survey conducted in this study comprises the aim of giving an opinion as a general guideline for the developments in the field of refinery catalysts.

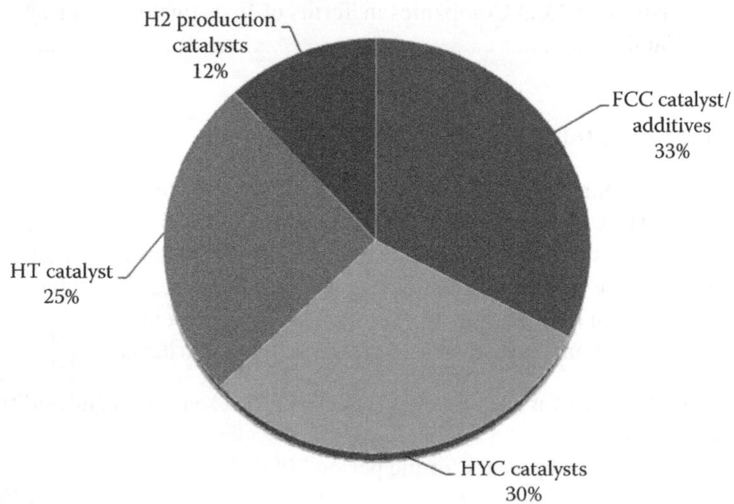

FIGURE 2.1 Distribution of the investigated patent topics.

Patents for refinery catalysts, namely FCC catalyst and additives, HYC catalysts, HT catalysts, and H_2 production catalysts, have been investigated between 2008 and 2014. The distribution of the investigated patents (a total number of 858) by the type of the catalyst can be seen in Figure 2.1. Catalysts and additives for FCC process, where the target product is majorly gasoline, and catalysts for HYC process, where the target product is middle distillates (diesel and kerosene), have the largest number of patents by 33% and 30%, respectively, of the total number. Patents for HT catalysts and H_2 production catalysts have 25% and 12% of the total number, respectively.

2.2 ANALYSIS BASED ON TOTAL NUMBER OF THE PATENTS BY YEAR

Figure 2.2 shows the total number of patents by years between 2008 and 2014. There is mainly an increasing trend in the total numbers. The distribution of the numbers in terms of the catalyst type is given in Figure 2.3 with respect to the years. Between 2008 and 2011, FCC catalyst and additives have the largest number of patents when compared to the other catalyst types. However, starting from 2012, the number of patents for HYC catalysts takes the lead in terms of numbers when compared to other types of investigated refinery catalysts. Moreover, there is also a sharp increase in the number of patents for HT catalysts in 2013. There is mainly a gradual increasing trend by years for the number of patents for H_2 production.

2.3 ANALYSIS BASED ON GEOGRAPHICAL DISTRIBUTION

For the investigated patents of refinery catalysts, the origin of the patent applicants based on the continents can be seen in Figure 2.4 where more than half of the patents (53%) are from Asian countries including China, Japan, India, Saudi

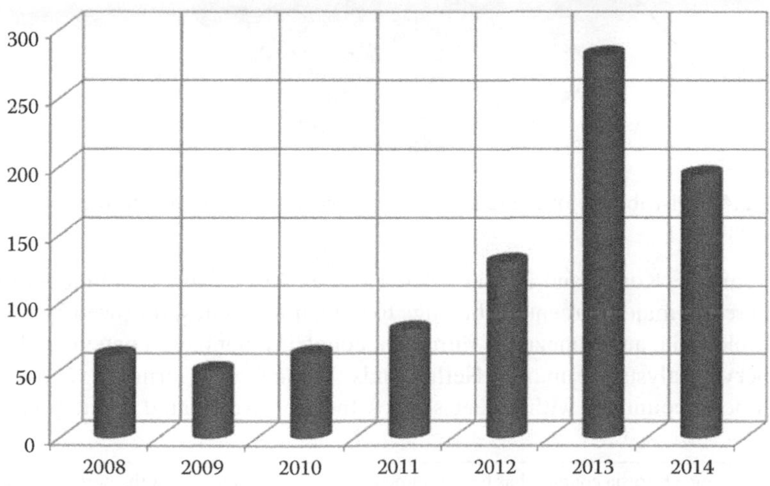

FIGURE 2.2 Distribution of the total number of patents by years.

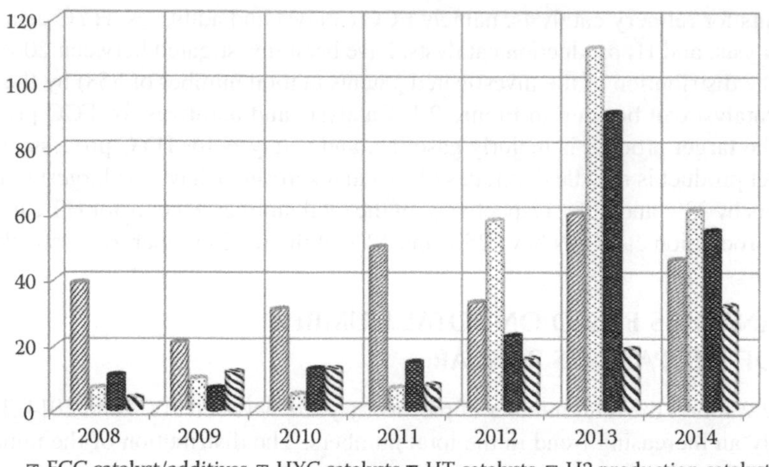

FIGURE 2.3 Distribution of the number of patents for each of the topics by years.

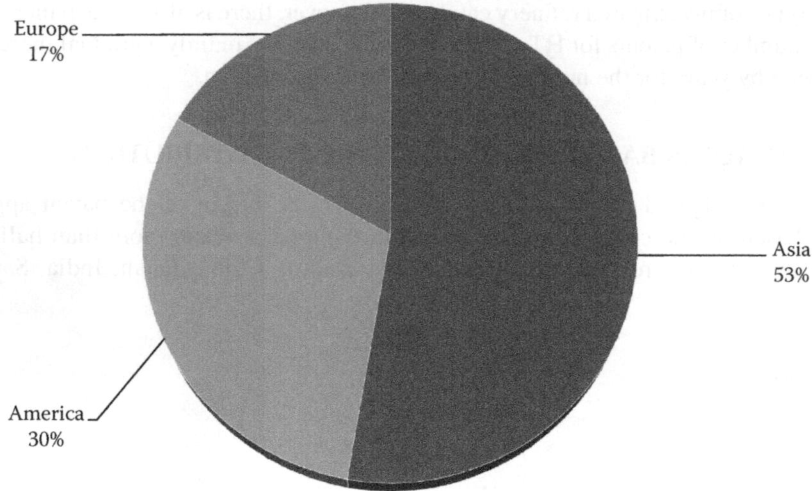

FIGURE 2.4 Distribution of the total number of patents with respect to the continents of the applicants.

Arabia, Taiwan, Korea, and Russia.* America has 30% of the total number of patents, where the major percentage belongs to the United States, followed by Mexico, Brazil, Colombia, and Venezuela. European countries that have a patent application for refinery catalysts are mainly Netherlands, France, and Germany. The remaining European countries with patent studies for the investigated refinery catalysts

* Russia, as being a Eurasia country, has been grouped as an "Asian" country for the percentage analysis of the patents, and there is no effect in terms of percentage whether the corresponding patents of Russia are grouped as Asian or European.

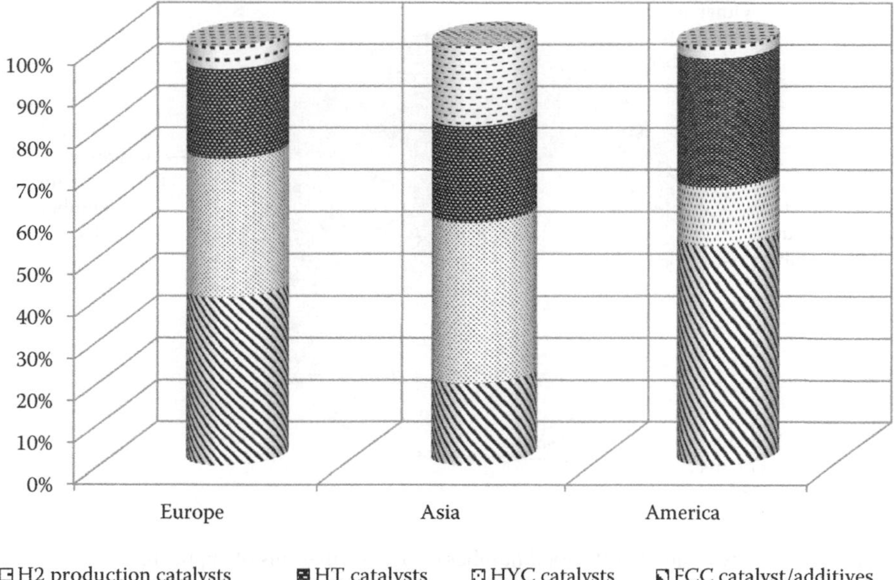

FIGURE 2.5 Distribution of the investigated patent topics with respect to the continents of the applicants.

are Italy, England, Spain, Denmark, Switzerland, Belgium, United Kingdom, and Czech Republic.

The continental distribution for each of the catalyst types is given in Figure 2.5. Patents for FCC catalyst and additives are the most investigated topic when compared to the other investigated refinery catalyst patents in America. Since the demand of America for gasoline is higher, the result given in Figure 2.5 for America meets the expectation as having the highest percentage of the patents for FCC catalyst and additives. Although Europe has a higher demand for diesel as an automotive fuel, which is produced mainly by using HYC catalysts, there are also many studies conducted there for the catalyst of the FCC process where the main product is gasoline. Asia has a focus toward HYC and hydrotreating catalysts. Moreover, H_2 production catalysts are mainly studied in Asia when compared to studies in America and Europe.

2.4 ANALYSIS BASED ON PATENT APPLICANTS

Patent applicants, for FCC catalyst and additives, HYC catalysts, HT catalysts, and H2 production catalysts, are divided into four main categories where the percentages can be seen in Figure 2.6, namely "universities, research institutes and R&D companies," "refineries," "catalyst manufacturers," and "other and individuals." Each category includes the total number of patent applicants including collaborative studies between the applicant categories.

Universities, research institutes, and R&D companies have the highest number of patents with 38% of the total studies. Patent applicants of the corresponding category

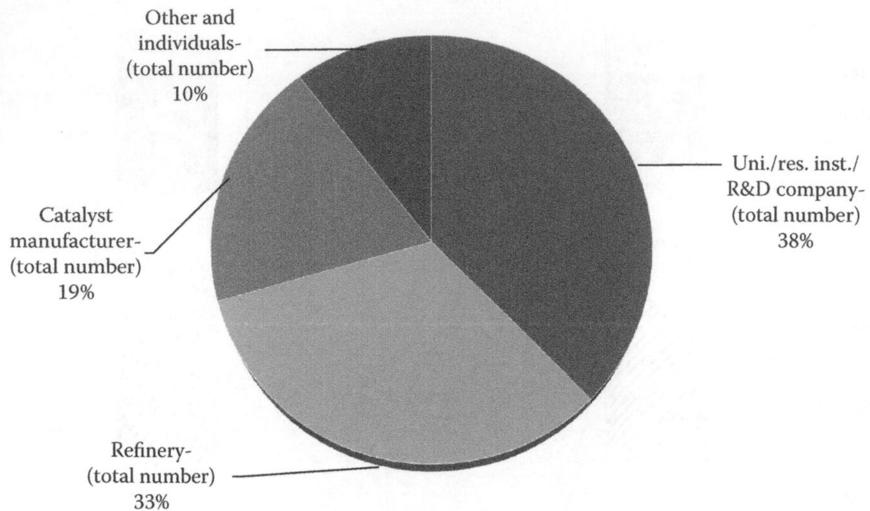

FIGURE 2.6 Distribution of the total number of patents in terms of applicants.

consist of SINOPEC Research Institute of Petroleum Processing, IFP Energies Nouvelles, Shell International Research Maatschappij B.V., SINOPEC Fushun Research Institute of Petroleum and Petrochemical, China University of Petroleum, Petroleum Energy Center, Korea Institute of Science and Technology, Instituto Mexicano Del Petroleo, Tianjin University, Dalian Institute of Chemical Physics, Chinese Academy of Sciences, King Fahd University of Petroleum and Minerals, Seoul National University Industry Foundation, Tokyo Institute of Technology, and East China University of Science and Technology.

Refineries, having patent applications (33%) for catalyst developments, include mainly ExxonMobil, Chevron, China Petroleum & Chemical, Petrochina Company Limited, Shell Oil Company, Cosmo Oil Company, Nippon Oil, ENI, Indian Oil Corporation Limited, Saudi Arabian Oil Company, China National Petroleum Corporation, Idemitsu Kosan, and Petrobras.

Grace, UOP, Albemarle, BASF, Rive Technology, Johnson Matthey, and Sued Chemie are the most featured patent applicants as catalyst manufacturers forming 19% of the total patent applicants for the investigated topics.

Other companies and individual applicants are grouped in an overall category where it comprises 10% of the total applicants. Several companies within this category are mainly Kior, China National Offshore Oil Corporation, Headwaters Heavy Oil, Japan Petroleum Exploration Company, NGK Insulators, Yueyang Dingge Yuntian Environmental Protection Technology Co. Ltd, and Asahi Kasei Chemicals Corporation.

The main focus for each of the applicants in terms of the investigated catalyst types can be seen in Figure 2.7. Studies of refineries in terms of patent applications of refinery catalysts are mainly for HT and HYC catalysts. Developments for FCC catalyst and additives are the governing portion (>75%) of the patents of catalyst

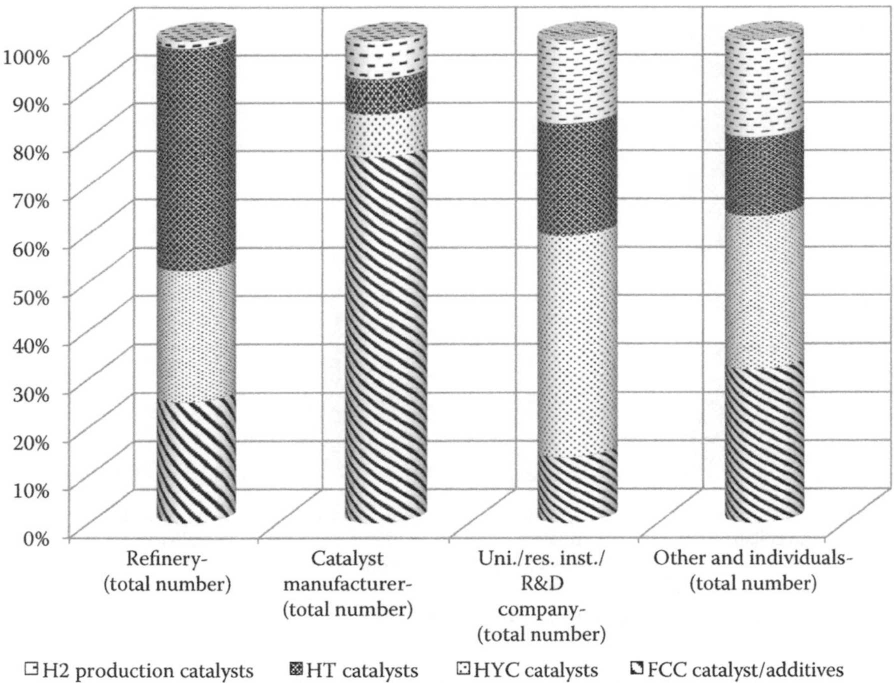

FIGURE 2.7 Distribution of the number of patents for each of the topic in terms of applicants.

manufacturers. Patents of HYC catalyst have the highest percentage within the applicant category of universities/research institutes and R&D companies.

2.4.1 Patent Studies and Collaborations of Refineries in Terms of Investigated Refinery Catalysts

Patents of the refineries were further classified whether the patents were applied with or without collaborators, as shown in Figure 2.8. The data indicate that refineries applied for patents mostly without collaborators. The rest of the data indicate that refineries mainly collaborate with universities/research institutes and R&D companies.

2.4.2 Patent Studies and Collaborations of Catalyst Manufacturers in Terms of Investigated Refinery Catalysts

Statistics for the patents of catalyst manufacturers in terms of collaborations are given in Figure 2.9. There are only a few patents in terms of collaborations of catalyst manufacturers for the investigated catalyst type. Almost all the patents of catalyst manufacturers are individual applications, without collaborations with other applicant categories.

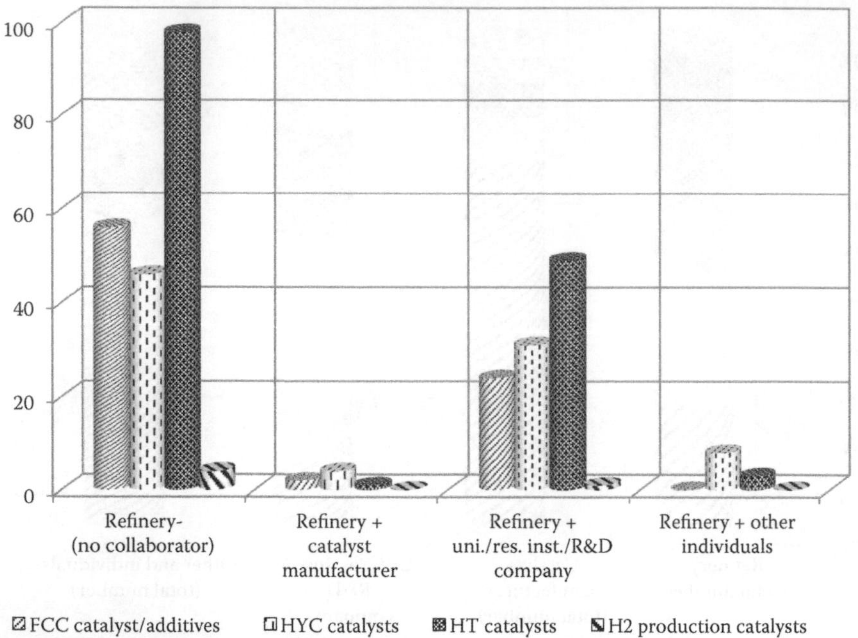

FIGURE 2.8 Patents of refineries together with collaborations.

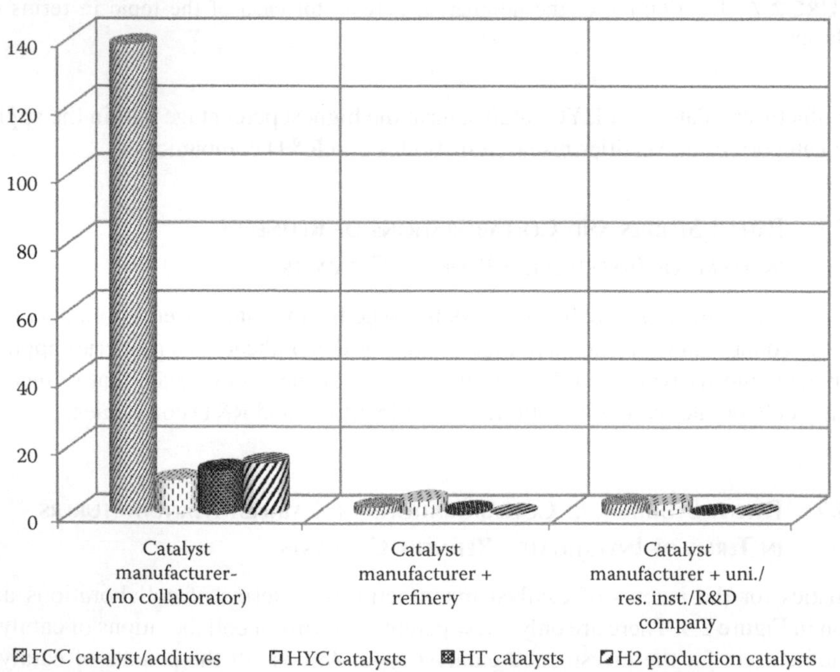

FIGURE 2.9 Patents of catalyst manufacturers together with collaborations.

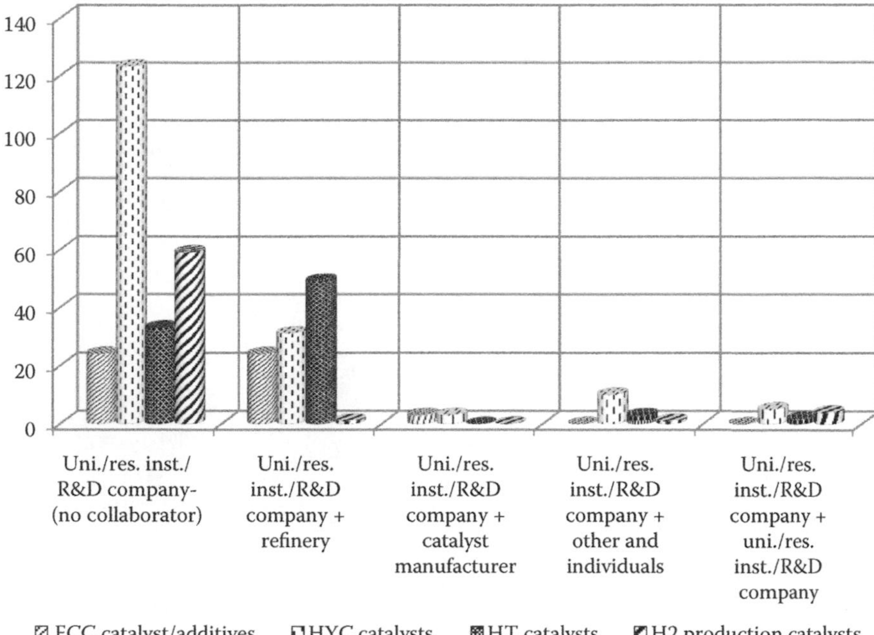

FIGURE 2.10 Patents of universities/research institutes/R&D companies together with collaborations.

2.4.3 Patent Studies and Collaborations of Universities/Research Institutes/R&D Companies in Terms of Investigated Refinery Catalysts

Figure 2.10 shows the patent applications of universities/research institutes/R&D companies together with the collaborations. Universities/research institutes and R&D companies have mainly collaborations with refineries followed by other companies and individuals. There are also some patent applications including collaboration of universities/research institutes/R&D companies with other universities/research institutes/R&D companies.

The nature of the engineering problems changes with the change in the landscape of the oil industry. The statistics above are a good indicator of the flow of information among refineries, catalyst manufacturing companies, R&D institutes, and universities.

Section II

Advances in Conventional Refining Technologies

3 Recent Developments in Hydrocracking Catalysts—Patent and Open Literature Survey*

Melek Bardakcı Türkmen

CONTENTS

* The review for the historical overview of the hydrocracking technology, process chemistry, and the developments of the catalysts has been performed by eight literature resources, namely *Petroleum Refining Process* (Speight, J.G. and Ozum, B.), *Handbook of Petroleum Processing* (David, S.J.J. and Pujado, P.R.), *Petroleum Refining Vol. 3, Conversion Processes* (Leprince, P.), *Petroleum Refining Technology and Economics* (Gary, J.H., Handwerk, G.E., and Kaiser, M.J.), *Practical Advances in Petroleum Processing* (Hsu, C.S. and Robinson, P.), *Hydroprocessing of Heavy Oils and Residua* (Ancheyta, J. and Speight, J.G.), *Hydrocracking Science and Technology* (Scherzer, J. and Gruia, A.J.), and *Diesel Fuels Technical Review* (Chevron) [1–8].

3.1 HYDROCRACKING TECHNOLOGY IN PERSPECTIVE

Hydrocracking is a refining technology within the scope of hydroprocessing that is used for the conversion of a variety of feedstocks to a range of products by adding hydrogen, removing impurities in the presence of a catalyst. Hydrocracking feeds, with higher molecular weights and lower hydrogen/carbon ratios, can range from heavy vacuum gas oils and coker gas oils to atmospheric oils, where products, having a lower molecular weight with higher hydrogen content and a lower yield of coke, usually range from heavy diesel to light naphtha [1,2].

Hydrocracking technology for conversion of coal to liquid fuels was developed in Germany as early as 1915. The forerunner of hydrocracking was the Bergius process, which may be considered as the first commercial plant and brought on stream in Germany in 1927 for hydrogenation of distillates-derived brown coal [2,3]. During World War II, hydrocracking processes played an important role in producing aviation gasoline. Nevertheless, after World War II, the emergent availability of Middle Eastern crude removed the incentive to convert coal to liquid fuels, which caused the development of hydrocracking technology to become less important as newly developed fluid catalytic cracking processes were much more economical than hydrocracking for converting high-boiling petroleum oils to fuels. In the mid-1950s, manufacturing of high-performance cars with high-compression ratio engines of the automobile industry required high-octane gasoline, where the switch of railroads from steam to diesel engines and the introduction of commercial jet aircraft in the late 1950s increased the demand for diesel fuel and jet fuel [2]. Thus, in the early 1960s, with the increasing demand for gasoline, diesel, and jet fuel, by-product hydrogen at low cost and in large amounts from catalytic reforming operations, and environmental concerns limiting sulfur and aromatic compound concentrations, hydrocracking technology gained importance again. And in 1958, the first modern distillate hydrocracker had been put into commercial operation by Standard Oil of California (now Chevron). It grew in other parts of the world, starting in the 1970s, primarily for the production of middle distillates, while hydrocracking was used in the United States primarily in the production of high-octane gasoline [2,4].

Hydrotreating and hydrocracking employ process flow schemes and similar catalysts where hydrocrackers tend to operate at more severe operational conditions as hydrotreaters are not conversion units through which breaking of carbon to carbon bonds is minimal. Hydrocrackers with higher severity in terms of process conditions have lower liquid hourly space velocity (LHSV), which means the given volume of feed requires more catalyst and higher pressure and temperature [5].

Hydrocracking is used to convert heavy fractions to lighter cuts similar to catalytic cracking, but unlike catalytic cracking, it is at relatively low temperatures and under high hydrogen partial pressures leading to inhibit rapid catalyst deactivation, which requires continuous regeneration of the catalyst during operation.

Since product balance is of major importance in any petroleum refinery, there are also relatively few operations that offer the versatility of the hydrocracking process, which has the ability to process high-boiling aromatic stocks produced by catalytic cracking or coking [1,4]. In general, hydrocracking is an in-between process that must be integrated in the refinery with other processes to take full advantage of the operations. As liquid, feed comes from the atmospheric/vacuum distillation, fluid catalytic cracking (FCC) bottom, delayed coker, or visbreakers, and the required hydrogen comes from catalytic reformers or steam/methane reformers or both. The outputs from a hydrocracker such as middle distillates usually meet finished-product specifications where heavy naphtha from a hydrocracker must be further processed in a catalytic reformer in the production of high-octane gasoline. The fractionator bottoms from a hydrocracker can be sent to an FCC, olefin, or lube plant [1,6].

Furthermore, conversion of heavy vacuum gas oil fraction into middle distillates requires not only a reduction in the number of carbon atoms but also an increase in the H/C ratio to get the desired product specifications, which also depend on the type of process. Naphtha cut, for example, can have a wide range of H/C ratio, where middle distillates must have a highly saturated structure to meet market quality requirements such as smoke point for jet fuel and cetane number for diesel cut [3].

Depending on the objectives chosen with respect to the type of feedstock, desired product quality, and degree of conversion, operating conditions are determined at the following range of conditions: liquid hourly space velocity (LHSV), 0.5 to 2.0 h^{-1}; H_2 circulation, 850–1700 Nm^3/m^3; H2PP, 10,300–13,800 kPa; and SOR temperatures ranging between 630.15 and 658.15 K [2]. In general, hydrocracker flow schemes can be grouped into two major categories: single-stage and two-stage operations. A single-stage once-through hydrocracking unit is the simplest configuration for a hydrocracking unit that resembles the first stage of the two-stage plant. The feed mixes with hydrogen before going to the reactor and the effluent goes to the fractionation. This type of hydrocracking unit has the lowest cost, but feedstock is not completely converted and highly refined heavy oil is required [1,2]. The once-through configuration is used when the fractionator bottoms are also the desired products. However, middle distillates obtained are of lower quality having higher aromatic content due to being operated at high-severity conditions in order to obtain high overall conversions [7]. In a single-stage recycle unit, which is the most widely found hydrocracking unit, unconverted feed is recycled by sending it back to the reactor section for further conversion [2]. This type of unit is more economical than a more complicated two-stage unit when the plant capacity is less than about 10,000–15,000 bbl/day. However, it is less selective for liquid products as compared with a two-stage configuration [1]. The two-stage hydrocracking process is also widely used especially for units with large capacities. In this type of operation in the first stage, hydrotreating and some cracking reactions take place where the effluent from the first stage is fractionated and the unconverted oil goes to the second stage [2]. This type of unit offers more product yield flexibility than a single-stage operation as different hydrocracking catalysts can be used in each stage. The use of a two-stage unit rather than a single-stage unit is advantageous when middle distillates are the desired products, which results in a complete hydrogenation with a significant reduction in aromatics and an increase in the smoke point of the middle distillate fractions [7].

3.2 HYDROCRACKING PROCESS CHEMISTRY AND CATALYSTS IN PERSPECTIVE

The hydrocracking process, which converts heavy feedstock to lower-molecular-weight, high-value products, is the result of combining catalytic cracking reactions with hydrogenation, and removes heteroatoms and saturates olefins and aromatics according to the complex reaction patterns. Essentially all of the initial reactions of catalytic cracking occur, but some of the secondary reactions are inhibited or stopped by the presence of hydrogen. Besides desirable reactions, such as treating, saturation, and cracking, and undesirable reactions, such as contaminant poisoning like coking of the catalyst, the reactions can be categorized into two main groups: hydrotreating reactions such as hydrodesulfurization, hydrodenitrogenation, and aromatic hydrogenation, and hydrocracking reactions such as breaking C–C bonds or C–C rearrangement reactions (hydroisomerization) [2,3,6].

Significant C–C bond breaking reactions within hydrocracking reactions comprising dealkylation of aromatic rings, opening of naphthene rings, and hydrocracking of paraffins are given by Equations 3.1, 3.2, and 3.3, respectively:

$$\varphi\text{–}CH_2\text{–}R + H_2 \rightarrow \varphi\text{–}CH_3 + RH \quad \Delta H_R = -1.3 \text{ to } -1.7 \text{ kJ/Sm}^3{*} \tag{3.1}$$

$$\text{Cyclo–}C_6H_{12} \rightarrow C_6H_{14} \quad \Delta H_R = -1.3 \text{ to } -1.7 \text{ kJ/Sm}^3{*} \tag{3.2}$$

$$R\text{–}R' + H_2 \rightarrow RH + R'H \quad \Delta H_R = -1.3 \text{ to } -1.7 \text{ kJ/Sm}^3{*} \tag{3.3}$$

Minimal C–C breaking reactions involve saturation of aromatics and olefins with hydrotreatment reactions that comprise hydrodesulfuration, hydrodenitrogenation, hydrodeoxygenation, and hydrodemetallation. Illustrations of the reactions that represent saturation of aromatics and saturation of olefins are seen in Equations 3.4 and 3.5, respectively:

$$C_{10}H_8 + 2\,H_2 \rightarrow C_{10}H_{12} \quad \Delta H_R = -5.5 \text{ kJ/Sm}^3{*} \tag{3.4}$$

$$R{=}R' + H_2 \rightarrow HR + R'H \quad \Delta H_R = -5.5 \text{ kJ/Sm}^3{*} \tag{3.5}$$

The rearrangement of C–C reactions also stated as hydroisomerization is given by the following reaction:

$$\text{n-RH} \rightarrow \text{i-RH} \tag{3.6}$$

The reactions of coke formation and mercaptan formation, which are causes of deactivation, are given by

$$2\,\varphi H \rightarrow \varphi\varphi + 2\,H_2 \quad \Delta H_R = +3 \text{ kJ/Sm}^3{*} \tag{3.7}$$

$$RR' + H_2S \rightarrow HS\text{–}R\text{–}R'H \quad \Delta H_R = -3 \text{ kJ/Sm}^3{*} \tag{3.8}$$

where ΔH_R is the heat of the reaction and is negative for the exothermic reactions, * is kilojoules per standard cubic meter of H_2 consumed, φ represents aromatics, and R represents alkyl [5].

The increasing interest of clean-burning fuels due to environmental constraints has led to research into the chemistry of the catalyst and engineering of heteroatom removal, with sulfur removal being the major focus [1]. All the hydrotreating (so called pretreating) reactions in which hydrogen is consumed, sulfur removal, nitrogen removal, (organo) metallic compound removal, olefin saturation, oxygen removal, and halides removal take place. The first three types of compounds are always present in varying amounts depending on the source of feedstock, whereas the others are not always present. In general, the ease of the removal of these heteroatoms by hydrotreating reactions is in the following descending order: (organo) metal removal, olefin saturation, sulfur removal, nitrogen removal, oxygen removal, and halide removal [2]. In these types of reactions, the feed desulfurization is nearly complete due to the high hydrogen pressure and the operating temperature; however, nitrogen content must be controlled by denitrogenation reactions before the feed contacts the catalyst, due to the impact of nitrogenated compounds on the acid sites of the hydrocracking catalyst, even at low concentrations [3]. The organic sulfur is transformed into H_2S, the nitrogen is transformed into NH_3, and oxygen compounds are transformed into H_2O within the scope of hydrotreating reactions. The aromatic hydrogenation reaction is also a requisite, as it is not possible to crack an aromatic compound, whereas it is possible to convert the naphthenes resulting from the hydrogenation of aromatic rings under hydrocracking conditions [2]. Apart from heteroatoms, another challenge to overcome is the contaminant poisoning of the catalyst like deposition of coke and metals onto the catalyst, which diminishes the cracking activity of the hydrocracking catalysts [1,2]. Polynuclear aromatics (PNA), sometimes called polycyclic aromatics or polyaromatic hydrocarbons, and coke formation are observed due to some aromatic saturation occurring in some treating and cracking sections. Aromatic saturation leading to PNA is observed toward the end of the catalyst life cycle because of equilibrium limitation when reactor temperature has to be increased to compensate the loss in catalyst activity resulting from coke formation and deposition. PNAs are compounds containing at least two benzene rings in the molecule, whereas the feed to a hydrocracker can normally contain PNAs with up to seven benzene rings in the molecule. Hydrocracking units operating with recycle through which unconverted feed is sent back to the reactor section for further conversion cause the creation of PNA with more than seven benzene rings in the molecule. These kinds of molecules containing more than seven benzene rings are called heavy PNA (HPNA), and the HPNA produced on the catalyst may cause downstream fouling; or their presence results in plugging of equipment; or they may deposit on the catalyst and form coke, which deactivates the catalyst [2].

Depending on the position of the C–C bonds, there are three types of reactions in terms of splitting these C–C bonds: (1) simple hydrocracking reactions, where a C–C bond in a chain is cracked; (2) hydrodealkylation, where a C–C bond adjacent to a ring is cracked; and (3) ring opening reactions, where a C–C bond in a ring is cracked during hydrocracking reactions in which a reduction of average molecular weight is achieved. In terms of C–C rearrangement reactions, hydroisomerization and hydroalkylation reactions improve the quality of the products [3].

In terms of product quality parameters, the octane number is a measurement of a fuel's resistance to knock, which indicates the efficiency of the work. A high octane number is associated with a high ratio of i/n paraffins and high concentration of aromatics and naphthenes within the chemical basis, and it is achieved by isomerization and alkylation processes. Accordingly, cloud and pour points in terms of cold behavior characteristics of fuel, indicating the cold weather performance of diesel due to the relative coagulation of wax in the oil and longer paraffin molecules in the fuel, precipitate as wax, and within the scope of product quality, low concentrations of n-paraffins are also necessary for low cloud and pour point. Similarly, fuels with low contents of n-paraffins and unsubstituted aromatics have lower freezing points compared to the melting points of isoparaffins and naphthalenes of the same carbon number, as the freezing point strongly depends on the molecular shape and the molecules that fit into a crystal structure easily. However, the more paraffinic a fuel is, the higher its cetane number will be, which is a measurement of the ignition delay for a diesel fuel [2,3,8].

In general, hydrocracking develops in the steps of reaction sequence that consist of the olefin dehydrogenation–hydrogenation reactions on a metal site and carbonium ion formation on an acid site, followed by isomerization with cracking of the carbonium ion and the generation of an olefin or cyclic olefin which occur simultaneously on the metal site of the catalyst. In other words, the acid site of the catalyst assists in the production of carbonium ion by adding protons to the olefin, and then the carbonium cracks to a smaller olefin and/or carbonium ion. These ions quickly hydrogenated preventing the adsorption of olefins on the catalyst, and thus their subsequent hydrogenation, which leads to coke formation. These are primary hydrocracking products and can produce still smaller molecules by reacting further. Besides, hydrocracking is a selective reaction for cracking of higher carbon number paraffins because the relative strength of adsorption is greater for large molecules. Thus, the equilibrium for olefin formation is more favorable for large molecules, so the conversion of large molecules tends to be favored by hydrocracking reactions leading to relatively stable single rings, low C_1 and C_3 formation, and highly isomerized products [1,2].

Hydrocracking reactions proceed through a bifunctional mechanism, which is one that requires two distinct types of catalytic sites. These two functions that catalyze separate steps in the reactions are the acid function, which provides for the cracking and isomerization, and the metal function, which provides for the olefin formation and hydrogenation. Overall, there is heat released in hydrocracking, as the cracking reaction requires heat while the hydrogenation reaction generates heat. The acidic support, on which cracking and isomerization reactions take place, consists of amorphous oxides (silica–alumina) and crystalline zeolite (mostly modified Y zeolite) plus binder, or a mixture of them. The number and strength of the acid sites designate the acidic characteristics and thus the cracking function of the hydrocracking catalyst. The metals providing hydrogenation function, consisting of noble metals (palladium, platinum) or non-noble (base) metal sulfides from group VIA (molybdenum, tungsten) and group VIIIA (cobalt, nickel), make the feedstock more reactive for cracking and heteroatom removal, as well as reduce the coking rate. The ranking of hydrogenation for metal combinations is in the following ascending order: Ni–W > Ni–Mo > Co–Mo > Co–W [1,2,5]. With the metal ratio, metal amount, degree of metal dispersion, and type of metal, metal–support interaction and location of metal on support are also important in order to avoid undesirable secondary reactions as long as a rapid molecular transfer can be achieved by having the hydrogenation sites located in the proximity of the cracking sites [2,3].

The ratio between the treating and cracking function is also important to optimize activity and selectivity. Aside from the optimization of the activity and selectivity, there are other performance parameters measured: initial activity is measured by the temperature required to obtain the desired product at the start of the run, stability is measured by the rate of increase in temperature required to maintain conversion, product selectivity is a measure of the ability of a catalyst to produce the desired product slate, and product quality is a measure of the ability of the process to produce products with the desired use specifications, such as pour point, smoke point, or cetane number [2,3]. For instance, a strong hydrogenation activity with a weak acid function causes a catalyst to operate at a higher temperature, which means low activity and a lower LHSV, but having better selectivity for middle distillates. Conversely, a weak hydrogenation function with a strong acid function makes the catalyst more active, but less selective for middle distillates [8]. Moreover, the type of feedstock, process configuration, PNA, catalyst design, operating variables, and process flexibility are also the criteria that affect the overall yield and product quality [7].

3.3 ANALYSIS BASED ON PATENT APPLICANTS AND COUNTRIES OF THE APPLICANTS

This chapter is about hydrocracking patents in terms of catalyst developments between 2008 and 2014. Throughout the investigation, the total number of analyzed patents is 315; among these patents, 259 are related to "hydrocracking catalysts" and the rest of them relevant to process developments of hydrocracking indicating that most of them are focused on hydrocracking catalyst-related investigations rather than process developments.

The survey for this and upcoming chapters is based on the latest issue date of each patent in order to cover the most recent technologies relevant to the development. The annual distribution of hydrocracking-related patents is given in Figure 3.1 in terms of two main topics: as hydrocracking catalyst and hydrocracking process-related developments. The number of patents related to relevant developments has a tendency to increase and seems to be the highest in 2013 since the latest version of each patent has been taken into account during the survey. The rest of the analysis within Section II has mainly focused on hydrocracking catalyst-related patents rather than process-related topics.

Hydrocracking catalyst-related patent distributions with respect to applicants are seen in Figure 3.2. Universities/research institutes/R&D companies are the major participants of the studies with 49%. Refineries ranked second with 26% of the total cited patents. Refinery collaboration with university/research institute/R&D companies has 9% of the total number, whereas both refineries with catalyst manufacturers

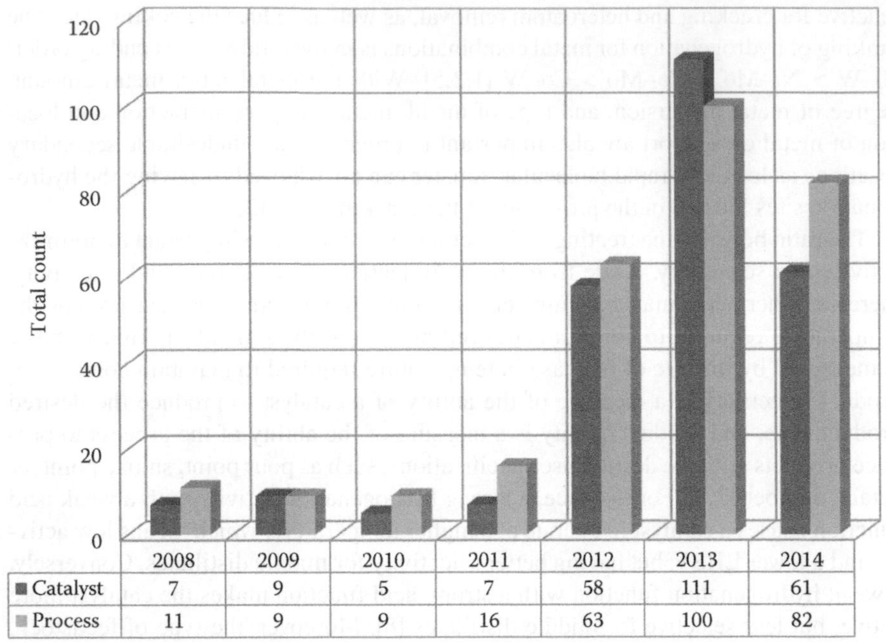

FIGURE 3.1 Annual distribution of hydrocracking patents between 2008 and 2014.

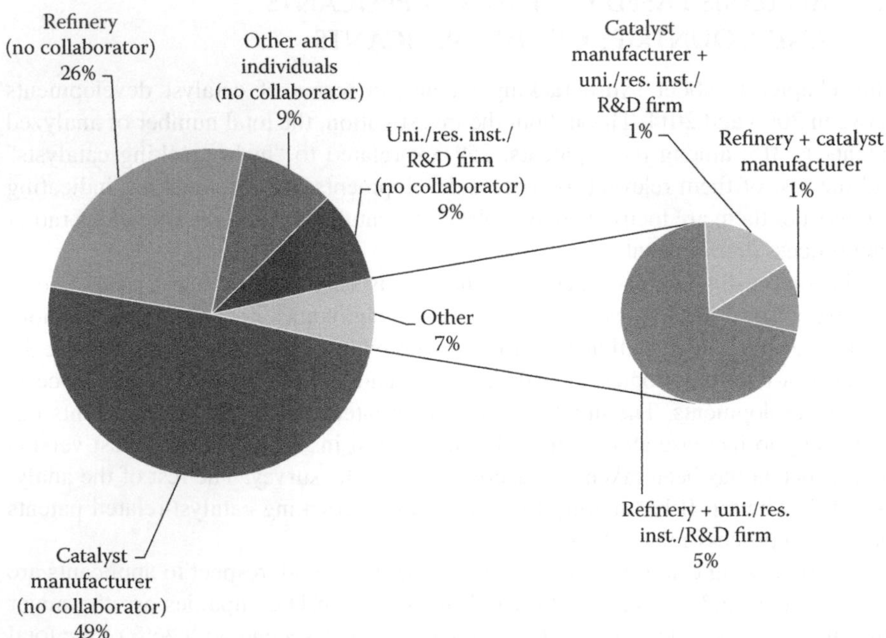

FIGURE 3.2 Patent distributions of hydrocracking catalyst based on applicants.

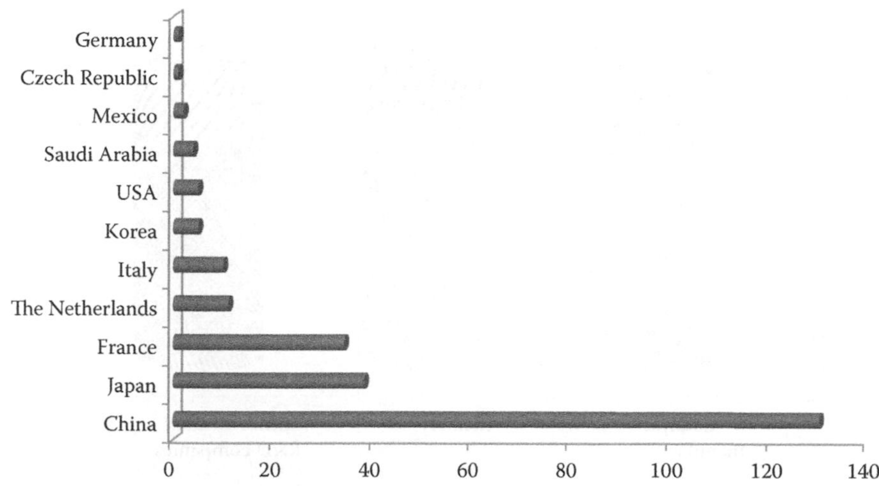

FIGURE 3.3 Patent distributions of hydrocracking catalysts in terms of total numbers by countries (note that if the corresponding country is in collaboration with another company, their collaborative studies is taken into account while making the percentage calculations).

and catalyst manufacturers with university/research institute/R&D companies have been found as just 1% of the total number. Therefore, refineries have collaborated with university/research institute/R&D companies much more than catalyst manufacturers for publication of the relevant patents, and most of the studies that catalyst manufacturers have published are without collaboration.

The patent distribution of hydrocracking catalysts is shown in Figure 3.3 in terms of total numbers by countries of applicants. Most of the patents are from China followed by Japan and France, respectively. The leading factor for these countries to be major within the number of published patents is the interest of universities and research institutes in the topic.

Catalyst manufacturers are found to have a higher number of patents in America, whereas refineries and university/research institute/R&D companies play a prominent role in Asia to the contribution of relevant patent publications. Moreover, catalyst manufacturers located in Europe have not been found to have any patent within the analysis as seen in Figure 3.4.

3.3.1 Involvement of Refineries

Hydrocracking catalyst patent applicants as refineries mainly consist of companies such as ENI SPA, ExxonMobil, Chevron, and others; the remaining 67% of the investigated patents are issued by JX Nippon Oil, Idemitsu Kosan, Cosmo Oil, Japan Oil & Gas, and China Petroleum & Chemical as seen in Figure 3.5. Furthermore, applicants that contribute less than 4% to the total count of patents are grouped together, and as a result, the "others" category consists of 18% of the total count.

According to further investigation of Asian companies that contribute 67% to the total number, China Petroleum & Chemical, which includes China National

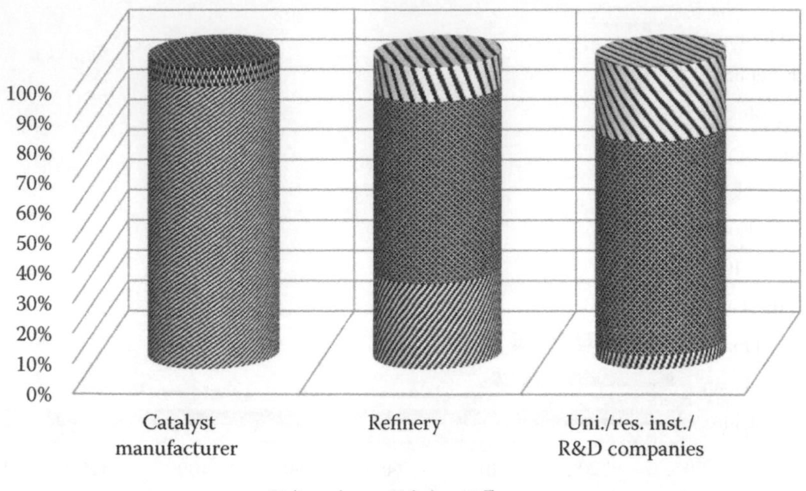

FIGURE 3.4 Patent distributions of hydrocracking catalysts in terms of total numbers by continents.

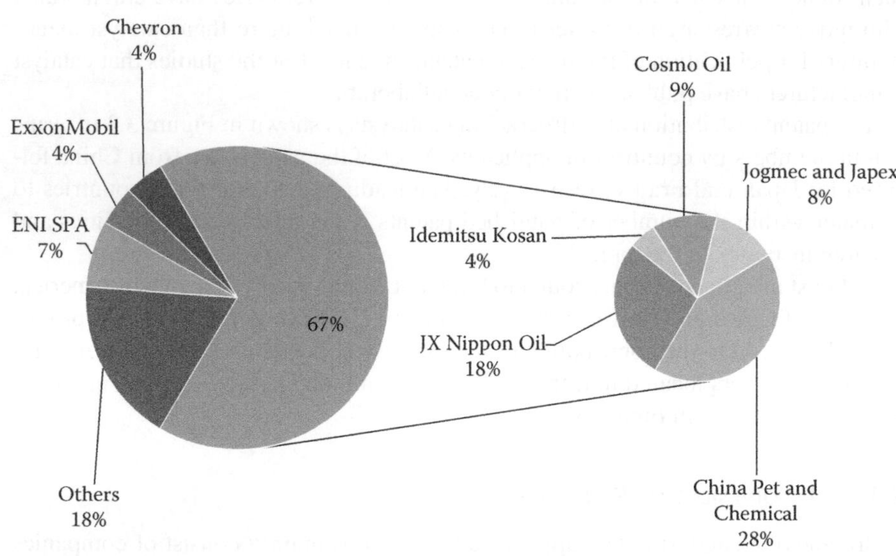

FIGURE 3.5 Patent applicants of hydrocracking catalysts as refineries (note that if the corresponding company is in collaboration with another company, their collaborative studies is taken into account while making the percentage calculations).

Offshore Oil Corporation (CNOOC), China National Petroleum Corp. (CNPC), and China Petrochemical Corp. (Sinopec), has the most total number of patents in refineries group with 28%. JX Nippon Oil & Energy Co. has the second most total number of patents in Asian refineries group with 18%.

3.3.2 Involvement of Catalyst Manufacturers

As seen in Figure 3.6, UOP LLC is the major contributor for patents of hydrocracking catalysts with 33%, which is followed by JGC Catalysts & Chemicals Ltd with 22% and both BASF Corp. and BASF Catalyst LLC (named as BASF in the figure) with 17%. Other catalyst manufacturer companies contributed similar percentages, each having 5% or 6%.

3.3.3 Involvement of Universities/Research Institutes/R&D Companies

In Figure 3.7, the applicants that contribute less than 4% to the total count of patents are grouped together as "others," which is 37% consisting of all universities except China University of Petroleum. The second plot, which comprises 36% of the total, consists of China University of Petroleum and the Sinopec group. The Sinopec group, which comprises Research Institute of Petroleum Processing, Fushun Research Institute of Petroleum and Petrochemicals, and Shanghai Research Institute

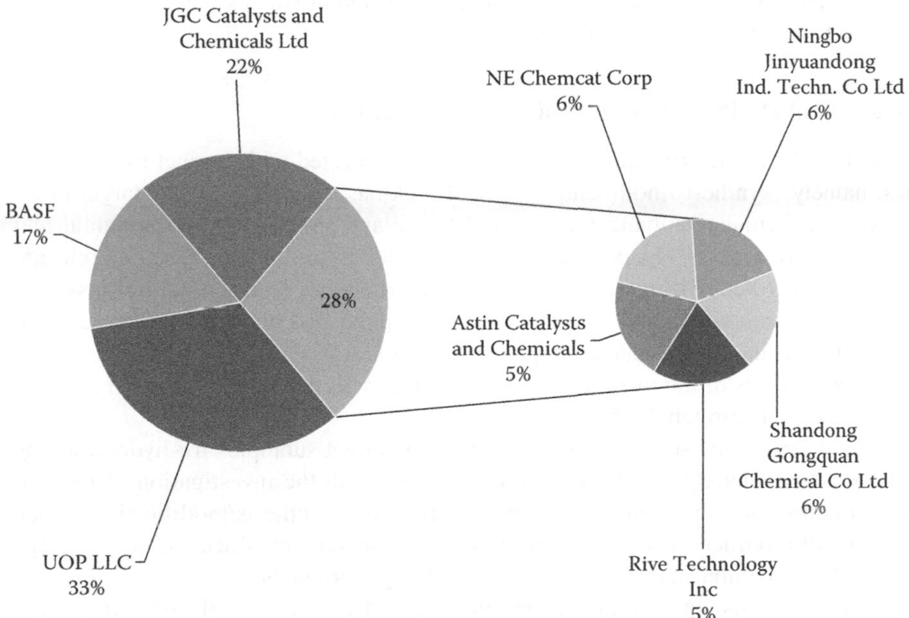

FIGURE 3.6 Patent applicants of hydrocracking catalysts as catalyst manufacturers (note that if the corresponding company is in collaboration with another company, their collaborative studies is taken into account while making the percentage calculations).

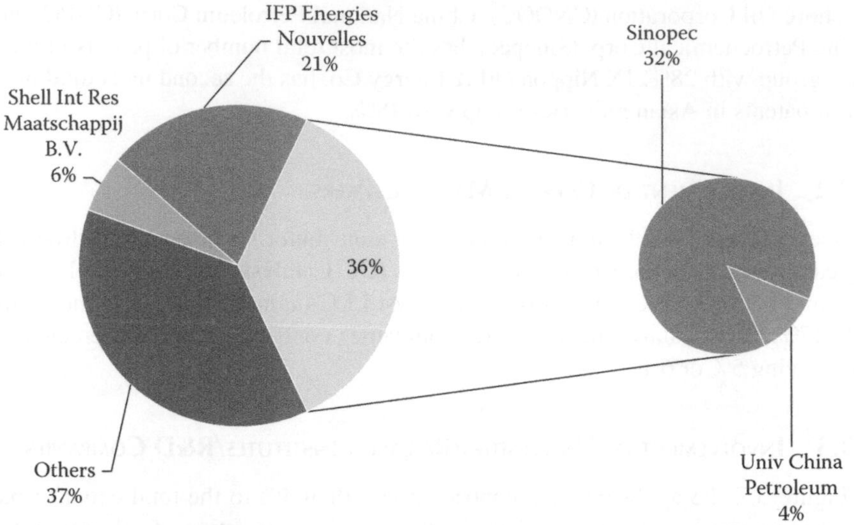

FIGURE 3.7 Patent applicants of hydrocracking as universities/research institutes/R&D companies (note that if the corresponding company is in collaboration with another company, their collaborative studies is taken into account while making the percentage calculations).

of Petrochemical Technology, has the major partition of studies with 32% followed by IFP Energies Nouvelles with 21%.

3.4 ANALYSIS BASED ON RESEARCH AREA

Patents of hydrocracking catalysts have been categorized with respect to main topics, namely "synthesis/modification," "feedstock-based technologies," "target products," "performance enhancements," and "ancillary processes, environmental and cost effective aspects." Each topic has been divided into subtopics based on relevant targets for further investigations. The topics and subtopics represent target classifications for hydrocracking catalyst-related patents published between 2008 and 2014. The referenced patents related to topics in terms of targets are given in Table 3.1. The detailed analysis of the most distinctive subtopics will be given in the next sections (Sections 3.4.1 through 3.4.5).

According to the survey, it was seen that the target subtopics for hydrocracking catalysts are related to each other in compliance with the investigation of the patents. For instance, the subtopic named support-related synthesis/modification, which consists of modified mesoporous structure for improvement of accessibility to large pores, leads to enhancement of selectivity for target products.

The synthesis and modification method-related patents of the survey are focused on support and metal parts of the hydrocracking catalysts under the heading of "synthesis/modification." The support-related synthesis and modification section comprises a modifying agent and different synthesis and modification methods where the metal-targeted section is composed of impregnation (wet and

TABLE 3.1
Referenced Patent Numbers Related to the Target Topics

Topic	Target	Referenced Patents
Synthesis/ modification methods	Support related: – Modifying agents – Carbonization methods	[9,10,15–17,19,21,23–25,35–37,40–42,44,46, 50–53,60,62–64,69,73,77,78,80–83,89–96,101, 109,114,115,120–131,133–136,143–150, 153–155,157,161,165,169,173,186,190,200–204, 208,209,212–220,224,226,229,232,235,237–244, 247–250,253,259–262,264]
	Metal related: – Impregnation (wet, dry) methods – Sol-gel methods – Dipping methods – Reduction treatment methods	[18,21,27,28,38,44,46,47,57–59,61,63,65,73,86, 89,105,112,114,121,132,136,138,140,142,143, 146–149,154,159–164,169,184,191–193,202, 206,208,209,215,219,220,223,224,232,239,244, 247,248,251–253,259,262]
Feedstock-based technology	Petroleum based Biomass and natural gas based Coal based Synthetic oil based	[11,12,18–21,24,26,35,36,46,50,54,55,61,63,64, 66–69,71,75,76,84,88,91–94,98,102,103,107, 108,114–116,120,121,134,137,139,143,148, 149,152,156,164,166–169,171,172,174–179, 182,183,190,191,206,207,219,221,222, 225–228,230,231,235–244,246–250,256–258]
Target product	Middle distillate (diesel/ kerosene) Naphtha/gasoline Base oil BTX Wax Food-level white oil Others (alkyl, aromatics)	[11,12,18–21,24,26,33,36,46,50,52,54,55,61,63, 64,67–69,71,75,76,84,88,92–94,98,102,103,107, 108,114,116–118,120,121,134,137,139,143,145, 148,149,152,156,164,166–169,171,174–179,182, 183,190,191,206,207,219,221,222,225–228,231, 235–244,246–250]
Performance enhancement	Activity Selectivity Stability Product quality: – Octane number – Cetane number – Pour point – Other product quality improvements	[9,10,13,14,16,19,22,25,26,29–32,38–41,44–50, 57–60,62,63,68,70,74,77,78,85,87,91–96,99–101, 104–106,108–114,118–121,130–144,148,151, 154–157,164,169,170,180,185,187–189,192–202, 205,210–216,218,225,238,241,251,252,254,255, 263]
Ancillary processes, environmental and cost-effective aspects	Ancillary processes: – Passivation (sulfuration, nitrogenation) – Regeneration – Reclamation – Utilization of waste catalyst Environmental aspects Cost-effective issues	[9,10,24,33,34,43,55,56,61,72,79,90,97–99,134, 158,164,181,194,195,210,217,233,234,238,245]

dry), sol-gel, dipping, and reduction treatment methods. The support-related synthesis and modification studies are a major focal point of the investigations since the features of the hydrocracking catalyst support designate both the structure of support and acidic characteristics of the hydrocracking catalyst, which plays a prominent role in getting the desired product slate and selectivity. The support contains amorphous alumina–silica components and/or various molecular sieves such as beta, Y, MCM and ZSM series, their modified versions, and/or composites of them.

In order for synthesis and modification methods to achieve a hydrocracking catalyst that has superior porous structured and acidic supports, a number of preparation methods and reagents are used, such as dealumination, dispersion of amorphous silica–alumina into porous inorganic support matrix, enrichment of mesoporous channels, forming intracrystalline structures, and utilization of heteropoly acids. The metal is related to the hydrogenation function of the catalyst. For the types of metals from Group VIB and Group VIII, different metal loading methods and dispersion state of active metals are applied to meet product specifications.

The patents surveyed in the scope of feedstocks used in the hydrocracking process are named "feedstock-based technology." The topics investigated according to feedstock-based technologies are grouped as petroleum, biomass–natural gas, coal, and synthetic oil.

The patents are classified with respect to the "target product" since meeting the market demands is directly relevant to the product slate obtained by the process. The related subtopics consist of middle distillates such as diesel/kerosene and naphtha/gasoline; benzene–toluene–xylene (BTX), base oil, wax, and food-level white oil; and others such as alkyl and aromatic hydrocarbons.

The performance parameters of hydrocracking catalysts that are studied in the patents are under the heading of "performance enhancements" including activity, selectivity, stability, and product quality as subtopics. The three main catalytic properties related to the performance of a hydrocracking catalyst, which are evaluated, are activity, selectivity, and stability. Activity can be determined by comparing the temperatures at which various catalysts are utilized under other constant conditions of hydrocracking process providing the same feedstock so as to produce a percentage, normally 65%, of products boiling in the desired range. The lower the temperature required for a given catalyst, the more active the relevant catalyst in relation to a catalyst requiring higher temperature. Selectivity may be determined during the preceding activity test measuring the percentage of the fraction of the product boiling to confirm if it is in the desired range of targeted products. Stability is a measure of deactivation over an extended time period when processing a determined hydrocarbon feedstock under the conditions of the activity test. Stability is generally a measure in terms of the change in temperature required per day to maintain a given conversion, generally 65%. The product quality is also an important term to meet desired product specifications and is composed of the octane number, the cetane number, and the pour point with other product quality improvements.

The issues under the topic "ancillary process, environmental and cost-effective aspects" are directly process relevant. The subtopic ancillary process involves passivation such as sulfuration and nitrogenation, regeneration, reclamation of metals, and utilization of waste catalysts. In terms of environmental aspects, it is aimed to overcome the pollution problems in the catalyst preparation methods with the help of different synthesis methods such as the use of ultrasonic wave environment. The cost-effective aspects consist of methods such as the use of a simple device during catalyst production, hence reducing energy consumption, cost, and time.

The distribution of hydrocracking patents related to mostly stated topics is seen in Figure 3.8. According to the survey, major topics are synthesis/modification of catalyst-related patents with 41% and performance enhancements, which ranks second, with 23% followed by target products. This is followed by developments for feedstock-based technologies and target products, each having 17% and 14%, respectively.

The distribution of main topics in terms of applicants is given in Figure 3.9, confirming that the major partition of the studies for each of the investigated main topics has been conducted by universities, research institutes, and R&D companies, where catalyst manufacturers have minimum participation. Studies of the universities/research institutes/R&D companies concern mostly the topic of "synthesis/modification," and the second major applicant for the corresponding topic is refineries within all of the hydrocracking patent applicants. Since the studies for each of the investigated topics conducted by universities/research institutes/R&D companies are mainly located in Asia, similar implications of Figure 3.9 can also apply to Figure 3.10.

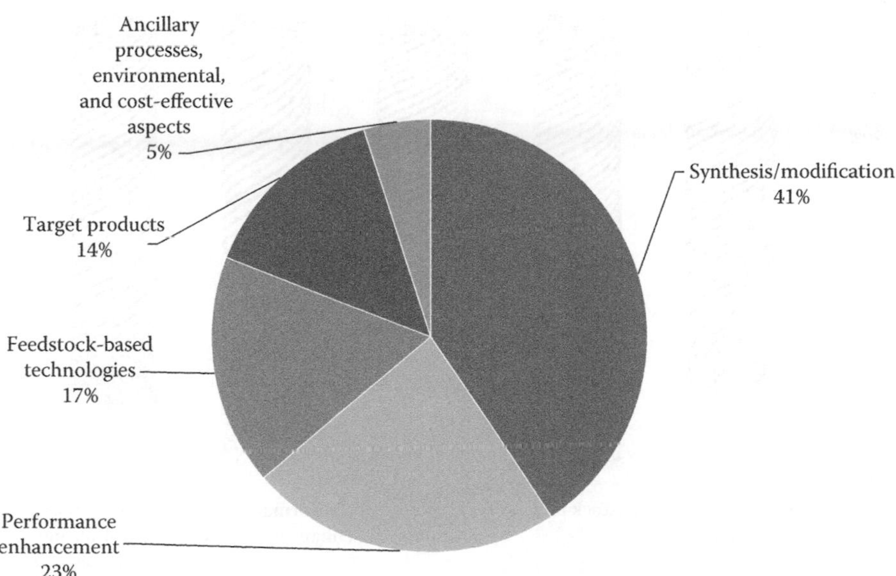

FIGURE 3.8 Patent distributions of hydrocracking catalysts based on mostly stated topics.

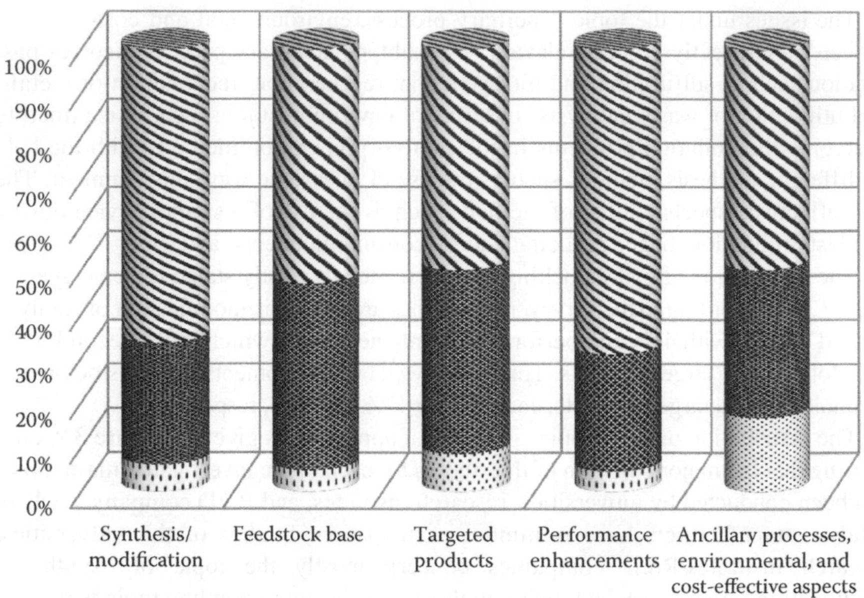

FIGURE 3.9 Topic analyses with respect to applicants.

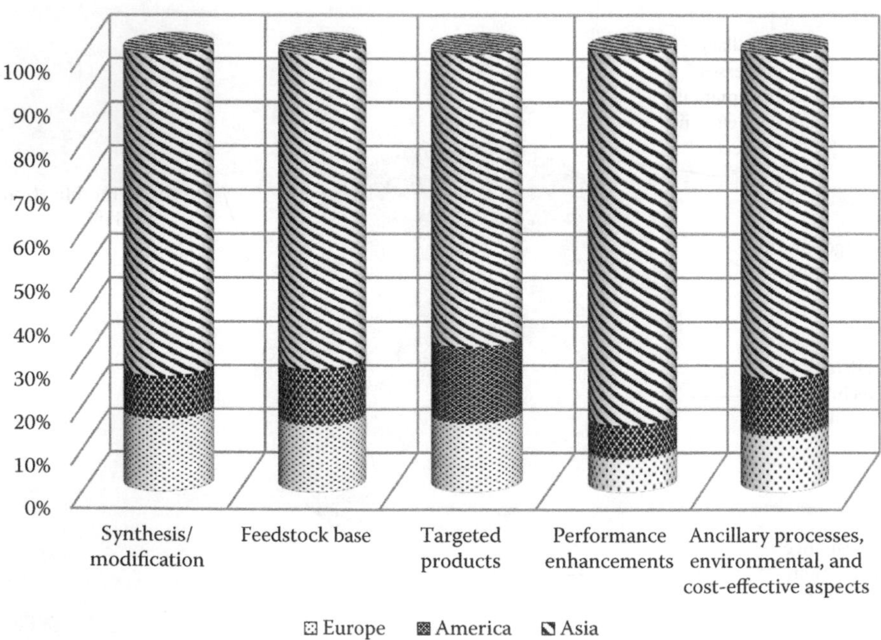

FIGURE 3.10 Topic analyses with respect to continents of the applicants.

3.4.1 Developments in Terms of "Synthesis/Modification"

In this section, patents of hydrocracking catalysts in terms of catalyst synthesis/modification are examined in two parts: "support related" and "metal related." The "support-related" part involves "synthesis/modification" and a "modifying agent," whereas the "metal-related" part consists of "synthesis/modification" with "impregnation," "dipping," "sol-gel," and "reduction treatment" methods as shown in the second plot in Figure 3.11. "Support-related" patents are more than "metal-related" ones with 67% of the catalyst synthesis/modification. Having support-related patents more in terms of number is consistent with the fact that support, in other words "carrier," is associated with the cracking function of the hydrocracking catalysts. IFP, Total, UOP, ExxonMobil, China Nat. Offshore Oil. Corp., China Petroleum & Chemical, Sinopec Fushun Res. Inst. Pet., Sinopec Res. Inst. Petroleum, PetroChina, and China University of Petroleum are the prominent contributors for the subtopic "support-related synthesis/modification."

Hydrocracking catalysts are bifunctional, combining an acidic function with a hydrotreating function. The acidic function, which plays a prominent role in cracking, is supplied by acidic supports, where the surface area varies from 150 to 800 m^2/g[1] such as alumina that can be halogenated, chlorinated, or fluorinated and is composed of combinations of oxides of aluminum and boron, and often amorphous silica–alumina and zeolites in combination with a binder to enhance the strength of the catalyst, which can be nonacidic such as clay and alumina.

According to the survey, most of the subtopics within the "support-related synthesis/modification" topic are about the improvement of modified zeolites to be obtained,

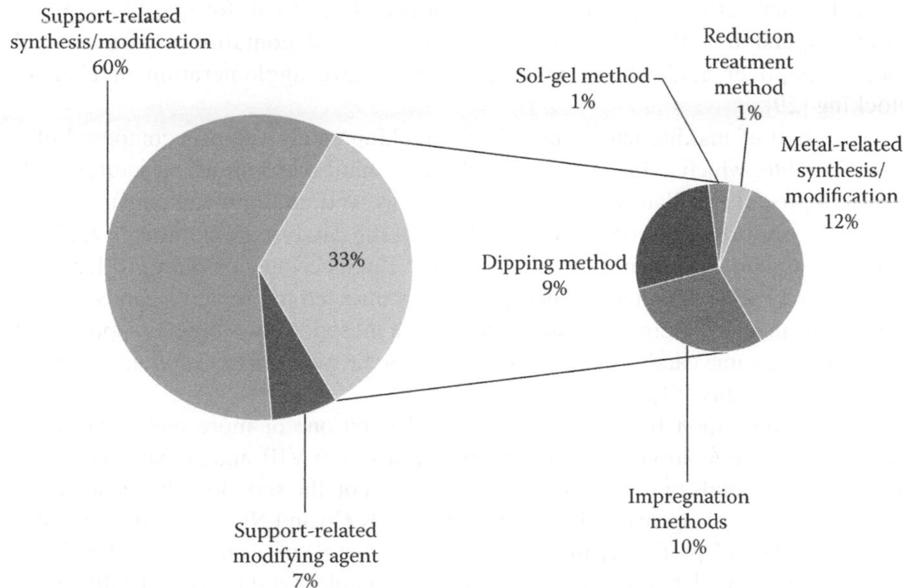

FIGURE 3.11 "Synthesis/modification"–related patents for hydrocracking catalysts.

i.e., mesoporous structures such as mesoporous channels, secondary pores to control diffusion in the scope of selectivity and conversion, specific surface area, silica–alumina ratio, type in terms of content of acidic structures, and distribution of acidic sites to achieve desired cracking activities. The zeolites of support, in particular their microporous structures, determine the factor for catalytic activity, stability, and/or selectivity, and as stated in drawbacks, the presence of the micropores causes inaccessibility to large molecules; undesired absorption of reagents and/or of products during operation leads to poor access of molecules in the support structure reducing the accessibility of the microporous volume of the support structure during reaction. According to the envisioned solutions, such solids undergo some treatments such as dealumination, desilication, and recrystallization processes in order to improve the efficiency of the catalyst. For instance, a system of secondary pores 2–50 nm in size is created in the microporous structures by introducing mesopores into the crystal structure of support by dealumination using hydrothermal treatment, acidic leaching techniques, or chemical treatments. Another challenge is developing appropriate catalysts having the slightest diffusional constraints and thereby mass transfer where mesoporous silicas having high surface area (1000 m^2/g) and a mesoporous structure with pores of uniform size would overcome the stated constraints relating to the diffusion of large molecules.

In some studies of Shell Internationale Research Maatschappij B.V., PetroChina Co Ltd, China University of Petroleum, and China National Offshore Oil Corporation, organic acids such as citric acid, tartaric acid, and oxalic acid are reported to be used as modifying agents of supports that enhance activity and stability of the catalyst improving secondary pore volume, crystallinity, and silica–alumina ratio [15,22,23,89,204]. Sinopec Fushun Research Institute of Petroleum and Petrochemicals also published some studies related to uniformity of alumina–silica dispersion with the usage of starting material containing extrusion aid such as peptizer acid which avoids molecular sieve agglomeration or channel blocking [29].

A method of making and using a hydrocracking catalyst that is composed of a new Y zeolite, which exhibits exceptionally low small mesoporous structures leading to improved distillate yield and selectivity as well as improved conversions at lower temperatures than conventional hydrocracking catalysts containing Y zeolites, has been studied by ExxonMobil Research and Engineering Company [10,135]. An application of Rive Technology Incorporation is directed to prepare a catalyst composition from a mesoporous zeolite by utilizing a mesoporous zeolite as support and incorporating some catalytic nanoparticles within the mesoporous zeolite, which has remarkable stability [51].

The hydrogenation function is supplied either by one or more metals of group VIB or by a combination of at least one metal of group VIB and at least one group of group VIII metals or oxides or sulfides deposited on the support. The most widely used metals for hydrogenation function are typically Co and Ni, optionally in combination with Mo/W, preferably in sulfided form, as metals of group VIB and Pt/Pd as metals of group VIII. One of the advantages of the noble metal usage is the direct use in the reduced metallic form. Secondly, sulfuration in order to keep the catalyst in its sulfided form is not required as it is typically the case with the catalysts including Co/Ni and W/Mo [21,161,202,248].

Furthermore, through the bifunctionality of the catalyst, the ratio, the strength, the balance, and the distance of the two acidic and hydrogenation functions also play important roles in terms of catalyst performance. For instance, a catalyst having a weak acid function and a strong hydrogenating function has low activity and requires low severe operating conditions but with good selectivity for middle distillates. Conversely, a catalyst having a strong acid function and a weak hydrogenation function has high activity but lower selectivity for middle distillates.

Metal-related synthesis/modification-related patents as seen in Figure 3.8 disclose mostly impregnation and dipping methods for metal loading. The most widely selected metal component compounds comprise group VIB elements particularly Mo and/or W and group VIII elements especially Ni and/or Co.

In a study of Shell International Research Maatschappij B.V. and Shell Oil Company, impregnation solution comprising C_3–C_{12} polyhydric compound, and any further additive such as phosphoric acid and fluorine, is used and the impregnated carrier is sulfided to obtain a sulfided catalyst [21]. According to another study of Shell International Research Maatschappij B.V., it can be difficult to impregnate a high amount of metals using conventional impregnation solutions, and the use of an organic compound having at least two moieties selected from carboxyl, carbonyl, and hydroxyl, but especially from carboxyl groups, citric acid or malic acid for the relevant study, assists in the impregnation [252].

China Petroleum and Sinopec Fushun Research Institute of Petroleum and Petrochemicals have an invention comprising the steps of dipping a formed porous support by using a water solution containing metal salts from groups VIB and VIII elements and/or a cosolvent wherein the number of dippings is not particularly limited and once or multiple impregnation operations to obtain the final catalyst with respect to the metal element content can meet the requirements [57]. In another study of China Petroleum and Sinopec Fushun Research Institute of Petroleum and Petrochemicals, a similar method comprises the step of impregnation by using water solution containing metal salts without a cosolvent. The catalyst with higher catalytic activity in hydrocarbon oil hydrocracking is provided by the relevant invention [138].

3.4.2 Developments in Terms of Feedstock-Based Technologies

The subtopic "petroleum-based feedstocks," which consists of titles such as heavy vacuum gas oil (HVGO), deasphalted oil (DAO), light cycle oil (LCO), oxygen-containing compounds, olefins, heavy naphtha, vacuum residues, wax, coker oil, tail oil, and base oil, is 57% of the related topic as seen in Figure 3.12 [11,18–21,24,46,50,63,64,67–69,71,84,102,108,114,121,139,143,149,164,167,175–179,183,191,206,219,222,226,231,235–237,239–243,246,249,250].

Other than the petroleum-based feedstock, there are also "synthetic oil based" such as synthetic wax and paraffinic Fischer–Tropsch synthesis oils with 11%; "biomass and natural gas based" with 17% such as fatty acid/esters/minerals, sunflower

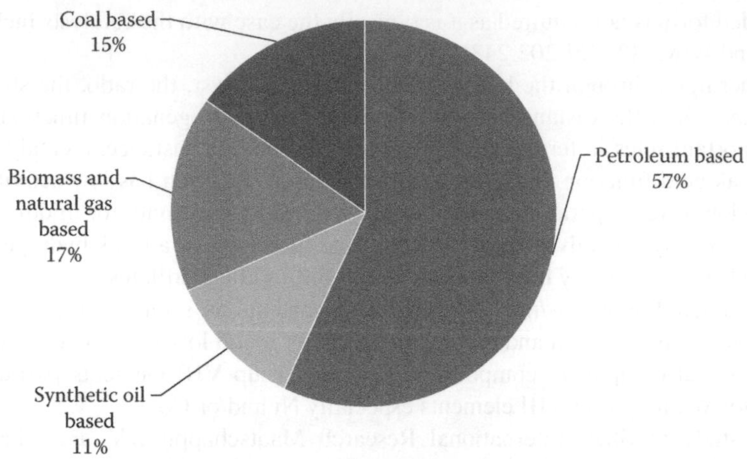

FIGURE 3.12 "Feedstock-based technologies"–related patents for hydrocracking catalysts.

oil, lignin, and sorbitol; and "coal based" such as brown coal, coal tar, and coal pitch with 15% of the total value of feedstock-based technologies.

Synthetic-based feedstocks mostly obtained by Fischer–Tropsch synthesis can be used both with hydrocarbon oils and without them to get predominantly biodiesels and food-level white oils [18,84,107,137,171,175,244,248–250]. Biomass-derived fuels such as biodiesel, bioaviation kerosene, or chemicals like BTX-type aromatics are mostly targeted products by thermochemical and/or catalytic upgrading processes of biomass [20,61,76,88,98,116,120,152,207,228,230]. Coal-based feedstocks can be processed in two or three phase systems, which comprise particles of coal components [26,36,55,67,75,134,156,168,169,190,231,238].

In an invention of Headwaters Heavy Oil LLC, methods and systems for hydrocracking of heavy oil and/or coal by using a colloidal or molecularly dispersed catalyst, which can lead to more efficient use of a supported catalyst if used in combination with the colloidal or molecular catalyst, are studied. The present systems and processes aforementioned within the study can be used to upgrade a coal feedstock and/or mixtures of heavy oil and coal feedstock as well as liquid heavy oil feedstocks [68].

3.4.3 Developments in Terms of Target Products

Developments in terms of target products are investigated categorizing them into two main groups: fuels such as kerosene and chemicals such as BTX, food-level white oil, base oil, and wax.

As presented in Figure 3.13, middle distillate involving both diesel and kerosene cuts is the major focus product with 46% and only naphtha/gasoline target production with 21%. The chemicals as target products, with 33%, are composed of base oil, BTX, wax, and food-level white oil. Base oil is the most investigated product of the chemicals as the target product, while food-level white oil and wax are the topics less subjected in the related patents.

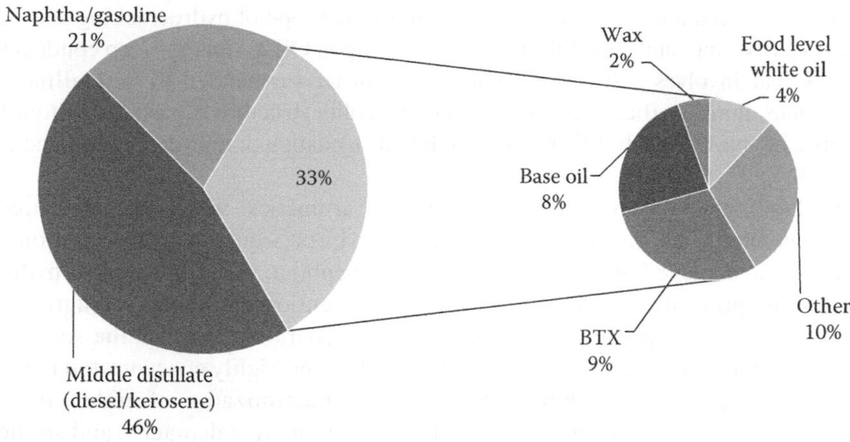

FIGURE 3.13 "Developments in terms of target product"–related patents for hydrocracking catalysts.

3.4.4 Developments in Terms of Performance Enhancements

Hydrocracking catalyst-related patents in terms of performance enhancements have been grouped into four main categories as activity, selectivity, stability, and product quality, as seen in Figure 3.14. The major focus topic is selectivity with 42%. Activity that involves hydrogenation, hydrocracking, hydrodesulfurization, and hydrodenitrogenation activities consists of 31% of the total number. The product quality with 11% of the performance enhancements consists of mainly octane number and cetane number. The remaining 5% is related to other product quality parameters such as pour point.

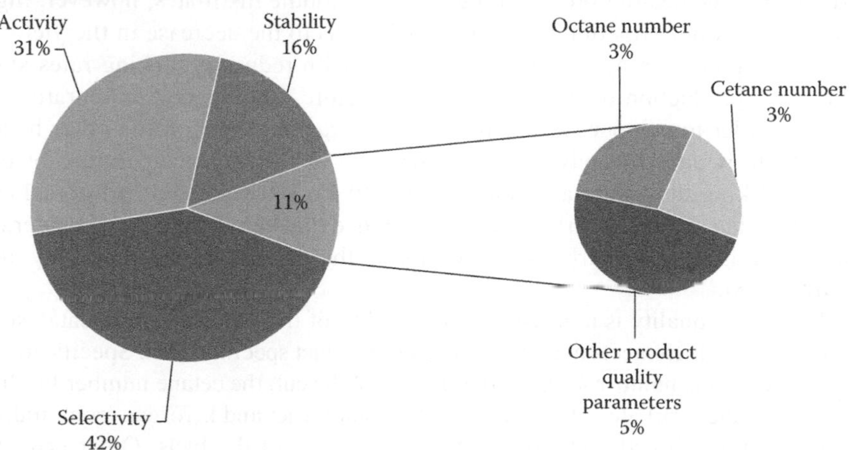

FIGURE 3.14 "Performance enhancements"–related patents for hydrocracking catalysts.

Hydrogenation and cracking activities are in the scope of hydrocracking activity term as a relevant catalyst is bifunctional. JX Nippon Oil & Energy Corp conducted a study that involves active hydrogen–metal cluster surrounded by crystalline or amorphous aluminosilicate matrix in which a zeolite structure is partially or totally collapsed to have a catalyst that shows high hydrogenation or dehydrogenation activity [108].

Naphtha has a relatively higher amount of aromatics, which provides high-octane numbers for gasoline derived from it. Less aromatic hydrogenation is necessary to achieve high aromatic contents in naphtha, which means low hydrogen consumption during the process. In an invention by Shell Internationale Research Maatschappij B.V., the process for the preparation of a naphtha-selective catalyst includes the usage of citric acid to stabilize the highly concentrated metal-containing impregnation solution [22]. However, maximization of middle distillates is the main target research area with respect to market demands, and studies aboutenhancement of middle distillate selectivity consist of catalysts having some features such as secondary pore, low acid density, and high medium-strong acid proportion for inhabitation of excessive cracking and secondary cracking via controlling the C–C bond cracking [25,29–31,45,49,50,68,101,108,136,150,151,169, 214,255].

One of the major concerns that is associated with cost-effective aspects and also indirectly related to environmental aspects is stability. From the point of environmental aspects, the primary problems are the wastes that occur during the production of the catalyst and disposal of the spent catalyst that cannot be regenerated. In terms of cost-effective aspects, enhancement of catalyst life provides economic benefit since catalyst purchase and loss of unit shutdown process are significant cost items. The metals within feedstocks and coke that occur from carbonaceous components under high-temperature process are poisons that shorten the catalyst life. The activity term is related to cracking hydrogenation and cracking rates. As for the cracking rate, a high activity catalyst is a necessity to count bottoms into the desired product range such as middle distillates; however, high activity causes light fraction increase, which leads to the decrease in the yield of the desired product range since lower activity with reduced cracking rates suppresses the production of light products. Therefore, the reaction temperature is raised in order to compensate for the reduced cracking rate; on the other hand, catalyst life is unfortunately shortened because of the high temperature of the process. Taking those into account, Cosmo Oil Co Ltd studied a hydrocracking catalyst having a derived carbonaceous substance that has an activity of moderate hydrocracking of hydrocarbons that optimize the balance between activity and stability [218].

The product quality is a measure of the ability of the process and its catalyst to yield product, which meets market demands of product specifications. Specifications such as the octane number for gasoline from naphtha cut, the cetane number for diesel from middle distillate cut, and the smoke point for jet and kerosene from middle distillate cut are directly relevant to the performance of the fuels. Other primary product quality parameters are composed of API gravity, sulfur content, pour point, aniline point, freeze point, cloud point, refractive index, viscosity, total acid number, and nitrogen content [5,29,41,61,62,75,77,78,95,99,104,108,113,139,141,201,202,205, 226,236,237,241].

3.4.5 Developments in Terms of "Ancillary Processes, Environmental, and Cost-Effective Aspects"

The distribution of developments in the catalysts in terms of subjects within the category "ancillary processes, environmental, and cost-effective aspects" is shown in Figure 3.15. Thirty-seven percent of the patents state that the proposed catalysts are improved with respect to environmental pollution reduction, while 32% of them are improved with respect to cost reduction. Ancillary processes with 31%, which are proposed to be investigated based on passivation, regeneration, and reclamation, seemed to just involve regeneration and utilization of waste catalysts within the topic "ancillary processes."

According to the investigation, most of the studies within the topic of ancillary process involve regeneration process [24,56,72,97,234]. In an invention of IFP Energies Nouvelles, pre-refining and hydroconversion in fixed bed of a heavy crude oil, which comprises removing metals in hydrodemetallization section and then hydrocracking and fractionating of the effluent, are performed [234]. Ferro Duo Ltd. has a study about recovery/reclamation that consists of recovering lanthanum from zeolite in which an aqueous acid containing lanthanum is added to zeolite compounds so that there is a pH value lower than or equal to 3. It is possible to recover lanthanum from zeolite, which is catalyst waste with the present invention [217].

The environmental aspects consist of synthesis of nonsulfur catalysts, clean preparation methods, and reduction of wastes during preparation [43,61,90,98,238]. In terms of cost-effective aspects, improvement of efficiency with conversions at lower temperatures and pressures of the hydrocracking process and reduction in period, raw materials, and energy consumption via simplifying production technology with simple devices are major focal points [10,34,90,98,238].

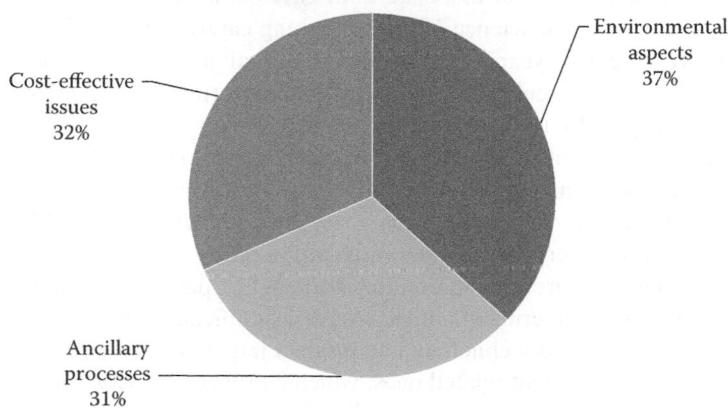

FIGURE 3.15 "Ancillary processes, environmental, and cost-effective aspects"–related patents for hydrocracking catalysts.

3.5 FUTURE PROSPECTS IN HYDROCRACKING TECHNOLOGIES

As transportation is the dominant sector in the final energy consumption, meeting the demand and specifications of the fuels has led to the growth in demand for the consumption of low-sulfur, high-quality products. However, the density of available crudes with low atomic hydrogen/carbon ratios has been rising as well as sulfur contents, as feedstocks get heavier year by year. Therefore, the requirement of having to cope with not only heavier feedstocks but also stricter product quality specifications due to environmental constraints would create a renewed interest in hydrocracking technology.

The refineries will have to process higher amounts of heavier feedstocks in the years ahead, and for the process of heavier feedstocks, higher hydrogenation activity, superior metal uptake capacity, and increased resistance to sulfur and nitrogen, which are directly associated with the improvement of the stability of the catalyst as well as the protection of the downstream catalyst, are essential. The longer cycle lengths and optimized hydrogen consumption for the economic balance, stability of the product quality, and operational conditions despite heavy feedstocks are also with respect to the optimization of hydrogenation and hydrocracking activity of the catalyst.

In terms of the stricter product quality, oil product specifications are shaped to reduce the air pollution caused by greenhouse gas emissions from the exhaust gas [265]. The sulfur content of diesel, for example, has dramatically reduced from 5000 to 10 ppm, almost sulfur free, during the last 35 years. A sulfur content of 2000 ppm is allowed by Euro I in 1993 and 500 ppm with Euro II in 1996. A limit of 350 ppm is designated by the beginning of the twenty-first century with Euro III, and it is further reduced to a limit of 50 ppm with the introduction of Euro IV in 2005. Euro V standard has been applied since 2009 with a maximum content of 10 ppm, so as Euro VI in 2014 [265–271]. In order to achieve the almost sulfur-free diesel, Rive Technology Inc. has published an invention presenting that the hydrogenation activity also provides sterically hindered sulfur compounds that are able to be removed [15]. Similar technologies that associate both environmental constraints and market demands with material science of hydrocracking catalysts will gain much more importance from year to year. The parameters of activity, stability, and yield are major focal points to have enhanced hydrocracking performance as the improvement of both the activity and selectivity ensures meeting market demand, where hydrogenation activity and stability promote the process of heavier feedstocks.

Remarkable developments are noticed through the patent survey based on the classified main patent topics between the years 2008 and 2014 that had been mentioned in previous chapters. The annual distribution of the patents in terms of main topics can be seen in Figure 3.16. According to the figure, particularly environmental aspects-related patents in terms of numbers have a significant growth over the years. Moreover, feedstock-based technology and product targeted-related patents, as well as performance enhancement-related ones, which consist of activity, selectivity, stability, yield, and product quality parameters, seem to have an increment, despite the fact that there is no clear trend in the total count by years.

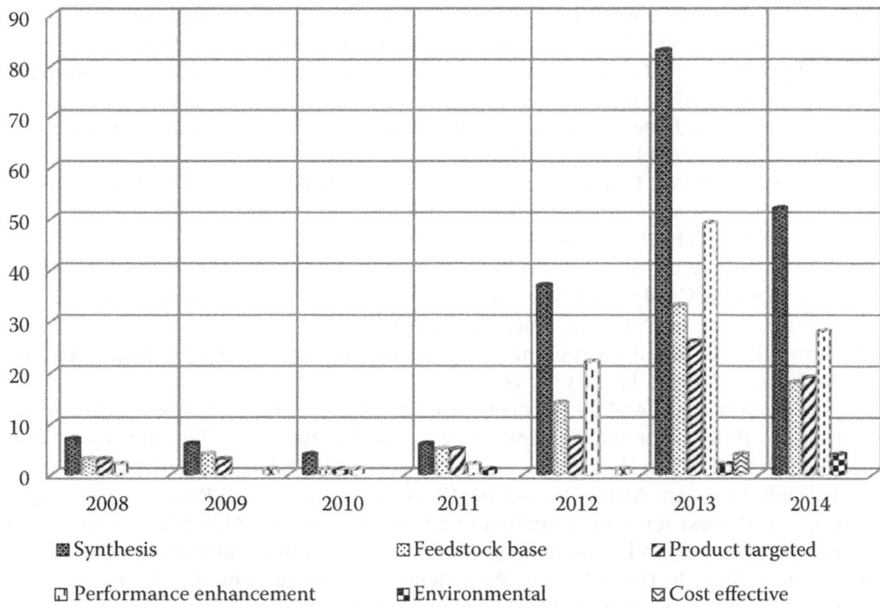

FIGURE 3.16 Annual distributions of the investigated patents in terms of the topic classification.

Hydrocracking technologies with catalyst innovations have taken on an increasingly prominent role in many refinery configurations. As the flexibility of hydrocracking processes and the diversity of catalysts used allow a variety of feedstocks and product ranges, the understanding and development of both complex process designs and material science of hydrocracking catalysts while processing difficult feedstocks and producing excellent product qualities are an outstanding requirement of the century [7,266]. As a result, the catalyst or the catalyst in conjunction with other types of catalysts used in the hydrocracking operation will have more significance as well as operational conditions and process design. Thus, pore structure and acidity of the catalysts, optimization of amount of precious metal, and highly dispersed metal incorporation methods for appropriate hydrocracking and hydrogenation activities are also the focus points of research fields [5,267].

REFERENCES

1. James G. Speight, Baki Özüm, *Petroleum Refining Process*, Marcel Dekker Inc., Alberta, Canada, 2002, pp. 485–499.
2. S.J. Jones David, Peter R. Pujado, *Handbook of Petroleum Processing*, Springer, Illinois, U.S.A., 2006, pp. 287–300, 308–306.
3. Pierre Leprince, *Petroleum Refining*, vol. 3, Conversion Processes, Institut Français du Pétrole, IFP, Paris, France, 2001, pp. 333–348.

4. James H. Gary, Glenn E. Handwerk, Mark J. Kaiser, *Petroleum Refining Technology and Economics*, 5th ed., CRC Press, Boca Raton, FL, 2007, pp. 161–169.
5. Chang S. Hsu, Paul Robinson, *Practical Advances in Petroleum Processing*, Springer, Texas, U.S.A., 2006, pp. 28–34, 200–217.
6. Jorge Ancheyta, James G. Speight, *Hydroprocessing of Heavy Oils and Residua*, CRC Press, Boca Raton, FL, 2007, pp. 287–295.
7. Julius Scherzer, A.J. Gruia, *Hydrocracking Science and Technology*, CRC Press, Boca Raton, FL, 1996, pp. 200–214, 174–198.
8. Chevron, Diesel Fuels Technical Review, 2007.
9. Jianxin Jason Wu, Ajit B. Dandekar, Christopher G. Oliveri, Mesoporous Zeolite-Y Hydrocracking Catalyst and Associated Hydrocracking Processes, WO2014098820 A1 filed December 19, 2012, and issued January 26, 2014.
10. Hiroyuki Seki, Masahiro Higashi, Processes for Production of Liquid Fuel, filed July 24, 2008, and issued January 11, 2008.
11. Thibault Alphazan, Audrey Bonduelle, Christele Legens, Pascal Raybaud, Christophe Coperet, Process for the Preparation of a Catalyst Based on Tungsten for Use in Hydrotreatment or in Hydrocracking, filed October 30, 2014, and issued April 30, 2014.
12. Thibault Alphazan, Audrey Bonduelle, Christele Legens, Pascal Raybaud, Christophe Coperet, Process for the Preparation of a Catalyst Based on Molybdenum for Use in Hydrotreatment or in Hydrocracking, filed October 31, 2014, and issued May 01, 2014.
13. Min Kee Choi, Ju Hwan Im, Do Woan Kim, Do Kyoung Kim, Tae Jin Kim, Seung Hoon Oh, Tae Hong Seok, Catalyst Containing Metal Cluster in Structurally Collapsed Zeolite, and Use Thereof, filed October 30, 2014, and issued April 22, 2014.
14. Yan Zifeng, Liu Xinmei, Chang Xingwen, Zhang Zhihua, Qın Lihong, Dai Baoqin, Gao Xionghou, Zhang Zhongdong, Method for Modifying Mesoporous-Rich Usy (Ultra-Stable Y) Molecular Sieve in Combined Manner, filed June 18, 2013, and issued December 24, 2014.
15. Qi Bangfeng, Gan Weğmin, Meng Fanmei, Silicon-Aluminum Carrier Containing Two Modified Molecular Sieves, Preparation Method and Application Thereof, filed July 22, 2013, and issued December 31, 2014.
16. Audrey Bonduelle, Emmanuelle Guillon, Magalie Roy-Auberger, Method for Preparing A Catalyst Usable in Hydroconversion and İncluding at Least One Nu-86 Zeolite, filed November 23, 2012, and issued November 06, 2014.
17. Takuya Niitsuma, Marie Iwama, Diesel Fuel or Diesel Fuel Base and Process for Manufacturing Same, filed October 09, 2014, and issued March 27, 2013.
18. Kwan Young Lee, Hee Jun Eom, Seong Min Kim, Young Gul Hur, Method for Preparing Light Distillate from Extra-Middle Distillate by Using Heteropolyacid Catalyst Substituted with Cesium, and Method for Regenerating Catalyst, A Method for Preparing Light Oil from Extra Heavy Oil Using Cesium Substituted Heteropolyacid Catalyst and Method for Regenerating the Catalyst, WO2014163252 A1 filed August 14, 2013, and issued October 09, 2014.
19. Antoine Daudin, Alain Quignard, Olivier Thinon, Optimized Method for Recycling Bio-Oils into Hydrocarbon Fuels, filed February 26, 2013, and issued January 21, 2015.
20. Cornelis Ouwehand, Marcello Stefano Rigutto, Welsenes Jan Arend Van, Process for Preparing Hydrocracking Catalyst, filed December 20, 2012, and issued June 27, 2013.
21. Wiebe Sjoerd Kijlstra, Ferry Winter, Process for Preparing Hydrocracking Catalyst Compositions, filed December 20, 2012, and issued June 27, 2013.
22. Zhang Zhihua, Yan Zifeng, Dai Baoqin, Liu Xinmei, Tian Ran, Li Haiyan, Sun Famin, Qin Lihong, Zhang Wencheng, Lyu Qian, Xie Bin, Wang Dongqing, Wang Fucun, Guo Shuzhi, Zhang Qingwu, Wang Yan, Yu Xiujuan, Bing Shuqiu, Usy Molecular Sieve Modification Method, filed April 03, 2013, and issued October 15, 2014.

23. Kwan Young Lee, Hee Jun Eom, Seong Min Kim, Young Gul Hur, Method for Preparing Light Distillate from Extra-Middle Distillate by Using Heteropolyacid Catalyst Substituted with Cesium, and Method for Regenerating Catalyst, filed August 14, 2013, and issued October 09, 2014.
24. Dong Songtao, Zhao Yang, Xin Jing, Hu Zhihai, Nie Hong, Porous Support Used for Hydrogenation Catalyst, Preparation Method of Porous Support and Hydrocracking Catalyst and Hydrocracking Catalysis Method, filed March 30, 2013, and issued October 01, 2014.
25. Xiang Wenyu, Xiang Yuqiao, Hu Yibo, Wang Hua, Catalyst for Coal Tar All-Distillate Hydrocracking as well as Preparation Method and Application Method Thereof, filed December 18, 2013, and issued April 02, 2014.
26. Marius Vaarkamp, Catalyst, Catalyst Support and Process for Hydrogenation, Hydroisomerization, Hydrocracking and/or Hydrodesulfurization, filed April 18, 2011, and issued March 06, 2012.
27. Yoshii Kiyotaka, Yamada Atsushi, Hydrocracking Catalyst, Method for Producing Same, and Method for Producing Hydroxy Compound Using Said Catalyst, filed August 22, 2012, and issued February 28, 2013.
28. Sujatha Krishnamurthy, Cassandra Schoessow, Douglas Piotter, Anand Subramanian, Systems and Methods for Temporary Deactivation of Hydrocracking Catalyst, filed March 13, 2014, and issued September 18, 2014.
29. Dong Songtao, Dong Jianwei, Nie Hong, Shi Yahua, Li Dadong, Hydrocracking Catalyst, Preparation and Application Thereof, filed March 15, 2013, and issued September 17, 2014.
30. Xin Jing, Yang Ping, Dong Songtao, Mao Yichao, Zhang Yuying, Li Mingfeng, Nie Hong, Hydrocracking Catalyst and İts Preparation Method and Use, filed March 13, 2013, and issued September 17, 2014.
31. Xin Jing, Mao Yichao, Yang Ping, Dong Songtao, Zhang Yuying, Li Mingfeng, Nie Hong, Hydrocracking Catalyst and Use Thereof, filed March 13, 2013, and issued September 17, 2014.
32. Ren Xiangkun, Kenji Iguchi, Koji Sakawaki, Cui Yongjun, Preparation Method of İron Sulfide for Catalyzing Heavy Oil Hydrocracking, and Technology Using İron Sulfide, filed March 06, 2013, and issued September 10, 2014.
33. Ji Jianbing, Yu Yunliang, Liu Xuejun, Lu Meizhen, Yu Fengwen, Zhu Cuihan, Reaction Device for Preparing Bio-Aviation Kerosene Through Hydrocracking, filed June 13, 2014, and issued September 10, 2014.
34. Shen Guoliang, Song Yuhong, Xu Tiejun, Cong Lanbo, Chen Yuannan, Tong Qingjie, Wang Yu, Gao Bing, Vulcanizing Agent of Hydrocracking Catalyst and Preparation Method Thereof, filed June 20, 2014, and issued September 03, 2014.
35. Zhang Qianwen, Zhou Houfeng, Sun Jinchang, Coal Tar Hydrocracking Pretreatment Catalyst and Preparation Method Thereof, filed May 14, 2014, and issued August 27, 2014.
36. Bonduelle Audrey, Guillon Emmanuelle, Roy-Auberger Magalie, Catalyst İncluding At Least One Nu-86 Zeolite, At Least One Usy Zeolite, and A Porous İnorganic Matrix, and Method for the Hydroconversion of Hydrocarbon Feedstocks Using Said Catalyst, filed November 23, 2012, and issued August 20, 2014.
37. Deng Wenan, Li Chuan, Zhang Qianqian, Li Shufeng, Wen Ping, Mu Baoquan, Oil Soluble Molybdenum-Based Precursor of Catalyst, Preparation Method and Application Thereof, filed May 16, 2014, and issued August 13, 2014.
38. Huang Peng, Zhang Xiaojing, Mao Xuefeng, Wang Yong, Li Peilin, Yan Bingfeng, Zhao Yuan, Du Shufeng, Zhong Jinlong, Hu Fating, Zhao Peng, Chang Qiulian, Wang Guangyao, Zhang Fan, Oil Solubility Compound-Type Suspended Bed Hydrocracking Catalyst As Well As Preparation Method Thereof, filed May 21, 2014, and issued August 13, 2014.

39. Yan Zifeng, Qiao Ke, Liu Xinmei, Xu Benjing, Wang Yingying, He Lifeng, Zhang Zhihua, Dai Baoqin, Wang Fucun, Compound Modification Method for Usy Molecular Sieve, filed April 02, 2014, and issued June 25, 2014.
40. Li Haiyan, Qin Lihong, Lyu Qian, Wang Dongqing, Xie Bin, Tian Ran, Sun Famin, Zhang Wencheng, Yu Chunmei, Dai Baoqin, Wang Yan, Ma Baoli, Dong Chunming, Xu Tiega, Ma Shoutao, Zhang Guojia, Yang Xiaodong, Wang Liang, Heteropoly Acid-Containing Hydrocracking Catalyst and Use Thereof, filed December 10, 2012, and issued June 18, 2014.
41. Tang Tiandi, Ma Yuli, Hu Jianbo, Liu Taotao, Fu Wenqian, Xiang Mei, Jin Huile, Mesoporous Zsm-5 Zeolite, Mesoporous Zsm-5 Zeolite-Loaded Metal Sulfide Catalyst and Application, filed May 14, 2014, and issued December 30, 2013.
42. Du Yanze, Fan Dandan, Wang Fenglai, A Clean Preparation Method of a Hydrocracking Catalyst, filed November 03, 2012, and issued May 14, 2014.
43. Jiang Hong, Yang Zhanlin, Tang Zhaoji, Wang Jifeng, Wen Derong, Wei Dengling, High-Activity Hydrocracking Catalyst and Preparation Method Thereof, filed November 01, 2012, and issued May 14, 2014.
44. Jiang Hong, Yang Zhanli, Tang Zhaoji, Wang Jifeng, Wei Dengling, Wen Derong, Preparation Method of Middle Distillate-Type Hydrocracking Catalyst, filed November 01, 2012, and issued May 14, 2014.
45. Gu Mingdi, Wang Gang, Cao Chunqing, Bu Yan, Bai Zhenmin, Huang Wei, Preparation Method of Hydrocracking Catalyst, filed November 01, 2012, and issued May 14, 2014.
46. Gu Mingdi, Bu Yan, Cao Chunqing, Wang Gang, Bai Zhenmin, Huang Wei, Hydrocracking Catalyst, and Preparation Method and Application Thereof, filed November 01, 2012, and issued May 14, 2014.
47. Gu Mingdi, Bai Zhenmin, Bu Yan, Wang Gang, Cao Chunqing, Huang Wei, Metal Nitride Hydrocracking Catalyst, and Preparation Method and Application Thereof, filed November 01, 2012, and issued May 14, 2014.
48. Gu Mingdi, Wang Gang, Bai Zhenmin, Bu Yan, Huang Wei, Cao Chunqing, Preparation Method for Middle Distillate Type Hydrocracking Catalyst Containing Metal Nitride, filed November 01, 2012, and issued May 14, 2014.
49. Omer Refa Koseoglu, Adnan Al-Hajji, Ali Mahmood Al-Somali, Ali H. Al-Abdul'Al, Mishaal Al-Thukair, Masaru Ushio, Ryuzo Kuroda, Takashi Kameoka, Koji Nakano, Yuichi Takamori, Hydrocracking Catalyst for Hydrocarbon Oil, Method for Producing Hydrocracking Catalyst, and Method for Hydrocracking Hydrocarbon Oil with Hydrocracking Catalyst, filed August 02, 2011, and issued July 11, 2013.
50. Garcia-Martinez Javier, Mesoporous Zeolite Catalyst Supports, filed November 08, 2013, and issued May 08, 2014.
51. Yuichi Tanaka, Takuya Niitsuma, Kazuhiko Tasaka, Marie Iwama, Hydrocracking Catalyst and Method for Producing a Hydrocarbon Oil, filed March 26, 2012, and issued May 08, 2014.
52. Patricia Rayo Mayoral, Juarez Jorge Ancheyta, Marroquin Sanchez Gustavo Jesus, Centeno Nolasco Guillermo, Jorge Fernando Ramirez Solis, Mesoporous Composite of Molecular Sieves for Hydrocracking of Heavy Crude Oils and Residues, filed November 06, 2013, and issued May 08, 2014.
53. Hommeltoft Sven Ivar, Elomari Saleh, Lacheen Howard S, An Acid Catalyst Composition Comprising a Conjunct Polymer and A Lewis Acid, filed November 17, 2009, and issued April 02, 2014.
54. Xiang Wenyu, Xiang Yuqiao, Hu Yibo, Wang Hua, Catalyst for Coal Tar Hydrocracking and Preparation Method Thereof, filed December 18, 2013, and issued April 02, 2014.

55. Tanaka Yuichi, Niitsuma Takuya, Tasaka Kazuhiko, Iwama Marie, Regenerated Hydrocracking Catalyst and Method for Producing a Hydrocarbon Oil, filed March 26, 2012, and issued March 27, 2014.
56. Zhang Le, Nie Hong, Long Xiangyun, Liu Qinghe, Mao Yichao, Liu Xuefen, Li Dadong, Wang Kui, Wang Zhe, Catalyst with Hydrogenation Catalysis Effect, Preparation Method and Application of Catalyst and Hydrocracking Method, filed August 29, 2012, and issued March 26, 2014.
57. Pierre Dufresne, Mickael Bremaud, Pauline Galliou, Kirumakk Sharath, Passivation Method of Zeolite Catalyst, Particularly Hydrocracking Catalyst by Using Nitrogen-Containing Compound, filed August 09, 2013, and issued February 24, 2014.
58. Zhang Le, Liu Qinghe, Mao Yichao, Zhao Xinqiang, Long Xiangyun, Liu Xuefen, Nie Hong, Shi Yahua, Gao Xiaodong, Catalyst with Hydrogenation Catalytic Action, Preparation Method, Applications and Hydrocracking Method, filed August 29, 2012, and issued March 12, 2014.
59. Maesen Theodorus, Kuperman Alexander E, Fong Darren P, Hydrocracking Catalyst and Process Using a Magnesium Aluminosilicate Clay, filed October 03, 2008, and issued April 08, 2010.
60. Liu Can, Rong Long, Zhao Xiaomeng, Yang Mingfeng, Wang Yan, Li Juntao, Chang Mingming, Huang Tiran, Zhang Jing, Method for Preparing Alkane Type Biodiesel by Using Tung Oil, filed November 11, 2013, and issued February 05, 2014.
61. Qi Bangfeng, Gan Weimin, Meng Fanmei, Composite Molecular Sieve-Containing Medium Oil Type Hydrocracking Catalyst, Preparation Method and Application Thereof, filed July 22, 2013, and issued February 05, 2014.
62. Shen Baojian, Wang Yandan, Ren Shenyong, Guo Qiaoxia, Wang Wennian, Li Jiangcheng, Sun Famin, Li Haiyan, Zhang Zhihua, Hydrocracking Catalyst Containing Ti-P-Beta Molecular Sieve and Preparation Method of Hydrocracking Catalyst, filed October 18, 2013, and issued January 22, 2014.
63. Shen Baojian, Wang Yandan, Ren Shenyong, Guo Qiaoxia, Wang Wennian, Li Jiangcheng, Sun Famin, Li Haiyan, Zhang Zhihua, Hydrocracking Catalyst Containing Ti-P-Y Molecular Sieve and Preparation Method of Hydrocracking Catalyst, filed October 18, 2013, and issued January 22, 2014.
64. Yu Haibin, Zhu Jinjian, Zhang Jingcheng, Xiao Han, Zhang Guohui, Zhang Yuting, Geng Shan, Li Xiaoguo, Preparation Method of Hydrocracking Catalyst Controlling Active Metal Distribution, filed September 18, 2013, and issued September 18, 2013.
65. Hermanus Jongkind, Wiebe Sjoerd Kijlstra, Bart Pelgrim, Marcello Stefano Rigutto, Ferry Winter, Process for the Preparation of A Hydrocracking Catalyst, filed September 16, 2013, and issued March 20, 2014.
66. Chang Yu-Hwa, Methods for increasing Catalyst Concentration in Heavy Oil and/or Coal Resid Hydrocracker, filed October 28, 2008, and issued August 11, 2010.
67. Masahiro Higashi, Hiroyuki Seki, Sumio Saito, Kuroda Ryuzo, Kameoka Takashi, Hydrocracking Catalyst and Process for Producing Fuel Base Material, filed March 22, 2007, and issued July 08, 2010.
68. Eric D Joseck, Michael B Carroll, David Mentzer, Teck Mui Hoo, Michel A Daage, Ajit B Dandekar, Production of Lubricating Oil Basestocks, filed July 12, 2012, and issued January 24, 2013.
69. Mayoral Patricia Rayo, Juarez Jorge Ancheyta, Sanchez Gustavo Jesus Marroquin, Nolasco Guillermo Centeno, Solis Jorge Fernando Ramirez, Mesoporous Composite of Molecular Meshes for the Hydrocracking of Heavy Crude Oil and Residues, filed November 06, 2012, and issued October 16, 2013.

70. Marchand Karin, Chaumonnot Alexandra, Bonduelle Audrey, Dufaud Veronique, Lefebvre Frederic, Bader Manuela, Lopes Silva Susana, Process for the Hydrocracking of Hydrocarbon Cuts Using a Catalyst Based on Heteropolyanions Trapped in A Mesostructured Oxide Support, filed December 15, 2011, and issued December 19, 2013.
71. Tanaka Yuichi, Niitsuma Takuya, Tasaka Kazuhiko, Iwama Marie, Hydrocracking Catalyst and Method for Producing Hydrocarbon Oil, filed March 26, 2012, and issued December 18, 2013.
72. Zhu Jinjian, Yu Haibin, Zhang Jingcheng, Xiao Han, Zhang Guohui, Zhang Yuting, Geng Shan, Li Xiaoguo, Method for Preparing Hydrocracking Catalyst for Controlling Acidic Site Distribution, filed September 18, 2013, and issued December 18, 2013.
73. Liu Baijun, Pang Xinmei, Wang Rui, Liu Shaopeng, Zhang Yin, Wang Xiaohua, Li Fayong, Hydrocracking Catalyst Containing Y-Type Molecular Sieve, and Preparation Method Thereof, filed June 01, 2012, and issued December 18, 2013.
74. Gu Zhihua, Lang Ying, Shang Xinli, Zhang Jing, Gao Min, Wu Jian, Hai Jun, Process Method and Catalyst for Preparing Diesel Oil by Hydrogenating Coal Tar, filed September 09, 2013, and issued December 11, 2013.
75. Quignard Alain, Thinon Olivier, Dulot Hugues, Huyghe Raphael, Optimized Process for Upgrading Bio-Oils of Aromatic Bases, filed May 23, 2013, and issued December 05, 2013.
76. Li Haiyan, Zhang Zhihua, Dai Baoqin, Qin Lihong, Tian Ran, Sun Famin, Wang Weizhong, Si Chaoxia, Zhang Wencheng, Ma Jianying, Yu Chunmei, Lv Qian, Wang Dongqing, Xiao Yong, Xie Bin, Yu Hongwei, Guan Xu, Cong Liru, Zhang Quanguo, Li Shujie, Wang Yan, Ma Shout Aao, Song Jinhe, Hydrocracking Catalyst Containing Heteropoly Acid and Application Thereof, filed May 15, 2012, and issued December 04, 2013.
77. Qi Bangfeng, Gan Weimin, Meng Fanmei, Light-Oil-Type Hydrocracking Catalyst Carrier, Preparation Method Thereof and Applications Thereof, filed July 22, 2013, and issued November 27, 2013.
78. Kagami Shigeari, Inamura Kazuhiro, Takahashi Nobuyuki, Hydrodesulfurization Apparatus Using Hydrocracking Catalyst and Hydroprocessing Method of Heavy Oil, filed March 30, 2012, and issued October 17, 2013.
79. Inamura Kazuhiro, Watabe Mitsunori, Eura Shinya, Crystalline Aluminosilicate, Hydrocracking Catalyst for Heavy Oil, and Method for Manufacturing Same, filed March 30, 2012, and issued October 17, 2013.
80. Qi Bangfeng, Gan Weimin, Meng Fanmei, Light Oil Type Hydrocracking Catalyst with Composite Molecular Sieve as well as Preparation Method and Application Thereof, filed July 22, 2013, and issued November 20, 2013.
81. Bulut Metin, Kenmogne-Gatchuissi Regine, Fajula Francois, Dath Jean-Pierre, Van Donk Sander, Finiels Annie, Hulea Vasile, Method of Preparing A Hydroconversion Catalyst Based on Silica or Silica–Alumina Having an İnterconnected Mesoporous Texture, filed December 23, 2011, and issued November 14, 2013.
82. Ying Jackie Y, Garcia-Martinez Javier, Mesostructured Zeolitic Materials Suitable for Use in Hydrocracking Catalyst Compositions and Methods of Making and Using the Same, filed July 09, 2013, and issued November 07, 2013.
83. Grezaud Aline, Heraud Jean Philippe, Dulot Hugues, Bouchy Christophe, Calemma Vincenzo, Optimized Method for Producing Middle Distillates from a Feedstock Originating from the Fischer–Tropsch Process Containing a Limited Quantity of Oxygenated Compounds, filed April 11, 2013, and issued October 17, 2013.
84. Li Chuan, Deng Wenan, Xia Zhitong, Mu Baoquan, Oil-Soluble Self-Vulcanizing Molybdenum Catalyst, and Preparation Method, Use Method and Application of Oil-Soluble Self-Vulcanizing Molybdenum Catalyst, filed July 26, 2013, and issued October 16, 2013.

85. Milam Stanley Nemec, Reynolds Michael Anthony, Wellington Scott Lee, Buhrman Frederik Arnold, Hydrocracking of A Heavy Hydrocarbon Feedstock Using a Copper Molybdenum Sulfided Catalyst, filed December 08, 2011, and issued June 14, 2012.
86. Qiu He, Zhou Bing, Highly Stable Hydrocarbon-Soluble Molybdenum Catalyst Precursors and Methods for Making Same, filed March 06, 2013, and issued September 26, 2013.
87. Radlein Desmond, J Wang, Y Yuan, Quignard Alain, Methods of Upgrading Biooil to Transportation Grade Hydrocarbon Fuels, filed September 12, 2011, and issued March 22, 2012.
88. Xiao Han, Yang Jianguo, Li Xiaoguo, Yu Haibin, Zhu Jinjian, Shi Fang, Zhang Guohui, Li Jia, Preparation Method of Catalyst Carrier for Hydrotreating Pretreatment of Wax Oil, filed June 21, 2013, and issued December 17, 2014.
89. Lu Yinhua, Fan Lichuang, Liu Feipeng, Hao Kun, Tao Zhichao, Yang Yong, Li Yongwang, Non-Sulfur Catalyst as well as Preparation Method and Application Thereof, filed July 11, 2013, and issued September 18, 2013.
90. Lott Roger K, Lee Lap Keung, Method for Upgrading Ebbulated Bed Reactor and Upgraded Ebbulated Bed Reactor, filed April 18, 2013, and issued September 12, 2013.
91. Gao Liang, Zong Baoning, Wen Langyou, Yu Fang, Mu Xuhong, Dong Minghui, Pyrite-Containing Heavy Oil Hydrogenation Catalyst, and Preparation and Application Thereof, filed February 22, 2012, and issued September 11, 2013.
92. Gao Liang, Zong Baoning, Wen Langyou, Yu Fang, Mu Xuhong, Dong Minghui, Bauxite-Containing Heavy Oil Hydrogenation Catalyst, and Preparation and Application Thereof, filed February 22, 2012, and issued September 11, 2013.
93. Gao Liang, Zong Baoning, Wen Langyou, Yu Fang, Mu Xuhong, Dong Minghui, iron-Oxide-Ore-Containing Heavy Oil Hydrogenation Catalyst, and Preparation and Application Thereof, filed February 22, 2012, and issued September 11, 2013.
94. Yuan Xiaoliang, Lan Ling, Hu Sheng, Tan Qingfeng, Liu Hongtao, Wang Shuqin, He Hao, Zhang Peng, Wu Pingyi, Zhao Qinfeng, Wang Peng, Hydrocracking Catalyst Containing Mesoporous–Microporous Molecular Sieve and Preparation Method of Catalyst, filed March 01, 2012, and issued September 11, 2013.
95. Koseoglu Omer Refa, Al-Hajji Adnan, Al-Somali Ali Mahmood, Al-Abdul Al Ali H, Al-Thukair Mishaal, Ushio Masaru, Kuroda Ryuzo, Kameoka Takashi, Nakano Kouji, Takamori Yuuichi, Hydrocracking Catalyst for Hydrocarbon Oil, Method for Producing Hydrocracking Catalyst, and Method for Hydrocracking Hydrocarbon Oil with Hydrocracking Catalyst, filed August 02, 2011, and issued August 28, 2013.
96. Myers Daniel N, Palmas Paolo, Myers David N, Process for Regenerating Catalyst in a Fluid Catalytic Cracking Unit, filed March 01, 2012, and issued September 11, 2013.
97. Schuetzle Robert, Schuetzle Dennis, Devilliers David, Catalytic Process for the Direct Production of Hydrocarbon Fuels from Syngas, filed August 22, 2013, and issued March 08, 2013.
98. Wei Xianyong, Qi Shichao, Li Zhanku, Zhao Changlin, Zong Zhimin, Zhao Yunpeng, Fan Xing, Zhao Wei, Cao Jingpei, Preparation Device of Load Type Magic Acid Catalyst, filed January 31, 2013, and issued August 21, 2013.
99. Lv Qian, Sun Famin, Dai Baoqin, Qin Lihong, Zhang Zhihua, Zhao Dongyuan, Tian Ran, Niu Guoxing, Xiao Yong, Song Mingjuan, Wang Dongqing, Li Haiyan, Sun Bo, Sun Honglei, Xia Endong, Chen Shuxian, Ma Shoutao, Wang Gang, Xiong Yuping, Liu Liying, Method of Producing Middle Distillate Oil through Hydrocracking, filed February 16, 2012, and issued August 21, 2013.
100. Lv Qian, Sun Famin, Dai Baoqin, Zhang Zhihua, Zhao Dongyuan, Xiao Yong, Qin Lihong, Niu Guoxing, Sun Bo, Song Mingjuan, Wang Dongqing, Tian Ran, Li Haiyan, Zhao Ye, Xia Endong, Li Ruifeng, Xie Bin, Zhang Lijun, Sun Shengbo, Shao Wei, Wang Yan, Composite Zeolite Hydrocracking Catalyst Carrier, filed February 16, 2012, and issued August 21, 2013.

101. Kim Do Woan, Koh Jae Hyun, Lee Sang ii, Lee Seung Woo, Oh Seung Hoon, Koh Jae Suk, Kim Yong Seung, Kim Gyung Rok, Choi Sun, Kim Hong Chan, Oh Sang Hun, Hydrocracking Catalyst for Preparing Valuable Light Aromatic Hydrocarbons from Polycyclic Aromatic Hydrocarbons, filed October 21, 2011, and issued August 15, 2013.
102. Gao Liang, Zong Baoning, Wen Langyou, Yu Fang, Mu Xuhong, Dong Minghui, iron Ore-Containing Heavy Oil Hydrogenation Catalyst, and Preparation Method and Application Thereof, filed February 03, 2012, and issued August 14, 2013.
103. Yuan Xiaoliang, Lan Ling, Hu Sheng, Wu Pingyi, He Hao, Zhao Qinfeng, Zhang Peng, Lu Xu, Lv Zhongwu, Wang Shuqin, Wang Peng, Hydrocracking Catalyst, and Preparation and Applications Thereof, filed February 08, 2012, and issued August 14, 2013.
104. Liu Dong, Lv Renqing, Du Hui, Liu Zhangyong, Zong Zhiqiao, Super-Dispersing Nanometer Catalyst for Hydrocracking Process of Suspended Bed, filed April 27, 2013, and issued August 07, 2013.
105. Mao Yichao, Nie Hong, Dong Jianwei, Xiong Zhenlin, Hu Zhihai, Shi Yahua, Li Dadong, A Catalyst for Hydrocracking Process, the Preparing Method and the Use Thereof, filed April 26, 2006, and issued June 07, 2013.
106. Iwama Marie, Tasaka Kazuhiko, Tanaka Yuichi, Process for Producing Hydrocarbon Oil and System for Producing Hydrocarbon Oil, filed August 12, 2011, and issued July 31, 2013.
107. Van Donk Sander, Lakrua Maksim, Kenmon'-Gatchuissi Rezhin, Fazhula Fransua, Bjulju Meten, Dat Zhan-P'Er, De Jong Krijn Piter, De Jong Petra Ehlizabet, Zechevich Jovana, Van Lak Andrianus Nikolas Kornelis, Hydrocracking Catalyst and Process for Producing Liquid Hydrocarbon, filed March 10, 2005, and issued July 24, 2013.
108. Donk Sander Van, Maxime Lacroix, Régine Kenmogne-Gatchuıssı, François Fajula, Métin BULUT, Jean-Pierre Dath, Jong Krijn Pieter De, Jongh Petra Elisabeth De, Jovana ZEČEVIĆ, Laak Adrianus Nicolaas Cornelis Van, Modified Zeolites Y with Trimodal İntracrystalline Structure, Method for Production Thereof and Use Thereof, filed December 22, 2009, and issued January 27, 2013.
109. Domokos Laszlo, Ouvekhend Kornelıs, Hydrocracking Catalyst, filed April 15, 2010, and issued June 10, 2013.
110. Ju Xueyan, Zhang Yuying, Jiang Donghong, Hu Zhihai, Wang Jinye, Nie Hong, Li Dadong, Methanation Method of Hydrocarbon, filed December 06, 2011, and issued June 17, 2013.
111. Fang Lei, Feng Xiufang, Guo Jintao hang Zhihua, Tian Ran, Wang Weizhong, Yu Chunmei, Sun Famin, Wang Gang, Xu Weichi, Wang Dan, Wen Guangming, Tian Chunguang, Liu Qinghua, Jin Yanchun, Li Ruifeng, Wang Fucun, Yu Hongwei, He Yulian, Guan Xu, Liu Lying, Method for Preparing Hydrocracking Catalyst, filed November 08, 2012, and issued July 10, 2013.
112. Diehl Fabrice, Bonduelle Audrey, Hydrocracking Method for Producing Gasoline Blending Component with High Octane Value, filed December 31, 2011, and issued July 03, 2013.
113. Yu Weiqiang, Xu Jie, Lu Fang, Yang Yanliang, Gao Jin, Miao Hong, Zhang Junjie, inferior Coker Gas Oil Hydrotreatment Method, filed December 28, 2011, and issued July 03, 2013.
114. Fabrice Diehl, Audrey Bonduelle, Method for Hydrocracking a Hydrocarbon Feedstock in the Presence of a Sulfide Catalyst Prepared Using a Cyclic Oligosaccharide, filed June 24, 2011, and issued May 22, 2013.
115. Kong Dejin, Zheng Junlin, Li Xuguang, Qian Bin, Chen Yan, Application for Catalyst In Hydrocracking for Biological Polyol, filed December 08, 2011, and issued June 19, 2013.
116. Li Shijun, Wang Jiankang, Yan Liang, Chen Xiaoming, Li Jing, Method for Producing Liquid Fuel, filed January 23, 2013, and issued May 02, 2013.

117. Liu Chang, Wang Fenglai, Du Yanze, Zhao Hong, Guan Minghua, Process for Producing Btx from a C5–C12 Hydrocarbon Mixture, filed June 04, 2013, and issued December 12, 2013.
118. Jiang Guang An, Zhang Ye, Fang Xiangchen, Wang Jifeng, Method for Preparing Arene by Mixture Containing Hydrocarbon with Condensed Rings, filed November 18, 2011, and issued May 29, 2013.
119. Sun Xiaoyan, Wang Jifeng, Fan Hongfei, Biomass Heavy-Oil Hydrocracking Catalyst and Preparation Method Thereof, filed November 18, 2011, and issued May 29, 2013.
120. Sun Xiaoyan, Fan Hongfei, Wang Zhanyu, Beta Molecular Sieve and Preparation Method Thereof, filed November 18, 2011, and issued May 15, 2013.
121. Jiang Guang An, Zhang Ye, Fang Xiangchen, Wang Jifeng, Preparation Method of Catalyst Carrier Material Containing Molecular Sieve and Amorphous Silica–Alumina, filed November 11, 2011, and issued May 15, 2013.
122. Jiang Guang An, Zhang Ye, Fang Xiangchen, Wang Jifeng, Preparation Method of Carrier Material, filed November 11, 2011, and issued May 15, 2013.
123. Jiang Guang An, Zhang Ye, Fang Xiangchen, Wang Jifeng, Carrier Material Containing Molecular Sieve and Amorphous Silica–Alumina and Preparation Method Thereof, filed November 11, 2011, and issued May 15, 2013.
124. Sun Xiaoyan, Fan Hongfei, Wang Zhanyu, Preparation Method of Catalyst Carrier Material, filed November 11, 2011, and issued May 15, 2013.
125. Wang Fenglai, Liu Chang, Guan Minghua, Du Yanze, Zhao Hong, Preparation Method of Catalyst Carrier Material Containing Molecular Sieve and Alumina, filed November 11, 2011, and issued May 15, 2013.
126. Jin Hao, Sun Suhua, Liu Jie, Zhu Huihong, Yang Guang, Preparation Method of Catalyst Carrier Dry Gel Powder, filed November 11, 2011, and issued May 15, 2013.
127. Ruan Caian, Wang Jifeng, Chen Song, Preparation Method of Composite Carrier Material, filed November 11, 2011, and issued May 15, 2013.
128. Wei Xianyong, Qi Shichao, Li Zhanku, Zhao Changlin, Zong Zhimin, Zhao Yunpeng, Fan Xing, Zhao Wei, Cao Jingpei, Preparation Method of Hydrogenation Catalyst, filed November 11, 2011, and issued May 15, 2013.
129. Wu Jianxin Jason, Dandekar Ajit B, Oliveri Christopher G, Hydrocracking Catalyst Carrier and Preparation Method Thereof, filed August 21, 2012, and issued November 21, 2012.
130. Wang Li, Hydrocracking Catalyst Carrier Containing Beta Molecular Sieve and Preparation Method Thereof, filed November 09, 2011, and issued May 15, 2013.
131. Aoki Nobuo, Seki Hiroyuki, Higashi Masahiro, Ikeda Masakazu, Waku Toshio, Hydrocracking Catalyst and Preparation Method Thereof, filed November 09, 2011, and issued May 15, 2013.
132. Ward Andrew Mark, Preparation Method of Hydrocracking Catalyst Containing in-situ Y Zeolite, filed November 09, 2011, and issued May 15, 2013.
133. Fan Hongfei, Wang Jifeng, Sun Xiaoyan, Method and Device for Preparing Loaded Magic Acid Catalyst, filed January 31, 2013, and issued May 15, 2013.
134. Bangfeng Qi, Lizhi Liu, Chun Chen, Jingjing Wu, Mesoporous Y Hydrocracking Catalyst and Associated Hydrocracking Processes, filed December 19, 2012, and issued May 16, 2013.
135. Laszlo Domokos, Cornelis Ouwehand, Hydrocracking Catalyst Containing Beta and Y Zeolites, and Process for İts Use to Make Jet Fuel or Distillate, filed November 02, 2006, and issued February 21, 2013.
136. Nuno Miguel Rocha Batalha, Ludovic Pinard, Francisco Manuel Da Silva Lemos, Fernando Manuel Ramoa Ribeiro, Emmanuelle Guillon, Christophe Bouchy, Method of Hydrocracking and Oligomerizing A Paraffinic Product of Fischer–Tropsch Synthesis Using a Dispersed Zeolite Beta Based Catalyst, filed October 26, 2012, and issued May 02, 2013.

137. Zhang Le, Mao Yichao, Liu Qinghe, Zhao Xinqiang, Long Xiangyun, Liu Xuefen, Nie Hong, Shi Yahua, Gao Xiaodong, Catalyst and Preparation Method and Application Thereof, and Hydrocracking Method, filed September 30, 2011, and issued April 10, 2013.
138. Tanaka Yuichi, Niitsuma Takuya, Tasaka Kazuhiko, Iwama Marie, Hydrocracking Method for Preparation of High-Octane Naphtha, filed April 17, 2014, and issued May 01, 2013.
139. Zhang Le, Liu Xuefen, Long Xiangyun, Liu Qinghe, Mao Yichao, Nie Hong, Li Dadong, Wang Kui, Chen Ruolei, Catalyst, Preparation Method and Application Thereof and Hydrocracking Method, filed October 25, 2011, and issued May 01, 2013.
140. Du Yan-Ze, Guan Ming-Hua, Wang Feng-Lai, Liu Chang, Hydrocracking Catalyst and İts Manufacturing Method and Useage, filed July 04, 2011, and issued January 16, 2013.
141. Yoshii Kiyotaka, Yamada Atsushi, Catalyst for Hydrocracking and Method of Manufacturing Hydroxy Compound Using the Catalyst, filed September 02, 2011, and issued March 21, 2013.
142. Wang Haitao, Ma Tao, Xu Xuejun, Wang Jifeng, Liu Dongxiang, Feng Xiaoping, Preparation Method of Rare-Earth-Containing Y Molecular Sieve, filed October 21, 2011, and issued April 24, 2013.
143. Fan Feng, Ling Fengxiang, Wang Shaojun, Zhang Huicheng, Chen Xiaogang, Yang Chunyan, Hydrocracking Catalyst as well as Preparation Method and Application Thereof, filed October 24, 2011, and issued April 24, 2013.
144. Xu Xuejun, Wang Haitao, Wang Jifeng, Liu Dongxiang, Feng Xiaoping, Preparation Method of Hydrocracking Catalyst Composition, filed October 21, 2011, and issued April 24, 2013.
145. Wang Haitao, Xu Xuejun, Liu Dongxiang, Wang Jifeng, Feng Xiaopin, Preparation Method of Bulk Phase Hydrocracking Catalyst, filed October 21, 2011, and issued April 24, 2013.
146. Qi Bangfeng, Chen Chun, Liu Lizhi, Wu Jingjing, Medium Oil Type Hydrocracking Catalyst and Preparation Method Thereof, filed December 17, 2012, and issued April 17, 2013.
147. Haizmann Robert, Leonard Laura E, Reforming Process with integrated Fluid Catalytic Cracker Gasoline and Hydroprocessed Cycle Oil, filed September 07, 2012, and issued April 11, 2013.
148. Simon Laurent, Guillon Emmanuelle, Hydrocracking Process Using a Zeolite Modified by Basic Treatment, filed November 23, 2012, and issued April 11, 2013.
149. Dong Songtao, Xin Jing, Nie Hong, Shi Yahua, Li Dadong, Porous Carrier, Preparation Method and Application Thereof as well as Catalyst and Hydrocracking Method, filed September 30, 2011, and issued April 10, 2013.
150. Dong Songtao, Xin Jing, Nie Hong, Shi Yahua, Li Dadong, Porous Support, Preparation Method and Application Thereof, Catalyst, and Hydrocracking Method, filed September 30, 2011, and issued April 10, 2013.
151. Daudin Antoine, Quignard Alain, Thinon Olivier, Optimized Method for Recycling Bio-Oils into Hydrocarbon Fuels, filed February 26, 2013, and issued September 19, 2013.
152. Ge Shaohui, Lan Ling, Ma An, Liu Kunhong, Hou Yuandong, Liu Yan, Zhang Peng, Wu Pei, Wu Pingyi, Zhao Qinfeng, Yuan Xiaoliang, Wang Peng, Sun Honglei, He Hao, Ma Jianbo, Lv Zhongwu, Wang Shuqin, Lu Xu, Method for Producing İntermediate Distillate Oil Through Heavy Oil Medium Pressure Hydrocracking, filed September 15, 2011, and issued March 27, 2013.

153. Saxton Robert J, Kibby Charles L, Jothimurugesan Kandaswamy, Das Tapan, Process of Synthesis Gas Conversion to Liquid Hydrocarbon Mixtures Using Alternating Layers of Synthesis Gas Conversion Catalyst and Hydrocracking Catalyst, filed June 10, 2011, and issued March 14, 2013.
154. Xiao Han, Yang Jianguo, Zhang Jingcheng, Yu Haibin, Shi Fang, Li Xiaoguo, Li Jia, Method for Preparing Hydrocracking Catalyst Containing Hierarchical Pore Beta Molecular Sieve, filed September 20, 2012, and issued March 06, 2013.
155. Wu Zhiguo, Wang Yamin, Shen Haiping, Wang Weiping, Wang Yun, Wang Pengfei, Carbonyl Composite Catalyst and İts Application Method, filed August 11, 2011, and issued February 13, 2013.
156. Yang Jianguo, Xiao Han, Zhang Jingcheng, Yu Haibin, Li Xiaoguo, Li Jia, Zhao Xunzhi, Preparation Method for Middle Oil Type Hydrocracking Catalyst Carrier, filed September 20, 2012, and issued February 06, 2013.
157. Deng Wenan, Li Chuan, Que Guohe, Mu Baoquan, Li Shufeng, Liu Dong, Wen Ping, Shang Meng, Qiao Peng, Oil-Soluble Binary Compound Catalyst for Hydrocracking High-Sulfur Low-Quality Heavy-Oil Slurry Bed, filed August 03, 2011, and issued February 06, 2013.
158. Gu Mingdi, Huang Wei, Preparation Method for Hydrocracking Catalyst, filed August 01, 2011, and issued February 06, 2013.
159. Gu Mingdi, Huang Wei, Preparation Method of Metal Phosphide-Containing Medium Oil Type Hydrocracking Catalyst, filed August 01, 2011, and issued February 06, 2013.
160. Gu Mingdi, Huang Wei, Preparation Method of Metal Phosphide Type Hydrocracking Catalyst, filed August 01, 2011, and issued February 06, 2013.
161. Gu Mingdi, Huang Wei, Preparation Method for Middle Distillate-Type Hydrocracking Catalyst, filed August 01, 2011, and issued February 06, 2013.
162. Wang Yaobin, Preparation Method of Catalyst, filed October 30, 2012, and issued February 06, 2013.
163. Hou Wenjie, Yu Zhaoxiang, Xia Jianzhong, Zhu Zhirong, Zhu Chunyan, Chen Liang, Wu Qian, High Activity Catalyst Used for Hydrocracking and Upgrading Reactions of Pro Residual Oil and Preparation Method Thereof, filed June 20, 2011, and issued February 06, 2013.
164. Bonduelle Audrey, Guillon Emmanuelle, Roy-Auberger Magalie, Method for Preparing A Catalyst Usable in Hydroconversion and İncluding at Least One Nu-86 Zeolite, filed November 23, 2012, and issued June 27, 2013.
165. Bonduelle Audrey, Guillon Emmanuelle [Fr], Roy-Auberger Magalie, Catalyst including At Least One Nu-86 Zeolite, At Least One Usy Zeolite, and a Porous İnorganic Matrix, and Method for the Hydroconversion of Hydrocarbon Feedstocks Using Said Catalyst, filed November 23, 2012, and issued June 27, 2013.
166. Shih Stuart S, Hydroprocessing of Gas Oil Boiling Range Feeds, filed March 31, 2011, and issued January 23, 2013.
167. Iguchi Kenji, Sakawaki Koji, Han Jue, Composite Type Coal Tar Hydrogenation Catalyst and Preparation Method Thereof, filed July 07, 2011, and issued January 09, 2013.
168. Wang Xiaoying, Coal Tar Hydrocracking Catalyst and Preparation Method Thereof, filed August 27, 2012, and issued January 02, 2013.
169. Kijlstra Wiebe Sjoerd, Winter Ferry, Process for Preparing Hydrocracking Catalyst Compositions, filed December 20, 2012, and issued June 27, 2013.
170. Bouchy Christophe, Grezaud Aline, Heraud Jean-Philippe, France-Dulot Hugues, Calemma Vincenzo, Method for Producing Middle Distillates İn Which the Feedstock From the Fischer-Tropsch Process and the Hydrogen Stream Have Limited Oxygen Levels, filed October 12, 2012, and issued May 01, 2013.

171. Bloch Michel, Fedou Stephane, Chapus Thierry, Bouchy Christophe, Method for Producing Middle Distillates from a Mixture of a Feedstock from Renewable Sources and a Paraffinic Effluent, filed October 08, 2012, and issued April 17, 2013.
172. Batalha Nuno, Pinard Ludovic, Da Silva Lemos Francisco Manuel, Guillon Emmanuelle, Bouchy Christophe, Ramora Ribeiro Fernando Manuel, Oligomerising Hydrocracking Method Using a Catalyst Based on Dispersed Beta Zeolite, filed July 13, 2012, and issued January 31, 2013.
173. Bhattacharyya Alakananda, Mezza Beckay J, Bricker Maureen L, Bauer Lorenz J, Composition of Supported Molybdenum Catalyst and Process for Use in Slurry Hydrocracking, filed June 01, 2011, and issued April 17, 2013.
174. Takuya Niitsuma, Marie Iwama, Diesel Fuel or Diesel Fuel Base Stock and Production Method Thereof, filed March 27, 2013, and issued March 05, 2015.
175. Sakoda Hisao, Konno Hirofumi, Hydrocracking Catalyst and Method of Producing Liquid Hydrocarbon, filed September 23, 2003, and issued November 14, 2012.
176. Hiroyuki Seki, Masahiro Higashi, Process for Hydrogenation of Wax and Process for Production of Fuel Base, filed February 05, 2007, and issued November 14, 2012.
177. Hiroyuki Seki, Masahiro Higashi, Method of Hydrotreating Wax and Process for Producing Fuel Base, filed March 01, 2007, and issued October 15, 2012.
178. Hiroyuki Seki, Masahiro Higashi, Method for Hydrocracking Wax, filed March 13, 2007, and issued September 14, 2012.
179. Hiroyuki Seki, Masahiro Higashi, Sumio Saito, Ryuzo Kuroda, Takashi Kameoka, Hydrocracking Catalyst, and Method for Production of Fuel Base Material, filed March 14, 2007, and issued August 15, 2012.
180. Takahashi Shinya, Tasaka Kazuhiko, Tanaka Yuichi, Iwama Marie, Method for Washing Reactor, filed August 12, 2011, and issued February 23, 2012.
181. Shih Stuart S, Hydroprocessing of Gas Oil Boiling Range Feeds, filed March 31, 2011, and issued October 30, 2012.
182. Hiroyuki Seki, Masahiro Higashi, Method of Hydrocracking Wax, filed August 08, 2012, and issued November 01, 2012.
183. Mingdi Gu, Wei Huang, Special Hydrocracking Catalyst and Preparation Method and Application Thereof, filed May 23, 2011, and issued November 28, 2012.
184. Bangfeng Qi, Lizhi Liu, Chun Chen, Jingjing Wu, Medium-Oil-Type Hydrocracking Catalyst and Preparation Method Thereof, filed August 21, 2012, and issued November 14, 2012.
185. Kiyoomi Kaneda, Hirokazu Matsuda, Method for Producing 1,2-Propanediol, filed March 10, 2011, and issued October 04, 2012.
186. Bushi Kristof, Gijon Ehmmanjuehl', Method of Obtaining Middle Distillates by Hydrocracking of Raw Material, Obtained in Fischer–Tropsch Process, in Presence of Catalyst, Which Contains Solid izm-2, filed July 29, 2009, and issued September 20, 2012.
187. Gijon Ehmmanjuehll', Simon Loran, izm-2 Zeolite-Based Catalyst and Method for Hydroconversion/Hydrocracking of Hydrocarbon Material, filed July 29, 2009, and issued September 20, 2012.
188. Deju Wang, Youdi Guo, Hui Wang, Zhongneng Liu, Catalyst for Yield İncreases of Btx (Benzene, Toluene and Xylol) Aromatic Hydrocarbons and Trimethylbenzene through Hydrocracking Heavy Aromatic Hydrocarbons, filed April 20, 2011, and issued October 24, 2012.
189. Chunshan Li, Tao Kan, Hongyan Wang, Hongxing He, Suojiang Zhang, Coal Tar Hydrocracking Catalyst with Pillared Clay as Vector and Preparation Method Thereof, filed April 13, 2011, and issued October 17, 2012.

190. Fengyun Ma, Qiuyan Chen, Huiling Hu, Jingmei Liu, Xuejiao Liu, Zhiqiang Sun, Hydrocracking Catalyst for Vacuum Residue Suspension Bed and Preparation Method and Use Method Thereof, filed June 08, 2012, and issued October 10, 2012.
191. Jeremy Francis, Laurent Simon, Emmanuelle Guillon, Nicolas Bats, Avelino Corma, Christophe Pichon, Hydrocracking Method Using Zeolite Catalyst Containing Two Different Hydrogenation Functions, filed January 06, 2012, and issued August 02, 2012.
192. Sajiki Hironao, Sawama Yoshinari, Yabe Yuki, Yamada Tsuyoshi, Selective Contact Reduction Catalyst, Method for Producing the Same, and Selective Hydrogenation Contact Reduction Method Using the Catalyst, filed October 04, 2011, and issued August 02, 2012.
193. Ryu Jae Wook, Jeong il Yong, Kim Gyung Rok, Park Sung Bum, Kim Do Woan, Kim Eun Kyoung, Choi Sun, Lee Chang Ha, Lee Jae Hyuk, Viet Tran Tan, Kim Yo Han, Method For Hydro-Cracking Heavy Hydrocarbon Fractions Using Supercritical Solvents, KR20120075368 (A), filed in December 20, 2011, issued in July 06, 2012.
194. Lee Jae Hyuk, Viet Tran Tan, Kim Yo Han, Method for Hydro-Cracking Heavy Hydrocarbon Fractions Using Supercritical Solvents, filed December 20, 2011, and issued July 06, 2012.
195. Dongjing Xu, Chunliang An, Jianjun Zhao, Chunxiang Cao, Xinwen Hao, Continuous Dryer Charging Machine for Preparing Hydrocracking Catalyst, filed December 31, 2011, and issued August 29, 2012.
196. Stockwell David M, Lerner Bruce A, Hydrocracking Catalyst Comprising an In Situ Produced Y-Fauajasite and Hydrocracking Process, filed February 26, 2007, and issued November 14, 2012.
197. Inamura Kazuhiro, Ilno Akira, Takahashi Nobuyuki, Sunagawa Yoji, Watabe Mitsunori, Yamahata Yuichi, Eura Shinya, Shirahama Yuji, Hydrocracking Catalyst for Heavy Oil and Method for Hydrotreating Heavy Oil Using Same, filed August 02, 2010, and issued August 02, 2012.
198. Xianyong Wei, Zhimin Zong, Xiaoming Yue, Yinghua Wang, Xing Fan, Yunpeng Zhao, Bing Sun, Yu Qing, Yugao Wang, Peng Li, Bo Chen, Chang Liu, Lili Huang, Zhou Jun, Yao Lu, Method for Preparing Loaded Solid Super Acidic Catalyst Directly by Microwave Method, filed March 12, 2012, and issued August 01, 2012.
199. Famin Sun, Baojian Shen,, ian Lv, Baoqin Dai, Qiaoxia Guo, Ran Tian, Zhihua Zhang, Liang Wang, Bojun Shen, Lihong Qin, Yandan Wang, Fucun Wang, Dongqing Wang, Haiyan Li, Shuzhi Guo, Endong Xia, Ruifeng Li, Method for Preparing Chemical Material by Hydrocracking of Heavy Hydrocarbon Oil, filed January 25, 2011, and issued July 25, 2012.
200. Kongyuan Zhang, Chenguang Liu, Kaichang Song, Luyan Xu, Zhaolin Fu, Hydrotreatment Catalyst and Preparation Method Thereof, filed March 05, 2012, and issued July 25, 2012.
201. Dongqing Wang, Gang Wang, Famin Sun, Wencheng Zhang, Quanzhi Li, Baoqin Dai, Jianying Ma, Yuhong Chang, Lihong Qin, Fucun Wang, Xuguang Li, Qian Lv, Haiyan Li, Shuzhi Guo, Xiaodong Yang, Guojia Zhang, Yao Huang, Shuyan Zhang, Double-Microporous-Mesoporous Composite Molecular Sieve Hydrocracking Catalyst, filed December 17, 2010, and issued July 11, 2012.
202. Haiyan Li, Zhihua Zhang, Baoqin Dai, Ran Tian, Lihong Qin, Famin Sun, Wencheng Zhang, Dongqing Wang, Qian Lv, Fucun Wang, Ye Zhao, Shuzhi Guo, Dongmei Ge, Hong Jie, Medium Oil Hydrocracking Catalyst Carrier and Preparation and Application Thereof, filed December 17, 2010, and issued July 11, 2012.

203. Famin Sun, Yasong Zhou, Baoqin Dai, Qiang Wei, Zhihua Zhang, Guiyuan Jiang, Ran Tian, Jianying Ma, Qian Lv, Ye Zhao, Lihong Qin, Dongqing Wang, Zhonghua Yang, Haiyan Li, Shoutao Ma, Aixian Gou, Method for Producing Chemical Raw Material by Hydrocracking, filed December 23, 2010, and issued July 04, 2012.
204. Famin Sun, Yasong Zhou, Baoqin Dai, Qiang Wei, Zhihua Zhang, Ran Tian, Qian Lv, Aijun Duan, Ye Zhao, Dongqing Wang, Fucun Wang, Haiyan Li, Lihong Qin, Shoutao Ma, Zhonghua Yang, Shuqiu Bing, Method for Selective Hydrocracking of Light Oil, filed December 23, 2010, and issued July 04, 2012.
205. Ran Tian, Wenyong Liu, Zhihua Zhang, Ye Zhao, Zhibo Han, Shoutao Ma, Wencheng Zhang, Famin Sun, Chunmei Yu, Beibing Yin, Ruifeng Li, Jinling Zhu, Endong Xia, Lili Jin, Quanfu Liu, Tiezhen Zhang, Liwei Liang, Aixian Gou, Preparation Method for Pretreating Catalyst by Hydrocracking, filed December 10, 2010, and issued July 04, 2012.
206. Zhihua Zhang, Lei Fang, Jintao Guo, Ran Tian, Chunmei Yu, Famin Sun, Wencheng Zhang, Gang Wang, Xiufang Feng, Weichi Xu, an Wang, Guangming Wen, Jinling Zhu, Shoutao Ma, Fengxuan Li, Dongmei Ge, Bingquan Tian, Poor Quality Wax Oil Hydrotreatment Method, filed December 03, 2010, and issued June 06, 2012.
207. Tao Zhang, Changzhi Li, Mingyuan Zheng, Aiqin Wang, Application of Tungsten-Based Catalyst in Lignin Catalytic Hydrogenation for Producing Aromatic Compound, filed December 14, 2010, and issued May 30, 2012.
208. Guang Chen, Yulan Gao, Youliang Shi, Ziming Wu, Hongfei Fan, Shike Sun, Sulfuration Method of Catalyst for Hydrocracking Technology, filed November 04, 2010, and issued May 23, 2012.
209. Fenglai Wang, Yanze Du, Minghua Guan, Chang Liu, Medium Oil Type Hydrocracking Catalyst and İts Preparation Method, filed November 04, 2010, and issued May 23, 2012.
210. Baijun Liu, Xinmei Pang, Qinglei Meng, Youdong Gai, Yanlin Chen, Xiaohui Gao, Xiaohua Wang, Zhiyuan Zhou, Jiarui Piao, Lijun Yan, Hydrocracking Catalyst Containing Y-Shaped Molecular Sieve/Amorphous Silicon-Aluminum, and Preparation Method Thereof, filed October 14, 2010, and issued May 16, 2012.
211. Xiaoyan Sun, Hongfei Fan, Zhanyu Wang, Preparation Method of Alumina Containing Titanium and Silicon, filed October 15, 2010, and issued May 16, 2012.
212. Caian Ruan, Jifeng Wang, Xiaoping Zhang, Yanze Du, Modification Method of İn-Situ Y Zeolite, filed October 13, 2010, and issued May 09, 2012.
213. Mingdi Gu, Hydrocracking Catalyst, Preparation Method and Application Thereof, filed October 13, 2010, and issued May 09, 2012.
214. Xuejun Xu, Jifeng Wang, Haitao Wang, Dongxiang Liu, Xiaoping Feng, Composite Molecular Sieve-Containing Hydrocracking Catalyst and Preparation Method Thereof, filed October 13, 2010, and issued May 09, 2012.
215. Mingdi Gu, Wei Huang, Preparation Method for Modified B-Beta Zeolite-Containing Hydrocracking Catalyst, filed October 13, 2010, and issued May 09, 2012.
216. Xuejun Xu, Jifeng Wang, Dongxiang Liu, Haitao Wang, Xiaoping Feng, Hydrocracking Catalyst with High Activity and High-Medium Oil Selectivity and Preparation Method of Hydrocracking Catalyst, filed October 13, 2010, and issued May 09, 2012.
217. Alexander Kehrmann, Method for Recovering Lanthanum from Zeolites Containing Lanthanum, filed October 05, 2011, and issued April 12, 2012.
218. Abe Masaki, Takaya Akira, Munakata Hiroshi, Support for Hydrocracking Catalyst of Hydrocarbon Oil, Hydrocracking Catalyst, and Method for Hydrocracking of Hydrocarbon Oil, filed July 16, 2010, and issued April 12, 2012.
219. Mezza Beckay J, Bhattacharyya Alakananda, Ringwelski Andrzej Z, Process for Determining Presence of Mesophase in Slurry Hydrocracking, filed December 19, 2011, and issued April 12, 2012.

220. Laszlo Domokos, Cornelis Ouwehand, Hydrocracking Catalyst, filed April 15, 2010, and issued April 11, 2012.
221. Marius Vaarkamp, Catalyst, Catalyst Support and Process for Hydrogenation, Hydroisomerization, Hydrocracking and/or Hydrodesulfurization, filed December 02, 2011, and issued March 29, 2012.
222. Koichi Matsushita, Production Method for Alkyl Benzenes, and Catalyst Used in Same, filed March 03, 2010, and issued March 07, 2012.
223. Du Yanze, Guan Minghua, Wang Fenglai, Liu Chang, A Hydrocracking Catalyst, Process for Preparing the Same and Use Thereof, filed July 06, 2011, and issued January 13, 2012.
224. Murata Kazuhisa, Ryu Genyu, Takahara İsao, Inaba Hitoshi, Method of Manufacturing Hydrocarbon, filed July 07, 2010, and issued January 26, 2012.
225. Inui Tomoyuki, Inui Masayuki, Kimura Takuma, Al-Saleh Mohammad Abdullah, Redhwi Halim Hamid, Ali Mohammad Ashraf, Ahmed Shakeel, Catalyst Exhibiting Hydrogen Spillover Effect, filed October 20, 2011, and issued February 16, 2012.
226. Chen Cong-Yan, Miller Stephen J, Ziemer James N, Liang Ann J, Multi-Stage Hydroprocessing for the Production of High Octane Naphtha, filed June 16, 2011, and issued February 02, 2012.
227. Marchand Karin, Chaumonnot Alexandra, Bonduelle Audrey, Dufaud Veronique, Lefebvre Frederic, Bader Manuela, Lopes Silva Susana, Method for the Hydrocracking of Hydrocarbon Fractions Using a Catalyst Based on Heteropolyanions Trapped in a Mesostructured Oxide Support, filed December 15, 2011, and issued June 28, 2012.
228. Yao Jianhua, Ghonasgi Dhananjay B, Xu Xiaochun, Bao Yun, Sughurue ii Edward L, Hydrocracking Process for Making Renewable Diesel from Oils and/or Fats, filed May 02, 2011, and issued January 26, 2012.
229. Van Donk Sander, Lacroix Maxime, Kenmogne-Gatchuissi Regine, Fajula Francois, Bulut Metin, Dath Jean-Pierre, De Jong Krijn Pieter, De Jongh Petra Elisabeth, Zecevic Jovana, Van Laak Adrianus Nicolaas Cornelis, Modified Y-Type Zeolites Having a Trimodal İntracrystalline Structure, Method for Making Same, and Use Thereof, filed December 22, 2009, and issued January 26, 2012.
230. Jinglai Zhang, Lei Zhou, Chenmin He, Minghua Jin, Min Song, Duan Zhang, Qing Liu, Ce Guan, Catalyst for Catalyzing Hydrocracking of Biomass Oil and Preparation Method and Application Thereof, filed July 13, 2011, and issued January 18, 2012.
231. Lott Roger K, Chang Yu-Hwa, Systems for increasing Catalyst Concentration in Heavy Oil and/or Coal Resid Hydrocracker, filed September 19, 2011, and issued January 12, 2012.
232. Hommeltoft Sven Ivar, Lacheen Howard S, Elomari Saleh, Acid Catalyst Composition Having a High Level of Conjunct Polymer, filed September 08, 2011, and issued January 05, 2012.
233. Kijlstra Wiebe Sjoerd, Rigutto Marcello Stefano, Van Welsenes Arend Jan, Hydrocracking Catalyst Composition, filed September 13, 2011, and issued March 22, 2012.
234. Guibard Isabelle, Quignard Alain, Pre-Refining and Hydroconversion In Fixed-Bed of A Heavy Crude Oil of Hydrocarbons, Comprises Removing Metals in Hydrodemetallation Section, Hydrocracking at Least Part of the Effluent, and Fractionating a Portion of the Effluent, filed January 18, 2011, and issued July 20, 2012.
235. Bhattacharyya Alakananda, Mezza Beckay J, Bricker Maureen L, Bauer Lorenz J, Process for Using Supported Molybdenum Catalyst for Slurry Hydrocracking, filed June 10, 2010, and issued December 15, 2011.
236. Yuichi Tanaka, Shinya Takahashi, Yoshifumi Chiba, Process for Production of Fuel Base, filed August 21, 2008, and issued August 15, 2011.

237. Abe Masaki, Takaya Ken, Munakata Hiroshi, Process for Treating Petroleum-Based Hydrocarbon, filed September 03, 2009, and issued March 17, 2011.
238. Fujuan Li, Zhirong Zhu, Qian Wu, Method for Coal Tar Pitch Hydrocracking Lightening Reaction Technology, filed November 29, 2010, and issued March 23, 2011.
239. Hayasaka Kazuaki, Ono Hideki, Nagayasu Yoshiyuki, Process for Producing Lube Base Oil, and Lube Base Oil, filed June 25, 2010, and issued January 06, 2011.
240. Seki Khirojuki, Method of Producing Liquid Fluid, filed August 30, 2007, and issued December 27, 2010.
241. Kagami Shigeari, Matsumoto Yuei, Hirano Tomoaki, Kajima Kazuhiro, Manufacturing Method for High Octane Number Gasoline Fraction, filed March 13, 2009, and issued September 30, 2010.
242. Kagami Shigeari, Hirano Tomoaki, Kajima Kazuhiro, Method and Apparatus for Producing Ultra Low-Sulfur Fuel Oil, filed March 31, 2008, and issued October 22, 2009.
243. Iino Akira, Inamura Kazuhiro, Method for Decomposing Atmospheric Distillation Residual Oil, filed March 28, 2008, and issued October 22, 2009.
244. Bijlsma Focco Kornelis, Dierickx Jan Lodewijk Maria, Hoek Arend, Process for Hydrocracking and Hydro-isomerisation of a Paraffinic Feedstock, filed September 10, 2008, and issued August 13, 2009.
245. Chen Guang, Gao Yulan, Du Yanze, Huang Xinlu, Cao Fenglan, Li Chonghui, Xu Liming, Start Working Method for Hydrocracking Process, filed January 23, 2008, and issued July 29, 2009.
246. Konno Hirofumi, Higashi Masahiro, Seki Hiroyuki, Method for Hydrocracking Wax and Method for Producing Fuel Base Material, filed March 12, 2007, and issued April 15, 2009.
247. Ling Fengxiang, Zhang Xiwen, Sun Wanfu, Li Ruifeng, Zhang Zhizhi, Fan Hongfei, Ma Jinghong, Yin Zequn, Fang Xiangchen, Hydrocracking Catalyst and Preparation Thereof, filed July 09, 2007, and issued January 14, 2009.
248. Euzen Patrick, Gueret Christophe, Method for Producing Middle Distillates by the Hydroisomerization and Hydrocracking of Feedstocks Obtained from the Fischer–Tropsch Method, filed April 11, 2006, and issued June 25, 2008.
249. Guan Minghua, Quan Hui, Yao Chunlei, Liu Ping, Lin Zhenfa, Method for Producing Food-Level White Oil by Hydrogenation Technique, filed September 20, 2006, and issued March 26, 2008.
250. Guan Minghua, Quan Hui, Yao Chunlei, Liu Ping, Lin Zhenfa, One-Stage Hydrogenation Method for Producing Food-Level White Oil, filed September 20, 2006, and issued March 26, 2008.
251. Domokos Laszlo, Jongkind Hermanus, Rigutto Marcello S, Jurriaan Stork Willem H, Stork-Blaisse Beatrijs A, Carola Van De Voort Esther H, Rigutto Marcello Stefano, Stork Willem Hartman Jurriaan, Stork-Blaisse, Legal Representative Beatrijs Anna, Van De Voort Esther Hillegarda Carola, Catalyst Composition for Hydrocracking and Process for İts Preparation, filed September 08, 2005, and issued January 19, 2010.
252. Edward Julius Creyghton, Laurent Georges Huve, Cornelis Ouwehand, Veen Johannes Anthonius Ro Van, Catalyst Carrier and Catalytical Composition, Methods of Their Preparation and Application, filed March 01, 2005, and issued October 29, 2008.
253. Stuart Leon Soled, Kenneth Lloyd Riley, Gary P. Schleicher, Richard A. Demmin, Ian Alfred Cody, William L. Schuette, Hydrocracking Process Using Bulk Group Viii/Group Vib Catalysts, filed January 14, 2000, and issued April 07, 2009.
254. Hye Kyung C. Timken, Extremely Low Acidity Ultra Stable Y Zeolite and Homogeneous, Amorphous Silica–Alumina Hydrocracking Catalyst and Process, filed October 10, 2003, and issued February 03, 2010.

255. Christophe Bouchy, Making Middle Distillates from Paraffin Charge Produced by Fischer–Tropsch Synthesis Comprises İmplementing Hydrocracking Catalyst Comprising Hydrodehydrogenating Metal and Composite Support Formed by Y-Type Zeolite and Silicon Carbide, filed October 06, 2009, and issued April 08, 2011.
256. Christophe Bouchy, Emmanuelle Guillon, Process for Producing Middle Distillates by Hydrocracking of Feedstocks Obtained by the Fischer–Tropsch Process in the Presence of a Catalyst Comprising An izm-2 Solid, filed July 29, 2009, and issued July 28, 2011.
257. Patrick Euzen, Vincenzo Calemma, Method for Producing Middle Distillates by Hydroisomerization and Hydrocracking of Feeds Derived from a Fischer–Tropsch Process Using a Multifunctional Guard Bed, filed July 17, 2006, and issued February 09, 2010.
258. Vincenzo Calemma, Cristina Flego, Luciano Cosimo Carluccio, Wallace Parker, Roberto Giardino, Giovanni Faraci, Process for the Preparation of Middle Distillates and Lube Bases Starting from Synthetic Hydrocarbon Feedstocks, filed June 28, 2004, and issued May 19, 2009.
259. Alberto Delbianco, Nicoletta Panariti, Process for the Conversion of Heavy Hydrocarbon Feedstocks to Distillates with the Self-Production of Hydrogen, filed January 08, 2009, and issued June 17, 2008.
260. Christophe Bouchy, Alexandra Chaumonnot, Method for Producing Medium Distillates by Hydrocracking of Charges Arising from the Fischer–Tropsch Process with an Amorphous Material Catalyst, filed May 13, 2009, and issued January 21, 2010.
261. Patrick Euzen, Christophe Gueret, Vincenzo Calemma, Method of Producing Middle Distillates, Comprising the Hydroisomerisation and Hydrocracking of Feedstocks Resulting from the Fischer–Tropsch Method, Using a Doped Catalyst Based on Mesoporous Alumina–Silica Having a Controlled Macropore Content, filed June 27, 2006, and issued July 02, 2008.
262. Novak Vaclav, Ricanek Milan, Goluch Vaclav, Process for Producing Aluminosilicate Hydrocracked Catalyst, filed May 14, 2010, and issued December 22, 2010.
263. Jeffrey T. Miller, Ronald B. Fisher, Tracy L. Marshbanks, Resid Hydroprocessing Method Utilizing a Metal-Impregnated, Carbonaceous Particle Catalyst, filed November 24, 1997, and issued January 15, 2008.
264. Rafael L. Espinoza, Keith H. Lawson, Kandaswamy Jothimurugesan, Combination of Amorphous Materials for Hydrocracking Catalysts, filed July 13, 2005, and issued January 29, 2008.
265. Cooperation on International Tracebility in Analytical Chemistry (CITAC), Specifications for Unleaded and Low Sulphur Fuels: Global and Africa Trends, Paper presented at the meeting of Workshop on Leaded Gasoline Phase-Out in North Africa, Tunisia, 2008.
266. Vasant Thakkar, Jill M Meister, Richard J Rossi and Li Wang, Process and Catalyst Innovations in Hydrocracking to Maximize High Quality Distillate Fuel, UOP LLC, 2008.
267. James G. Speight, *The Refinery of the Future*, Elsevier, Oxford, UK, 2010, pp. 275–341.
268. Chris Beddoes, Statistical Report 2014, Fuels Europe, 2014, Retrieved from https://www.fuelseurope.eu/uploads/Modules/Resources/statistical_report_fuels_europe-_v25_web.pdf
269. Ministry of Business, Innovation & Employment, Reviewing Aspects of the Engine Fuel Specifications Regulations 2011, New Zealand, 2015.
270. Orbital Australia Pty Ltd, Review of Sulphur Limits in Petrol, 2013.
271. Lars Skyum, Next Generation, Haldor Topsoe, Denmark, 2015, Retrieved from http://www.topsoe.com/sites/default/files/topsoe_next_generation.ashx_.pdf

4 Hydrodesulfurization Catalysis Fundamentals

Arzu Kanca and Basar Caglar

CONTENTS

4.1 INTRODUCTION

Hydrodesulfurization (HDS) is one of the most important industrial operations used to reduce sulfur level of highway and nonroad fuels in the presence of a catalyst. It is the application of catalytic technology, which processes the largest volume of material [1,2]. HDS catalysts provide an energetically more favorable pathway, in which sulfur compounds are converted to H_2S and hydrocarbon. HDS is widely practiced in petroleum refineries, but it is also used as a postprocess to clean gases produced by coal and biomass gasification.

Desulfurization of fuels is becoming more important since the current refined crude oils contain higher sulfur content, while the regulated sulfur level of fuels is being reduced due to environmental concerns. Since the combustion of sulfur-containing fuels releases SO_X, thereby causing atmospheric pollution and acid rain, the sulfur content in fuels has to be reduced to meet emission standards. Sulfur regulation is also important for reducing other pollutants such as nitrogen oxides (NO_X) and particulate matter (PM) emitted by motor vehicles. Since catalysts used in exhaust aftertreatment technologies are sensitive to sulfur, an ultralow sulfur level is required to achieve drastic reduction in NO_X and PM emission [3,4]. A further reason for sulfur removal is that catalysts used for processing of oil products have low tolerance to sulfur. For example, selective ring opening (SRO) is a process that improves the quality of diesel fuel by decreasing fuel density and increasing the cetane number [5]. Since sulfur compounds easily poison catalysts used for SRO, all SRO feedstocks must be desulfurized. Moreover, the growing interest in fuel cell–based transportation systems appeals advanced desulfurization processes for hydrogen production with a very low level of sulfur-containing compounds (e.g., H_2S, volatile organosulfur compounds). Since solid oxides and polymer electrolytes in fuel cells are very sensitive to sulfur, hydrogen produced by hydrocarbon reforming should possess a very low level of sulfur (<10 ppmw for solid oxide fuel cell and <0.1 ppmw for PEMFC [6]).

Sulfur limits in gasoline and diesel for different countries based on their own emission standards are shown in Table 4.1 [7,8]. As seen from the table, regulations for fuel sulfur content become more stringent, and the current sulfur content in fuels is limited to ultralow sulfur level (10–15 wppm). This introduces major challenges to refineries for meeting the fuel sulfur specification. In order to achieve ultralow sulfur

TABLE 4.1
Sulfur Limit in Fuels Applied in Different Countries

	Sulfur Limit (wppm)		
Countries	**Gasoline**	**Diesel (On-Road)**	**Emission Standards**
USA	30	15	EPA Tier 2 (2006)
	10	15	EPA Tier 3 (2017)
EU	50	50	Euro IV (2005)
	10	10	Euro V (2009)
China	150	350	China III (2011)
	50	50	China IV (2015)
	10	10	China V (2018)
Russia	150	350	2013
	50	50	2015
	10	10	2016
India	500	500	Bharat II (2005)
	150	350	Bharat III (2010)
	NA	50	Bharat IV

Source: K.O. Blumberg et al., Low-sulfur gasoline & diesel: The key to lower vehicle emissions, in: *United Nations Environment Programme*; J.D. Miller, C. Facanha, The state of clean transport policy—A 2014 synthesis of vehicle and fuel policy developments, Washington, 2014.

level in fuels, the performance of HDS units should be improved. This requires a better understanding of HDS catalysts and processes, and the development of advanced technologies, which can be applied to industry in a short time.

In this chapter, we will investigate HDS catalysts and processes in different aspects. We mainly focus on HDS processes used in petroleum refineries and provide information about HDS of coal- and biomass-derived gases. We will first describe the types of sulfur compounds present in different fuels (e.g., petroleum, coal, and biomass) and reaction classes of HDS. Then we will go through conditions that satisfy these reactions to occur and the reaction kinetic in the presence of an accelerating agent, i.e., a catalyst. The effects of active species, support, particle size, porosity, and shape of catalysts will be included. Finally, HDS processes and technologies will be evaluated, and the advanced technologies applied in industrial scale will be discussed.

4.2 SULFUR COMPOUNDS

Sulfur compounds in fuels are present in organic and inorganic forms. Inorganic sulfur compounds are pyritic and sulfatic $\left(SO_4^{2-}\right)$, and they can generally be found in solid fuels, such as coal and biomass. Since inorganic sulfur compounds are minor in coal and biomass and they are not present in petroleum, we focus on organic sulfur compounds. Organic sulfur compounds can be categorized into two groups: (1) aliphatic organosulfur compounds and (2) aromatic organosulfur compounds. Aliphatic organosulfur compounds are mercaptans (thiols), sulfides, and disulfides, whereas aromatic organosulfur compounds are thiophenes, benzothiophenes (BTs), dibenzothiophenes (DBTs), and their alkylated derivatives [9–11]. The molecular formulas of organosulfur compounds commonly observed are shown in Table 4.2.

TABLE 4.2
Typical Organic Sulfur Compounds in Fuels

Sulfur Compounds	Structure
Mercaptane (thiols)	RSH
Sulfides	RSR
Disulfides	RSSR
Thiophene	 [structure: thiophene ring, S]
Benzothiophene	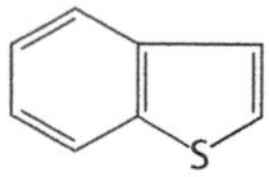 [structure: benzothiophene, S]
Dibenzothiophene	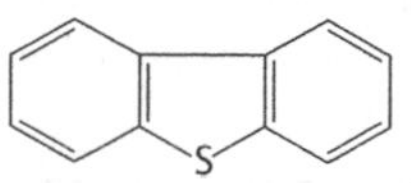 [structure: dibenzothiophene, S]

Crude oil has mainly organic sulfur compounds, and the sulfur content in crudes varies between 0.1 and 5 wt% depending on the origin of crudes. A lighter fraction of crude oil (e.g., naphtha, gasoline) contains mainly thiols, sulfides, and disulfides along with a minor amount of thiophenes and its alkylated derivatives, whereas in a heavier fraction of crudes (e.g., kerosene, gas oil), thiophenes, BTs, DBTs, and their alkyl derivatives are primary sulfur compounds [1,12–15]. Thiols are the typical organic sulfur compounds in coal and biomass [16].

4.3 HDS CHEMISTRY AND THERMODYNAMICS

In HDS, the removal of organosulfur compounds is achieved by hydrogenation, which generally results in the formation of saturated hydrocarbon and H_2S. The HDS reactions of different classes of organosulfur compounds and their reaction enthalpies are listed in Table 4.3. HDS of all organosulfur compounds are exothermic. They are thermodynamically favored under ordinary reaction conditions (e.g., 330–410°C, 35–150 atm) [17–19]. This is indicated by the logarithms of the equilibrium constants for the reduction sulfur compounds to saturated hydrocarbons over a wide temperature range (Figure 4.1). As shown in the figure, as temperature increases, the equilibrium constants decrease consistent with the exothermicity of the reactions. The equilibrium constants are higher than 1 over a wide temperature range, indicating that the reactions proceed to completion in the presence of a stoichiometric amount of hydrogen. The equilibrium constants are lower than 1 for only thiophene and disulfide at temperatures considerably higher than those required in practice.

Thermodynamic analysis indicates that HDS of organosulfur compounds are favorable under ordinary operation conditions. The next stage will be the investigation of reaction kinetics and HDS catalyst in different aspects.

TABLE 4.3
Reaction Enthalpies of Hydrodesulfurization Reactions

Organosulfur Compounds	Reactions	ΔH°_{rxn} (kj/mol)
Methane thiol	$CH_3–SH + H_2 = CH_4 + H_2S$	−72
Dimethyl sulfide	$CH_3–S–CH_3 + 2H_2 = 2CH_4 + H_2S$	−134
Dimethyl sulfide	$CH_3–S–CH_3 + 2H_2 = C_2H_6 + H_2S$	−59
Thiophene	$C_4H_4S + 4H_2 = C_4H_{10} + H_2S$	−262
Benzothiophene	$C_8H_6S + 3H_2 = C_8H_{10} + H_2S$	−203
Dibenzothiophene	$C_{12}H_8S + 2H_2 = C_{12}H_{10} + H_2S$	−148

Source: M.L. Vrinat, *Applied Catalysis*, 6 (1983) 137–158; A. Syed Ahmed, Thermodynamics of hydroprocessing reactions, in: *Hydroprocessing of Heavy Oils and Residua*, CRC Press, Boca Raton, FL, 2007, pp. 51–69.

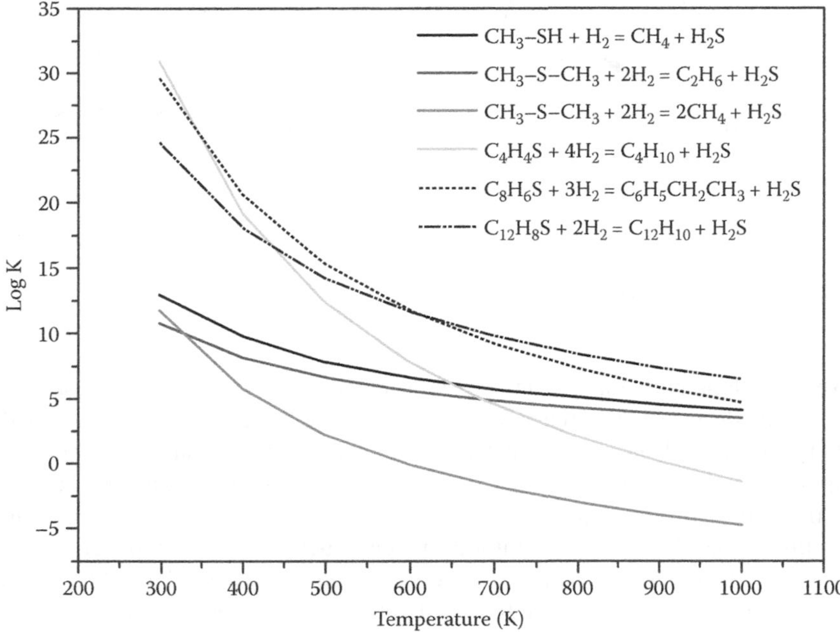

FIGURE 4.1 Change in logarithms of equilibrium constants (calculated by using Van't Hoff equation) with respect to temperature for the HDS reactions of different sulfur compounds.

4.4 HDS CATALYSIS AND KINETICS

4.4.1 Reactivities

Organosulfur compounds show different HDS reactivity depending on their structure. Desulfurization of aliphatic organosulfur compounds (e.g., alkyl and aryl thiols [RSH], thioethers [RSR'], and disulfides [RSSR']) is easier than the aromatic thiophenes [14]. The reactivities of aromatic organosulfur compounds differ in the following order: thiophene > alkylated thiophene > BTs > alkylated BT > DBTs and alkylated DBT without substituents at the 4 and 6 positions > alkylated DBT with one substituent at either the 4 or 6 position > alkylated DBT with alkyl substituents at the 4 and 6 positions [15,20,21]. The general trend is that low-molecular-weight compounds are more readily desulfurized than high-molecular-weight compounds. Furthermore, for DBTs, methyl substitution at the 4 or/and 6 positions results in reactivity decrease. This is attributed to a strong steric effect, arising from the methyl substitution at the 4 or/and 6 positions [17].

4.4.2 Reaction Mechanism

Organosulfur compounds show different mechanisms depending on the nature of the sulfur compounds. For sulfur compounds without a conjugation structure

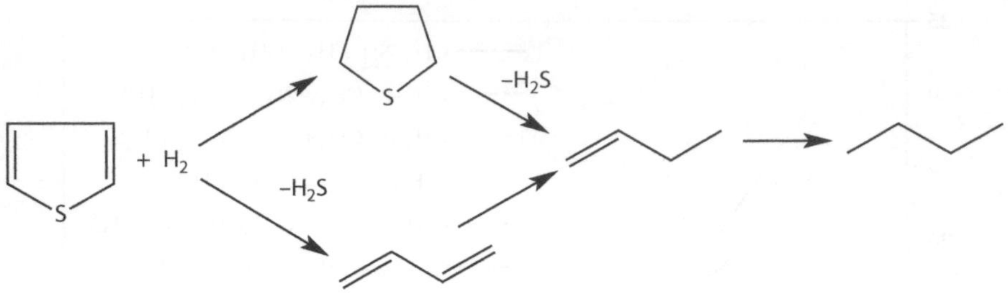

FIGURE 4.2 Reaction pathways for thiophene HDS.

between the lone pairs on S atom and the p-electrons (e.g., thiols, sulfides, disulfides, tetrahydrothiophene), the HDS reaction proceeds via a hydrogenolysis pathway [1]. In this pathway, since electrons are localized on the sulfur atoms, these compounds first lose their sulfur atoms upon adsorption on a catalyst surface and then they are hydrogenated. On the other hand, for aromatic sulfur compounds (e.g., thiophene, BT, DBT, naphthothiophene, benzo-naphthothiophene), both hydrogenolysis and a new pathway called hydrogenation drive HDS [22–25]. In the hydrogenation pathway, since the conjugation structure between the lone pairs on S atom and the p-electrons on the aromatic ring leads to electron delocalization, the adsorbed sulfur compounds are first hydrogenated and then the C–S bond is broken. The schematic representation of both mechanisms for thiophene (as a representative compound) is shown in Figure 4.2. Wang and Iglesia [26] proposed two different adsorption configurations for hydrogenolysis and hydrogenation pathways based on the study of thiophene desulfurization on supported Ru clusters. They suggest that for the hydrogenolysis pathway, thiophene binds onto sulfur vacancies via lone pairs in its sulfur atom (σ-bond) followed by C–S bond cleavage due to the interaction between adsorbed thiophene and vicinal adsorbed hydrogen. This also results in the formation of an adsorbed thiolate, which rapidly breaks the remaining C–S bond to form butadiene and its derivatives. On the other hand, for the hydrogenation pathway, thiophene binds onto sulfur vacancies via aromatic ring (π-bond) and reacts with the adjacent S–H intermediate to form protonated species. These species further react with surface hydrogen atoms to complete hydrogenation and unstable dihydrothiophene intermediate forms. This is followed by C–S bond cleavage, producing butadiene and its derivatives. Literature also suggests that hydrogenolysis and hydrogenation occur on separate active sites of the catalyst used [17,27].

4.4.3 Kinetics

Several research groups have extensively studied the reaction kinetic of HDS on different catalyst surfaces where the reaction takes place [11,18,23,28–33]. There are no fundamental differences in the reaction mechanism observed to date. The main difference among different catalysts is related to their activity and stability. In order to describe the HDS kinetics, a general rate equation can be applied. The general rate equation, which is consistent with Langmuir–Hinshelwood kinetics, is given by

$$r_A = \frac{k_A P_A}{\left(1+\sum K_i P_i\right)^n} f(P_H) \tag{4.1}$$

where r_A is the rate of disappearance of reactant A, k_A is the reaction rate constant, P_A is the partial pressure of A, K_i is the equilibrium constant for adsorption of species, P_i is the partial pressure of species in equilibrium, and n is a constant (1 or 2), reflecting the number of adsorbed species involved in the rate determining step [23]. Based on the quasi-equilibrium assumption, one elementary step is considered as being slow and it determines the rate, while other steps are fast and they are all in equilibrium. The rate-limiting step is generally the reaction between the adsorbed organosulfur compounds and the adsorbed hydrogen. The $f(P_H)$ term usually changes linearly with hydrogen pressure, but when hydrogen adsorbs on a different site from that of the organosulfur compounds, it depends on both hydrogen pressure and the equilibrium constant of hydrogen adsorption (K_H). The general rate equation includes the effect of strong inhibition by the adsorption of the organosulfur compounds and H_2S that is formulized by the term of K_iP_i. If these compounds adsorb on the surface strongly, this will lead to the decrease in the HDS rate [23].

The general rate equation gives an insight about how reactants and products affect the reaction kinetic. However, it is impractical to derive a single rate equation applicable for all HDS reactions due to the complex structures of sulfur compounds and the resulting steric effects [34,35]. Several equilibrium stages involved in HDS and the site occupancy of reactants also contribute to the complexity of HDS kinetics [36].

We refer to literature studies for a detailed kinetic analysis and complete rate equations for HDS of thiophene [18,37–39], BT [18,40–42], and DBT [18,43,44]. Some findings of literature studies on thiophene and DBT are summarized in Tables 4.4 and 4.5, respectively. The tables indicate that the HDS reaction rates are directly proportional to hydrogen pressure. The tables also suggest that organosulfur compounds and H_2S compete for the same sites. H_2S blocks access of some sites for the adsorption of thiophene or DBT and hydrogen, thereby inhibiting the HDS reaction. The HDS reaction takes place on the remaining available surface sites. In some cases, especially for DBT, hydrogen is assumed to adsorb on different surface sites from sulfur compounds and H_2S, indicated by the term of $(K_HP_H)^n/(1+K_HP_H)^n$. Furthermore, it is clearly seen that the activation energy of thiophene HDS is lower than that of DBT HDS.

The kinetic analysis for HDS of organosulfur compounds, summarized in Tables 4.4 and 4.5, combines all experimental findings in a single rate expression. However, Wang and Iglesia [26] suggested two separate kinetic expressions for thiophene HDS due to the fact that HDS is driven by two different pathways, hydrogenation and hydrogenolysis (Section 4.2). The rate equations proposed by authors

TABLE 4.4
Kinetic Expressions of Thiophene Hydrodesulfurization

Thiophene Source	T (K)	P (atm)	Catalysts	Activation Energy E_a (kJ/mole)	Recommended Rate Equation	Ref.
	508–538	1	Co–Mo/Al_2O_3	15.5	$R_{HDS} = k \times \frac{K_T P_T \times K_H P_H}{(1 + K_T P_T + K_S P_S)^2} \times P_H$	[18,37]
	580–623	1	Co–Mo/Al_2O_3	90.3	$R_{HDS} = k \times \frac{K_T P_T}{(1 + K_T P_T)^2} \times P_H$	[18]
In benzene	523–623	1	Co–Mo/Al_2O_3	83.6	$R_{HDS} = k \times \frac{K_T P_T}{(1 + K_T P_T + K_S P_S)^2} \times P_H$	[18,38]
In naphtha	510–563	1	Ni–Mo/Al_2O_3	12.5	$R_{HDS} = k \times \frac{K_T P_T}{(1 + K_T P_T)^2} \times P_H$	[18]
In n-hexane	543–623	1	Ni–Mo/Al_2O_3	66.9	$R_{HDS} = k \times \frac{K_T P_T}{(1 + K_T P_T + K_S P_S)^2} \times \frac{K_H P_H}{\left(1 + \sqrt{K_H P_H}\right)^2}$	[18,39]

TABLE 4.5
Kinetic Expressions of Dibenzothiophene Hydrodesulfurization in the Presence of Co–Mo/Al_2O_3 Catalyst

DBT Source	T (K)	P (atm)	Activation Energy (kJ/mole)	Recommended Rate Equation	Ref.
	513	0.06–1	68.1	$R_{HDS} = k \times \frac{K_{DBT}P_{DBT}}{(1 + K_{DBT}P_{DBT} + K_S P_S)} \times \frac{(K_H P_H)^m}{1 + (K_H P_H)^m}$ where $m = 0.5$ or 1	[18]
	473–520	5–50	96.1	$R_{HDS} = k \times \frac{K_{DBT}P_{DBT}}{(1 + K_{DBT}P_{DBT} + K_S P_S)} \times \frac{K_H P_H}{(1 + K_H P_H)}$	[18,43]
In n-hexadecane	548–595	34–160	125.4	$R_{HDS} = k \times \frac{K_{DBT}P_{DBT}}{(1 + K_{DBT}P_{DBT} + K_S P_S)^2} \times \frac{K_H P_H}{(1 + K_H P_H)}$	[18,44]
In cetane	583–648	10–80	71.1	$R_{HDS} = k \times \frac{K_{DBT}P_{DBT}}{\left(1 + K_{DBT}P_{DBT} + K_S P_S + \sum K_i P_i\right)} \times P_H$	[18]
Tetralin	558–623	7–26	246.6	$R_{HDS} = k \times \frac{K_{DBT}P_{DBT}}{1 + K_{DBT}P_{DBT} + K_P P_P} \times \frac{K_H P_H}{(1 + K_H P_H)}$	[18]

for hydrogenation (Equations 4.2 and 4.4) and hydrogenolysis (Equations 4.3 and 4.4) pathways based on the pseudo steady-state assumption for all intermediates are shown below:

$$R_{\text{hydrogenolysis}} = k_1 \times \frac{K_T P_T K_{H2} P_{H2}}{\alpha^2} \tag{4.2}$$

$$R_{\text{hydrogenation}} = k_2 \times \frac{K'_T P_T (K_{H2} P_{H2} + K_{H2S} P_{H2S})}{\alpha^2} \tag{4.3}$$

$$\alpha = 1 + K'_T P_T + K_T P_T + K_{H2} P_{H2} + K_{H2S} P_{H2S} + K_S \frac{(K_{H2S} P_{H2S})}{K_{H2} P_{H2}} \tag{4.4}$$

The $K'_T P_T$ and $K_T P_T$ terms represent thiophene concentration in two different adsorption configurations explained in Section 4.2. Each term in Equation 4.4 represents the concentration of surface intermediates relative to the concentration of sulfur vacancies. The rate equations show that for both pathways, reaction rate is directly proportional to hydrogen and thiophene concentrations. Moreover, hydrogenation reaction is less sensitive to H_2S inhibition than hydrogenolysis reaction due to the term $K_{H2S} P_{H2S}$ in the numerator of the rate equation for hydrogenation. Thus, as H_2S pressure increases, the ratio of hydrogenation to hydrogenolysis increases.

In summary, kinetic analysis of HDS reactions shows that the HDS reaction rate increases with hydrogen pressure, and H_2S (the reaction product) has the inhibiting effect on the HDS activity. This suggests that for high HDS activity, hydrogen pressure should be increased, and the HDS catalyst should be designed to have active sites on which H_2S adsorbs weakly.

4.4.4 Catalyst

An HDS catalyst consists of a transition metal and a solid support. The activity of a catalyst is mainly attributed to the metal. The most commonly used metals are molybdenum (Mo) and tungsten (W) promoted by cobalt (Co) and nickel (Ni). Since W is more expensive than Mo, it is only applied for HDS under harsh conditions [45]. The metal concentration in the catalyst is usually 1–4 wt% for Co and Ni, 8–16 wt% for Mo, and 12–25 wt% for W [23]. Alumina (Al_2O_3), silica (SiO_2), magnesia (MgO), zirconia (ZrO_2), titania (TiO_2), mixed metal oxides, and zeolites are generally used as a support material. The surface area of support materials is in the range of 100–300 m^2/g.

In the following section, we discuss the effect of active species and support of the catalyst on the HDS reactivity. We also present discussion about how the physical properties of the catalyst (e.g., shape, size, and porosity) affect HDS reactions and how the catalyst loses its activity.

4.4.4.1 Active Species

The selection of a catalyst is made depending on the properties of the feed treated in HDS units. CoMo (on γ-Al_2O_3) is the conventional HDS catalyst, and it shows a superior HDS activity for light feed containing lower molecular weight sulfur compounds. CoMo catalyst has high selectivity toward C–S bond scission without hydrogenation of sulfur compounds. Therefore, it requires low hydrogen consumption, and it is the preferable catalyst for hydrogen economy. In contrast, the NiMo catalyst has high hydrogenation activity, and so it is useful for HDS of heavy feeds that require extensive hydrogenation [15]. The amount of sulfur (S), nitrogen (N), oxygen (O), and metals, such as Ni, V, Ti, and Fe, in feedstock must also be taken into account for a suitable catalyst design. The NiMo catalyst is effective for N, O, and metal removal from feeds due to its high hydrogenation activity. This is important particularly for the HDS of coal- and biomass-derived feeds that contain high amount of nitrogen and oxygen and petroleum-derived heavy feeds that contain high amount of nitrogen and metals. Therefore, NiMo should be the choice of catalyst for these feeds.

An HDS catalyst usually consists of Co- or Ni-promoted molybdenum disulfide (MoS_2) on a support since the catalyst in a sulfide form is more active than the one in a metallic form or in an oxide form [45]. In order to prepare a catalyst in the sulfide form, aqueous solutions of Co and Mo precursors are first impregnated onto a support (usually γ-Al_2O_3). The resulting oxidic catalyst is then converted into the sulfide state by either applying external sources containing sulfur (e.g., H_2S, carbon disulfide, dimethyl sulfide, etc.) or by the initial contact between the catalyst and the sulfur-containing compounds [46]. Mo and promoters (i.e., cobalt or nickel) are converted into the sulfide state more or less simultaneously [45].

In order to understand how catalysis works for HDS, it is essential to know the description of structure and active sites. In this study, we only focus on the CoMo catalyst since NiMo, CoW, or NiW catalysts are highly similar to the CoMo catalyst in terms of structure and active species. There are different models on structures and active sites of the conventional CoMo/Al_2O_3 catalyst reported in the literature. We describe four models in the present study.

In the monolayer model, molybdenum species bond to the alumina surface via oxygen bridges [23,47–51]. The charge balance is achieved by a "capping layer" of O^{2-} ions on top of the monolayer. Upon sulfiding, oxygen ions are replaced by sulfide ions, and some of the sulfur ions are removed under hydrogen pressure during HDS operation. This leads to the formation of sulfur vacancies and the reduction of adjacent molybdenum ions from Mo^{4+} to Mo^{3+}. Mo^{3+} ions are assumed to be the catalytically active species for HDS. In the presence of a promoter, cobalt (Co^{2+}), Co cations replace aluminum cations (Al^{3+}) in the surface layer adjacent to the Mo monolayer, which increases the stability of Mo and so HDS activity [23].

The monolayer model is good at describing the catalyst structure during the initial stage of the sulfidation, but for the further stage where sulfidation is completed, the intercalation model describes the catalyst structure better than the monolayer model. According to the intercalation model, a molybdenum sulfide (MoS_2) or tungsten sulfide (WS_2) layer is sandwiched between two hexagonal, close-packed planes of sulfur atoms [52,53]. The promoters, Co or Ni, accommodate on the edge of the MoS_2 layer.

The promotion effect of cobalt is linked to an increase in the concentration of Mo^{3+} ions, which form due to the Co–Mo interaction ($Co^0 + 2Mo^{4+} = 2Mo^{3+} + Co^{2+}$) [23].

The contact synergy model suggests the formation of Co_9O_8 and MoS_2 on the supported catalyst [54–56]. The promotional effect of Co is associated with the interaction between the Co_9O_8 and MoS_2 phases. Due to this interaction, hydrogen spillover takes place from Co_9O_8 to MoS_2, leading to increase in the activity of the MoS_2 species [23].

The Co–Mo–S model includes several Co–Mo–S structures with different Co:Mo:S stoichiometry. At maximum Co concentrations, MoS_2 edges are fully covered by Co atoms [57,58]. The properties of Co atoms in Co–Mo–S structures vary depending on edge-site geometries [58,59], Co–Co interactions [60,61], and sulfur coordination [23]. The catalytic activity is attributed to the edge sites on which Co atoms facilitate the formation of sulfur vacancies. The resulting anion vacancies create Lewis acid sites that are active in HDS. In this connection, the catalytic activity of HDS depends on how easy sulfur is removed from the catalyst surface to create a vacancy. Therefore, the catalytic activity correlates with the sulfur–metal bond strength. This partly explains why the Co-promoted MoS_2 catalyst shows higher activity compared to the unpromoted MoS_2 catalyst.

Recently, we summarize surface species that are responsible for catalytic activity and models to describe them. The models suggest that the HDS catalyst is active in the sulfide state, and the catalytic activity is attributed to the presence of sulfur vacancies (or Lewis acid sites) in the MoS_2 layer. Moreover, Co (as a promoter) promotes the formation of sulfur vacancies, thereby increasing the HDS activity. In Section 4.4.4.2, we will describe properties of a suitable support, which enables to obtain high HDS activity.

4.4.4.2 Support

Support has an important effect on the catalyst activity. It plays roles in dispersing active components and promoters, and altering the catalytic functionalities via metal–support interaction. Al_2O_3 is the traditional support material for Co–Mo, Ni–Mo, and Ni–W active phases. It provides a high surface area, and it has high thermal and mechanical stabilities. However, the strong interaction of Al_2O_3 with active metal species (e.g., Mo, W) and promoters (e.g., Ni, Co) decreases the activity and causes low metal dispersion. Furthermore, Al_2O_3 shows low acidity, which has a negative impact on HDS activity. In order to overcome these drawbacks, many materials have been studied as alternative support materials, such as SiO_2 [62], MgO [63], ZrO_2 [64,65], TiO_2 [66,67], zeolites like Na–Y [62], mesoporous materials like MCM-41 [68–70], HMS [62,71], SBA-15 [62,72], and mixed metal oxides.

Among the studied support materials, TiO_2 and ZrO_2 show outstanding HDS activities [64–67]. Since metal–support interaction is weak on a ZrO_2- and TiO_2-supported catalyst, the presence of Zr or Ti in the support leads to the increase in reducibility and dispersion of Mo. The increase in reducibility and consequent increase in anion vacancies give rise to HDS activity increase. However, their low surface area, limited thermal stability, and unsuitable mechanical properties prevent commercial exploitation. With an aim to overcome these drawbacks and improve catalytic activities, TiO_2 and ZrO_2 are mixed with γ-Al_2O_3 [73–75]. These mixed oxides can be prepared by different techniques, such as coprecipitation, chemical vapor deposition, and impregnation. Preparation methods influence the physical and chemical properties of support materials. For the interested reader, we refer to the literature for detailed information about preparation techniques [12,62,73,76–78].

Both Mo and CoMo supported on ZrO_2–Al_2O_3 and TiO_2–Al_2O_3 show higher activity than an Al_2O_3-supported catalyst. When ZrO_2 or TiO_2 is mixed with Al_2O_3 with the ratio of 1/1 (wt/wt), the highest HDS reactivity is obtained [62]. This is attributed to high surface area and high metal dispersion. In addition, for a TiO_2–Al_2O_3 mixed oxide supported catalyst, the promoter effect of Ni on HDS process was found higher than Co [62].

Silica is also a potential support material, which is assumed to improve the HDS activity when it is mixed with Al_2O_3 due to an increase in the acidity of the support. However, since the presence of SiO_2 causes the decrease in the number of OH groups on the surface, metal dispersion and HDS activity decrease considerably [79,80]. Moreover, a high amount of SiO_2 in the support material leads to high acidity, thereby enhancing deactivation of the catalyst via coke formation [12].

4.4.4.3 Physical Properties of the Catalyst

The catalytic activity also depends on physical properties of the catalyst, such as particle size, the shape of the particle, and pore size. The particle size of the catalyst is an important parameter that influences the resistance for mass transfer of reactants to the catalyst surface or mass transfer of products from the catalyst surface. In HDS operation, the external mass transfer resistance does not affect the reaction rate significantly, whereas the effect of intraparticle mass transfer resistance on the HDS rate is significant [47]. In order to minimize this resistance, small particles are preferred due to their small diffusion length. However, small particles cause a high-pressure drop throughout the catalyst bed and lead to activity loss. Therefore, an optimum value of particle size should be determined to maintain a low diffusion limitation and to minimize the pressure drop. For the same particle size, the pressure drop can be also minimized by using particles with different shapes and pore structures. The pressure drop buildup in the presence of various shaped catalysts was studied in the literature [23]. It was shown that among catalyst particles in different geometries (e.g., trilobes, pentalobes, rings, cylinders, etc.), rings create lower pressure drop than the others due to the increased voidage in the catalyst bed at the expense of catalyst activity (Figure 4.3). On the other hand, the traditional cylinder particles create high-pressure drop and show highest HDS activity per unit volume. The pore size of the catalyst has also strong influence on the activity and the pressure drop. Larger pores are required for lower pressure drop, but in this case, low activity is obtained due to the loss of surface area. Therefore, an optimum pore size should be determined to obtain the highest activity [81].

The selection of pore size of the catalysts must take into account feed types. For the HDS of light feed, catalysts with micropores (<50 nm) are preferred to increase the surface area and so the catalytic activity [82]. The built-up back pressure due to small pore size is in the acceptable limit and is offset by the increased activity. On the other hand, the pore size of the catalyst is critical for the effective HDS of heavy feed. Since heavy feed contains large molecules with high molecular weight,

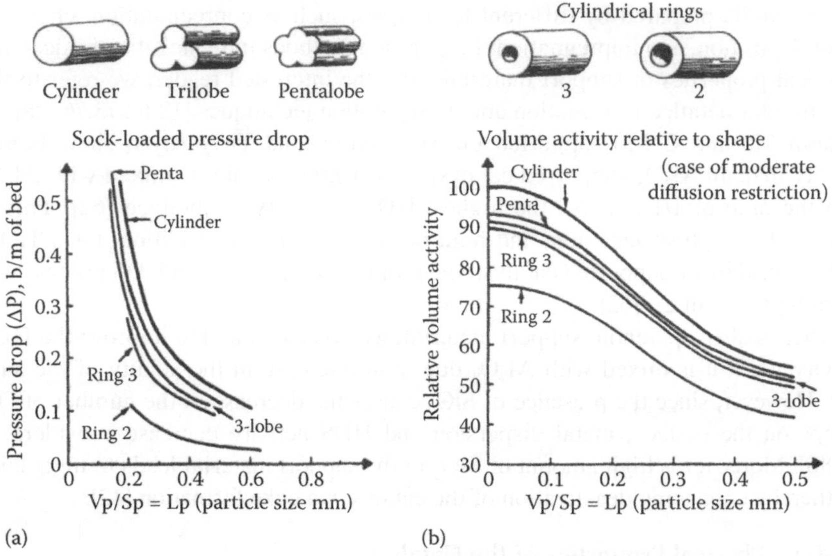

FIGURE 4.3 Effect of catalyst particle size and shape on (a) pressure drop and (b) HDS activity. (Adapted from B.H. Cooper et al., *Oil & Gas Journal*, 8 (1986) 39–44.)

small pores restrict the diffusion of molecules in and out pores. This results in low activity and high-pressure drop. High metal content in heavy feed also causes the same effects to happen. Since metals plug the pore mouth, they prevent the catalytic HDS reaction to occur [82,83]. Therefore, for the HDS of heavy feed, the catalyst should have mesopores (50–1000 nm) and macropores (>1000 nm) for efficient catalyst utilization [82].

4.4.4.4 Catalyst Deactivation

Aromatic hydrocarbons, nitrogen-containing compounds, saturated hydrocarbons that are present in fuel, and H_2S (the reaction product) are known inhibitors for HDS. The inhibition effect of these compounds increases in the following order: aromatic hydrocarbons < H_2S < nitrogen-containing compounds [84].

Among aromatic hydrocarbons present in diesel and jet fuels, naphthalene and its derivatives are dominant species. Studies in the literature reveal that naphthalene and its derivatives inhibit the conversion and selectivity of HDS of thiophene [85] and DBT [1]. It was reported that naphthalene reduces the selectivity of the catalyst for the hydrogenation pathway since it poisons the active sites that are responsible for hydrogenation [34,86].

Hydrogen sulfide has also an inhibiting effect on the HDS activity due to equilibrium and kinetic considerations. Since it is a product of HDS reactions, the high amount of hydrogen sulfide pushes the equilibrium toward the reactants. In terms of kinetics, H_2S blocks active sites on the catalyst, thereby decreasing the HDS activity. It is reported that H_2S poisons mainly hydrogenolysis sites on a sulfided Co–Mo/γ-Al_2O_3 catalyst [23].

The inhibition effect of nitrogen-containing organic compounds on HDS activity is dramatic due to their low reactivity. The nitrogen content of crude oil is between 0.1 and 1 wt%. About a third of nitrogen-containing compounds are basic compounds containing the pyridine nucleus (e.g., carbozole and indole), while the rest is present mainly in the form of relatively nonbasic compounds containing the pyrrole nucleus (e.g., quinolone) [23]. Basic and nonbasic compounds inhibit the HDS activity of thiophene and DBT and its alkyl derivatives [15,85,87]. The basic nitrogen-containing compounds are more poisonous than the nonbasic ones. This is supported by the study, indicating that the degree of poisoning of nitrogen compounds is correlated to their gas phase proton affinities for the HDS of DBT on NiMo catalyst [88]. Nitrogen-containing compounds poison different active sites on the catalyst. The poisonous effect differs depending on the type of sulfur compounds. To illustrate, nitrogen-containing compounds poison mainly the hydrogenation pathway for the HDS of DBT, while they poison the hydrogenolysis pathway for BT [15].

Metal deposition and coke formation are other reasons for catalyst deactivation. Metals, such as V and Ni, are deposited within the pores of catalyst particles and make the movement of sulfur compounds in the pores difficult or impossible. In order to increase the tolerance of catalyst particles for metal deposition, large pore diameter and pore volume are preferred. The disadvantage of this design is the low surface area of catalyst particles, leading to low intrinsic catalyst activity. The other deactivating agent, coke, forms on the catalyst surface during HDS operation and poisons catalytically active sites [89]. Coke accumulates by time and decreases the activity slowly [23]. The coked catalyst can be regenerated by an oxidative burn-off of carbon deposits under oxygen environment [90–92]. The original activity of the catalyst can be recovered by this way only when the burn-off is carefully controlled to prevent overheating of catalyst particles. Otherwise, the active phase of the catalyst can change irreversibly [93].

4.4.4.5 Industrial Outlook–Advanced HDS Catalyst

The utilization of advanced catalysts enables HDS reactions to be carried out under mild conditions with high selectivity and stability. The expected properties of catalysts must be as follows [15]: (1) high desulfurization activity to meet environmental regulations, (2) high nitrogen and metal atoms removal ability for the long lifetime of the catalyst, and (3) high selectivity for desulfurization to improve the fuel quality. There are many successfully applied HDS catalysts possessing these properties in industry. Some of known HDS catalysts along with their manufacturer and specifications are listed in Table 4.6. As shown in the table, companies have developed different HDS catalysts, which have high activity, selectivity, and stability at specific conditions due to their well-designed chemical and physical structures. We will mention some of them here.

Akzo Nobel developed STARS technology by using super active CoMo (KF 757) and NiMo (KF 848) catalysts [15,94]. CoMo-STARS catalysts are used for high-level sulfur streams (100–500 ppm), while NiMo-STARS catalysts are preferred for the removal of the low sulfur streams (<100 ppm). They are highly effective for removing sulfur from sterically hindered compounds. Sulfur content of the feedstock can be decreased up to 1–2 ppm level in the presence of STARS catalysts at the common

TABLE 4.6
Industrially Applied HDS Catalysts, Their Manufacturer, and Specification

Manufacturer	Catalyst Name	Some Specifics	Ref.
ConocoPhillips	S Zorb™ SRT	Metal-based adsorbent 10 ppm in gasoline In 2001	[95] [96]
Haldor Topsoe Denmark	TK-558 BRIM™	CoMo 10 ppm sulfur in gasoline	[97]
	TK-559 BRIM™	NiMo 10 ppm sulfur in gasoline	[97]
	TK-568 BRIM®	CoMo, ultralow sulfur diesel (ULSD) and kerosene applications	[98]
	TK-570 BRIM®	CoMo, ULSD Low to medium pressure	[99]
	TK-575 BRIM®	NiMo, ULSD High pressure (45+ bar) In 2005	[100]
	TK-576 BRIM®	CoMo, ULSD Medium pressure (27 bar) In 2004	[100]
	TK-578 BRIM®	CoMo, ULSD production (less than 10 wt ppm sulfur)	[101]
	TK-609 HyBRIM™	NiMo, ULSD Medium to high pressure	[102]
Axens France	HR 606	CoMo on Al_2O_3 Desulfurization and denitrification of gasoline and naphtha	[103]
Johnson Matthey	$HYTREAT_{JM}$	Cobalt–molybdenum-sulfide particles on Al_2O_3 ULSD (<15 ppm S)	[104]
Criterion Catalysts & Technologies, USA	CENTINEL DC-2118 catalyst	CoMo <50 ppm S diesel	[105]
Süd-Chemie Germany	HDMax® 213	Stabilized alumina extrusion impregnated with cobalt oxide and molybdenum oxide	[106]
Grace Davison UK	SuRCA® FCC	Achieving on average 30–40% reduction of sulfur in full-range FCC naphtha	[107]
Albemarle–Akzo Nobel The Netherlands	STARS catalyst KF 757	Co–Mo on Al_2O_3 <50 ppm S diesel First announced in 1998 Medium to high pressure	[108] [95]
	KF848-STARS	Ni–Mo, ULSD In 2002 (commercialized)	[95,109]

(*Continued*)

TABLE 4.6 (CONTINUED)
Industrially Applied HDS Catalysts, Their Manufacturer, and Specification

Manufacturer	Catalyst Name	Some Specifics	Ref.
ExxonMobil Corporation USA	SCANfining™	SCANfiningI™ for mercaptanes SCANfiningII™ for ULSD	[110]
Akzo Nobel with ExxonMobil	NEBULA	Bulk catalyst no support In 2001 Low sulfur diesel in most high-pressure units for the production of 500 ppm S	[109] [95]

HDS conditions. The high HDS activities of these catalysts are attributed to Type II reaction sites, which contain stable Co–MoS_2 nanostructures. Long-term stability is the common property of this type of catalysts, thanks to their optimized pore structures [111].

NEBULA catalysts, also produced by Akzo Nobel, are useful for the diesel hydrogenation at high pressure and temperature conditions [15]. The NEBULA catalyst is a bulk base metal catalyst without using a porous support, and it shows higher HDS and HDN activities than the NiMo-STARS catalyst [111]. The catalyst is active in the sulfided state and has a different structure from the conventional HDS catalyst. The catalyst works at harsh conditions, but it requires high hydrogen consumption.

Criterion Catalysts and Technologies produced CENTINEL catalysts by modifying the preparation method of the traditional HDS catalysts. CENTINEL catalysts are 80% more active than the conventional HDS catalyst. This is associated with the better metal dispersion on the catalyst support [111]. CoMo-CENTINEL catalysts are employed for the high sulfur level stream at low H_2 pressure. On the other hand, NiMo-CENTINEL catalysts are used for the low sulfur level stream at high H_2 pressures [15].

Henrik Topsoe and his team proposed two types of a molybdenum disulfide–based BRIM catalyst in order to meet the ultralow sulfur emission standards. During BRIM catalyst discovery, Topsoe's Research and Development Department collaborated with Denmark's Technical University and the University of Aarhus, Denmark. They used molybdenum disulfide crystals in nanosize and tested hydrodesulfurization performance via a scanning tunneling microscope. The results of the experiments revealed the presence of new catalytic active sites. These more reactive sites are located on the top surface, close to the molybdenum disulfide nanocrystal edges. They are called BRIM sites [97,112,113]. During the process, thiophene is first hydrogenated and C–S bond is broken on BRIM sites, and the final ring-opened molecule is transferred to the cluster edge for another C–S bond breaking [112]. Initially, Type I of BRIM catalysts was developed to increase the density of the catalytic reaction cites. Additionally, in order to improve the catalytic activity, the discovery of Type II catalysts was carried out [113]. TK-558 BRIM™ (CoMo catalyst) and TK-559 BRIM™ (NiMo catalyst) are the first commercial catalysts providing 10 ppm sulfur level. TK-558 BRIM™ (CoMo catalyst) makes the low sulfur product formation possible by preserving the yields. This catalyst shows better HDS activity than TK-559 BRIM™, while the HDN activity of TK-559 BRIM™ is higher when pressure is higher than 50 bars [97].

In addition to the currently used industrial catalyst mentioned above, there are some published patents on the HDS catalyst. We list some of them in Table 4.7. The table shows that different active species and support material have been applied to produce ultralow sulfur fuels. The sulfide forms of nickel, cobalt, and molybdenum are generally used for obtaining a high desulfurization activity. As shown in the table, these species are either promoted by metals such as tungsten (W), iron (Fe), ruthenium (Ru), copper (Cu), magnesium (Mg), titanium (Ti), barium (Ba), and metal oxides like zinc oxide (ZnO_x) and niobium oxide (NbO_x), or replaced by them. In addition, alumina, silica, and zeolites with different textural properties are commonly used as a support material. A list of patented HDS catalysts are given in Table 4.7.

The combination of the new type of active species and support materials became new alternatives for the advanced HDS. Pt and Pt–Pd catalysts were found very active for the deep desulfurization of prehydrotreated straight-run gas oil. They decrease the sulfur content up to 6 ppm. These catalysts also cause 75% decrease in aromatic contents. When amorphous silica–alumina (ASA) is used as a support material, PtPd/ASA catalysts show better sulfur removal performance for the feed with low aromatics and medium and/or low sulfur level. When aromatic content is high, the desulfurization ability of Pt on ASA is higher than that of PtPd/ASA. Since the presence of sulfur leads to catalyst poisoning, when sulfur is present in high level, ASA-supported NiW catalysts are preferred. Sulfur tolerance of the active species (noble metals in this case) is the limiting parameter for the advanced HDS process [114,115]. In this regard, bifunctional catalysts can improve the sulfur resistance of the catalysts by providing the combination of two active sites with sulfur resistance and support materials with bimodal pore size distribution. In this model, the first active sites in the larger pores remove the organosulfur compounds, and the second active sites placed in the small pores can be used for the transportation of dissociated hydrogen between the pores. The design and synthesis of the bifunctional catalysts are not easy, and no industrial application has been presented yet [15].

4.5 HDS PROCESSES AND TECHNOLOGIES

4.5.1 General Description

HDS processes are generally employed in refineries to produce low-sulfur fuel and to prevent catalyst deactivation, which occurs due to sulfur poisoning. The operating conditions of HDS strongly depend on the desired product specification and feedstock properties. For processing light feeds, mild processing conditions (e.g., 330–360°C, 5–35 atm) are applied to remove sulfur via hydrogenation. On the other hand, the HDS of heavy feeds requires higher pressures (50–150 atm) and temperatures (360–410°C) since HDS reaction rates of these compounds are relatively slow. Due to the high temperature and pressure treatments, heavy distillates also decompose to

TABLE 4.7
Published Patents on HDS Catalyst

Patent Application Number	Patent Name	Company or Inventor	Publication Date	Some Details on Desulfurization Materials
US 8697598	Hydrogenation catalyst and use thereof	China Petroleum & Chemical Corporation SINOPEC	April 15, 2014	Ni, Mo, W on two types of carriers.
US 20050059545	Molybdenum sulfide/carbide catalysts	Alonso Gabriel, Chianelli Russell R., Fuentes Sergio	March 17, 2005	MoS_2 and $MoS_{2-x}C_x$ promoted with Co, Ni, Fe, and/or Ru sulfides.
US 20150306585	Supported catalysts for producing ultra-low sulfur fuel oils	Universida Nacional Autónoma De México	October 29, 2015	CoMo and NiMo supported on Y-type zeolite.
US 20120292231	A high activity hydrodesulfurization catalyst, a method of making a high activity hydrodesulfurization catalyst, and a process for manufacturing an ultralow sulfur distillate product	Shell Oil Company (Houston, TX, US)	November 22, 2012	Gamma-alumina produced from aluminum hydroxide.
WO 2004002620	FCC catalyst for reducing the sulfur content in gasoline and diesel	Albemarle Netherlands B.V.	January 8, 2004	5–55 wt% metal-doped anionic clay, 10–50 wt% zeolite, 5–40 wt% matrix alumina, 0–10 wt% silica, 0–10 wt% of other ingredients, and balance kaolin.
WO 2005077498	Sulfur oxide adsorbents and emissions control	Battelle, Memorial Institute Liyu LI. King, David L. (WA, 99352, US)	August 25, 2005	Manganese-based octahedral molecular sieve as a SO_2 adsorbent.

(*Continued*)

TABLE 4.7 (CONTINUED)
Published Patents on HDS Catalyst

Patent Application Number	Patent Name	Company or Inventor	Publication Date	Some Details on Desulfurization Materials
WO 2013043629	Low-temperature adsorbent for removing sulfur from fuel	ExxonMobil Research and Engineering Company (NJ, 08801-0900, US) NOVAK, William, J. (NJ, 07921, US) GATT, Joseph, E. (NJ, 08801, US)	March 28, 2013	An active copper component disposed on a zeolitic and/or mesoporous support.
US 6482314	Desulfurization for cracked gasoline or diesel fuel	Phillips Petroleum Company (Bartlesville, OK)	November 19, 2002	A mixture of zinc oxide, silica, alumina, and a substantially reduced valence cobalt.
US 5882614	Very low sulfur gas feeds for sulfur sensitive syngas and hydrocarbon synthesis processes	Exxon Research and Engineering Company (Florham Park, NJ)	March 16, 1999	The gas first with zinc oxide and then with nickel metal.
US 7709412	Bulk metal hydrotreating catalyst used in the production of low sulfur diesel fuels	ExxonMobil Research and Engineering Company (Annandale, NJ, US)	May 4, 2010	A bulk metal hydrotreating catalyst. MoxCoyNbz
WO 2007127022	Method for removing sulfur compounds from gasoline or diesel fuel using molecularly imprinted polymers	LEONE, Anna, Madeleine (CA, 94608, US)	November 8, 2007	A plurality of molecularly imprinted polymer beads.
WO 2001088064	Antistatic lubricity additive for ultralow sulfur diesel fuels	The Lubrizol Corporation (OH, 44092-2298, US) Wilkes, Mark F. (1UX, GB) Duncan, David A. (DE63 AW, GB) Carney, Shaun P. (DE56 OLE, GB)	November 22, 2001	0.001–1 ppm of a hydrocarbyl monoamine or hydrocarbyl-substituted poly(alkyleneamine) 10–500 ppm of at least one fatty acid containing 8–24 carbon atoms, or an ester.

(Continued)

TABLE 4.7 (CONTINUED)
Published Patents on HDS Catalyst

Patent Application Number	Patent Name	Company or Inventor	Publication Date	Some Details on Desulfurization Materials
US 20080060977	Catalyst and process for the manufacture of ultralow sulfur distillate product	Bhan, Opinder Kishan (Katy, TX, US)	March 13, 2008	The calcined mixture comprises molybdenum trioxide, a Group VIII metal compound, and an inorganic oxide material.
US 8394735	Catalyst for ultradeep desulfurization of diesel via oxidative distillation, its preparation, and desulfurization method	Dalian Institute of Chemical Physics, Chinese Academy of Sciences (Dalian, CN)	March 12, 2013	Amphiphilic oxidative catalyst. Quaternary ammonium cation; X is P, Si, As or B; and M is Mo or W.
US 20130296163	A catalyst and process for the manufacture of ultralow sulfur distillate product	Shell Oil Company	November 7, 2013	A calcined mixture of inorganic oxide material, a high concentration of a molybdenum component, and a high concentration of a Group VIII metal component.
WO 2009073469	Catalyst to attain low sulfur diesel	CHOI, Ki-Hyouk (Dhahran, 31311, SA) MOCHIDA, Isao (Kasuga Fukuoka, 816-8580, JP)	June 11, 2009	One of the metals selected from molybdenum, cobalt, and nickel, and a silicon dioxide support via the spray pyrolysis technique.
US 4719196	Process for producing a supported catalyst	Hoechst AG (DE)	May 13, 1987	A noble metal onto a ceramic honeycomb structure coated with an aluminum oxide layer.

(*Continued*)

TABLE 4.7 (CONTINUED)
Published Patents on HDS Catalyst

Patent Application Number	Patent Name	Company or Inventor	Publication Date	Some Details on Desulfurization Materials
US 8992768	Method for the desulfurization of fuels and highly active nickel carrier catalyst based on aluminum oxide suitable for said method	Süd-Chemie IP GmbH & Co. KG (München, DE)	March 31, 2015	A nickel catalyst based on aluminum oxide and a promoter, selected from the compounds of Mg, Ti, Pb, Pt, Ba, Ca, and/or Cu.
US 8216958	Selective catalysts having silica supports for naphtha hydrodesulfurization	ExxonMobil Research and Engineering Company (Annandale, NJ, US)	July 10, 2012	A Co/Mo metal hydrogenation component loaded on a silica or modified silica support in the presence of organic ligand.
US 8637423	Selective catalysts having high-temperature alumina supports for naphtha hydrodesulfurization	ExxonMobil Research and Engineering Company (Annandale, NJ, US)	January 28, 2014	A Co/Mo metal hydrogenation component loaded on a high-temperature alumina support.

lower-boiling products via C–C bond scission concurrent with hydrogenation. High hydrogen pressure also minimizes coke formation and promotes high desulfurization reaction rates [46].

The simplified scheme of the HDS process is shown in Figure 4.4. In HDS processes, hydrocarbon feedstock is first pressurized and mixed with recycle gas stream. Then the final mixture is preheated and fed to the reactor. Desulfurized products in liquid phase are removed from the recycle gases in the separator, while H_2S and light hydrocarbon gases are separated from the recycle gas in the scrubber. The remaining gases are mixed with fresh hydrogen and recycled back to the system to increase the total yield and to minimize hydrogen losses. The fresh hydrogen can be produced on-site by steam reforming or other reforming reactions. In the HDS process, sulfur removal is usually in the level of 90–95% [46].

Hydrogen requirements of the HDS process depend on the nature of the feedstock. Heavy feedstocks, containing less reactive aromatic organosulfur compounds, e.g., BT, DBT, and their alkylated derivatives, require more hydrogen to produce a product with a desirable sulfur level than the one containing aliphatic organosulfur compounds. The oxygen and nitrogen contents of feedstock also lead to more hydrogen consumption. O_2 and N_2 react with hydrogen to produce water and ammonia, respectively. In addition to oxygen- and nitrogen-containing compounds, metals in feedstock are the reason for extra hydrogen requirement since metal deposition on the catalyst enhances hydrogenation reactions [46].

In the HDS reactor, the reaction medium contains corrosive compounds like hydrogen sulfide, hydrogen, as well as other corrosive agents arising from feedstock. In order to have a safe operation, the reactor wall should have certain thickness and composition, providing strength to harsh condition and resistance to corrosion [46]. Moreover, since desulfurization reactions are generally exothermic, a great amount of heat is released. Therefore, the reactor temperature has to be controlled. One of the desirable features of the reactor is low-pressure drop, which can lead to substantial decrease in reaction yield.

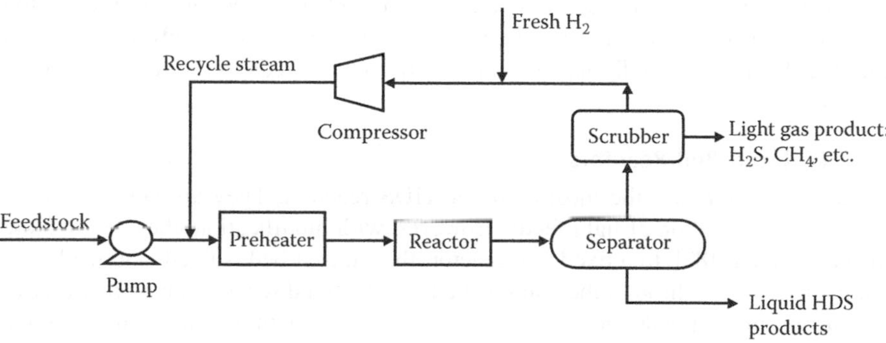

FIGURE 4.4 Simplified scheme of the HDS process. (Adapted from J.G. Speight, Desulfurization chemistry, in: *The Desulfurization of Heavy Oils and Residua*, CRC Press, Boca Raton, FL, 1999, pp. 180–217.)

The process variables of HDS reactions are reactor temperature, hydrogen pressure, flow rate of reactants, and recycle ratio. The reactor temperature is a very important control parameter for an efficient operation. An increase in the reactor temperature gives rise to higher reaction rate (e.g., a 10°C increase doubles the reaction rate), thereby leading to less catalyst consumption. However, there is a limit for temperature increase, above which the process efficiency can be adversely affected. At temperatures above 410°C, the reaction rate of thermal cracking of hydrocarbons becomes considerably high, resulting in the formation of low molecular weight hydrocarbons and hydrogen [46]. The resulting hydrogen could lead to a substantial increase in temperature due to excessive heat generated by thermal cracking and desulfurization reactions. The increase in temperature causes catalyst sintering and damages the reactor wall.

Hydrogen pressure has also important influence on the HDS reaction rate. The rate increases with the partial pressure of hydrogen. Nevertheless, excessive increase in the partial pressure of hydrogen can lead to the formation of a high amount of NH_3 and H_2S, which poison the catalyst. A high amount of H_2S also causes corrosion.

The reaction rate can be increased by decreasing space velocity, which is the ratio of the volumetric flow rate to the catalyst volume. This results in a lower production rate, but it can be compensated by temperature increase.

In the HDS process, there are two major challenges for having an efficient operation: (1) the development of catalysts with high resistance to deactivation and (2) designing three-phase reactors capable of processing large quantities at high temperatures and pressures [46]. We already discussed the desired properties of catalysts for efficient HDS operation in Section 4.4.5. Now, we present reactor technologies applied to the HDS process and advanced technologies in this field.

4.5.2 Reactor and Its Configuration

Reactor design for HDS varies depending on the arrangement of the catalyst bed and the means by which feedstocks are introduced to the reactor. In terms of the arrangement of the catalyst bed, HDS reactors are divided into two groups: (1) fixed-bed reactors and (2) fluidized-bed reactors. In the following sections, we present the description of these reactor technologies with advantages and disadvantages when utilizing them. We also discuss the ways to introduce feeds to the reactor in these sections.

4.5.2.1 Fixed-Bed Reactors

Fixed-bed reactors are the most common HDS reactors. They are generally used for the desulfurization of light feeds. However, with modification, they can operate on heavier feeds [82]. In a fixed-bed reactor, the catalyst bed is fixed, and feedstock and hydrogen pass through the catalyst bed. A fixed-bed reactor can work either in a cocurrent mode (feeds move in the same direction) or in a countercurrent mode (feeds move in the opposite direction). In the cocurrent design, it is easy to control gas–liquid mixing and to provide contact between the catalyst and feeds in liquid and gaseous states. The drawback of this design is the lowest hydrogen concentration at the outlet of the reactor, which causes the ineffective desulfurization at the outlet. Furthermore, a high concentration of H_2S at the outlet suppresses the activity of the catalyst. These problems can be overcome by the countercurrent design, where hydrogen is fed from the bottom of the reactor and liquid feedstock is introduced from the top. However, in this case, a poor gas–liquid contact is present, and liquid flooding and hot spots occur in the reactor.

In the fixed-bed reactor, the temperature rises inside the reactor due to the exothermic nature of HDS. In order to compensate for the temperature rise, several catalyst beds can be incorporated within the reactor, and cold quench gas can be introduced in between these beds. One of the disadvantages of the fixed-bed reactor is the high pressure drop. The pressure drop can be minimized by using a radial flow reactor due to its shorter bed length [23]. The radial flow reactor also provides a larger catalyst cross-sectional area than the fixed-bed reactor. However, the occurrence of hot spots is more probable, and it is more difficult to remove contaminants [46]. Therefore, the radial flow reactors are only used for processing light feedstock, such as naphtha and kerosene. For heavy distillates, fixed-bed reactors are mostly preferred.

The main limitation of fixed-bed reactors is the metal and coke deposition in the pores of catalysts. In the beginning of the operation, metals, especially Ni and V, block access for HDS and lead to a gradual activity loss. At the end of the operation, carbon deposition also occurs and reduces the activity further [23,46]. As the catalyst is poisoned by metal and coke deposition, the reactor temperature is increased to maintain a constant conversion [116]. However, the temperature rise during the operation also causes deactivation due to sintering. At the end of the catalyst lifetime, which is mainly dictated by its metal holding capacity, the catalyst beds should be replaced. This is an expensive operation since the reactor should be shut down.

The HDS activity of the fixed-bed reactors can be increased by dense catalyst loading [82,83,117]. In this way, more catalyst loading per unit volume of the reactor is achieved. Since catalyst particles are loaded slowly, they settle into bed without creating large voids. This provides better contact between the feedstock and the catalyst and reduces channeling. Dense loading increases activity per reactor volume and decreases the start of run temperature [117]. The negative effect of this type of loading is high pressure drop. However, positive effects mentioned above compensate for the negative effect.

4.5.2.2 Fluidized-Bed Reactors

In order to eliminate high pressure drop and catalyst deactivation by metal deposition, fluidized-bed reactors are used. In fluidized-bed reactors, hydrogen flows upward and feed flows upward (cocurrent) or downward (countercurrent). Catalyst particles are suspended freely in a gas–liquid mixture, so they move throughout the reactor. Due to the nature of the catalyst bed, a high pressure drop or plugging problem is not encountered in the fluidized-bed reactor. Since the pressure drop can be easily avoided, small catalyst particles can be used. The ability to use small catalyst particles in the fluidized-bed reactor provides more efficient operation due to the low intraparticle mass transfer resistance [47]. Furthermore, catalyst particles in motion allow frequent withdrawal and replacement of these particles without the necessity of shutting down the reactor. The random motion of catalyst particles also eases the dissipation of heat produced by the exothermic HDS reactions, suggesting that a uniform temperature distribution can be obtained.

There are some drawbacks of fluidized-bed reactors. A high degree of mixing of reactants in the flow direction requires lower space velocities and higher temperatures compared to fixed-bed reactors [47]. The resulting high-temperature operation leads to undesired reactions and high hydrogen consumptions. Furthermore, it is difficult to control the reactor stability during the operation, especially for exothermic reactions involving hydrogen at high temperature and pressure. Fluidized-bed reactors are particularly favored for processing heavy feeds.

4.5.3 Advanced HDS Technologies

New reactor design and technology are required for achieving high HDS activity and stability. Fixed-bed reactors with cocurrent flow of feedstock and H_2 streams are the conventional reactors. However, nonhomogenous H_2 concentration through the reactor and high H_2S concentration of the exit stream prevent the deep desulfurization. In order to increase the HDS reactivity and to improve the cocurrent flow reactor configuration, multistep reactor system is considered as an alternative approach. In this reactor configuration, two or more separated catalyst beds in a single reactor with cocurrent and countercurrent flows are employed. Since nonhomogenous H_2 concentration through the reactor is the main drawback of the cocurrent flow, to avoid the problem, H_2 (bottom part) and distillation products (top part) are mixed in countercurrent flow. SynAlliance patented this reactor design, called SynTechnology [118]. In this single reactor configuration, a cocurrent reactor is placed in the initial part of the reactor, while countercurrent flow is placed in the last part of the reactor. The conversion of the sulfur compounds is followed by the H_2S removal from the reactants. During the countercurrent flow stage, the favorable H_2 and H_2S concentration profiles through the reactor are obtained. Additionally, utilization of noble metal-based catalysts in this reactor configuration provides more activity. However, the poisoning effect of sulfur limits the applicability of this type of catalysts in the countercurrent part of the reactor [1,15,82].

A swing reactor type is another alternative for replacing the fixed-bed reactor. It is employed when metal content reaches 350 ppm. The flexible reactor design makes the deep desulfurization possible by switching the catalyst and stream places.

New bed arrangements can be also developed to have higher performance. Satterfield [119] reported that the void fraction along the wall of the reactor results in the deviation from the plug flow and prevents full wetting of the catalyst with the liquid. Consequently, the conversion of HDS decreases. Carruthers and DiCamillo [120] suggested to use inert diluent particles along with catalyst particles. This helps in narrowing liquid flow channels, improves liquid holdup, and increases the contact time and conversion.

As mentioned before, catalyst deactivation, especially due to metal deposition, is one of the important problems in the HDS process, which causes activity loss and high pressure drop. In order to avoid these problems, catalyst particles should be replaced regularly. Since the conventional way to replace catalysts is costly, a new strategy should be developed. Literature suggests that the use of a multiple catalyst-bed system in fixed-bed reactors can be one of the solutions to the catalyst deactivation by metal deposition [82,121]. A dual or triple catalyst system can be chosen depending on metal content in the feed [82,122]. For feeds containing between 25 and 50 ppm metals, a dual catalyst system is preferred, whereas a triple catalyst system is suitable for feeds containing between 50 and 100 ppm metals. In these systems, catalyst particles in the front end have high metal tolerance, while those in the tail end possess high HDS activity. Each bed contains a different chemical composition and pore size of the catalyst. For the front and tail ends, a CoMo catalyst with small pore size and a NiMo catalyst with large pore size, respectively, are potential candidates. This design improves catalyst stability and lifetime. Jacobsen et al. [123] showed that the catalyst particles in the multiple bed are much more stable than those in the single bed.

In order to load and unload catalyst particles into multiple beds, Royal Dutch Shell developed the quick replacement reactor [122]. The reactor is a conical fixed-bed reactor. Catalyst-feed mixture is transported to the top of the reactor, where catalyst particles and liquid feed are separated and then they are fed to the reactor. In this design, catalyst particles are easily released and replaced.

HYVAHL S process developed by the ASVAHL consortium provides an alternative strategy for easy catalyst replacement [82]. This strategy is based on the swing reactor concept, in which several reactors are employed. Front reactors or guard reactors have high hydrodemetallization (HDM) activity, and they are placed before the main fixed-bed reactor with high HDS activity. Front reactors are switchable in the operation, and they can be switched on and off depending on the metal contents of feeds. This arrangement can provide 90% HDS and 98% HDM conversion for a wide range of heavy feeds [124].

Fluidized-bed reactors can be also employed for enabling easy catalyst replacement. Shell developed the bunker reactor [125], which combines fixed-bed and ebullated bed technologies. In bunker reactors, a fresh catalyst is loaded from the top part of the reactor. As metals deposit on the fresh catalyst, catalyst particles gain weight and move downward throughout the reactor. Then they are released from the bottom of the reactor. Since feeds are always faced with the fresh catalyst particles, the metal tolerance of the bunker reactor is higher than that of the fixed-bed reactor. Therefore, the bunker reactor can work for the HDS of heavy feeds.

4.6 SUMMARY AND FUTURE DIRECTIONS

More stringent environmental regulation for low-sulfur fuels and the increasing use of heavy feeds increase the importance of the HDS process. Sulfur in fuels is mainly present in organic compounds, such as aliphatic sulfur compounds (thiols, sulfides, disulfides) and aromatic sulfur compounds (thiophene, BT, DBT, and their alkylated substitutes). HDS is mainly applied for processing petroleum-derived feeds. The light fraction of petroleum-derived feeds contains aliphatic organosulfur compounds with low molecular weight, whereas the heavy fraction of petroleum-derived feeds contains aromatic organosulfur compounds with high molecular weight. The HDS of all organosulfur compounds is thermodynamically favorable at operation conditions (330–410°C and 35–150 atm). Aliphatic organosulfur compounds show high HDS reactivity, and so the sulfur removal from these compounds can be easily achieved. In contrast, the desulfurization of aromatic organosulfur compounds is difficult due to their conjugation structure between the lone pairs on S atom and the p-electrons. Therefore, the HDS of heavy feed containing aromatic sulfur compounds requires extensive hydrogenation and high-temperature treatment.

The chemical identity and physical properties of the catalyst play a significant role in the HDS activity. The conventional catalyst for HDS is CoMo or NiMo supported on alumina. CoMo catalysts show HDS activity, and they are usually preferred for the HDS of light feed. However, CoMo catalysts are not suitable for the HDS of heavy feeds since heavy feeds contain high amount of aromatic compounds, nitrogen compounds, and metals that block catalytic active sites and deactivate them. NiMo catalysts are generally used for the HDS of heavy feeds due to their high hydrogenation, hydrodenitrogenation, and HDM activities. The HDS of heavy feeds also requires small catalyst particles with high porosity to reduce mass transfer resistance and to prevent pore plugging caused by metal and coke deposition.

HDS reactions are usually carried out in fixed-bed reactors, which provide a good contact between the catalyst and reactants and easy control of reactor stability. However, for processing heavy feeds, fluidized-bed reactors can be favorable since fixed-bed reactors create high pressure drop, and the replacement of deactivated catalysts is not easy.

Conventional HDS technologies are still operational for the HDS of light feeds since it is quite straightforward to desulfurize low molecular weight aliphatic organosulfur compounds. However, advanced HDS technologies are required for the HDS of heavy feeds due to the low reactivity of high molecular weight aromatic sulfur compounds and the ease of catalyst deactivation. Future research interest will be on modification and improvements of HDS catalysts and process to achieve high sulfur removal from heavy feeds and a stable operation. In this respect, bifunctional catalysts with optimized pore structures, providing high sulfur and metal resistance, can be chosen as a research topic. It can provide high HDS activity and stability. Noble metals and TiO_2–ZrO_2 mixed with alumina are proven active metals and supports, respectively. With further studies on these materials, they can replace conventional catalysts. In terms of process innovations, multilayer beds and multistage reactor systems show higher HDS activity and stability compared to conventional systems. By combining suitable catalyst replacement technologies, they can be used to achieve deep desulfurization required for upcoming environmental regulations.

REFERENCES

1. L.V. Angel, A.P. Mark, O.H. John et al., Hydrotreating catalysts and processes, in: *Encyclopedia of Chemical Processing* (print version), CRC Press, Boca Raton, FL, 2005, pp. 1357–1365.
2. R. Prins, Catalytic hydrodenitrogenation, in: *Advances in Catalysis*, Academic Press, San Diego, USA, 2001, pp. 399–464.
3. EPA, Control of diesel fuel quality, in: Fed. Reg., 1999, pp. 26142–26158.
4. M.o.E.C. Association, The impact of sulfur in diesel fuel on catalyst emission control technology, in: *Manufacturers of Emissions Controls Association*, M.o.E.C Association, Washington, DC.

5. G.B. McVicker, M. Daage, M.S. Touvelle et al., Selective ring opening of naphthenic molecules, *Journal of Catalysis*, 210 (2002) 137–148.
6. C. Song, Fuel processing for low-temperature and high-temperature fuel cells: Challenges, and opportunities for sustainable development in the 21st century, *Catalysis Today*, 77 (2002) 17–49.
7. K.O. Blumberg, M.P. Walsh, G. Pera, Low-sulfur gasoline & diesel: The key to lower vehicle emissions, United Nations Environment Programme, Napa, CA.
8. J.D. Miller, C. Facanha, The state of clean transport policy—A 2014 synthesis of vehicle and fuel policy developments, International Council on Clean Transportation, Washington, 2014.
9. S. Hatanaka, M. Yamada, O. Sadakane, Hydrodesulfurization of catalytic cracked gasoline. 1. Inhibiting effects of olefins on HDS of alkyl(benzo)thiophenes contained in catalytic cracked gasoline, *Industrial & Engineering Chemistry Research*, 36 (1997) 1519–1523.
10. W.C. Cheng, G. Kim, A.W. Peters et al., Environmental fluid catalytic cracking technology, *Catalysis Reviews*, 40 (1998) 39–79.
11. U. Alkemade, T.J. Dougan, New catalytic technology for FCC gasoline sulfur reduction without yield penalty, in: J.B. M. Absi-Halabi, A. Stanislaus (Eds.) *Studies in Surface Science and Catalysis*, Elsevier, Amsterdam, 1996, pp. 303–311.
12. S. Brunet, D. Mey, G. Perot, C. Bouchy, F. Diehl, On the hydrodesulfurization of FCC gasoline: A review, *Applied Catalysis A—General*, 278 (2005) 143–172.
13. R.J. Angelici, Heterogeneous catalysis of the hydrodesulfurization of thiophenes in petroleum—An organometallic perspective of the mechanism, *Accounts of Chemical Research*, 21 (1988) 387–394.
14. R.J. Angelici, An overview of modeling studies in HDS, HDN and HDO catalysis, *Polyhedron*, 16 (1997) 3073–3088.
15. I.V. Babich, J.A. Moulijn, Science and technology of novel processes for deep desulfurization of oil refinery streams: A review, *Fuel*, 82 (2003) 607–631.
16. B.P. Baruah, P. Khare, Pyrolysis of high sulfur Indian coals, *Energy & Fuels*, 21 (2007) 3346–3352.
17. M.J. Girgis, B.C. Gates, Reactivities, reaction networks, and kinetics in high-pressure catalytic hydroprocessing, *Industrial & Engineering Chemistry Research*, 30 (1991) 2021–2058.
18. M.L. Vrinat, The kinetics of the hydrodesulfurization process—A review, *Applied Catalysis*, 6 (1983) 137–158.
19. A. Syed Ahmed, Thermodynamics of hydroprocessing reactions, in: *Hydroprocessing of Heavy Oils and Residua*, CRC Press, Boca Raton, FL, 2007, pp. 51–69.
20. R. Shafi, G.J. Hutchings, Hydrodesulfurization of hindered dibenzothiophenes: An overview, *Catalysis Today*, 59 (2000) 423–442.
21. B.C. Gates, H. Topsøe, Reactivities in deep catalytic hydrodesulfurization: Challenges, opportunities, and the importance of 4-methyldibenzothiophene and 4,6-dimethyldibenzothiophene, *Polyhedron*, 16 (1997) 3213–3217.
22. G.H. Singhal, R.L. Espino, J.E. Sobel, Hydrodesulfurization of sulfur heterocyclic compounds: Reaction mechanisms, *Journal of Catalysis*, 67 (1981) 446–456.
23. H. Topsøe, B. Clausen, F. Massoth, Hydrotreating catalysis, in: J. Anderson, M. Boudart (Eds.) *Catalysis*, Springer, Berlin, 1996, pp. 1–269.
24. X. Ma, K. Sakanishi, I. Mochida, Hydrodesulfurization reactivities of various sulfur compounds in diesel fuel, *Industrial & Engineering Chemistry Research*, 33 (1994) 218–222.
25. P.T. Vasudevan, J.L.G. Fierro, A review of deep hydrodesulfurization catalysis, *Catalysis Reviews—Science and Engineering*, 38 (1996) 161–188.
26. H. Wang, E. Iglesia, Thiophene hydrodesulfurization catalysis on supported Ru clusters: Mechanism and site requirements for hydrogenation and desulfurization pathways, *Journal of Catalysis*, 273 (2010) 245–256.

27. B.C. Gates, J.R. Katzer, G.C.A. Schuit, Chemistry of catalytic processes, *AIChE Journal*, 25 (1979) 734–734.
28. F.E. Massoth, Studies of molybdena–alumina catalysts: VI. Kinetics of thiophene hydrogenolysis, *Journal of Catalysis*, 47 (1977) 316–327.
29. P. Fott, P. Schneider, Mass transport and a complex reaction in porous catalyst pellets: Thiophene hydrodesulphurization, *Chemical Engineering Science*, 39 (1984) 643–650.
30. S.K. Ihm, S.J. Moon, H.J. Choi, Hydrodesulfurization of thiophene over cobalt–molybdenum, nickel–molybdenum, and nickel–tungsten/alumina catalysts: Kinetics and adsorption, *Industrial & Engineering Chemistry Research*, 29 (1990) 1147–1152.
31. I.A. Van Parijs, L.H. Hosten, G.F. Froment, Kinetics of the hydrodesulfurization on a cobalt–molybdenum/gamma-alumina catalyst. 2. Kinetics of the hydrogenolysis of benzothiophene, *Industrial & Engineering Chemistry Product Research and Development*, 25 (1986) 437–443.
32. I.A. Van Parijs, G.F. Froment, Kinetics of hydrodesulfurization on a cobalt–molybdenum/gamma-alumina catalyst. 1. Kinetics of the hydrogenolysis of thiophene, *Industrial & Engineering Chemistry Product Research and Development*, 25 (1986) 431–436.
33. T.C. Ho, J.E. Sobel, Kinetics of dibenzothiophene hydrodesulfurization, *Journal of Catalysis*, 128 (1991) 581–584.
34. T. Isoda, S. Nagao, X. Ma, Y. Korai, I. Mochida, Hydrodesulfurization of refractory sulfur species. 1. Selective hydrodesulfurization of 4,6-Dimethyldibenzothiophene in the major presence of naphthalene over $CoMo/Al_2O_3$ and Ru/Al_2O_3 blend catalysts, *Energy & Fuels*, 10 (1996) 482–486.
35. T. Isoda, S. Nagao, X. Ma, Y. Korai, I. Mochida, Hydrodesulfurization of refractory sulfur species. 2. Selective hydrodesulfurization of 4,6-Dimethyldibenzothiophene in the dominant presence of naphthalene over ternary sulfides catalyst, *Energy & Fuels*, 10 (1996) 487–492.
36. J.G. Speight, Desulfurization chemistry, in: *The Desulfurization of Heavy Oils and Residua*, CRC Press, Boca Raton, FL, 1999, pp. 139–179.
37. C.N. Satterfield, G.W. Roberts, Kinetics of thiophene hydrogenolysis on a cobalt molybdate catalyst, *AIChE Journal*, 14 (1968) 159–164.
38. S. Morooka, C.E. Hamrin Jr, Desulfurization of model coal sulfur compounds by coal mineral matter and a cobalt molybdate catalyst—I: Thiophene, *Chemical Engineering Science*, 32 (1977) 125–133.
39. Y. Kawaguchi, I.G.D. Lana, F.D. Otto, Hydrodesulphurization of thiophene over a $NiO–MoO_3–Al_2O_3$ catalyst, *The Canadian Journal of Chemical Engineering*, 56 (1978) 65–71.
40. R. Bartsch, C. Tanielian, Hydrodesulfurization: I. Hydrogenolysis of benzothiophene and dibenzothiophene over $CoO–MoO_3/Al_2O_3$ catalyst, *Journal of Catalysis*, 35 (1974) 353–358.
41. D.R. Kilanowski, B.C. Gates, Kinetics of hydrodesulfurization of benzothiophene catalyzed by sulfided $Co–Mo/Al_2O_3$, *Journal of Catalysis*, 62 (1980) 70–78.
42. F.P. Daly, Hydrodesulfurization of benzothiophene over $CoO–MoO_3/Al_2O_3$ catalyst, *Journal of Catalysis*, 51 (1978) 221–228.
43. G.H. Singhal, R.L. Espino, J.E. Sobel, G.A. Huff Jr, Hydrodesulfurization of sulfur heterocyclic compounds: Kinetics of dibenzothiophene, *Journal of Catalysis*, 67 (1981) 457–468.
44. D.H. Broderick, B.C. Gates, Hydrogenolysis and hydrogenation of dibenzothiophene catalyzed by sulfided $CoO–MoO_3/\gamma\text{-}Al_2O_3$: The reaction kinetics, *AIChE Journal*, 27 (1981) 663–673.
45. I. Chorkendorff, J.W. Niemantsverdriet, Oil refining and petrochemistry, in: *Concepts of Modern Catalysis and Kinetics*, Wiley-VCH Verlag GmbH & Co. KGaA, Weinheim, 2005, pp. 349–376.

46. J.G. Speight, Desulfurization chemistry, in: *The Desulfurization of Heavy Oils and Residua*, CRC Press, Boca Raton, FL, 1999, pp. 180–217.
47. G.C.A. Schuit, B.C. Gates, Chemistry and engineering of catalytic hydrodesulfurization, *AIChE Journal*, 19 (1973) 417–438.
48. J.M.J.G. Lipsch, G.C.A. Schuit, The CoO–MoO_3/Al_2O_3 catalyst: I. Cobalt molybdate and the cobalt oxide molybdenum oxide system, *Journal of Catalysis*, 15 (1969) 163–173.
49. J.M.J.G. Lipsch, G.C.A. Schuit, The CoO–MoO_3/Al_2O_3 catalyst: II. The structure of the catalyst, *Journal of Catalysis*, 15 (1969) 174–178.
50. J.M.J.G. Lipsch, G.C.A. Schuit, The CoO–MoO_3/Al_2O_3 catalyst: III. Catalytic properties, *Journal of Catalysis*, 15 (1969) 179–189.
51. V.H.J. De Beer, T.H.M. Van Sint Fiet, J.F. Engelen et al., The CoO–MoO_3/Al_2O_3 catalyst. IV. Pulse and continuous flow experiments and catalyst promotion by cobalt, nickel, zinc, and manganese, *Journal of Catalysis*, 27 (1972) 357–368.
52. R.J.H. Voorhoeve, Electron spin resonance study of active centers in nickel–tungsten sulfide hydrogenation catalysts, *Journal of Catalysis*, 23 (1971) 236–242.
53. R.J.H. Voorhoeve, J.C.M. Stuiver, Kinetics of hydrogenation on supported and bulk nickel–tungsten sulfide catalysts, *Journal of Catalysis*, 23 (1971) 228–235.
54. P. Grange, Catalytic hydrodesulfurization, *Catalysis Reviews—Science and Engineering*, 21 (1980) 135–181.
55. G. Hagenbach, P. Courty, B. Delmon, Physicochemical investigations and catalytic activity measurements on crystallized molybdenum sulfide–cobalt sulfide mixed catalysts, *Journal of Catalysis*, 31 (1973) 264–273.
56. P. Grange, B. Delmon, The role of cobalt and molybdenum sulphides in hydrodesulphurisation catalysts: A review, *Journal of the Less Common Metals*, 36 (1974) 353–360.
57. H. Topsøe, R. Candia, N. Burriesci, B.S. Clausen, S. Mørup, The influence of the support on Co–Mo hydrodesulfurization catalysts, in: P.G.P.J. B. Delmon, G. Poncelet (Eds.) *Studies in Surface Science and Catalysis*, Elsevier, Amsterdam, 1979, pp. 479–492.
58. S.M.A.M. Bouwens, F.B.M. Vanzon, M.P. Vandijk et al., On the structural differences between alumina-supported CoMoS Type I and alumina-, silica-, and carbon-supported CoMoS Type II phases studied by XAFS, MES, and XPS, *Journal of Catalysis*, 146 (1994) 375–393.
59. H. Topsøe, B.S. Clausen, N.-Y. Topsøe, P. Zeuthen, Progress in the design of hydrotreating catalysts based on fundamental molecular insight, in: S.A.M.A.-H. D.L. Trimm, A. Bishara (Eds.) *Studies in Surface Science and Catalysis*, Elsevier, Amsterdam, 1989, pp. 77–102.
60. C. Wivel, R. Candia, B.S. Clausen, S. Mørup, H. Topsøe, On the catalytic significance of a Co–Mo–S phase in Co–Mo/Al_2O_3 hydrodesulfurization catalysts: Combined in situ Mössbauer emission spectroscopy and activity studies, *Journal of Catalysis*, 68 (1981) 453–463.
61. H. Topsøe, B.S. Clausen, R. Candia, C. Wivel, S. Mørup, In situ Mössbauer emission spectroscopy studies of unsupported and supported sulfided Co–Mo hydrodesulfurization catalysts: Evidence for and nature of a Co–Mo–S phase, *Journal of Catalysis*, 68 (1981) 433–452.
62. G.M. Dhar, B.N. Srinivas, M.S. Rana, M. Kumar, S.K. Maity, Mixed oxide supported hydrodesulfurization catalysts—A review, *Catalysis Today*, 86 (2003) 45–60.
63. K.V.R. Chary, H. Ramakrishna, K.S. Rama Rao, G. Murali Dhar, P. Kanta Rao, Hydrodesulfurization on MoS_2/MgO, *Catalysis Letters*, 10 (1991) 27–33.
64. K.C. Pratt, J.V. Sanders, V. Christov, Morphology and activity of MoS_2 on various supports: Genesis of the active phase, *Journal of Catalysis*, 124 (1990) 416–432.

65. S.K. Maity, M.S. Rana, B.N. Srinivas et al., Characterization and evaluation of ZrO_2 supported hydrotreating catalysts, *Journal of Molecular Catalysis A: Chemical*, 153 (2000) 121–127.
66. S. Srinivasan, A.K. Datye, C.H.F. Peden, The morphology of oxide-supported MoS_2, *Journal of Catalysis*, 137 (1992) 513–522.
67. A.K. Datye, S. Srinivasan, L.F. Allard et al., Oxide supported MoS_2 catalysts of unusual morphology, *Journal of Catalysis*, 158 (1996) 205–216.
68. A. Wang, Y. Wang, T. Kabe et al., Hydrodesulfurization of dibenzothiophene over siliceous MCM-41-supported catalysts: I. Sulfided Co–Mo catalysts, *Journal of Catalysis*, 199 (2001) 19–29.
69. K.M. Reddy, B. Wei, C. Song, Mesoporous molecular sieve MCM-41 supported Co–Mo catalyst for hydrodesulfurization of petroleum resids, *Catalysis Today*, 43 (1998) 261–272.
70. C. Song, K. Madhusudan Reddy, Mesoporous molecular sieve MCM-41 supported Co–Mo catalyst for hydrodesulfurization of dibenzothiophene in distillate fuels, *Applied Catalysis A: General*, 176 (1999) 1–10.
71. T. Chiranjeevi, P. Kumar, M.S. Rana, G. Murali Dhar, T.S.R. Prasada Rao, Physico-chemical characterization and catalysis on mesoporous Al-HMS supported molybdenum hydrotreating catalysts, *Journal of Molecular Catalysis A: Chemical*, 181 (2002) 109–117.
72. L. Vradman, M.V. Landau, M. Herskowitz et al., High loading of short WS2 slabs inside SBA-15: Promotion with nickel and performance in hydrodesulfurization and hydrogenation, *Journal of Catalysis*, 213 (2003) 163–175.
73. C. Pophal, F. Kameda, K. Hoshino, S. Yoshinaka, K. Segawa, Hydrodesulfurization of dibenzothiophene derivatives over TiO_2–Al_2O_3 supported sulfided molybdenum catalyst, *Catalysis Today*, 39 (1997) 21–32.
74. K. Segawa, S. Satoh, TiO_2-coated on Al_2O_3 support prepared by CVD method for HDS catalysts, in: G.F.F. B. Delmon, P. Grange (Eds.) *Studies in Surface Science and Catalysis*, Elsevier, Amsterdam, 1999, pp. 129–136.
75. S. Yoshinaka, K. Segawa, Hydrodesulfurization of dibenzothiophenes over molybdenum catalyst supported on TiO_2–Al_2O_3, *Catalysis Today*, 45 (1998) 293–298.
76. W. Zhaobin, X. Qin, G. Xiexian et al., Titania-modified hydrodesulphurization catalysts: I. Effect of preparation techniques on morphology and properties of TiO_2–Al_2O_3 carrier, *Applied Catalysis*, 63 (1990) 305–317.
77. E. Rodenas, T. Yamaguchi, H. Hattori, K. Tanabe, Surface and catalytic properties of TiO_2–Al_2O_3, *Journal of Catalysis*, 69 (1981) 434–444.
78. G.M.K. Abotsi, A.W. Scaroni, A review of carbon-supported hydrodesulfurization catalysts, *Fuel Processing Technology*, 22 (1989) 107–133.
79. G. Muralidhar, F.E. Massoth, J. Shabtai, Catalytic functionalities of supported sulfides. 1. Effect of support and additives on the CoMo catalyst, *Journal of Catalysis*, 85 (1984) 44–52.
80. F.E. Massoth, G. Muralidhar, J. Shabtai, Catalytic functionalities of supported sulfides. 2. Effect of support on Mo dispersion, *Journal of Catalysis*, 85 (1984) 53–62.
81. J.M. Oelderik, S.T. Sie, D. Bode, Progress in the catalysis of the upgrading of petroleum residue: A review of 25 years of R&D on Shell's residue hydroconversion technology, *Applied Catalysis*, 47 (1989) 1–24.
82. E. Furimsky, Selection of catalysts and reactors for hydroprocessing, *Applied Catalysis A—General*, 171 (1998) 177–206.
83. B.H. Cooper, B.B.L. Donnis, B.M. Moyse, Hydroprocessing conditions affect catalyst shape selection, *Oil & Gas Journal*, 8 (1986) 39–44.
84. H. Schulz, W. Böhringer, P. Waller, F. Ousmanov, Gas oil deep hydrodesulfurization: Refractory compounds and retarded kinetics, *Catalysis Today*, 49 (1999) 87–97.

85. V. La Vopa, C.N. Satterfield, Poisoning of thiophene hydrodesulfurization by nitrogen compounds, *Journal of Catalysis*, 110 (1988) 375–387.
86. M. Nagai, T. Kabe, Selectivity of molybdenum catalyst in hydrodesulfurization, hydrodenitrogenation, and hydrodeoxygenation: Effect of additives on dibenzothiophene hydrodesulfurization, *Journal of Catalysis*, 81 (1983) 440–449.
87. E. Furimsky, F.E. Massoth, Deactivation of hydroprocessing catalysts, *Catalysis Today*, 52 (1999) 381–495.
88. M. Nagai, T. Sato, A. Aiba, Poisoning effect of nitrogen compounds on dibenzothiophene hydrodesulfurization on sulfided NiMo–Al_2O_3 catalysts and relation to gas-phase basicity, *Journal of Catalysis*, 97 (1986) 52–58.
89. R. Ramachandran, F.E. Massoth, The effects of pyridine and coke poisoning on benzothiophene hydrodesulfurization over CoMo/Al_2O_3 catalyst, *Chemical Engineering Communications*, 18 (1982) 239–254.
90. Y. Yoshimura, E. Furimsky, Oxidative regeneration of hydrotreating catalysts, *Applied Catalysis*, 23 (1986) 157–171.
91. E. Furimsky, J. Houle, Y. Yoshimura, Use of boudouard reaction for regeneration of hydrotreating catalysts, *Applied Catalysis*, 33 (1987) 97–106.
92. Y. Yoshimura, N. Matsubayashi, H. Yokokawa et al., Temperature-programmed oxidation of sulfided cobalt-molybdate/alumina catalysts, *Industrial & Engineering Chemistry Research*, 30 (1991) 1092–1099.
93. D.S. Kim, K. Segawa, T. Soeya, I.E. Wachs, Surface structures of supported molybdenum oxide catalysts under ambient conditions, *Journal of Catalysis*, 136 (1992) 539–553.
94. S. Mayo, E. Brevoord, L. Gerritsen, F.L. Plantenga, Process ultra-low sulfur diesel, *Hydrocarbon Processing*, 2 (2001) 84A.
95. C.S. Song, An overview of new approaches to deep desulfurization for ultra-clean gasoline, diesel fuel and jet fuel, *Catalysis Today*, 86 (2003) 211–263.
96. ConocoPhillips, http://docslide.us/documents/s-zorb.html, viewed in 2015.
97. P. Zeuthen, FCC pre-treatment catalysts TK-558 BRIM and TK-559 BRIM for ULS gasoline using BRIM technology, in: *C. Haldor Topsøe A/S, Denmark* (Ed.), Haldor Topsoe A/S, Copenhagen, Denmark.
98. Haldor Topsoe, http://www.topsoe.com/products/tk-568, viewed in 2015.
99. Haldor Topsoe, http://www.topsoe.com/products/tk-570, viewed in 2015.
100. Haldor Topsoe, http://www.topsoe.com/sites/default/files/topsoe_clean_cat_he.ashx.pdf, viewed in 2015.
101. Haldor Topsoe, http://www.topsoe.com/products/tk-578, viewed in 2015.
102. Haldor Topsoe, http://www.topsoe.com/products/tk-609-hybrimtm, viewed in 2015.
103. Axens France, http://www.axens.net/product/catalysts-a-adsorbents/20068/hr-606.html, viewed in 2015.
104. Johnson Matthey, http://www.jmprotech.com/hydroprocessing-hytreat-refineries-johnson-matthey, viewed in 2015.
105. Criterion Catalysts & Technologies, http://s02.static-shell.com/content/dam/shell/static/criterion/downloads/pdf/2001-march-hce.pdf, viewed in 2015.
106. Süd-Chemie, http://www.sud-chemie-india.com/Sulfur-Recovery-Units/, viewed in 2015.
107. Grace Davison, https://grace.com/catalysts-and-fuels/en-us/fcc-catalysts/SuRCA, viewed in 2015.
108. P.H. Desai, L.A. Gerritsen, Y. Inoue, Low Cost Production of Clean Fuels with STARS Catalyst Technology, in: NPRA Annual Meeting, San Antonio, 1999.
109. S. Eijsbouts, F. Plantenga, B. Leliveld, Y. Inoue, K. Fukita, Keynote address: Stars and nebula—New generations of hydroprocessing catalysts for the production of ultra low sulfur diesel, *Abstracts of Papers of the American Chemical Society*, 226 (2003) U529–U529.

110. Exxon Mobil Research and Engineering, http://cdn.exxonmobil.com/~/media/global/files/catalyst-and-licensing/2014-1275-scanfininglartc.pdf, viewed in 2015.
111. C. Song, X. Ma, Desulfurization technologies, in: *Hydrogen and Syngas Production and Purification Technologies*, John Wiley & Sons, Inc., New Jersey, 2009, pp. 219–310.
112. J.V. Lauritsen, M. Nyberg, J.K. Norskov et al., Hydrodesulfurization reaction pathways on MoS_2 nanoclusters revealed by scanning tunneling microscopy, *Journal of Catalysis*, 224 (2004) 94–106.
113. Haldor Topsoe, http://www.topsoe.com/forums-research/research-activities-papers/hydroprocessing, viewed in 2015.
114. H.R. Reinhoudt, R. Troost, S. van Schalkwijk et al., Testing and characterisation of Pt/ASA for deep HDS reactions, *Fuel Processing Technology*, 61 (1999) 117–131.
115. H.R. Reinhoudt, R. Troost, A.D. van Langeveld et al., Catalysts for second-stage deep hydrodesulfurisation of gas oils, *Fuel Processing Technology*, 61 (1999) 133–147.
116. A.V. Sapre, B.C. Gates, Hydrogenation of aromatic hydrocarbons catalyzed by sulfided cobalt oxide–molybdenum oxide/alpha-aluminum oxide. Reactivities and reaction networks, *Industrial & Engineering Chemistry Process Design and Development*, 20 (1981) 68–73.
117. F.M. Nooy, Dense loading, *Oil & Gas Journal*, 82 (1984) 152–157.
118. F.A. Dautzenberg, A call for accelerating innovation, *Cattech*, 3 (1999) 54–63.
119. C.N. Satterfield, Trickle-bed reactors, *AIChE Journal*, 21 (1975) 209–228.
120. J.D. Carruthers, D.J. DiCamillo, Pilot plant testing hydrotreating catalysts: Influence of catalyst condition, bed loading and dilution, *Applied Catalysis*, 43 (1988) 253–276.
121. C.T. Adams, A.A. Del Paggio, H. Schaper, W.H.J. Stork, W.K. Shiflett, Hydroprocess catalyst selection: Tailoring and selecting catalyst systems for fixed-bed residue hydroprocessing made easier by a unified reactor mode, *Hydrocarbon Processing*, 68 (1989) 57–61.
122. W.C. van Zijll Langhout, C. Ouwerker, K.M.A. Pronk, New process hydrotreats metal-rich feedstocks, *Oil & Gas Journal*, 78 (1980) 120.
123. A.C. Jacobsen, B.H. Cooper, P.N. Hannerup, in: 12th World Petroleum Congress, 1987, p. 97.
124. J.P. Peries, P. Renard, T. Des Courieres, J. Rossarie, in: Fourth UNITAR/UNDP International Conference on Heavy Crudes and Tar Sands, Edmonton, Canada, 1988, p. 21.
125. K.W. Robschlager, W.J. Deelen, J.E. Naber, in: Fourth UNITAR/UNDP International Conference on Heavy Crudes and Tar Sands, Edmonton, Canada, 1988, p. 249.

5 Recent Developments in Hydrotreatment Catalysts—Patent and Open Literature Survey*

Burcu Yüzüak

CONTENTS

5.1 HYDROTREATMENT TECHNOLOGY IN PERSPECTIVE

The increasing attention for reducing emissions, which are derived from fuel sources, and the increasing quantity of crude oil with heavy components have resulted in improvement of hydrotreating technology. In order to ensure emission regulations and remove hazardous contents of fuels such as sulfur, nitrogen, olefins, and aromatics, hydrotreating technology has a wide range of applications. In addition to the aim of reducing emission levels, there is also a need for accelerated demand of hydrotreating technology since the processed feedstocks get heavier day by day.

* The brief history overview of the hydrotreating technology has been performed by using five literature sources, namely *Practical Advances in Petroleum Processing* (Robinson P. R. and Dolbear G. E.); *Handbook of Petroleum Processing* (Gruia A.); *Handbook of Refinery Desulfurization* (El-Gendy N. S. and Speight J. G.); *Heterogeneous Catalysis and Solid Catalysts* (Deutschmann O., Knözinger H., Kochloefl K., and Turek T.); and *Refining Processes Handbook* (Parkash S.) [1–5].

Hydrotreating technology is used to effect conversion of heavy feedstocks to products with lower molecular weights, to treat feedstocks for processes such as hydrocracking, and/or to enhance the quality of finished products. The main objectives of the hydrotreating process are removing impurities such as metal, sulfur, nitrogen, oxygen, and aromatic compounds from feedstocks, and decreasing the density of the feedstock by incorporating hydrogen to the structure of the feedstock.

Hydrotreating units in refineries are designed to run at a wide range of process conditions according to the type of feed, required cycle length, and desired quality of the products. In a hydrotreating process, the preheated feedstock is reacted with hydrogen with the partial pressure of 14.28–140.72 kg/cm^2, in the range of 290–425°C of temperature and 296–433 kg/cm^2 of pressure, while flowing through fixed beds of a hydrotreating catalyst bed that mainly contains cobalt–molybdenum (Co–Mo) or nickel–molybdenum (Ni–Mo) on γ-alumina (γ-Al_2O_3) support according to the severity of process conditions [1]. Since hydrotreating reactions are exothermic, in many commercial units, quench flows are introduced to cool down the stream and accelerate the desired reactions. The ranges of the process conditions of hydrotreating units to obtain expected productivity are given in Table 5.1. The limits of the operating conditions vary according to the feedstock type and the desired reaction efficiencies [2,3].

In 1878, thanks to the invention of Sabatier and Senderens, it has been found that unsaturated hydrocarbons in the vapor phase could be converted into saturated hydrocarbons over a nickel catalyst by using hydrogen. Following Sabatier and Senderens's study, in 1904, Ipatief realized that the range of applicable hydrogenation reactions can be expanded by introducing hydrogen in increased pressure levels. The possibility of removing sulfur species and metal (ferric oxide) contaminants from feedstock was identified by Bergius's invention in 1910. In 1927, the first commercial hydrogenation unit that is used for hydrogenating brown coal was implemented in Germany. Although hydrogenation technology was attracting petroleum industry's attention, commercial application was limited because of the high cost of hydrogen production and lack of high-pressure processing knowledge.

TABLE 5.1
Hydrotreating Process Operating Conditions

Liquid hourly space velocity (LHSV) (1/h)	0.2–10
H_2 circulation rate (Nm^3/m^3)	50–844
H_2 partial pressure (kg/cm^2)	14.28–140.72
SOR operating temperature (°C)	290–425

Sources: Adrian Gruia, *Handbook of Petroleum Processing*, Springer, Dordrecht, The Netherlands, 2006, pp. 321–322; Nour Shafik El-Gendy, James G. Speight, *Handbook of Refinery Desulfurization*, CRC Press, Boca Raton, FL, 2016, p. 188.

Thanks to the invention of catalytic reforming technology in the early 1950s, the attention on hydrotreating technology increased since reforming is the major hydrogen-producing process [2,4].

5.2 HYDROTREATMENT PROCESS CHEMISTRY AND CATALYSTS IN PERSPECTIVE

Catalysts for hydrotreating reactions, including hydrodesulfurization (HDS), hydrodenitrogenation (HDN), hydrodeoxygenation (HDO), and hydrodemetallization (HDM), mainly consist of dispersed active metal sulfides such as MoS_2 (15–20 wt.%), which are supported on γ-Al_2O_2. Alumina is known to be the best alternative due to its surface area of 200–300 m^2/g, high mechanical strength, and low cost. In some cases, a mixture of alumina and silica can also be used as a support material of hydrotreating catalysts. Besides molybdenum, nickel and cobalt (1–5 wt.%) are also used as promoters on the metal phase of hydrotreating catalysts. Molybdenum catalysts are typically promoted with nickel or cobalt. Hydrogenation activity of catalysts according to the metal phase combinations is as follows: Pd, Pt > Ni–W > Ni–Mo > Co–Mo. At the end of the catalyst preparation step, calcination, metals on the hydrotreating catalyst are in oxide form; therefore, the catalyst presulfiding stage is required to obtain metal sulfides, which are known to be the active phase on the hydrotreating catalyst [1].

Although the reactions on the hydrotreating catalyst are not completely clarified, various reactions have been proposed. The removal of sulfur compounds is one of the most required reactions in refinery operations, since the sulfur content of crude oil has important effects on refining. Furthermore, most catalysts that are used for the processing of fuel products cannot handle a compound that contains sulfur and metals. In HDS reactions, the sulfur content of organic sulfur compounds such as mercaptans, sulfides, disulfides, thiophenes, benzothiophenes, and dibenzothiophenes is converted into H_2S. The level of difficulty of HDS reactions depends on the type of sulfur compound. As the complexity of compounds decreases, the ease of sulfur removal increases. The ranking on the basis of ease of removal according to the type of sulfur compounds is as follows: mercaptans, sulfides, disulfides, thiophenes, benzothiophenes, and dibenzothiophenes. Removal of nitrogen is more difficult to overcome than HDS. HDN is also important for many petroleum streams, since nitrogen compounds inhibit the acidic function of the catalyst. Therefore, nitrogen removal efficiency effects further reactions such as hydrocracking. Besides nitrogen-containing compounds, metal contents also damage catalyst surface by being deposited on the hydrotreating catalyst. HDM reactions are used to remove metal contents such as vanadium and nickel. Like the metal-containing compounds, olefinic hydrocarbons affect the activity of the hydrotreating catalysts. Removal of the olefinic compounds are used to decelerate formation of coke deposits on the catalysts. Hydrodeoxygenation (HDO) reactions are used to remove phenols and/or peroxides in the feedstock by releasing water. Chemical reactions that take place in hydrotreating processes are shown in Table 5.2 [1,2,5].

TABLE 5.2
List of Hydrotreating Reactions

Reaction Type	Illustration	ΔH_R (kj/Nm³)
Hydrodesulfurization	$R\text{-}S\text{-}R' + 2H_2 \rightarrow RH + R'H + H_2S$	–2.5 to –3.0
Hydrodenitrogenation	$R{=}N\text{-}R' + 3H_2 \rightarrow RH + R'H + NH_3$	–2.5 to –3.0
Hydrodeoxygenation	$R\text{-}O\text{-}R' + 2H_2 \rightarrow RH + R'H + H_2O$	–2.5 to –3.0
Hydrodemetallization	$R\text{-}M + ½H_2 + A \rightarrow RH + M\text{-}A$	–3.0
Saturation of aromatics	$C_{10}H_8 + 2H_2 \rightarrow C_{10}H_{12}$	–3.0
Saturation of olefins	$R{=}R' + H_2 \rightarrow HR\text{-}R'H$	–5.5
Isomerization	$n\text{-}RH \rightarrow i\text{-}RH$	n/a

Source: Paul R. Robinson, Geoffrey E. Dolbear, *Practical Advances in Petroleum Processing*, Springer, New York, 2006, Volume 1, pp. 182, 196, 261–263, 335–336.

Note: A: metals-adsorbing material; M: Fe, Ni, or V; R: alkyl.

5.3 ANALYSIS BASED ON PATENT APPLICANTS

In this section, developments of hydrotreatment catalysts between 2008 and 2014 have been investigated. The total number of investigated patents related to "hydrotreatment" is 215; among these patents, 128 are related to "hydrotreatment catalysts," 15 are related to "regeneration of hydrotreating catalysts," and the rest, 72 of them, investigate process-based developments of hydrotreating technologies. The total numbers indicate that most of the studies related to hydrotreating focused on catalyst preparations rather than process-based technologies [171–220]. During the study, in order to cover the most updated technologies relevant to the development, date analysis has been conducted by using the latest issue date of each patent.

The annual distribution of hydrotreatment patents between 2008 and 2014 in terms of "HT catalyst preparation," "HT process," and "HT catalyst regeneration" is shown in Figure 5.1. Not including 2008, the ranking of the studies is as follows: "catalyst preparation," "process," and "regeneration." Additionally, it can be seen that the number of the patents related to the aforementioned HT studies has gained most of the attention in 2013. The patents that are related to "HT catalysts" are going to be thoroughly investigated for the following analyses.

Analysis of HT patent distributions in terms of applicants is given in Figure 5.2. The results indicate that the major portion of the studies have been conducted by refineries covering 28% of the total count. The second major portion of the studies (18%) have been conducted by universities, research institutes, and R&D companies. Patent applications in collaboration with each other consist of 49% of the total patents. The results indicate that most of the remaining studies involve collaboration of universities, research institutes, R&D companies, and refineries.

Distributions of HT patents in terms of total numbers by countries of the applicants can be seen in Figure 5.3. Most of the patents have been published in China, followed by the United States. France, Japan, and the Netherlands, which are the countries that contribute to the patents of HT catalysts substantially. As shown in

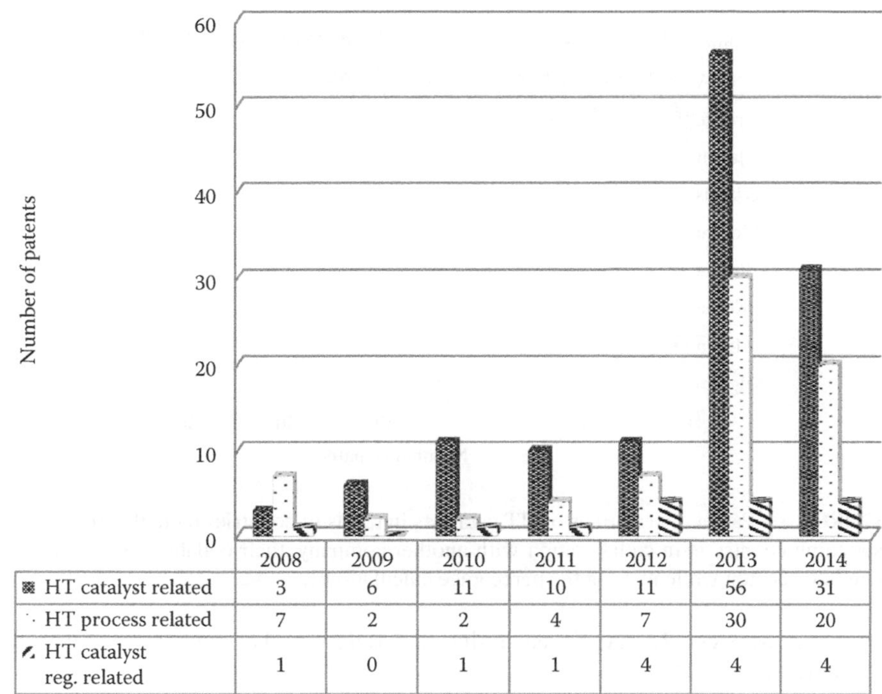

	2008	2009	2010	2011	2012	2013	2014
HT catalyst related	3	6	11	10	11	56	31
HT process related	7	2	2	4	7	30	20
HT catalyst reg. related	1	0	1	1	4	4	4

FIGURE 5.1 Annual distribution of HT patents between 2008 and 2014.

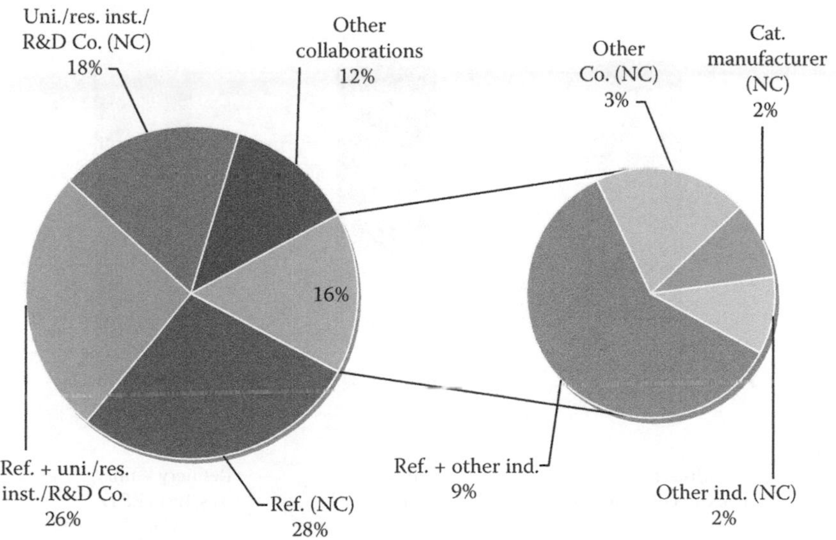

FIGURE 5.2 Patent distributions of HT catalysts based on applicants (ind.: ındividuals; NC: no collaborator; ref.: refinery).

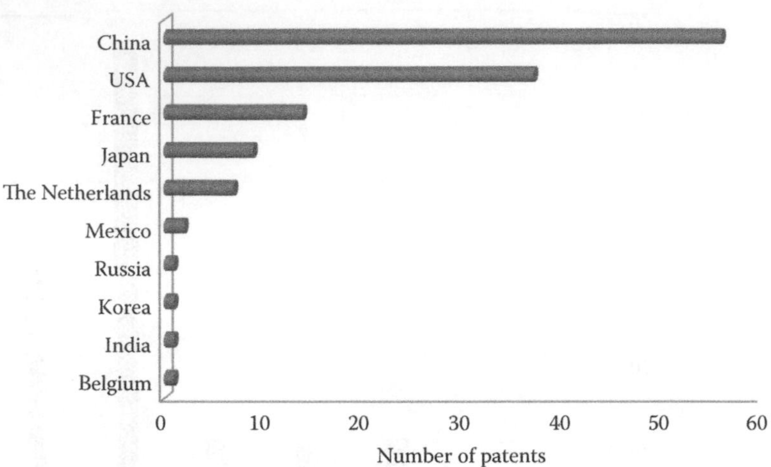

FIGURE 5.3 Patent distributions of HT catalysts in terms of countries (note that if the corresponding country is in collaboration with another company, their collaborative studies is taken into account while making the percentage calculations).

Figure 5.3, Mexico, Russia, Korea, India, and Belgium also published patents of HT catalysts.

Collaborative studies have been conducted in Asia among universities, research institutes, R&D companies, and refineries, as can be seen in Figure 5.4. Universities, research institutes, and R&D companies mainly conducted collaborative studies with refineries in Asia, whereas they tend to study without any collaborator in Europe.

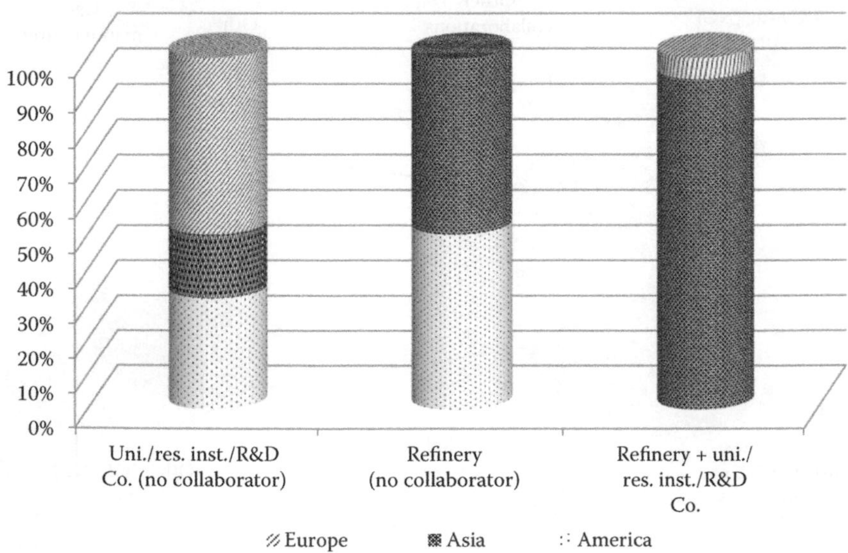

FIGURE 5.4 Continents of the applicants who have a patent in the general topic of HT catalysts.

Refineries that have patent application related to HT catalysts are located evenly in Asia and America.

5.3.1 Involvement of Refineries

Refineries that have published patents related to HT catalysts are mainly Sinopec, Shell, and Chevron, as can be seen in Figure 5.5. Sinopec, the company known also as China Petroleum & Chemical, China Petrochemical, and Petrochina, has published more patents compared to other companies with 45% of the total count of the patents.

5.3.2 Involvement of Universities/Research Institutes/R&D Companies

Patent distribution analyses of HT catalysts with respect to universities/research institutes/R&D companies are shown in Figure 5.6. Results show that the Research Institute of Petroleum Processing, Sinopec plays a dominant role in the patents of HT catalysts. The Research Institute of Petroleum Processing, Sinopec is followed by IFP Energies Nouvelles (17%), Shell Int. Research (14%), ExxonMobil Res. & Eng. Co. (8%), and Univ. China Petroleum (8%). The rest belongs to CNOOC Tianjin Chem Res. & Des., Univ. Fuzhou, Advanced Refining Technologies LLC, Beijing Res. Inst. Chem. Ind, CT DE Investigacion En Materiales Avanzados SC, Shanxi Coal Chem Inst., and Univ. Nac Autónoma De México.

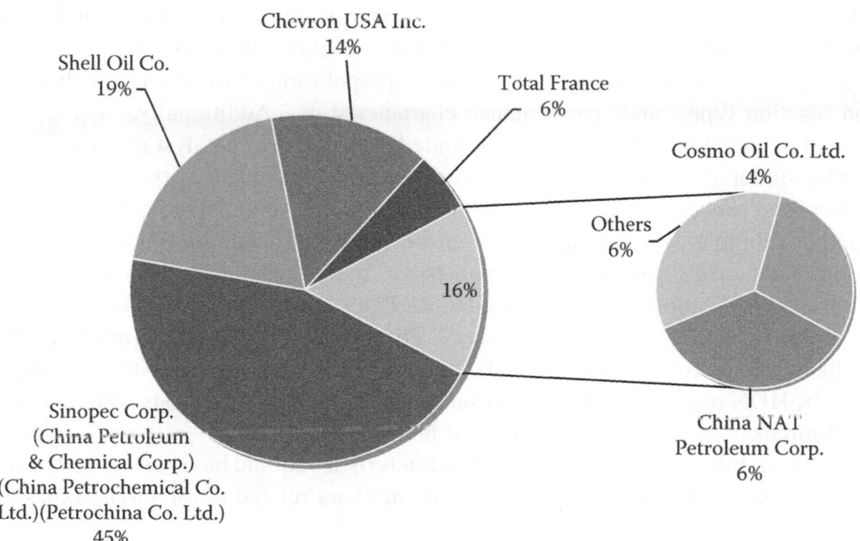

FIGURE 5.5 Parents applications of HT catalysts as refineries (note that if the corresponding company is in collaboration with another company, their collaborative studies is taken into account while making the percentage calculations).

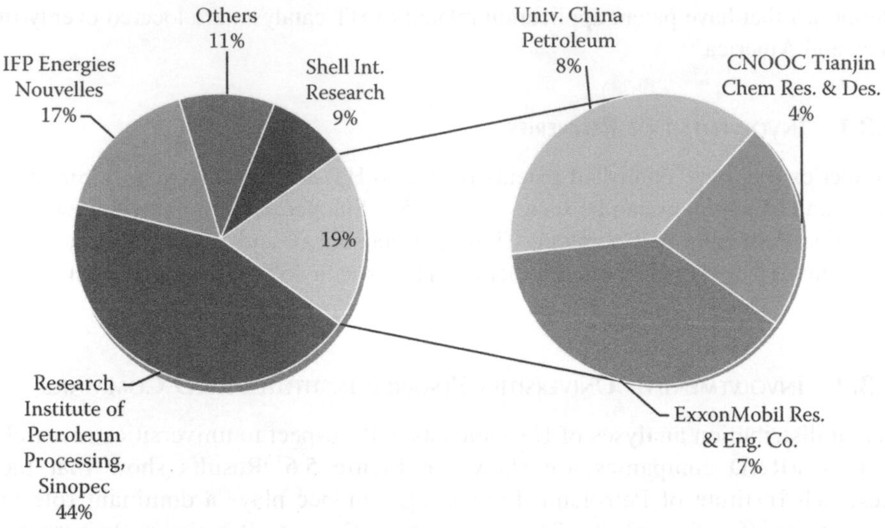

FIGURE 5.6 Patent applications of HT catalysts with respect to universities/research institutes/R&D companies (note that if the corresponding company is in collaboration with another company, their collaborative studies is taken into account while making the percentage calculations).

5.4 ANALYSIS BASED ON RESEARCH AREA

Patents of HT catalysts have been categorized in terms of main topics and stated targets as shown in Table 5.1. The main topics include "feedstock-based technology," "cost-effective, efficient, or alternative preparation of material," "hydrogenation reaction type," and "performance characteristics." Additionally, each topic is divided into subtopics based on mostly stated studies in HT catalyst patents.

The subtopic of "feedstock-based technology" is divided to the target names "heavy feed processing," "middle distillate processing," and "light feed processing." Another subtopic, which is named as "cost-effective, efficient, or alternative preparation of material," includes target names as "metal promoter addition," "support material modification," and "cost reduction." Patents are also classified according to the topic of "hydrogenation reaction type." The related target headings are defined as "reduction of sulfur components by HDS reactions," "reduction of nitrogen components by HDN reactions," and "reduction of metals by HDM reactions." The subtopic "performance characteristics" is divided into the target names "activity," "selectivity," and "stability," since performance characteristics should be investigated according to these parameters. Referenced patent numbers related to the target topics are given in Table 5.3.

During the classification of HT patents, it has been observed that each of the target subtopics can be interrelated. For instance, most of the catalyst preparation for heavy feed processing studies includes support material modification. Metal promoter addition affects the performance of HDS, HDN, and HDM reactions.

TABLE 5.3
Referenced Patent Numbers Related to the Target Topics

Topic	Target	Referenced Patents
Feedstock-based technology	Heavy feed processing	[6–91]
	Middle distillate processing	[12,19,21,24,28,30,33,38,40–44,48,49,51,52,56,70,72, 77,79,81,87–89,91–114]
	Light feed processing	[12,24,28,33,38,48,51,70,72,77,79,87,89,95,114–121]
Cost-effective, efficient, or alternative preparation of material	Metal promoter addition	[10,11,16,19,24,28,30–32,35,39,42,46,48,54,62,70,73, 74,79,89,92,94,99,100,103,104,110,115,119,120,122, 123–141]
	Support material modification	[7,9,12,17,18,22,24,27,29–32,34–37,39–41,46,47,50, 54,61–65,67,70–72,81–84,88,90,92,93,94,96,97, 99–101,104,106–108,110,114,115,119,120,122, 123–126,130–134,139–156]
	Cost reduction	[35,36,42,54,71,90,93,98,105,113,122,123,142,157–160]
Hydrogenation reaction type	Reduction of sulfur components by HDS reactions	[6,9,11,13–16,23,24,28–30,36,37,39–43,45–51,54, 56,60,61,65,67,69,70,72,73,75,77–79,81–82,84–88, 90,91,93,94,96–108,111,114,115,117,118,120,121, 123,127,128,132,137,141,143,152,155,161–165]
	Reduction of nitrogen components by HDN reactions	[9,13–15,19,24,29,32,35–37,39–43,46,48,49,51,56,60, 65,67,69,70,72–75,77,79,80,82,83,85,87,88,91,96, 98,100,101,104,118,120,127,136–138,141,152,161, 162,164]
	Reduction of metals by HDM reactions	[10,11,15–18,20,22,23,34,36,43,57,61,65,67,73, 84,89,114,115,146,155]
Performance characteristics	Activity	[6,7,9,13–16,18–21,23–33,35–43,46–51,53–56,59,62, 63,65–67,69–73,75–79,81,85,87–90,92–94,97–104, 107,108,110,111,113,115,117–120,122,123–128,130, 132,133,135–138,141,147–151,153,154,156,161, 162,164,169–171]
	Selectivity	[29,54,90,110,117,120,129,145]
	Stability	[6–8,10,13–15,17,19,25,26,32,34–36,40–42,50,53,55,59, 60,63,64,66,69,72,78,84,92,95,96,119–121,126,128,131, 133,142,143,146,149,152,156,166,167]

The distribution of patents for HT catalysts in terms of mostly stated topics is given in Figure 5.7. Developments related to activity, selectivity, and stability improvements almost cover most of the investigated patents for HT catalysts with 29%. This is followed by developments for feedstock-based technologies and hydrogenation reaction type targeted patents, each having 26%. Nineteen percent of the investigated patents for HT catalysts include technologies based on describing cost-effective, efficient, or alternative preparation of material.

Figure 5.8 shows studies of mostly encountered applicants in terms of the main topics. Developments for "feedstock-based technologies," "cost-effective, efficient,

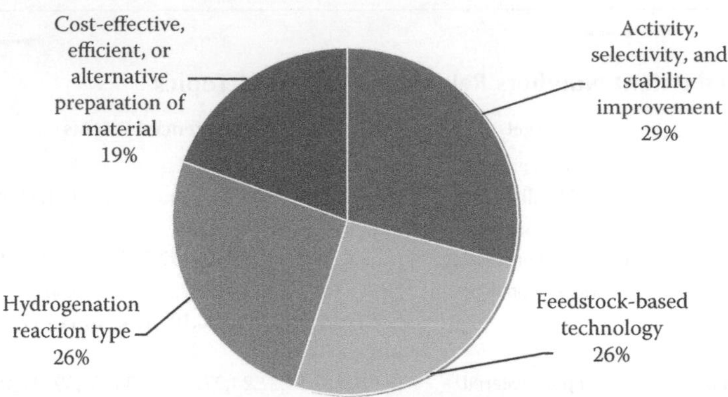

FIGURE 5.7 Patent distributions of HT catalysts based on mostly stated topics.

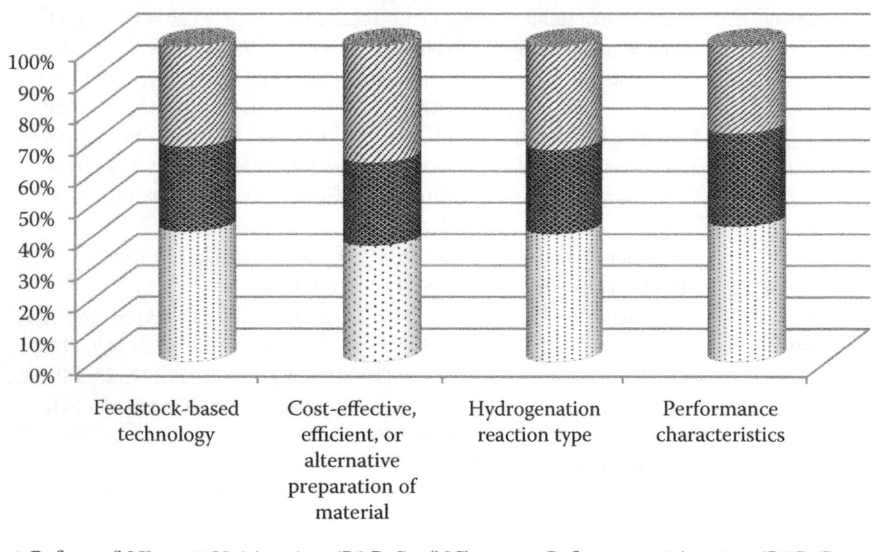

FIGURE 5.8 Topic analysis with respect to the applicants (NC: no collaborator; Co.: company).

or alternative preparation of material," "type of hydrogenation reaction," and "performance characteristics" are almost equally divided for each of the most encountered applicants, namely refineries without a collaborator, universities/research institutes/R&D companies without a collaborator, and collaborative studies of both of the applicants. Almost 40% of each of the three topics are the studies conducted by refineries. Approximately each remaining 30% of the total counts for three of the topics are patents of universities/research institutes/R&D companies without a collaborator and collaboration of refineries with universities/research institutes/R&D companies.

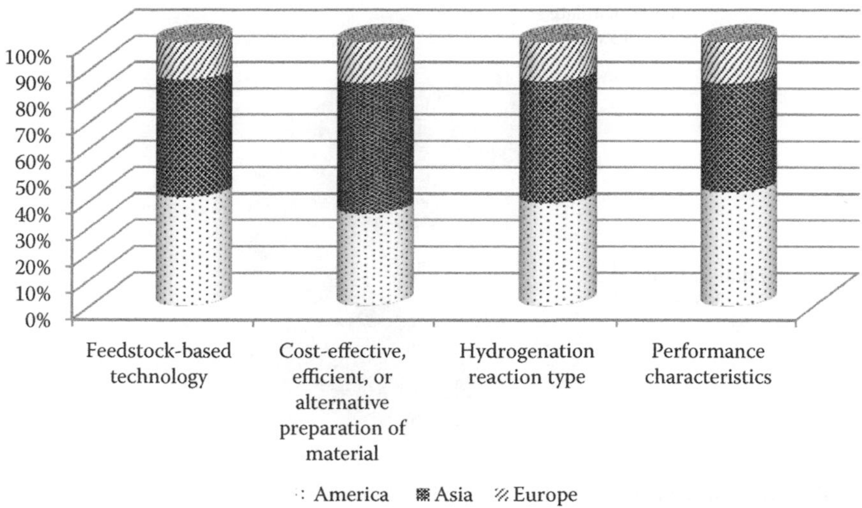

FIGURE 5.9 Topic analysis with respect to continents of the applicants.

The distribution of the main patent topics for HT catalysts, given in Figure 5.9, based on the continent of the patent applicant indicates that each of the topics is investigated in similar percentages for America, Asia, and Europe. Asia has the highest percentage for "feedstock-based technologies," "cost-effective, efficient, or alternative preparation of material," "reaction type, and "performance characteristics." Asia is followed by the studies of America in terms of patent numbers. When compared with Asia and America, the least number of patents is found in Europe for three of the investigated topics in terms of HT catalysts.

5.4.1 Developments in Terms of Feedstock-Based Technology

The subtopic distribution of patents related to feedstock-based technologies is given in Figure 5.10. Since the content of heavier feeds in terms of sulfur, nitrogen, and metal is at higher levels, most of the patent studies of HT catalysts focus on treatment of heavier feeds. Secondly, due to the depletion of lighter feeds, refineries are to process heavier feeds. As a combinatorial result, more than half of the studies (55%) aim at processing heavy feeds. Hydrotreatment of middle distillates constitutes 30% of the studies within the patents related to HT catalysts. The remaining 15% of the corresponding studies are for treatment of lighter distillates.

Patents with the corresponding applicants in the subtopic "heavy feed processing" are mainly applied by companies such as Chevron, ExxonMobil, Shell, IFP, and China Petroleum and Chemical. Shell, ExxonMobil, and Total France have a higher number of patents for hydrotreatment catalysts that are used for middle distillate processing.

Further breakdown given in Figure 5.11 shows the patent studies for processing heavy feeds in terms of the modifications applied to synthesize the desired HT catalyst. Pore diameters should be large enough for the heavier feeds to interact

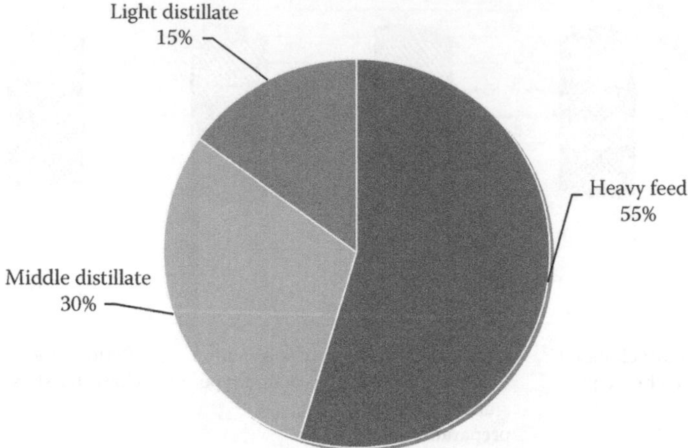

FIGURE 5.10 Developments in terms of feedstock-based technology.

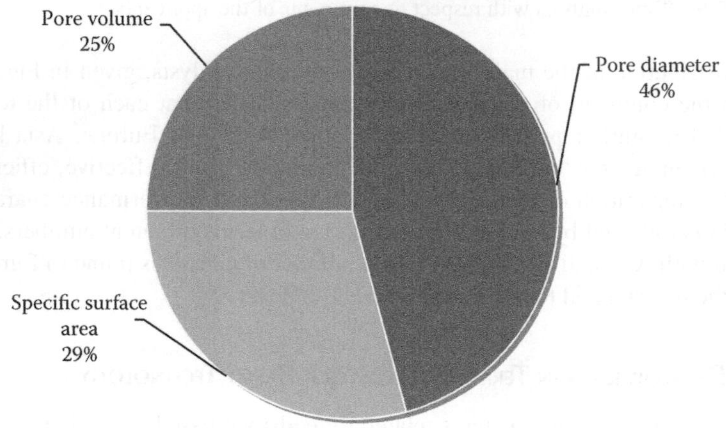

FIGURE 5.11 Catalyst modification studies to obtain heavy feed processing catalysts.

with the active sites for hydrogenation reactions. Therefore, most of the catalyst modification studies (46%) [11,12,17,18,22,24,29,36,46,47,50,61–64,67,72,73,82–84,99,107,114,120,127,134,136,137,139,142,156,158,159] within the investigated patents describe methods to enlarge the pore diameter. Moreover, the pore diameter expansion process has also positive effect on stability, since larger pores delay the coke accumulation, which results in blocking of the pore mouths. Twenty-nine percent of the studies focus on modifications for specific surface area [11,12,17,19,27,31,46,62,67,72,73,81,82,99,119,120,127,129,138,139,142,146,156]. Since larger specific surface area enables loading of more metals, which are known to be hydrogenation agents within the HT catalyst content, the hydrogenation activity of the catalyst increases. Pore diameter modifications also result in changes in the pore volume [11,12,17,18,36,46,61,62,72,73,88,99,114,120,136,142,146,156].

5.4.2 Developments in Terms of Cost-Effective, Efficient, or Alternative Preparation of Material

Subtopic distributions of patents that are related to material preparation in terms of cost-effective, efficient ways or alternative methods for synthesis of HT catalysts are given in Figure 5.12. The highest proportion of the patent studies is for synthesis of support. Since the number of patent studies is higher for heavy feeds and also the modifications mainly have a focus on support-related characteristics such as pore diameter, specific surface area, and pore volume, the percentage of the support modifications is more than half of the studies in the corresponding category (56%) [62–65,67,70–72, 81–84,88,90,92–94,96,97,99–101,104,106–108,110,114,115,119–122,126,127,129–132,136–140,142,145–160]. Since the hydrogenation active component of the catalyst is within the metal phase, there are many patents (34%) [10,11,16,19,24,28,30–32,35,39,42,46,48,54,62,70,73,74,79,89,92,94,99,100,103,104,110,115,119, 120,122,126,130–147] describing the process of synthesis in terms of metal phase.

Further breakdown for the applications described within the investigated patents in terms of metal phase and support material modifications is given in Figure 5.13. Specific surface area has high importance since it enables loading of active metal sites properly. Addition of phosphorous reduces the interaction between support and metal phase and increases the activity. It also enables uniform metal phase distribution on the catalyst surface, which is responsible for the enhancement of hydrogenation activity of the hydrotreating catalysts [19,24,30,35,46,50,62,71,73, 83,119,120,130,134,135,137,142,147,149,156]. Besides improving metal phase distribution, boron addition also increases acid sites [7,24,71,96,130,157]. Carbon addition is used for obtaining bimodal pore distribution, wherein the macropores are ensured by the incorporation of carbon-containing materials into support material [46,84,110,150,153]. Following the carbon addition process, carbon-containing substances are removed by heating or alternative processes. Mechanical strength

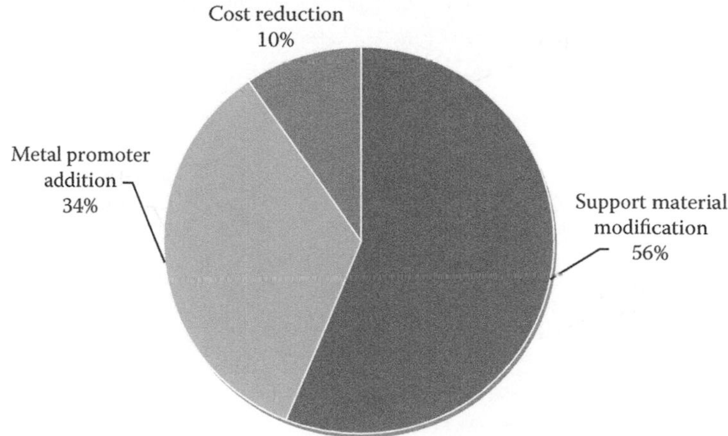

FIGURE 5.12 Target breakdown of "cost-effective, efficient, or alternative preparation of material" used for HT catalysts.

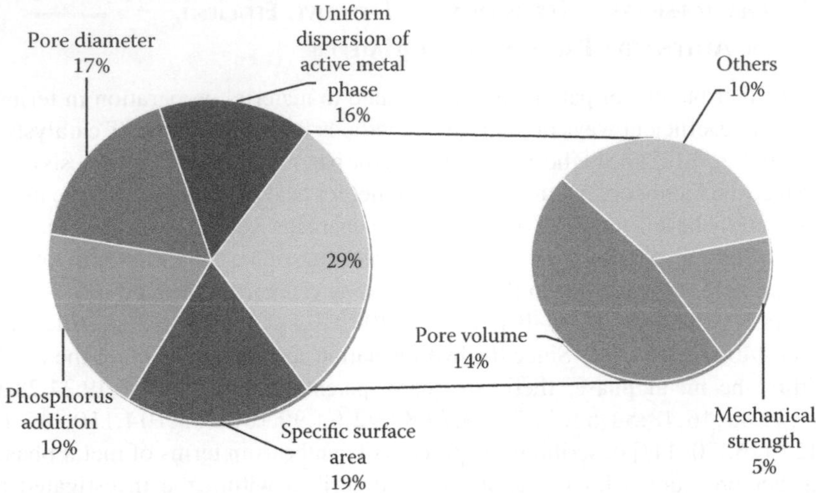

FIGURE 5.13 Applications in metal phase and support material modifications.

improvements result in the increase in stability of the catalyst [120,127,137,139,160]. Carbon and boron are the contents of the "others" group in Figure 5.13.

5.4.3 Developments in Terms of Hydrogenation Reaction Type

Figure 5.14 shows the distribution of the patents for HT catalysts in terms of hydrogenation reaction type. Half of the patent studies of the corresponding category describe catalyst formulation for HDS reaction (51%) due to the strict regulations for reduction of sulfur emissions. Thirty-two percent of the investigated patents within this category are related to HDN reactions and the remaining 17% with HDM reactions.

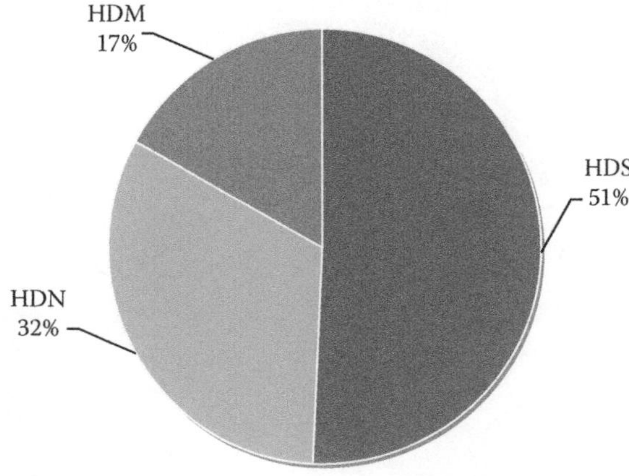

FIGURE 5.14 Distribution of HT catalyst studies on hydrogenation reactions: HDS, HDN, and HDM.

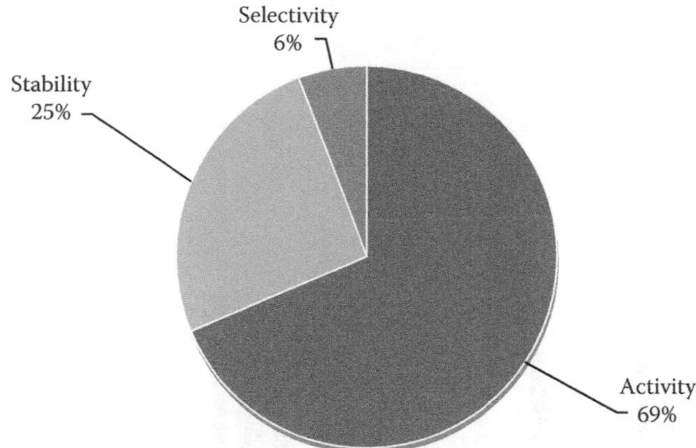

FIGURE 5.15 HT catalyst studies on activity, selectivity, and stability improvement.

5.4.4 Developments in Terms of Performance Characteristics

The distribution of the patent studies concerning the catalytic performance characteristics is given in Figure 5.15 in terms of breakdown for activity, selectivity, and stability. Most of the studies describe catalyst formulations for achieving higher activity. Selectivity-targeted studies [29,54,90,110,117,120,135,149] have the least percentage (6%) when compared to activity studies (69%) for increasing hydrogenation and stability studies (25%) [6–8,10,13–15,17,19,25,26,32,34–36, 40–42,50,53,55,59,60,63,64,66,69,72,78,84,92,95,96,119–121,127,129,132,134,137, 139,150,156,160,169,170] for increasing the cycle length of the HT catalyst.

5.5 FUTURE PROSPECTS IN HYDROTREATMENT TECHNOLOGIES

The trend of processing more heavy and sour crude oil and compelling regulations related to fuel qualities evokes developing high-efficiency hydrotreating catalysts and processes. In addition, environmental regulations promote refineries for producing fuels with close to zero content of sulfur. Therefore, industry is seeking for advanced hydrotreating catalysts to keep up with the competitive environment. Hydrotreating processes need a new catalyst with increased activity mainly for processing heavy feedstocks and ultradeep desulfurization of diesel fuel. The increasing content of heavier feedstocks accelerates studies on new active phases and new carriers. For example, in order to be able to process heavier feedstocks and suppress coke formations, studies mainly focus on surface modifications such as expansion of pore structure [11,12,17,18,22,24,29,36,46,47,50,61–64,67,72,73,82–84,99,107,114,120,127,134,136,137,139,142,156,158,159]. In addition to the studies on the modification of catalyst surface, alternative metal contents are introduced to the metal phase of the catalyst in order to obtain improved hydrogenation performance [10,11,16,19,24,28,30–32,35,39,42,46,48,54,62,70,73,74,79,89,91,94,99,100, 103,104,110,115,119,120,122,126,130–147]. The annual distribution of the investigated patents between 2008 and 2014 in terms of the main topics is given in Figure 5.16.

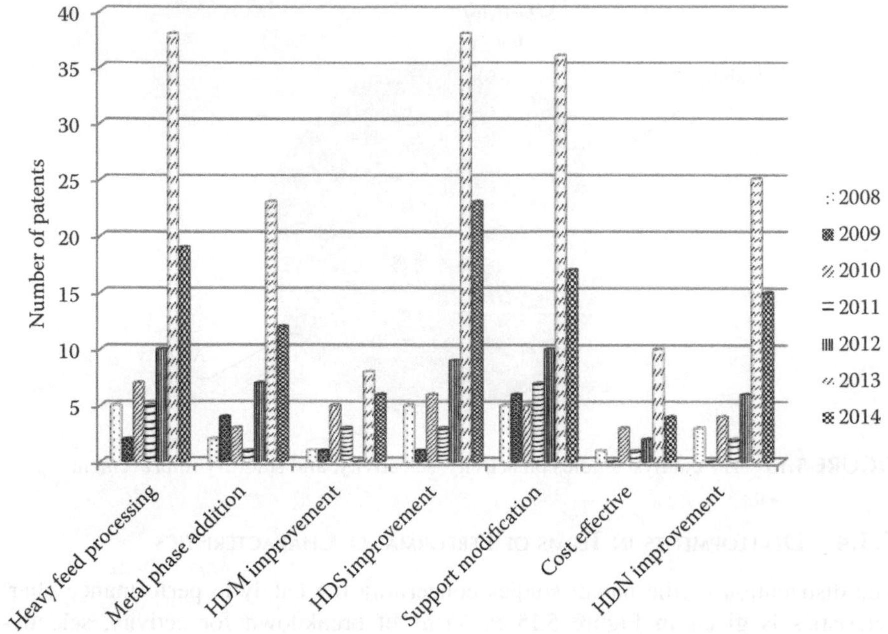

FIGURE 5.16 Annual distributions of the investigated patents between 2008 and 2014 in terms of the main topics.

REFERENCES

1. Paul R. Robinson, Geoffrey E. Dolbear, *Practical Advances in Petroleum Processing*, Springer, New York, 2006, Volume 1, pp. 182, 196, 261–263, 335–336.
2. Adrian Gruia, *Handbook of Petroleum Processing*, Springer, The Netherlands, 2006, pp. 321–322.
3. Nour Shafik El-Gendy, James G. Speight, *Handbook of Refinery Desulfurization*, CRC Press, Boca Raton, FL, 2016, p. 188.
4. Olef Deutschmann, Helmt Knözinger, Karl Kochloefl, Thomas Turek, Heterogeneous catalysis and solid catalysts, *Ullmann's Encyclopedia of Industrial Chemistry*, Wiley-VCH Verlag GmbH & Co. KGaA, Weinheim, Germany, 2009, pp. 5–6.
5. Surinder Parkash, *Refining Processes Handbook*, Elsevier, Burlington 2003, pp. 29–31.
6. Hiroshi Mizutani, Nobumasa Nakajima, Yoshihiro Seriguchi, Hydroprocessing catalyst and method of hydroprocessing vacuum-distilled gas oil, JP2008290030 A, filed May 28, 2007, and issued December 4, 2008.
7. Dawei Hu, Qinghe Yang, Bin Liu, Chuanfeng Niu, Tao Liu, Lishun Dai, Heavy oil hydrogenating treatment catalyst and preparation method thereof, CN101332430 A, filed June 27, 2007, and issued December 31, 2008.
8. Zhongneng Liu, Xiaoling Wu, Minbo Hou, Hongyuan Zong, Xinghua Jiang, Nickel catalyst for selective hydrogenation, CN101147871 A, filed September 20, 2006, and issued March 26, 2008.

9. Fujikawa Takashi, Osaki Takayuki, Kimura Hiroshi, Mizuguchi Hiroshi, Hashimoto Minoru, Tagami Hiroyasu, Kato Katsuhiro, Catalyst for hydrogenation treatment of gas oil and method for preparation thereof, and process for hydrogenation treatment of gas oil, JP2008105031 A, filed April 16, 2002, and issued May 8, 2008.
10. Iwamoto Ryuichiro, Kagami Shigeari, Demetallization catalyst and method for hydrotreating heavy oil by using the same, JP2008093493 A, filed October 5, 2006, and issued April 24, 2008.
11. Jie Liu, Suhua Sun, Huihong Zhu, Cheng Zhang, Guang Yang, Heavy oil hydrogenating treatment catalyst and preparation method thereof, CN101433848 A, filed November 15, 2007, and issued May 20, 2009.
12. Zhongneng Liu, Jianqiang Wang, Duo Zhao, Xiaoling Wu, Large hole nickel-based catalyst, CN101347734 B, filed July 18, 2007, and issued January 21, 2009.
13. Mironov Oleg, Kuperman Alexander, Hydroprocessing bulk catalyst and uses thereof, US7678730 B2, filed October 29, 2008, and issued March 16, 2010.
14. Mironov Oleg, Kuperman Alexander, Lopez Jaime, Brait Axel, Reynolds Bruce, Chen Kaidong, Hydroprocessing bulk catalysts and uses thereof, US7737072 B2, filed October 31, 2007, and issued June 15, 2010.
15. Mironov Oleg, Kuperman Alexander, Hydroprocessing bulk catalyst and uses thereof, US7678731 B2, filed October 29, 2008, and issued March 16, 2010.
16. Guichard Bertrand, Guillaume Denis, Hydrodemetallization and hydrodesulphurization catalysts, and use in a single formulation in a concatenated process, US20100155301 A1, filed December 17, 2009, and issued June 24, 2010.
17. Dingcong Wang, Jiduan Liu, Hydrotreating guard catalyst and application thereof, CN101890381 A, filed October 18, 2009, and issued November 24, 2010.
18. Lishun Dai, Dawei Hu, Dadong Li, Bin Liu, Jia Liu, Hong Nie, Chuanfeng Niu, Zhicai Shao, Yahua Shi, Qinghe Yang, Shuangqin Zeng, Hydrogenation catalyst and preparation method thereof, CN101745400 A, filed December 17, 2008, and issued June 23, 2010.
19. Oyama Shigeo, Transition metal phosphide catalysts, EP2218502 A2, filed September 28, 2000, and issued August 18, 2010.
20. Sakamoto Norio, Catalyst for hydrotreating heavy oil and method for producing the catalyst, JP2011245418 A, filed May 26, 2010, and issued December 8, 2011.
21. Haitao Wang, Xuejun Xu, Jifeng Wang, Dongxiang Liu, Xiaoping Feng, Hydrotreating catalyst and preparation method thereof, CN102049265 A, filed October 27, 2009, and issued May 11, 2011.
22. Dingcong Wang, Hydrotreating catalyst and preparation method thereof, CN102049309 A, filed October 27, 2009, and issued May 11, 2011.
23. Xinqiang Zhao, Xuefen Liu, Shuangqin Zeng, Qinghe Yang, Dawei Hu, Shuling Sun, Lishun Dai, Chuanfeng Niu, Tao Liu, Zhicai Shao, Qinghe Liu, Hong Nie, Dadong Li, Heavy oil hydrotreating catalyst and application thereof, CN102049263 A, filed October 27, 2009, and issued May 11, 2011.
24. Patrick Euzen, Alexandra Chaumonnot, Magalie Roy-Auberger, Patrick Bourges, Tivadar Cseri, Maryline Delage, Nathalie Lett, Doped alumino-silicate catalyst and improved process for treatment of hydrocarbon feeds, US7879224 B2, filed September 20, 2005, and issued February 1, 2011.
25. Shuwu Yang, Julie Chabot, Bruce Edward Reynolds, Bo Kou, Hydroprocessing catalysts and methods for making thereof, US20120000821 A1, filed July 21, 2009, and issued January 5, 2012.
26. Shuwu Yang, Julie Chabot, Bruce Edward Reynolds, Hydroprocessing catalysts and methods for making thereof, US20120004097 A1, filed July 21, 2009, and issued January 5, 2012.

27. Christine N. Elia, Mohan Kalyanaraman, Lei Zhang, Madhav Acharya, Michael A. Daage, Darden Sinclair, Jocelyn A. Kowalski, Jose G. Santiesteban, Ajit B. Dandekar, Hydroprocessing catalysts with low surface area binders, US8263517 B2, filed December 18, 2008, and issued September 11, 2012.
28. Ali Azghay, Geert Marten Bakker, Jandirk Maarten Dortmundt, Van An Du, Marcello Stefano Rigutto, Ulrich Schubert, Welsenes Arend Jan Van, Hydrotreating catalyst comprising a group VIII and/or group VIB metal silicide compound, WO2012156294 A1, filed May 11, 2012, and issued November 22, 2012.
29. Jeremy Francis, Laurent Simon, Emmanuelle Guillon, Nicolas Bats, Avelino Corma, Christophe Pichon, Hydrocracking process using a zeolite catalyst containing two distinct hydrogenating functions, US20120205286 A1, filed January 6, 2012, and issued August 16, 2012.
30. Shinichi Inoue, Yukitaka Wada, Akihiro Mutou, Takeo Ono, Hydrotreating catalyst for hydrocarbon oil and method for producing same, and hydrocarbon oil hydrotreating method using same, US20120318717 A1, filed March 22, 2011, and issued December 20, 2012.
31. Hong Jiang, Zhanlin Yang, Jifeng Wang, Zhaoji Tang, Dengling Wei, Derong Wen, Method for preparing hydrogenation catalyst, CN102451773 A, filed October 15, 2010, and issued May 16, 2012.
32. Min Wang, Lihua Liu, Fengxiang Ling, Wanfu Sun, Shaojun Wang, Tao Tang, Preparation method of hydrotreating catalyst, CN102728373 A, filed April 14, 2011, and issued October 17, 2012.
33. Xiangyun Long, Xuefen Liu, Qinghe Liu, Xinqiang Zhao, Le Zhang, Mingfeng Li, Qinghe Yang, Hong Nie, Hydroprocessing catalyst as well as preparation and application thereof, CN102580757 A, filed January 13, 2011, and issued July 18, 2012.
34. Stephane Kressmann, Magalie Roy-Auberger, Jean Luc Le Loarer, Denis Guillaume, Jean Francois Chapat, Irregularly shaped non-spherical supported catalyst, and a process for hydroconversion heavy oil fractions, US8618017 B2, filed May 3, 2012, and issued December 31, 2013.
35. Opinder Kishan Bhan, A low cost and high activity hydroprocessing catalyst, WO2013063219 A1, filed October 25, 2012, and issued May 2, 2013.
36. Opinder Kishan Bhan, Hydroprocessing catalyst and process for treating heavy hydrocarbon feedstocks, US2013284640 A1, filed April 24, 2013, and issued October 31, 2013.
37. Alexei Grigorievich Gabrielov, William Douglas Gillespie, Composition having an active metal or precursor, an amine component and a non-amine containing polar additive useful in the catalytic hydroprocessing of hydrocarbons, a method of making such catalyst, and a process of using such catalyst, US2013165317 A1, filed June 26, 2012, and issued June 27, 2013.
38. Long Xiangyun, Liu Qinghe, Liu Xuefen, Wang Kui, Li Mingfeng, Chen Ruolei, Nie Hong, Hydrogenation catalyst with silicon oxide as carrier and preparation and application thereof, CN103157487 A, filed December 15, 2011, and issued June 19, 2013.
39. Mironov Oleg, Kuperman Alexander, Han Jinyi, Hydroprocessing bulk catalyst and methods of making thereof, JP2013512091 A, filed July 23, 2012, and issued April 11, 2013.
40. Gabrielov Alexei Grigorievich, Ganja Ed, Torrisi Salvatore Philip, A chelant and polar additive containing composition useful in the hydroprocessing of hydrocarbon feedstocks and method of making and use thereof, CN103052445 A, filed August 5, 2011, and issued April 17, 2013.
41. Gabrielov Alexei Grigorievich, Ganja Ed, Torrisi Salvatore Philip, Hydroprocessing catalyst with a single step metal and chelant incorporation, JP2013537100 A, filed August 22, 2011, and issued September 30, 2013.
42. Gabrielov Alexei Grigorievich, Ganja Ed, Torrisi Salvatore Philip, A hydroprocessing catalyst prepared with waste catalyst fines and its use, CN103052443 A, filed August 5, 2011, and issued April 17, 2013.

43. Fabrice Diehl, Elodie Devers, Karin Marchand, Bertrand Guichard, Process for hydrotreating a hydrocarbon cut with a boiling point of more than 250°C in the presence of a sulphide catalyst prepared using a cyclic oligosaccharide, US2013186806 A1, filed June 24, 2011, and issued July 25, 2013.
44. Jaime Lopez, Janine Lichtenberger, Method of making high energy distillate fuels, JP5364711 B2, filed September 13, 2013, and issued December 11, 2013.
45. Vinod Ramaseshan, Ali Hasan Al-Abdulal, Yuv Raj Mehra, Hydrotreating and aromatic saturation process with integral intermediate hydrogen separation and purification, US2013112596 A1, filed November 2, 2012, and issued May 9, 2013.
46. Takashi Fujikawa, Masahiro Kato, Nobumasa Nakajima, Minoru Hashimoto, Hydrotreating catalyst for gas oil, process for producing the same, and method of hydrotreating gas oil, DK1577007 T3, filed December 17, 2003, and issued April 15, 2013.
47. Uzio Denis, Fernandes Georges, Ould Chikh Samy, Rouleau Loic, New catalyst comprising hydro-dehydrogenating metal and carrier comprising porous core and peripheral porous layer having a pore distribution measured by mercury porosimetry, used for hydrotreating heavy loads, FR2988306 A1, filed March 21, 2012, and issued September 27, 2013.
48. Soled Stuart, Miseo Sabato, Baumgartner Joseph, Nistor Iulian, Venkataraman Pallassana, Kliewer Chris, Chimenti Robert, Guzman Javier, Kennedy Gordon, Levin Doron, Hydroprocessing catalysts and their production, KR20130077844 A, filed June 1, 2011, and issued July 9, 2013.
49. Ghosh Prasenjeet, El-Malki Ei-Mekki, Wu Jason, Bai Chuansheng, Han Jun, Lowe David M, Joshi Prasanna, Elks Jeff, Giaquinta Daniel, Venkataraman Pallassana S, Volpe Anthony F Jr, Mcconnachie Jon M, Jacobs Peter W, Sokolovskii Valery, Hagemeyer Alfred, Hou Zhigou, Bulk group viii/vib metal catalysts and method of preparing same, CN101578352 B, filed June 10, 2009, and issued August 28, 2013.
50. Rong He, Stanislaw Plecha, Meenakshi S. Krishnamoorthy, Bharat M. Patel, improved resid hydrotreating catalyst containing titania, WO2013192394 A1, filed June 20, 2013, and issued December 27, 2013.
51. Chris E. Kliewer, Stuart L. Soled, Sabato Miseo, Jeffrey S. Beck, Hydroprocessing catalysts and their production, SG181546 A1, filed December 14, 2012, and issued April 25, 2013.
52. Stuart S. Shih, Christopher G. Oliveri, Mohan Kalyanaraman, Timothy Lee Hilbert, Stephen J. Mccarthy, Post dewaxing hydrotreatment of low cloud point diesel, WO2013085533 A1, filed December 9, 2011, and issued June 13, 2013.
53. Shuwu Yang, Julie Chabot, Bruce Edward Reynolds, Bo Kou, Hydroprocessing catalysts and methods for making thereof, WO2013039950 A2, filed September 12, 2012, and issued March 21, 2013.
54. Bhan Opinder Kishan, Highly stable catalyst for heavy-hydrocarbon hydrodesulfurization, method for production thereof, and use thereof, JP2013056333 A, filed October 2, 2012, and issued March 28, 2013.
55. Bruce E. Reynolds, Axel Brait, Concentration of active catalyst slurry, JP5372770 B2, filed April 22, 2010, and issued December 18, 2013.
56. Chen Kaidong, Reynolds Bruce, Highly active slurry catalyst composition, JP5101285 B2, filed September 2, 2008, and issued December 19, 2012.
57. Jia Yanzi, Yang Qinghe, Li Dadong, Wang Kui, Hu Dawei, Zhao Xinqiang, Dai Lishun, Liu Tao, Heavy oil hydrotreating method, CN103374390 A, filed April 26, 2012, and issued October 30, 2013.
58. Jia Yanzi, Yang Qinghe, Li Dadong, Wang Kui, Sun Shuling, Liu Xuefen, Nie Hong, Dai Lishun, Heavy oil hydrotreating method, CN103374391 A-2013-10-30, filed April 26, 2012, and issued October 30, 2013.

59. Jia Li, Yang Tao, Heavy oil hydro-upgrading method using catalyst grading, CN103059980 A, filed October 21, 2011, and issued April 24, 2013.
60. Shao Zhicai, Sun Shuling, Dai Lishun, Niu Chuanfeng, Liu Tao, Yang Qinghe, Dong Kai, Hydrotreating method for heavy raw material by fixed bed, CN103374411 A, filed April 18, 2012, and issued October 30, 2013.
61. Hu Dawei, Yang Qinghe, Liu Bin, Liu Jia, Sun Shuling, Nie Hong, Wang Kui, Li Dadong, Dai Lishun, Zhao Xinqiang, Liu Tao, Method for hydrotreating heavy oil with high nickel and vanadium contents, CN103374387 A, filed April 13, 2012, and issued October 30, 2013.
62. Xiao Han, Yang Jianguo, Li Xiaoguo, Yu Haibin, Zhu Jinjian, Shi Fang, Zhang Guohui, Li Jia, Preparation method of catalyst carrier for hydrotreating pretreatment of wax oil, CN103301888 A, filed June 21, 2013, and issued September 18, 2013.
63. Yang Zhanlin, Jiang Hong, Tang Zhaoji, Wang Jifeng, Wen Derong, Wei Dengling, Preparation method of hydrotreating catalyst, CN103100397 A, filed November 9, 2011, and issued May 15, 2013.
64. Yang Zhanlin, Tang Zhaoji, Jiang Hong, Wang Jifeng, Wei Dengling, Wen Derong, Preparation method of alumina carrier, CN103100426 A, filed November 9, 2011, and issued May 15, 2013.
65. Zhao Liang, Sun Suhua, Zhu Huihong, Liu Jie, Yang Gang, Wang Gang, Fang Xiangchen, Residual oil hydrotreating catalyst and preparation method thereof, CN103055932 A, filed October 24, 2011, and issued April 24, 2013.
66. Shao Zhicai, Sun Shuling, Dai Lishun, Niu Chuanfeng, Liu Tao, Yang Qinghe, Dong Kai, Fixed bed hydrotreating method of heavy oil, CN103059935 A, filed October 20, 2011, and issued April 24, 2013.
67. Yang Qinghe, Sun Shuling, Hu Dawei, Liu Jia, Wang Kui, Dai Lishun, Liu Tao, Nie Hong, Li Dadong, Heavy-petroleum hydrogenating deasphaltenizing catalyst and preparation and application thereof, CN103357445 A, filed March 31, 2012, and issued October 23, 2013.
68. Hu Dawei, Yang Qinghe, Liu Jia, Sun Shuling, Nie Hong, Wang Kui, Li Dadong, Dai Lishun, Liu Xuefen, Liu Tao, Hydrotreating method of heavy oil with high contents of iron and calcium, CN103374389 A, filed April 13, 2012, and issued October 30, 2013.
69. Zhao Yusheng, Zhao Yuansheng, Liu Yuandong, Zhang Zhiguo, Zhou Zhiyuan, Yu Shuanglin, Fan Jianguang, Cui Ruili, inferior heavy oil catalyst-combined hydrotreating technique capable of maximally enhancing utilization ratio of catalyst, CN103289736 A, filed March 1, 2012, and issued September 11, 2013.
70. Thierry Cholley, Jean-Pierre Dath, Claude Brun, Georges Fremy, Francis Humblot, Hydroprocessing catalyst, preparation method thereof and use of same, CA2595335 C, filed January 19, 2006, and issued January 8, 2013.
71. Ge Shaohui, Liu Kunhong, Lan Ling, Hou Yuandong, Wu Pingyi, Zhao Qinfeng, Ma Jianbo, Zhong Haijun, Yuan Xiaoliang, Ju Ya Na, Zhang Peng, Wu Pei, Sun Honglei, Wang Peng, Lv Zhongwu, He Hao, Wang Shuqin, Lu Xu, Heavy oil hydrotreating composite catalyst preparation method, CN102989493 A, filed September 15, 2011, and issued March 27, 2013.
72. Kazuyuki Kiriyama, Takashi Fujikawa, Masahiro Kato, Minoru Hashimoto, Catalyst for hydrotreating hydrocarbon oil, process for producing the same, and method for hydrotreating hydrocarbon oil, CA2560925 C, filed March 22, 2005, and issued April 16, 2013.
73. Bertrand Guichard, Mathieu Digne, Residue hydrotreatment catalyst comprising vanadium, and its use in a residue hydroconversion process, US2014166540 A1, filed December 13, 2013, and issued June 19, 2014.
74. Theodorus Ludovicus, Michael Maesen, Alexander E. Kuperman, Hydroconversion multi-metallic catalyst and method for making thereof, US8702970 B2, filed November 14, 2012, and issued April 22, 2014.

75. Yang Shuwu, Chabot Julie, Reynolds Bruce Edward, Kou Bo, Hydroprocessing catalysts and methods for making thereof, CN103906828 A, filed September 12, 2012, and issued July 2, 2014.
76. Reynolds Bruce, Brait Axel, integrated unsupported slurry catalyst preconditioning process, CN101600783 B, filed December 5, 2007, and issued June 4, 2014.
77. Gabrielov Alexei Grigorievich, Ganja Ed, Torrisi Salvatore Philip, Method for restoring activity to a spent hydroprocessing catalyst, a spent hydroprocessing catalyst having restored catalytic activity, and a hydroprocessing process, JP2014502205 A, filed April 11, 2013, and issued January 30, 2014.
78. Opinder Kishan Bhan, Self-activating hydroprocessing catalyst and process for treating heavy hydrocarbon feedstocks, US2014116924 A1, filed October 25, 2012, and issued May 1, 2014.
79. Laszlo Domokos, Hermanus Jongkind, Willem Martman Jurriaan Stork, Den Tol-Kershof Johanna Maria Helena Van, Catalyst composition, its preparation and use, CA2516691 C, filed February 24, 2004, and issued April 29, 2014.
80. William J. Novak, Kathryn Y. Cole, Patrick L. Hanks, Timothy L. Hilbert, Hydroprocessing of high nitrogen feed using bulk catalyst, US2014048448 A1, filed October 21, 2013, and issued February 20, 2014.
81. Kochappilly Ouseph Xavier, Alex Cheru Pulikottil, Mohan Prabhu Kuvettu, Brijesh Kumar, Santanam Rajagopal, Ravinder Kumar Malhotra, Hydrotreating catalyst and process for preparing the same, WO2014033653 A2, filed August 29, 2013, and issued March 6, 2014.
82. Jiang Lilong, Cao Yanning, Lin Ke, Ma Yongde, Wei Kemei, Wei Ying, Carrier, Residual oil hydrogenation catalyst based on carrier, and preparation method of carrier, CN104014328 A, filed June 20, 2014, and issued September 3, 2014.
83. Zhou Yasong, Wei Qiang, Liu Tingting, Ding Sijia, Luo Xiujuan, Zhang Tao, Heavy oil hydrogenation catalyst and preparation method thereof, CN104069884 A, filed June 20, 2014, and issued October 1, 2014.
84. Jiang Lilong, Lin Ke, Cao Yanning, Ma Yongde, Wei Kemei, Residual oil hydrogenation catalyst and preparation method thereof, CN104084222 A, filed June 23, 2014, and issued October 8, 2014.
85. Maesen Theodorus, Kuperman Alexander, Dykstra Dennis, Hydroconversion multimetallic catalyst and method for making thereof, JP5502192 B2, filed March 30, 2014, and issued May 28, 2014.
86. Gao Yixin, Gao Yiqiang, Li Xiaogang, Hydrorefining treatment process of aromatic hydrocarbon oil, CN104194827 A, filed September 11, 2014, and issued December 10, 2014.
87. Stuart L. Soled, Sabato Miseo, Sona Eijsbouts-Spickova, Robertus Gerardus Leliveld, Paul Joseph Maria Lebens, Frans Lodewijk Plantenga, Bob Gerardus Oogjen, Hank Jan Tromp, Hydrocarbon hydroprocessing using bulk catalyst composition, US2014027350 A1, filed December 2, 2008, and issued January 30, 2014.
88. Chai Yongming, Xue Chao, Liu Chenguang, Liu Bin, Liu Yunqi, Zhao Ruiyu, Zhao Huiji, Yin Changlong, Preparation method of high active site density hydrotreating catalyst, CN104248965 A, filed July 24, 2014, and issued December 31, 2014.
89. Bi-Zeng Zhan, Method for making a hydroprocessing catalyst, US2014151266 A1, filed December 9, 2013, and issued June 5, 2014.
90. Nuñez Gabriel Alonso, Partida Trino Armando Zepeda, Moyados Sergio Fuentes, Elena Smolentseva, De Leon Hernandez Jorge Noé Diaz, Supported catalysts for producing ultra-low sulphur fuel oils, WO2014088388 A1, filed November 14, 2013, and issued June 12, 2014.
91. Evgeny T. Kolev, Transition metal phosphides and hydrotreating process using the same, US7446075 B1, filed August 23, 2005, and issued November 4, 2008.

92. Zhanlin Yang, Shaozhong Peng, Hong Jiang, Xueling Liu, Jifeng Wang, Gang Wang, Preparation method of hydrogenation catalyst, CN101491766 A, filed January 23, 2008, and issued July 29, 2009.
93. Grigorievich Gabrielov Alexei, Anthony Smegal John, Philip Torrisi Salvatore, A composition useful in the catalytic hydroprocessing of hydrocarbon feedstocks, a method of making such catalyst, and a process of using such catalyst, CN101808732 A, filed August 4, 2008, and issued August 18, 2010.
94. Cholley Thierry, Dath Jean-Pierre, Brun Claude, Fremy Georges, Humblot Francis, Hydroprocessing catalyst, method for the preparation thereof and use of the same, US7741241 B2, filed January 19, 2006, and issued June 22, 2010.
95. Hao Wu, Bin Liu, Xiaodong Gao, Qinghe Yang, Hong Nie, Zongfu He, Ziwen Wang, Method for prolonging service life of coking gasoline and diesel hydrotreating catalyst, CN101683623 A, filed September 27, 2008, and issued March 31, 2010.
96. Laszlo Domokos, Hermanus Jongkind, Johannes Anthonius Robert Van Veen, Catalyst composition preparation and use, US20120065056 A1, filed November 17, 2011, and issued March 15, 2012.
97. Alexei Grigorievich Gabrielov, John Anthony Smegal, Oil and polar additive impregnated composition useful in the catalytic hydroprocessing of hydrocarbons, a method of making such catalyst, and a process of using such catalyst, US20120295786 A1, filed July 27, 2012, and issued November 22, 2012.
98. Stephen J. McCarthy, Chuansheng Bai, William G. Borghard, William E. Lewis, Hydroprocessing using rejuvenated supported hydroprocessing catalysts, US8128811 B2, filed April 3, 2009, and issued March 6, 2012.
99. Haitao Wang, Xuejun Xu, Dongxiang Liu, Xiaoping Feng, Jifeng Wang, Method for preparing hydrotreating catalyst composition, CN102451707 A, filed October 15, 2010, and issued May 16, 2012.
100. Cholley Thierry, Dath Jean Pierre, Brun Claude, Fremy Georges, Humblot Francis, Hydrotreatment catalyst, method for the preparation thereof, and use of the same, KR101189205 B1, filed August 17, 2007, and issued October 9, 2012.
101. Karl Marvin KRUEGER, Puneet Gupta, Selenium-containing hydroprocessing catalyst, its use, and method of preparation, US2013256195 A1, filed March 25, 2013, and issued October 3, 2013.
102. Rajesh Muralidhar Badhe, Alok Sharma, Brijesh Kumar, Santanam Rajagopal, Ravinder Kumar Malhotra, Anand Kumar, Catalytical hydrodesulfurization of kerosene in two steps on cobalt–molybdenum catalyst and intermediate stripping, US2013056391 A1, filed March 16, 2011, and issued March 7, 2013.
103. Opinder Kishan Bhan, A catalyst and process for the manufacture of ultra-low sulfur distillate product, US2013296163 A1, filed July 1, 2013, and issued November 7, 2013.
104. Banerjee Soumendra, Hoehn Richard, Hydrotreating catalyst comprising group viii and vib metals on a refractory oxide support comprises an organic compound comprising two thiol groups separated by a ketone- or ether-functional hydrocarbon group (hydrotreatment catalyst, method for production and use thereof), JP5200219 B2, filed February 22, 2013, and issued June 5, 2013.
105. Chen Guang, Zeng Ronghui, Shi Youliang, Gao Yulan, Du Yanze, Wu Ziming, Catalyst sulfurization method of diesel hydrotreating technique, CN103357449 A, filed April 4, 2012, and issued in October 23, 2013.
106. Wang Haitao, Xu Xuejun, Ma Tao, Wang Jifeng, Liu Dongxiang, Feng Xiaoping, Preparation method of hydrotreating catalyst composition, CN103055887 A, filed October 21, 2011, and issued April 24, 2013.

107. Hiroyuki Seki, Masanori Yoshida, Shogo Tagawa, Tomoyasu Kagawa, Hydrodesulfurization catalyst for hydrocarbon oil, process of producing same and method for hydrorefining, US2013153467 A1, filed June 21, 2011, and issued June 20, 2013.
108. Opinder Kishan Bhan, High activity hydrodesulfurization catalyst, a method of making a high activity hydrodesulfurization catalyst, and a process for manufacturing an ultra-low sulfur distillate product, US2014076783 A1, filed November 18, 2013, and issued March 20, 2014.
109. Peter Kokayeff, Suheil F. Abdo, Methods for producing diesel range materials having improved cold flow properties, US2014319024 A1, filed July 10, 2014, and issued October 30, 2014.
110. Yuichi Tanaka, Takuya Niitsuma, Kazuhiko Tasaka, Marie Iwama, Hydrogenation refining catalyst and method for producing a hydrocarbon oil, EP2692429 A1, filed March 26, 2012, and issued February 5, 2014.
111. Li Yanpeng, Liu Dapeng, Li Aiting, Li Feifei, Liu Chenguang, Nimo diesel hydrorefining catalyst with improved hydrogenation activity and preparation method, CN104117362 A, filed July 22, 2014, and issued October 29, 2014.
112. Chai Yongming, Xue Chao, Liu Chenguang, Liu Bin, Bai Rui, Zhao Jinchong, Yin Changlong, Liu Yunqi, Method for improving activity of distillate oil hydrogenation catalyst, CN104248995 A, filed July 24, 2014, and issued December 31, 2014.
113. Yuichi Tanaka, Takuya Niitsuma, Kazuhiko Tasaka, Marie Iwama, Regenerated hydrogenation refining catalyst and method for producing a hydrocarbon oil, US2014076782 A1, filed March 26, 2012, and issued March 20, 2014.
114. Qinghe Yang, Dawei Hu, Shuling Sun, Jia Liu, Hong Nie, Xinqiang Zhao, Xuefen Liu, Dadong Li, Lishun Dai, Zhicai Shao, Tao Liu, Process for hydrotreating heavy raw oils, US2014001090 A1, filed March 29, 2013, and issued January 2, 2014.
115. Do Kyoung Kim, Do Woan Kim, Sang Il Lee, Seung Hoon Oh, Woo Kyung Kim, Han Seung Pan, Woo Young Kim, Kyung Soo Jun, Sun Choi, Method of simultaneously removing sulfur and mercury from hydrocarbon material using catalyst by means of hydrotreating reaction, US2013313165 A1, filed February 8, 2012, and issued November 28, 2013.
116. Chen Lin, Chen Guang, Shi Youliang, Zeng Ronghui, Du Yanze, Wu Ziming, Gao Yulan, Catalyst sulfurization method of gasoline hydrotreating technique, CN103361111 A, filed April 4, 2012, and issued October 23, 2013.
117. Zhang Xuejun, Lan Ling, Zhao Qinfeng, Ju Yana, Zhong Haijun, Wang Shuqin, Lu Xu, Ma Jianbo, Liquefied gas hydrotreating method, CN103146429 A, filed September 28, 2012, and issued June 12, 2013.
118. Zeng Youfu, Zou Shiying, Tan Zhenming, Chao Huixia, Liao Bin, Yan Guiling, Zhu Linghui, Bie Ke, Wang Wenbo +, inferior gasoline hydrotreating catalyst and preparation method and application thereof, CN102872891 A, filed July 15, 2011, and issued January 16, 2013.
119. Liu Renjun, Xiao Leye, Li Xin, Li Jiancai, Crude benzene hydrorefining catalyst and preparation method thereof, CN104226325 A, filed August 20, 2014, and issued December 24, 2014.
120. Yu Haibin, Zhong Dule, Zhao Xunzhi, Li Xiaoguo, Shi Fang, Qu Xiaolong, Sui Zhiyu, Zhang Yonghui, Jiang Xuedan, Preparation method of hydrofining catalyst for high-activity coker gasoline, CN103521236 A, filed October 11, 2013, and issued January 22, 2014.
121. Ross April, Halbert Thomas, Novak William, Greeley John, Hydroprocessing methods utilizing carbon oxide-tolerant catalysts, JP2014511914 A, filed November 12, 2013, and issued May 19, 2014.

122. Hirano Tomoak, Kojika Hiromichi, Inamura Kazuhiro, Titanium-carrying fire-resistant inorganic oxide carrier, hydrotreating catalyst of hydrocarbon oil, and hydrotreatment method using the same, JP2008178843 A, filed January 26, 2007, and issued August 7, 2008.
123. Alexei Grigorievich Gabrielov, John Anthony Smegal, Salvatore Philip Torrisi, Composition useful in the catalytic hydroprocessing of hydrocarbon feedstocks, a method of making such catalyst, and a process of using such catalyst, US8431510 B2, filed August 4, 2008, and issued April 30, 2013.
124. Zhanlin Yang, Shaozhong Peng, Hong Jiang, Xueling Liu, Jifeng Wang, Preparation of hydrogenation catalyst, CN101491767 A, filed January 23, 2008, and issued July 29, 2009.
125. Zhanlin Yang, Shaozhong Peng, Hong Jiang, Xueling Liu, Jifeng Wang, Preparation method of hydrogenation catalyst, CN101434861 A, filed November 15, 2007, and issued May 20, 2009.
126. Shaozhong Peng, Zhenhui Lv, Preparation method of hydroprocessing catalyst, CN102463150 A, filed November 1, 2010, and issued May 23, 2012.
127. William Douglas Gillespie, Additive impregnated composition useful in the catalytic hydroprocessing of hydrocarbons, a method of making such catalyst, and a process of using such catalyst, US2013005566 A1, filed June 26, 2012, and issued January 3, 2013.
128. Yoshinori Kato, Hiroshi Kimura, Takashi Fujikawa, Kazuyuki Kiriyama, Hydrodesulfurization/dewaxing catalyst for hydrocarbon oil, process for producing the same, and method of hydrotreating hydrocarbon oil with the catalyst, EP2072127 B1, filed September 10, 2007, and issued in January 9, 2013.
129. Tomina Natal'ja Nikolaevna, Pimerzin Andrej Alekseevich, Antonov Sergej Aleksandrovich, Maksimov Nikolaj Mikhajlovich, Drjaglin Jurij Jur'evich, Catalyst for hydrofining of oil fractions and raffinates of selective purification and method of its preparation, RU2012104059 A, filed February 6, 2012, and issued August 20, 2013.
130. Zhang Cheng, Wang Yonglin, Yang Gang, Wang Gang, Preparation method of hydrotreating catalyst, CN103100390 A, filed November 9, 2011, and issued May 15, 2013.
131. Lv Zhenhui, Peng Shaozhong, Wang Gang, Yuan Shenghua, Preparation method of hydrotreating catalyst, CN103055908 A, filed October 21, 2011, and issued April 24, 2013.
132. Bao Xiaojun, Han Wei, Shi Gang, Yuan Pei, Fan Yu, Liu Haiyan, Hydrogenation catalyst and preparation method thereof, CN103143365 A, filed December 6, 2011, and issued June 12, 2013.
133. Cui Chenglai, Preparation method of hydrotreating catalyst, CN103055892 A, filed December 16, 2012, and issued April 24, 2013.
134. Bertrand Guichard, Laurent Simon, Sylvie Lopez, GRANDI Valentina DE, Delphine Minoux, Jean-Pierre Dath, Method for preparing a catalyst usable in hydroprocessing and hydroconversion, WO2013093228 A1, filed November 27, 2012, and issued June 27, 2013.
135. Liu Qinghe, Nie Hong, Long Xiangyun, Liu Xuefen, Zhang Le, Zhao Xinqiang, Auxiliary-metal-component-containing hydrotreating catalyst and preparation and application thereof, CN103285871 A, filed February 23, 2012, and issued September 11, 2013.
136. Xiao Han, Yang Jianguo, Yu Haibin, Li Xiaoguo, Zhu Jinjian, Shi Fang, Zhang Guohui, Li Jia, Preparation method for hydrogenation pretreatment catalyst containing silicon aluminum–phosphorus aluminum composite molecule sieve, CN103285914 A, filed June 21, 2013, and issued September 11, 2013.

137. Ornelas Gutierrez Carlos Elias, Alvarez Contreras Lorena, Farias Mancilla Jose Rurik, Aguilar Elguezabal Alfredo, Unsupported and supported promoted ruthenium sulfide catalyst with high catalytic activity for hydrocarbon hydrotreatments and its method., US2013157842 A1, filed April 11, 2012, and issued June 20, 2013.
138. Smegal John Anthony, A hydroprocessing catalyst and methods of making and using such a catalyst, CN103717305 A, filed June 20, 2012, and issued April 9, 2014.
139. Li Yanpeng, Liu Dapeng, Li Aiting, Li Feifei, Liu Chenguang, Aluminum oxide supported sulfurized molybdenum (tungsten) base hydrogenation catalyst activity phase sample preparing method, CN104122128 A, filed July 22, 2014, and issued October 29, 2014.
140. Jinyi Han, Alexander E. Kuperman, Theodorus Ludovicus Michael Maesen, Horacio Trevino, Hydroconversion multi-metallic catalysts and method for making thereof, US2014135207 A1, filed September 5, 2013, and issued May 15, 2014.
141. Maarten Vogelaar Bastiaan, Jacob Arie Bergwerff, Oene Johan Nan, Henk Jan Tromp, Supported hydrotreating catalysts having enhanced activity, WO2014056846 A1, filed October 7, 2013, and issued April 17, 2014.
142. Yan Chuang, Zhao Suyun, Li Jingbin, Tooth-spherical alumina carrier, tooth-spherical hydrotreating catalyst and preparation methods thereof, CN103285842 A, filed June 25, 2013, and issued September 11, 2013.
143. Li Xuekuan, Tang Mingxing, Lyu Zhanjun, Zhou Ligong, Du Mingxian, Ge Hui, Yang Ying, Fuel oil deep-adsorption desulfurizationn catalyst, preparation method and application thereof, CN104056632 A, filed June 10, 2014, and issued September 24, 2014.
144. Hong Jiang, Shaozhong Peng, Zhanlin Yang, Xuelıng Liu, Preparation method of hydrogenation catalyst, CN101279291 A, filed April 4, 2007, and issued October 8, 2008.
145. Bhan Opinder Kishan, Catalyst and process for the selective hydrodesulfurization of an olefin containing hydrocarbon feedstock, WO2009111720 A2, filed March 6, 2009, and issued September 11, 2009.
146. Birke Peter, Goerlitz Frankk Heinz, Himmel Wigbert Gerhard, Hunold Juergen, John Hans-Heino, Catalyst and hydrotreating process, US20100101979 A1, filed November 12, 2009, and issued April 29, 2010.
147. Wang Jinye, Zeng Shuangqin, Li Mingfeng, Nie Hong, Li Dadong, Yang Qinghe, Xia Guofu, Zhang Runqiang, Chen Ruolei, Gao Xiaodong, Wang Kui, Zhu Mei, Li Jian, Zhu Li, Hydrotreating catalyst and application thereof, CN102274731 A, filed June 10, 2010, and issued December 14, 2011.
148. Jing Xin, Songtao Dong, Kui Wang, Shuangqin Zeng, Mingfeng Li, Hong Nie, Li Zhu, Yuying Zhang, Zhihai Hu, Hydrotreating catalyst and application thereof, CN102188990 A, filed March 4, 2010, and issued September 21, 2011.
149. Zhanlin Yang, Shaozhong Peng, Hong Jiang, Xueling Liu, Preparation method of hydrotreating catalyst, CN101940929 A, filed July 9, 2009, and issued January 12, 2011.
150. Wang Jinye, Zeng Shuangqin, Nie Hong, Xia Guofu, Li Mingfeng, Yang Qinghe, Wang Kui, Li Dadong, Zhang Runqiang, Zhu Mei, Li Jian, Zhu Li, Hydrotreating catalyst and application thereof, CN102274730 A, filed June 10, 2010, and issued December 14, 2011.
151. Wang Jinye, Zeng Shuangqin, Nie Hong, Li Mingfeng, Wang Kui, Yang Qinghe, Xia Guofu, Li Dadong, Chen Ruolei, Zhang Runqiang, Gao Xiaodong, Zhu Mei, Li Jian, Zhu Li, Hydrotreating catalyst and application thereof, CN102274732 A, filed June 10, 2010, and issued December 14, 2011.
152. Daisuke Usui, Yoshinori Kato, Hydroprocessing catalyst for hydrocarbon oil, method for producing hydroprocessing catalyst for hydrocarbon oil and hydroprocessing method for hydrocarbon oil, JP2013027847 A, filed July 29, 2011, and issued February 7, 2013.

153. Long Xiangyun, Liu Xuefen, Liu Qinghe, Li Mingfeng, Yang Qinghe, Jiang Donghong, Hu Zhihai, Nie Hong, Hydrotreating catalyst, and preparation method and application thereof, CN103372449 A, filed April 26, 2012, and issued October 30, 2013.
154. Long Xiangyun, Liu Xuefen, Wang Kui, Xin Jing, Li Mingfeng, Nie Hong, Hu Zhihai, Li Dadong, Mao Yichao, Zhang Yuying, Molecular-sieve-containing hydrotreating catalyst, and preparation and application thereof, CN103372458 A, filed April 26, 2012, and issued October 30, 2013.
155. Wu Jianming, Preparation method and application of macroporous alumina supporter, CN104085909 A, filed July 31, 2014, and issued October 8, 2014.
156. Ma Junjian, Yang Qinghe, Zeng Shuangqin, Nie Hong, Li Dadong, Hydrated alumina forming matter and preparation method thereof, alumina forming matter, catalyst and preparation method and application thereof and hydrotreating method, CN103480249 A, filed June 11, 2012, and issued January 1, 2014.
157. Bhaduri Rahul, Stiksma John, Berezowsky Roman, Process for separating and recovering base metals from used hydroprocessing catalyst, US7837960 B2, filed December 19, 2008, and issued November 23, 2010.
158. Cheng Zhang, Yonglin Wang, Gang Yang, Guolin Wu, Method for regenerating inactivated hydrotreating catalyst, CN102451774 A, filed October 15, 2010, and issued May 16, 2012.
159. Bhaduri Rahul, Powers Christopher, Mohr Donald, Reynolds Bruce, Lopez Jose Guitian, Process for recovering base metals from used hydroprocessing catalyst, EA017665 B1, filed November 27, 2008, and issued February 28, 2013.
160. Alexander E. Kuperman, Theodorus Maesen, Dennis Dykstra, Ping Wang, Soy Uckung, Hydroconversion multi-metallic catalyst and method for making thereof, US8703641 B2, filed October 18, 2011, and issued April 22, 2014.
161. Yulan Gao, Xiangchen Fang, Gang Wang, Fenglan Cao, Chonghui Li, Hydroprocessing technique, CN101148609 B, filed September 20, 2006, and issued March 26, 2008.
162. Mccarthy Stephen J., Bai Chuansheng, Borghard William G., Lewis William, Regeneration and rejuvenation of supported hydroprocessing catalysts, US7906447 B2, filed April 3, 2009, and issued March 15, 2011.
163. Omer Refa Koseoglu, integrated isomerization and hydrotreating process, US2013062257 A1, filed July 27, 2012, and issued March 14, 2013.
164. Kiriyama Kazuyuki, Kimura Hiroshi, Kato Yoshinori, Method for regenerating catalyst for hydrotreating hydrocarbon oil, JP2013017999 A, filed September 6, 2012, and issued January 31, 2013.
165. Gabor Kiss, Lulian Nistor, John Zengel, Hydrotreating process, US8747659 B2, filed February 23, 2010, and issued June 10, 2014.
166. Hyung S. Woo, Jane C. Cheng, Teh C. Ho, Stephen H. Brown, Hydroprocessing of heavy hydrocarbon feeds using small pore catalysts, WO2013033293 A2, filed August 30, 2012, and issued March 7, 2013.
167. Hyung S. Woo, Jane C. Cheng, Teh C. Ho, Stephen H. Brown, Richard C. Dougherty, David T. Ferrughelli, Federico Barrai, Hydroprocessing of heavy hydrocarbon feeds, CA2845340 A1, filed August 30, 2012, and issued March 7, 2013.
168. Long Xiangyun, Liu Qinghe, Zhang Le, Li Mingfeng, Wang Zhe, Hu Zhihai, Nie Hong, Hydrofining catalyst, and preparation and application thereof, CN103372448 A, filed April 26, 2012, and issued October 30, 2013.
169. Long Xiangyun, Liu Xuefen, Liu Qinghe, Li Mingfeng, Jiang Donghong, Hu Zhihai, Nie Hong, Hydrogenation catalyst, and preparation and application thereof, CN103372455 A, filed April 26, 2012, and issued October 30, 2013.
170. William Douglas Gillespie, Alexei Grigorievich Gabrielov, Peter Wolohan, John Anthony Smegal, Composition and a method of making and use of such composition, US2014339135 A1, filed August 7, 2014, and issued in November 20, 2014.

171. Michael Glenn Hunter, Angelica Hidalgo Vivas, Lars Skov Jensen, Gordon Gongngai Low, Hydrotreating and hydrocracking process and apparatus, EP1931752 A1, filed September 12, 2006, and issued June 18, 2008.
172. Mayeur Vincent, Vergel Cesar, Mariette Laurent, Process for the hydrotreatment of a gas-oil feedstock, hydrotreatment reactor for implementing said process, and corresponding hydrorefining unit, WO2008012415 A2, filed July 18, 2007, and issued January 31, 2008.
173. Baozhong Li, Zhongqing Zhang, Ying Zhang, Minghua Guan, Method for deep desulfurization olefin hydrocarbon reduction of inferior gasoline, CN101492608 A, filed January 23, 2008, and issued July 29, 2009.
174. Kokayeff Peter, Leonard Laura Elise, Hydrodesulfurization Process, US7749375 B2, filed September 7, 2007, and issued July 6, 2010.
175. Kiss Gabor, Nistor Iulian, Zengel John, Miseo Sabato, Krycak Roman, Ho Teh, Hydrotreating process, WO2011106277 A2, filed February 21, 2011, and issued September 1, 2011.
176. Koseoglu Omer Refa, Bourane Abdennour, integrated hydrotreating and oxidative desulfurization process, WO2011123383 A1, filed March 28, 2011, and issued October 6, 2011.
177. Hassan Abbas, Anthony Rayford G, Borsinger Gregory, Hassan Aziz, Bagherzadeh Ebrahim, Process for hydrodesulfurization, hydrodenitrogenation, hydrofinishing, or amine production, EP2379678 A1, filed January 18, 2010, and issued October 26, 2011.
178. Cullen Mark, Process for removing sulfur from hydrocarbon streams using hydrotreatment, fractionation and oxidation, US20110226670 A1, filed May 17, 2011, and issued September 22, 2011.
179. Manuela Serban, Alakananda Bhattacharyya, Beckay J. Mezza, Bussche Kurt M. Vanden, Christopher P. Nicholas, Joseph A. Kocal, Warren K. Bennion, Process for removing nitrogen from vacuum gas oil, EP2519610 A2, filed December 16, 2010, and issued November 7, 2012.
180. Edward S. Ellis, John P. Greeley, Vasant Patel, Murali V. Ariyapadi, Selective hydrodesulfurization and mercaptan decomposition process with interstage separation, CA2593062 C, filed December 13, 2005, and issued January 3, 2012.
181. Mahmoud Bahy Noureldin, Ahmed Saleh Bunaiyan, Energy-efficient and environmentally advanced configurations for naptha hydrotreating process, US20120279900 A1, filed May 2, 2011, and issued November 8, 2012.
182. Soled Stuart L, Miseo Sabato, Eijsbouts Sonja, Plantenga Frans, Louwen Jacobus Nicolaas, Leliveld Robertus Gerardus, Cerfontain Marinus Bruce, Vogt Eelco Titus Carel, Ryley Kenneth, Hydroprocessing using bulk bimetallic catalysts, CA2627372 C, filed October 25, 2006, and issued July 17, 2012.
183. Cui Ruili, Zhao Yusheng, Ma An, Cheng Tao, Yu Shuanglin, Zhao Yuansheng, Tan Qingfeng, Zhou Zhiyuan, Fan Jianguang, Zhang Chunguang, Na Meiqi, Liu Yuandong, Zhang Zhiguo, Residual oil hydrotreating-catalytic cracking-solvent refining combined process, CN102816594 A, filed June 10, 2011, and issued December 12, 2012.
184. Anand Kumar, Brijesh Kumar, Sarvesh Kumar, Ravinder Kumar Malhotra, Santanam Rajagopal, Alok Sharma, A process for desulfurization of diesel with reduced hydrogen consumption, WO2012066574 A2, filed November 16, 2011, and issued May 24, 2012.
185. Kazuya Nasuno, Operation method of middle distillate hydrotreating reactor, and middle distillate hydrotreating reactor, US2012006721 A1, filed January 21, 2010, and issued January 12, 2012.
186. Daniele Molinari, Giuseppe Bellussi, Alberto Landoni, Paolo Pollesel, Catalytic system and process for the total hydroconversion of heavy oils, WO2013034642 A1, filed September 6, 2012, and issued March 14, 2013.

187. Doesburg Edmundo Steven Van, Process for hydrotreating a hydrocarbon oil, WO2013098336 A1, filed December 27, 2012, and issued July 4, 2013.
188. Jorge Ancheyta Juarez, Jose Antonio Domingo Munoz Moya, Luis Carlos Castaneda Lopez, Sergio Ramirez Amador, Gustavo Jesus Marroquin Sanchez, Guillermo Centeno Nolasco, Fernando Alonso Martinez, Rodolfo Antonio Aguilar Escalante, Process of hydroconversion–distillation of heavy and/or extra-heavy crude oils, US2013056394 A1, filed August 30, 2012, and issued March 7, 2013.
189. Beckay J. Mezza, Haiyan Wang, Alakananda Bhattacharyya, Christopher P. Nicholas, Process for removing refractory nitrogen compounds from vacuum gas oil, WO2013089915 A1, filed October 17, 2012, and issued in June 20, 2013.
190. John A. Petri, Vedula K. Murty, Peter Kokayeff, Two-stage hydrotreating process, US8608947 B2, filed September 30, 2010, and issued December 17, 2013.
191. Oluwaseyi Abiodun Odueyungbo, Thermal treatment system for spent hydroprocessing catalyst, US2013247406 A1, filed September 26, 2013, and issued May 21, 2013.
192. Gary P. Schleicher, Kenneth L. Riley, A hydrotreating catalyst system suitable for use in hydrotreating hydrocarbonaceous feedstreams, RU2011142844 A, filed March 30, 2010, and issued May 10, 2013.
193. Chuanfeng Niu, Lishun Dai, Yongcan Gao, Dadong Li, Yahua Shi, Hong Nie, Qinghe Yang, Combined process for hydrotreating and catalytic cracking of residue, US8529753 B2, filed December 27, 2007, and issued September 10, 2013.
194. Soumendra M. Banerjee, Richard K. Hoehn, Hydrotreating methods and hydrotreating systems, WO2013089893 A1, filed October 1, 2012, and issued June 20, 2013.
195. Terry E. Helton, Benjamin S. Umansky, William J. Tracy, III, Stephen J. McCarthy, Timothy L. Hilbert, Mohan Kalyanaraman, Christopher G. Oliveri, Activation of dual catalyst systems, US2013130893 A1, filed November 15, 2012, and issued May 23, 2013.
196. Zackory S. Akin, Craig R. Boyak, Abdenour Kemoun, Ralph Evan Killen, Krishniah Parimi, Steven Xuqi Song, Steven Alden Souers, Multiphase contact and distribution apparatus for hydroprocessing, US8597595 B2, filed November 27, 2012, and issued December 3, 2013.
197. Marchand Karin, Digne Mathieu, Process for preparing a hydrotreatment catalyst by impregnation with a phosphorus-containing compound, JP5362712 B2, filed May 30, 2011, and issued December 11, 2013.
198. Ding He, Song Yongyi, Niu Shikun, Catalytically cracked gasoline pretreatment method, CN102876375 A, filed July 11, 2011, and issued January 16, 2013.
199. Chen Liehang, Ju Yana, Lan Ling, Sun Honglei, Zhang Xuejun, Zhang Zhenli, Catalytic process for selective hydrodesulfurization of gasoline, CN101845321 B, filed May 12, 2010, and issued July 31, 2013.
200. Stephane Cyrille Kressmann, Raheel Shafi, Ali Hussain Alzaid, Esam Z. Hamad, Process for the sequential hydroconversion and hydrodesulfurization of whole crude oil, US2013068661 A1, filed November 8, 2012, and issued March 21, 2013.
201. César Vergel, Process for hydrotreating a diesel fuel feedstock, hydrotreating unit for implementing said process, and corresponding hydrorefining unit, US8541636 B2, filed February 12, 2008, and issued September 24, 2013.
202. Faiz Pourarian, Marc A. Portnoff, David A. Purta, Margaret A. Nasta, Jingfeng Zhang, Heather A. Elsen, Patricia A. Bielenberg, Interstitial metal hydride catalyst activity regeneration process, US8618010 B2, filed November 17, 2010, and issued December 31, 2013.
203. James J. Schorfheide, Sean C. Smyth, Bal K. Kaul, David L. Stern, Hydrotreating process with improved hydrogen management, US8518244 B2, filed January 23, 2006, and issued August 27, 2013.

204. Peter Kokayeff, Paul R. Zimmerman, Process for producing diesel, WO2014085278 A1, filed November 25, 2013, and issued June 5, 2014.
205. Jeffrey N. Daily, Controlling temperature within a catalyst bed in a reactor vessel, US2014083905 A1, filed November 26, 2013, and issued March 27, 2014.
206. Omer Refa Koseoglu, Abdulrahman Al-Bassam, Selective middle distillate hydrotreating process, EP2737022 A1, filed July 26, 2012, and issued June 4, 2014.
207. Benoit Touffait, Herve Innocenti, Jamil Zaari, Bal K. Kaul, Narasimhan Sundaram, Two stage hydrotreating of distillates with improved hydrogen management, CA2594498 C, filed January 23, 2006, and issued January 7, 2014.
208. Kazuhiko Tasaka, Yuichi Tanaka, Marie Iwama, Process for hydrotreating naphtha fraction and process for producing hydrocarbon oil, CA2779048 C, filed October 13, 2010, and issued September 9, 2014.
209. Soumendra M. Banerjee, Richard K. Hoehn, Hydrotreating methods and hydrotreating systems, US8911514 B2, filed December 15, 2011, and issued December 16, 2014.
210. Antonio O. Ramos, Chithranjan Nadarajah, Hans G. Korsten, Benjamin S. Umansky, Yi En Huang, Process for reactor catalyst loading, US2014037419 A1, filed August 2, 2013, and issued February 6, 2014.
211. Richard Hoehn, Soumendra Mohan Banerjee, Dave Bowman, Xin X. Zhu, Hydrotreating process and apparatus relating thereto, US8877040 B2, filed August 20, 2012, and issued November 4, 2014.
212. Reynolds Bruce, Lam Frederick, Chabot Julie, Antezana Fernando, Bachtel Robert, Gibson Kirk, Threlkel Richard, Leung Pak, Upflow reactor system with layered catalyst bed for hydrotreating heavy feedstock, JP5475510 B2, filed July 29, 2010, and issued April 16, 2014.
213. Frederic Bazer-Bachi, Mathieu Digne, Jan Verstraete, Nicolas Marchal, Cecile Plain, Process for hydrotreating heavy hydrocarbon feeds with switchable reactors including at least one step of progressive switching, US2014027351 A1, filed December 20, 2011, and issued January 30, 2014.
214. Frederic Bazer-Bachi, Christophe Boyer, Isabelle Guibard, Nicolas Marchal, Cecile Plain, Method for hydrotreating heavy hydrocarbon feedstocks using permutable reactors, including at least one step of short-circuiting a catalyst bed, US2014001089 A1, filed December 20, 2011, and issued January 2, 2014.
215. Laurent Georges Huve, Meng Loong Chua, Hydrotreating and dewaxing process, WO2014082985 A1, filed November 26, 2013, and issued June 5, 2014.
216. Ishihara Hisaya, Tagawa Shogo, Kagawa Tomoyasu, Regeneration method for hydrotreating catalyst, TW201400186 A, filed December 28, 2010, and issued January 1, 2014.
217. Craig Boyak, Ralph Evans Killen, Fixed-bed catalyst support for a hydroprocessing reactor, WO2014070325 A1, filed September 18, 2013, and issued May 8, 2014.
218. Hasan Dindi, Brian Paul Lamb, Luis Eduardo Murillo, Brian Boeger, Jeffrey D. Caton, Hydroprocessing process using uneven catalyst volume distribution among catalyst beds in liquid-full reactors, US8894838 B2, filed April 29, 2011, and issued November 25, 2014.
219. Xiangchen Fang, Suhua Sun, Huihong Zhu, Gang Wang, Jie Liu, Guang Yang, Shenghua Yuan, Li Cai, Ebullated bed hydrotreating process of heavy crude oil, EP2441817 B1, filed October 13, 2011, and issued July 2, 2014.

6 Catalytic Naphtha Reforming

Volkan Balci, İbrahim Şahin, and Alper Uzun

CONTENTS

6.1 INTRODUCTION

Catalytic naphtha reforming (CNR) process was pioneered by UOP in the late 1940s to meet the burgeoning demand for high-octane motor fuels and has been a pivotal unit in petroleum refineries all over the world since its inception. The CNR process is specifically designed to convert naphtha to high-octane gasoline blending components called *reformate*. The low-octane components that usually have octane number in the range of 40–65 in naphtha, such as normal paraffins (*n*-paraffins), are converted into isoparaffins (*i*-paraffins) and naphthenes, and naphthenes are converted to aromatics in catalytic reformers to enhance the octane number of gasoline

TABLE 6.1
Pure and Blending Octane Numbers of Various Hydrocarbons

Compound	Formula	RON	RON Blending[a]
	Paraffins		
n-Butane	C_4H_{10}	94.0	113
2-Methylpropane (isobutane)		102.1	122
n-Pentane	C_5H_{12}	61.8	62
2-Methylbutane (isopentane)		92.3	100
n-Hexane	C_6H_{14}	24.8	19
2-Methylpentane		73.4	82
2,2-Dimethylbutane (neohexane)		91.8	89
n-Heptane	C_7H_{16}	0.0	0.0
3-Methylhexane		52.0	56.0
2,3-Dimethylpentane		91.1	88.0
2,2,3-Trimethylbutane (triptane)		112.1	112
n-Octane	C_8H_{18}	<0	–18
3,3-Dimethylhexane		75.5	72
2,2,4-Trimethylpentane (isooctane)		100.0	100.0
n-Nonane	C_9H_{20}	<0	–18
2,2,3,3-Tetramethylpentane		116.8	122
n-Decane	$C_{10}H_{22}$	<0	–41
	Olefins		
1-Hexene	C_6H_{12}	76.4	96
1-Heptene	C_7H_{14}	54.5	65
2-Methyl-2-hexane		90.4	129
2,3-Dimethyl-1-pentene		99.3	139
	Naphthenes		
Cyclopentane	C_5H_{10}	101.6	141
Cyclohexane	C_6H_{12}	83.0	110
Methylcyclopentane		91.3	107
Methylcyclohexane	C_7H_{14}	74.8	104
trans-1,3-Dimethylcyclopentane		80.6	90
1,1,3-Trimethylcyclopentane	C_8H_{16}	87.7	94
Ethylcyclohexane		45.6	43
Isobutylcyclohexane	$C_{10}H_{20}$	33.7	38
	Aromatics		
Benzene	C_6H_6	–	98
Methylbenzene (toluene)	C_7H_8	119.7	124
Ethylbenzene	C_8H_{10}	111.2	124
1,2-Dimethylbenzene (*o*-xylene)		–	120
1,3-Dimethylbenzene (*m*-xylene)		117.5	145
1,4-Dimethylbenzene (*p*-xylene)		116.4	146
n-Propylbenzene	C_9H_{12}	111.0	127

(Continued)

TABLE 6.1 (CONTINUED)
Pure and Blending Octane Numbers of Various Hydrocarbons

Compound	Formula	RON	RON Blending[a]
Isopropylbenzene (cumene)		113.1	132
1-Methyl-3-ethylbenzene		112.1	162
1,3,5-Trimethylbenzene (mesitylene)		>120.3	170
n-Butylbenzene	$C_{10}H_{14}$	104.4	114
1-Methyl-3-isopropylbenzene		–	154
1,2,3,4-tetramethylbenzene		105.3	146

Source: Reprinted from Prestvik, R., Moljord, K., Grande, K., and Holmen, A. 2004. Compositional Analysis of Naphtha and Reformate. In *Catalytic Naphtha Reforming, Revised and Expanded*: CRC Press, Boca Raton, FL. With permission.

Note: RON: Research octane number.

[a] Obtained using a 20% hydrocarbon—80% mixture of isooctane and *n*-heptane (60:40).

blends up to 90–105. In order to elaborate the dependency of octane number on chemical structure, numerous hydrocarbons are compared in Table 6.1 with respect to their research octane numbers. In general, aromatics possess the highest octane number, followed by naphthenes, olefins, and *n*-paraffins having the lowest octane number among other hydrocarbons listed. One of the essential characteristics of the CNR process is that it is the primary source of aromatics, such as benzene, toluene, and xylene (BTX) with more than 50 vol.% of production volume on worldwide basis. Moreover, it produces hydrogen as a by-product (also called net gas as a mixture of hydrogen, methane, ethane, and trace propanes), which can be utilized in hydrogen-consuming processes (i.e., hydrocracking, hydrotreating, hydrogenation, etc.) refinery-wide.

TABLE 6.2
Typical Gasoline Pool Composition in a US Refinery

Gasoline Components	Percent of Pool Volume	Percent of Pool Sulfur
FCC naphtha	36	98
Reformate	34	–
Alkylate	12	–
Isomerate	5	–
Butanes	5	–
Light straight-run naphtha	3	1
Hydrocracked naphtha	2	–
MTBE	2	–
Coker naphtha	1	1

Source: Reprinted from Ali, S. A. 2004. Naphtha Hydrotreatment. In *Catalytic Naphtha Reforming, Revised and Expanded*: CRC Press, Boca Raton, FL. With permission.

As given in Table 6.2, reformate is of paramount importance in the gasoline pool among other gasoline blending components. It occupies the second place after fluid catalytic cracking (FCC) naphtha with approximately 34 vol.% contribution to the gasoline pool.

6.2 PROCESS OVERVIEW

The commercial CNR processes differ from each other by (1) type of the catalyst used, (2) catalyst regeneration mode, and (3) reactor configuration. According to these aspects, the commercially available CNR processes can be divided into three categories:

1. Semiregenerative catalytic reforming
2. Cyclic catalytic reforming
3. Continuous catalytic reforming (CCR)

More than 700 commercial CNR units are in operation worldwide, which accounts for an approximate reforming capacity of 11 million barrels per day. Of this capacity, almost 40% is resided in North America, 20% in western Europe, and 20% in the Asia-Pacific region (Aitani 2004). Among these reforming units, the semiregenerative CNR process is the most commonly used configuration with more than half of the total reforming capacity worldwide. Furthermore, continuous regenerative and cyclic processes share more than 1/4 and 1/10 of the total reforming capacity, respectively.

6.2.1 Semiregenerative Catalytic Reforming Process

The semiregenerative CNR process was developed as the first technology to meet the high-octane gasoline specification. It, therefore, occupies the majority among the total reformer installations worldwide. In this mode of operation, the catalyst loses its activity in time because of coke deposition on the surface. When the catalyst activity is decreased by the virtue of coke deposition, the purity of hydrogen produced and the yield of aromatics are reduced as well. The operation cycle of semiregenerative CNR reactors lasts from 6 months to 1 year. At the end of each cycle, reactors are periodically shut down for catalyst regeneration. The catalyst regeneration occurs *in situ* at approximately 8 bar in the presence of air. The catalyst used in the semiregenerative CNR reactors can be regenerated 5 to 10 times until it loses its economic feasibility; then it is replaced with a fresh one. To enhance the tolerance of the catalyst against coking, bimetallic and multimetallic reforming catalysts were developed. With these novel catalysts, reactors can be operated at 14 to 17 bar, and sustained at the same cycle length with that of monometallic catalysts operated at higher pressures.

As shown in Figure 6.1, three reactors are connected in series in a semiregenerative system. A fourth reactor is also added to boost the process severity or throughput in some of the units. The reformer feed is first pressurized with hydrogen before being fed to the first reactor. The feed is heated to the corresponding reactor temperatures up to 495–525°C prior to each reactor. At the end of the last reactor, hydrogen is separated from the products, and some portion of it is recycled back to the system

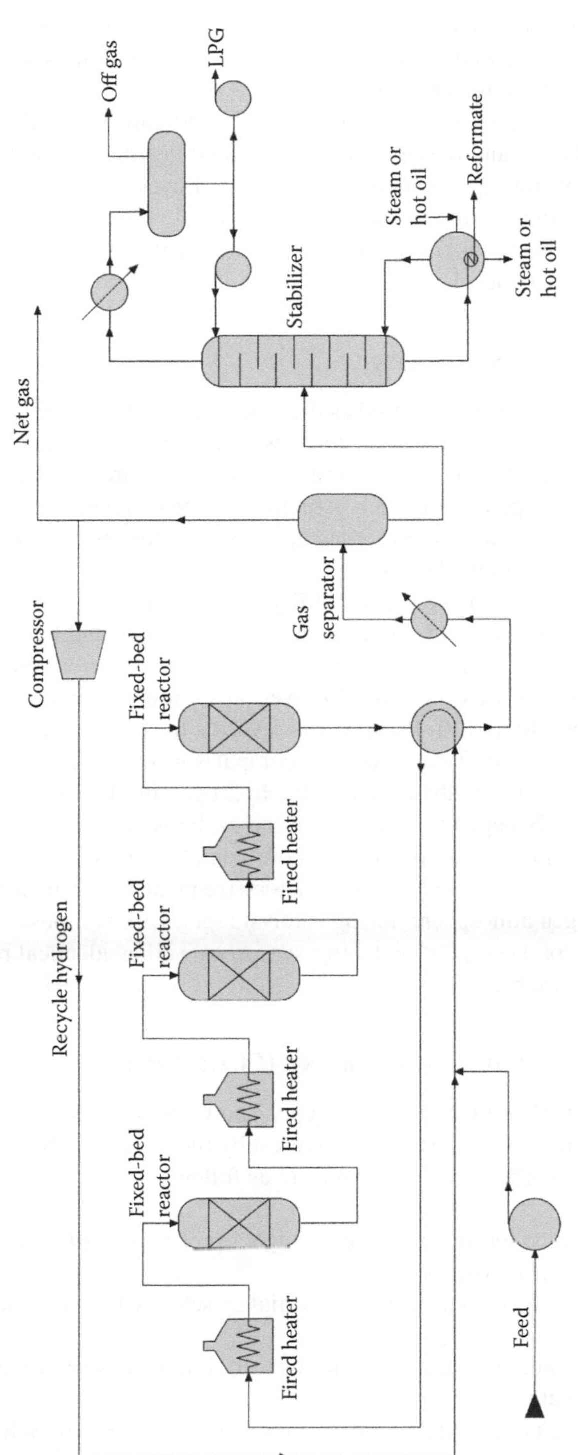

FIGURE 6.1 Semiregenerative CNR process.

while the major fraction is sent to the naphtha hydrotreater (NHT) unit. The product stream is subsequently passed through a series of separator columns to remove light gases; hence, high-octane reformate is produced.

In this process, research octane number of 85–100 can be acquired, which is controlled by feedstock and designated gasoline qualities, the amount of required gasoline, and the operating conditions for a designed cycle length. Once the cycle length is increased, the amount of the catalyst required is increased as well. When the catalyst starts to lose its activity, the reactor temperatures are increased to maintain the same conversion level.

6.2.2 Cyclic Reforming Process

The cyclic process is not as widely used as the other types of CNR processes. Similar to the semiregenerative CNR process, it consists of three reactors connected in series (Figure 6.2). However, it contains a swing reactor as a spare reactor, designed to replace any reactor in operation when it is required to be regenerated. Since reactors are identical in size, the catalyst preserves its activity at longer cycle lengths in the early stages than it does in the later stages.

One of the advantages of the cyclic CNR process is that it can be operated for a broader boiling point range of the feed, at lower pressures and hydrogen-to-feed ratios than the semiregenerative CNR process. In addition, reformate and hydrogen yields are higher in this process. However, at its low-pressure process condition and high-octane (RON 100–104) process severity, the catalyst activity is rapidly decreased less than a week to a month. In comparison with the semiregenerative process, the catalyst activity, the purity of the hydrogen produced, and the conversion alter much slightly depending on time in the cyclic process. On the other hand, the process design in the cyclic process is more complex to eliminate high level of safety risk. Because of the fact that the reactors frequently switch from hydrogen atmosphere to oxygen atmosphere during catalyst regeneration, it poses a high level of risk and a need for a complex switching system as well as identical reactor sizes as illustrated in Figure 6.2.

6.2.3 Continuous Catalytic Reforming (CCR) Process

CCR is the state-of-the-art CNR process type. The CCR process was developed in the early 1970s, and it has received great interest by the industry. The advantages of CCR as compared to other CNR processes are as follows:

- It can be operated for the same or even higher process severities in comparison with the cyclic process.
- It uses a less stable catalyst but with higher selectivity and yield toward reformate.
- It can be operated for a lower quality of feed, but it can still produce high-octane reformate.
- The CCR units are usually operated at lower pressures and recycle ratios.
- It produces more uniform reformate with higher aromatic content.

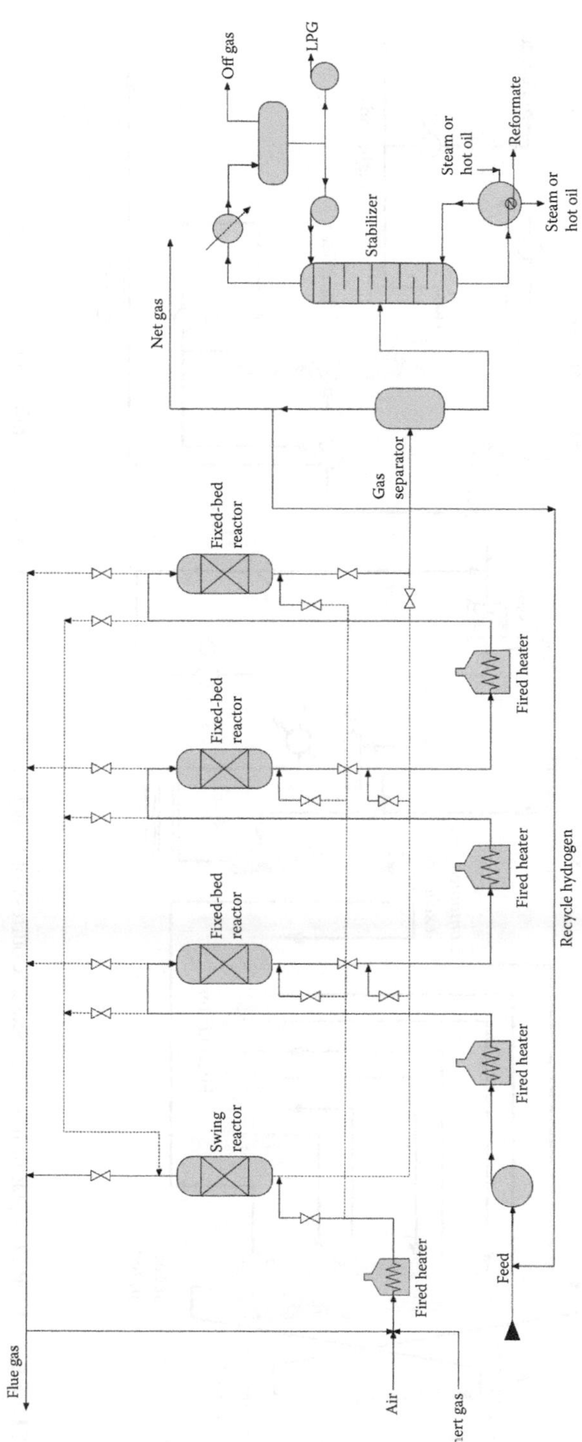

FIGURE 6.2 Cyclic CNR process.

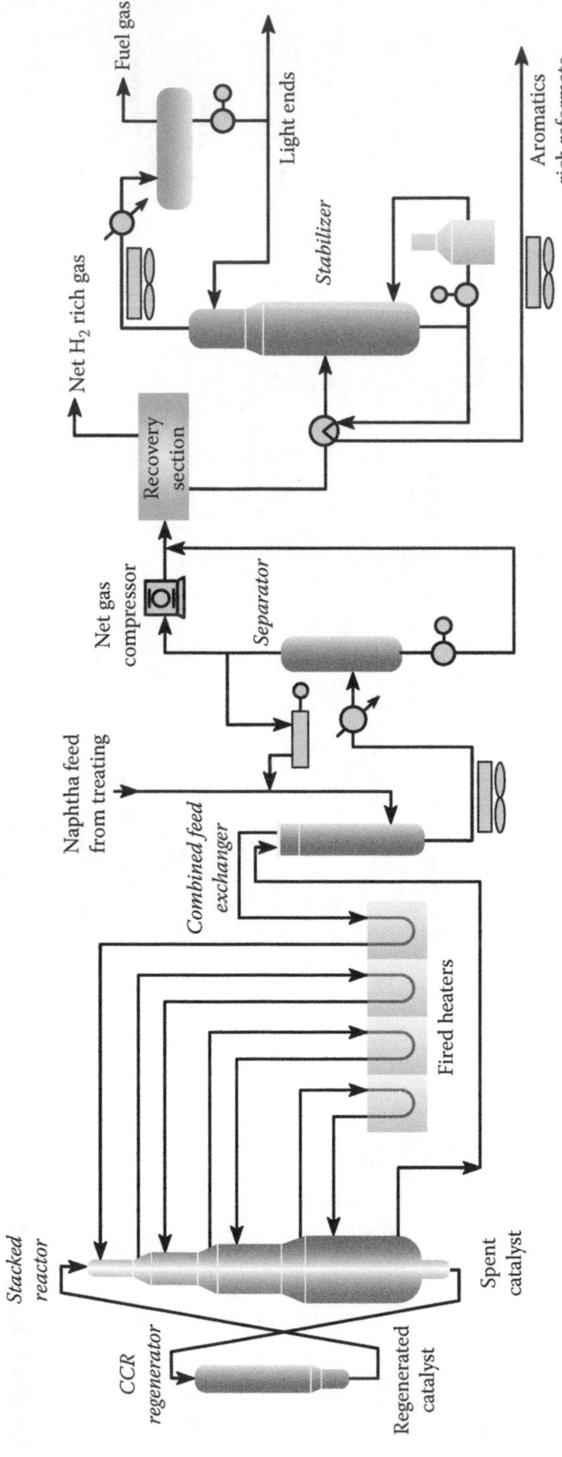

FIGURE 6.3 UOP CCR platforming © process. (Courtesy of Honeywell UOP, http://www.uop.com/reforming-ccr-platforming/.)

- A higher purity and yield of hydrogen can be achieved.
- The catalyst regeneration occurs continuously during normal operation mode so that the downtime for catalyst regeneration is eliminated.

There exist two different reactor configurations in the CCR process. In the first configuration (by UOP), moving bed reactors are used, and all of them are vertically stacked on top of each other as seen in Figure 6.3. The reactor system has a common catalyst bed that moves from top to bottom of the reactors. The deactivated catalyst is continuously withdrawn from the last reactor and sent to the regeneration unit to burn off the deposited coke. The catalyst circulation rate can be tuned in accordance with desired reformate and hydrogen yield. The reduction of the reformer catalyst is performed at the top of the first reactor or at the bottom of the regeneration unit. The catalyst is moved through the reactors by the virtue of gravity; however, it is transported by the gas lift method from the last reactor to the top of the regeneration unit. Similarly, the regenerated catalyst is transported from the bottom of the regeneration unit to the top of the reactor by the gas lift method.

In the second CCR configuration (by Axens), the reactors are connected in series as illustrated in Figure 6.4. The catalyst transportation system through the reactors is modified to allow the constant flow and easy operation. The deactivated catalyst is withdrawn from the last reactor and sent to the regeneration unit as in the case of the first CCR configuration. The regenerated catalyst is recycled back to the system at the top of the first reactor.

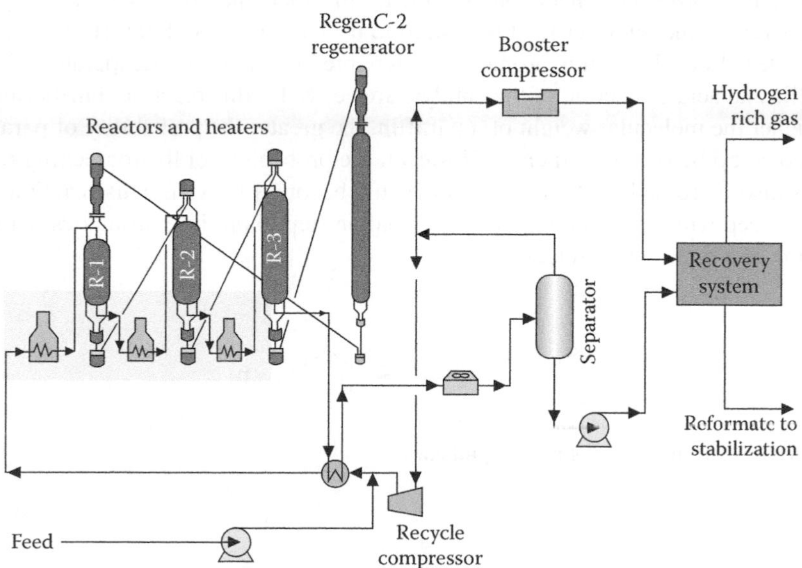

FIGURE 6.4 Axens CCR octanizing © process. (Courtesy of Axens, http://www.axens.net/document/703/octanizing-reformer-options/english.html.)

6.3 NAPHTHA REFORMING CHEMISTRY

The major reactions of the CNR process are dehydrogenation, dehydrocyclization, isomerization, and hydrocracking. For these reactions, the catalyst is required to encompass a bifunctional property *vis.* acidic and metallic. In addition to the catalyst properties, operating conditions and reformer feed composition predominantly affect both yield and purity of reformate and hydrogen, which will be discussed in detail in the following sections.

6.3.1 Dehydrogenation

Dehydrogenation is the fastest and the most octane-boosting reforming reaction, and it involves the direct conversion of naphthenes to aromatics (Figure 6.5). The composition of naphthenes in the reformer feed is, therefore, desired to be as high as possible. Accordingly, the naphthenes, such as cyclohexane, methylcyclohexane, and dimethylcyclohexane, are converted to benzene, toluene, and xylene via dehydrogenation, respectively. For each mole of naphthenes converted to aromatics, three moles of hydrogen is produced as a by-product. Thus, the yield of hydrogen produced by dehydrogenation reaction is the highest among other CNR reactions. The reaction is highly endothermic and occurs really fast at high temperature and low pressure. The metallic function of the catalyst is responsible for this reaction.

6.3.2 Dehydrocyclization

In comparison with other CNR reactions, dehydrocyclization is the least favored reaction. It involves the conversion of paraffins to naphthenes as illustrated in Figure 6.6. Therefore, the reformer feed is desired to be lean in terms of paraffinic components. Dehydrocyclization is promoted by low pressure and high temperature. Both metallic and acidic function of the catalyst are active for this reaction. Furthermore, the higher the molecular weight of a paraffin, the greater the probability of paraffin to become cyclic (i.e., naphthenic). However, the probability of hydrocracking reaction is also increased as the paraffin molecule becomes heavier. This reaction can also proceed with an additional dehydrogenation step to produce aromatics with an addition of 3 moles of H_2 release.

$$\text{(cyclohexane)} + \Delta H \longrightarrow \text{(benzene)} + 3H_2$$

FIGURE 6.5 Dehydrogenation of naphthenes.

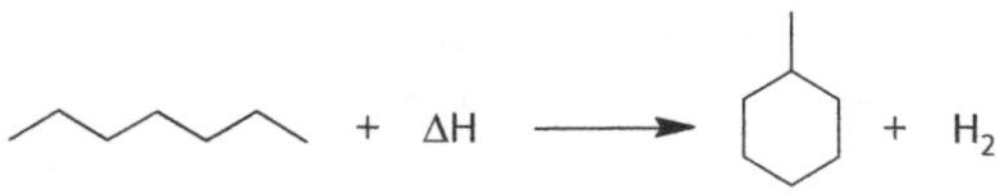

$$\text{(n-heptane)} + \Delta H \longrightarrow \text{(methylcyclohexane)} + H_2$$

FIGURE 6.6 Dehydrocyclization of *n*-paraffins.

6.3.3 Isomerization

Isomerization of *n*-paraffins can also occur during the CNR process. The *n*-paraffins are converted to branched paraffins, mostly *i*-paraffins (Figure 6.7), which increases the octane number of reformate. However, formation of more branched isomers is not thermodynamically favorable at the reaction conditions. The isomerization reaction is an exothermic reaction, and its rate depends slightly on the operating pressure. The acidic function of the catalyst is responsible for this reaction.

Moreover, naphthenes are subjected to isomerization reactions as well. For instance, in order to convert alkylcyclopentane to an aromatic (i.e., benzene), it is first required to be isomerized to alkylcyclohexane as given in Figure 6.8. During naphthene isomerization, it is also likely to form paraffin via ring opening.

6.3.4 Hydrocracking

Paraffins can also undergo hydrocracking reactions by the act of an acidic function to form lighter paraffins (Figure 6.9). As a result of this reaction, paraffins are deviated from the gasoline boiling range. This allows the aromatics to be concentrated in the liquid reformate stream; hence the octane number of reformate is increased. Due to the fact that hydrogen is consumed and liquid yield is decreased by the hydrocracking reactions, they are not desired in the CNR reaction network. High temperature and pressure operating conditions promote hydrocracking reaction.

6.3.5 Demethylation

Demethylation is not a regular CNR reaction, yet it can occur in some certain cases. During demethylation, methyl groups of paraffins or aromatics are cracked and react with hydrogen to form methane gas (Figures 6.10 and 6.11). This reaction may occur when a fresh or regenerated catalyst is loaded to the reactors operated at high process severity, such as at high temperature and pressure. The metallic function of the catalyst is responsible for this reaction. Therefore, this reaction can be hindered by reducing the metallic strength via sulfiding the catalyst or using a multimetallic catalyst. Similar to hydrocracking, hydrogen is consumed during demethylation; thus,

$+ \Delta H$

FIGURE 6.7 Isomerization of *n*-paraffins.

$+ \Delta H$

FIGURE 6.8 Isomerization of naphthenes.

$+ H_2 \longrightarrow + + \Delta H$

FIGURE 6.9 Hydrocracking of *n*-paraffins.

$$\text{n-paraffin} + H_2 \longrightarrow \text{n-paraffin} + CH_4 + \Delta H$$

FIGURE 6.10 Demethylation of *n*-paraffins.

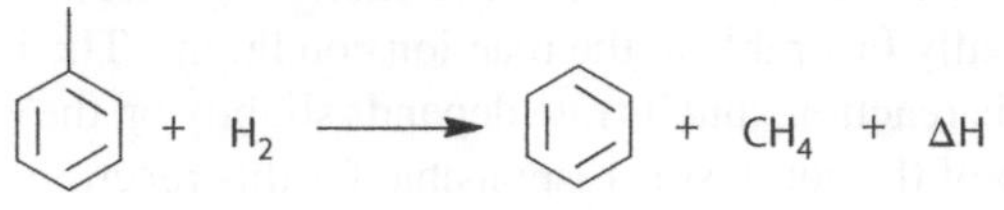

$$\text{Toluene} + H_2 \longrightarrow \text{Benzene} + CH_4 + \Delta H$$

FIGURE 6.11 Demethylation of aromatics.

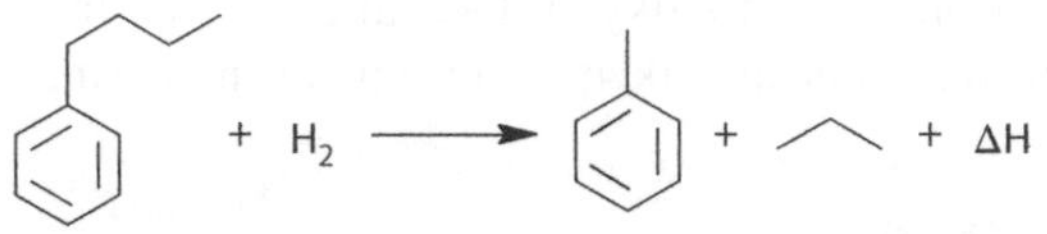

$$\text{Butylbenzene} + H_2 \longrightarrow \text{Toluene} + \text{Propane} + \Delta H$$

FIGURE 6.12 Dealkylation of aromatics.

this reaction decreases the hydrogen yield. Moreover, the cyclization probability of paraffins is reduced with decreasing chain length. Toluene may also be converted to benzene during demethylation, which is an undesired aromatic. Thus, demethylation is not wanted during CNR operation.

6.3.6 Dealkylation

Dealkylation is very similar to demethylation; the only difference is that longer alkyl chains on aromatic rings are cracked during dealkylation instead of methyl chains (Figure 6.12). When the alkyl chain on aromatic ring is long enough, the reaction takes place similar to hydrocracking. Dealkylation is promoted at high temperature and pressure, which is catalyzed by acidic function of the catalyst.

6.4 PROCESS VARIABLES

6.4.1 Feed

The boiling range of reformer feed is an important parameter, which affects the performance of the catalyst, and hence the yield of reformate. The maximum ASTM end point for reformer feed is determined to be 205°C, and C_6–C_{10} cut is specifically preferred. Hydrocarbons boiling higher than 205°C may form polycyclic aromatics leading to coke formation on the catalyst surface. As a rule of thumb, 13°C change in the ASTM end point of the feed boiling in the range of 190–218°C costs about 35% of the catalyst life.

The primary source of CNR feed is the straight-run naphtha. As discussed next, sometimes naphthas from visbreaker, coker, hydrocracker, and FCC units may also be used as reformer feedstock to a limited extent.

i. *Straight-run naphtha*: It is the naphtha distilled directly from the atmospheric crude distillation column. The primary feedstock for catalytic reformers is heavy straight-run naphtha boiling in the range of 85–205°C.

Since light straight-run naphtha boiling in the range of 35–85°C contains mainly light paraffins that are susceptible to hydrocracking, it is not a viable catalytic reformer feedstock. Furthermore, naphtha from FCC, hydrocracker, coker, and visbreaker units can also be mixed with straight-run naphtha and fed to the catalytic reformers. However, high sulfur, nitrogen, and olefin content of these alternative naphthas makes it difficult to be hydrotreated.

ii. *Visbreaker naphtha*: Depending on the source, the feed from a visbreaker unit might contain a high level of sulfur up to 15,000 ppm and nitrogen up to 500 ppm. Therefore, visbreaker naphtha requires a deep hydrotreatment process before being fed to the reformer. It is usually mixed with heavy straight-run naphtha as reformer feed in small amounts for this reason.

iii. *Coker naphtha*: Coker naphtha demonstrates similar properties with visbreaker naphtha. Nevertheless, the yield of a coker unit is higher so that more of the coker naphtha is to be processed into reformer feed.

iv. *Hydrocracked naphtha*: Hydrocracked naphtha is produced by hydrocracking of heavier petroleum fractions, such as vacuum gas oil or residues. Since it is composed of high content of naphthenes, it is an appropriate source of reformer feedstock.

v. *FCC naphtha*: It is produced by FCC of light gas oil from vacuum distillation column. However, it contains a high amount of sulfur contributing to about 98% of sulfur in the gasoline pool if the feed to the FCC is not hydrotreated beforehand. The 75–150°C fraction of FCC naphtha is appropriate to be processed in the reformer, but it requires a severe hydrotreatment operation prior to the reformer. However, high operation cost for severe hydrotreatment operation of the FCC naphtha does not compensate for the improvement of the octane number in the reformate.

Moreover, composition of a CNR feed varies depending on the source of naphtha as well as the blending ratio with another type of naphthas. Table 6.3 shows a typical hydrocarbon composition and the octane number of a typical reformer feed. Accordingly, *n*-paraffins constitute the major part of a reformer feed. Furthermore,

TABLE 6.3
Typical Properties of a CNR Feed

Property	Value
n-Paraffins (vol.%)	40–50
i-Paraffins (vol.%)	2–5
Naphthenes (vol.%)	30–40
Olefins (vol.%)	0–2
Aromatics (vol.%)	5–10
Hydrogen (vol.%)	0
Octane number (RON)	40–45

the quality of naphtha feed is characterized by N+A or N+2A value, where N and A represent volume percentages of naphthenes and aromatics in the feed, respectively. Accordingly, with the increasing values of N+A or N+2A, the yield of the reformate and the reformate octane number increase, while the coke deposition rate decreases.

6.4.2 Pressure

The operating pressure is a crucial parameter governing the catalyst activity and cycle as well as the reformate yield and composition. Here, the pressure is attributed to average reactor pressure that is usually at the inlet of the last reactor. In the CNR process, the operating pressure is usually in the range 3.5–35 bar depending on the process type, such as semiregenerative, cyclic, and CCR.

The operating pressure is able to control the CNR process in various aspects. For instance, the yield of reformate and hydrogen increases as the operating pressure decreases. Similarly, the operating temperature to sustain the same product yield also decreases with decreasing reactor pressure. Because of the fact that the coke deposition rate increases at lower pressures, the catalyst regeneration frequency is to be increased as well. Therefore, the CCR type of the CNR process enabling continuous catalyst regeneration is more adequate for low operating pressures.

6.4.3 Temperature

The operating temperature is a key process parameter controlling the yield, composition, and octane rating of the reformate. There are two different expressions used for CNR operating temperature, namely weighted-average inlet temperature (WAIT), and weighted-average bed temperature (WABT). The definitions for WAIT and WABT are expressed as follows:

$$\text{WAIT} = \sum_{i=1}^{N} T_i^{\text{in}} W_i^{\text{cat}}$$

where T^{in} is inlet temperature of the reactor, W^{cat} is the weight fraction of the catalyst bed, and the indices i and N represent the corresponding reactor number.

$$\text{WABT} = \sum_{i=1}^{N} T_i^{\text{avg}} W_i^{\text{cat}}$$

where T^{avg} is the average catalyst bed temperature, W^{cat} is the weight fraction of the catalyst bed, and the indices i and N represent the corresponding reactor number.

Since WAIT is the most commonly used expression for temperature, the temperature will be referred to WAIT throughout this chapter unless otherwise stated.

In general, CCR units are operated in the range of 525–540°C while semiregenerative units are operated in the temperature range of 490–525°C.

6.4.4 Hydrogen-to-Hydrocarbon Ratio

Because most of the reforming reactions are thermodynamically limited, hydrogen-to-hydrocarbon ratio (H_2/HC) is an important process variable. H_2/HC can be expressed as the ratio of moles of hydrogen in the recycle gas to moles of naphtha fed to the reformer. A typical H_2/HC ratio varies in the range of 4–6 mol/mol. Any change in this ratio can significantly affect the catalyst life and stability. It can be attributed to the fact that coke deposition on the catalyst active surface reversely depends on hydrogen partial pressure. Accordingly, coke formation can be reduced by increasing the hydrogen partial pressure through increasing the H_2/HC ratio. Thus, the catalyst life and stability as well as cycle length are enhanced overall.

6.4.5 Space Velocity

In CNR units, space velocity and operating temperature are modulated to set the octane number of the reformate. As the space velocity is increased, the CNR unit requires higher temperature in order to meet the octane set point in the reformate. The CNR process severity can be increased by reducing the space velocity or increasing the temperature. When the temperature is set to follow a constant process severity mode of operation, the effect of space velocity variation on the reformate and hydrogen yields is negligible. Slightly higher reformate yields may be achieved at higher space velocities because of the fact that dealkylation reactions are hindered at lower residence time of the naphtha in the reactors. However, hydrocracking reactions at higher temperatures counteract the gain of the latter case.

6.5 REFORMING CATALYSTS

In the beginning of the 1950s, commercial naphtha reforming catalysts consist of monometallic platinum catalysts supported on chlorinated alumina. These catalysts were capable of producing high octane reformate; however, the activity of catalysts rapidly vanished due to coke formation. Therefore, the CNR units containing these catalysts had to be operated at high-pressure and low-octane process severity. In the early 1970s, bimetallic catalysts were developed to meet the growing demand for high process severity operations. Bimetallic reforming catalysts are composed of platinum and a second metal (usually rhenium, tin, germanium, or iridium) supported on chlorinated alumina. In commercial CNR catalysts, platinum content changes between 0.3 and 0.7 wt.%, while chlorine between 0.1 and 1.0 wt.% is added to the alumina (γ or η phase) support material to provide acidity. Therefore, the CNR catalysts are regarded as bifunctional having both metallic and acidic active sites. Since the 1980s, trimetallic CNR catalysts were developed by adding a third metal to the bimetallic catalysts to enhance their performance and their tolerance to coke formation. In this context, Table 6.4 shows the evolution of the CNR catalysts over the years.

TABLE 6.4
Evolution of CNR Catalysts

Patent Holder	Catalyst	Year
UOP	Pt/Al_2O_3–Cl	1949
Chevron	Pt–Re/Al_2O_3–Cl	1968
Total	Pt–Sn/Al_2O_3	1969
UOP	Pt–Ge/Al_2O_3	1971
ExxonMobil	Pt–Ir/Al_2O_3	1976
UOP	Pt–In/Al_2O_3	1976
UOP	Pt–Re–Ge/Al_2O_3	1982
ExxonMobil	Pt–Re–Ir/Al_2O_3	1985
ExxonMobil	Pt–Ir–Sn/Al_2O_3	1993
UOP	Pt–Sn–In/Al_2O_3	2000

Source: Reprinted from *Applied Energy* 109, Rahimpour, M. R., Jafari, M., and Iranshahi, D., Progress in catalytic naphtha reforming process: A review, 79–93, Copyright 2013, with permission from Elsevier.

The general properties of multimetallic CNR catalysts supported on chlorinated alumina are listed below:

- High-purity alumina support material provides high resistance to erosion.
- Platinum used in conjunction with other metals provides high selectivity and stability.
- High regeneration ability of these catalysts complies with the need for CCR operation.

6.6 ISSUES AFFECTING THE PROCESS PERFORMANCE AND PRODUCT QUALITY

The yield and quality of gasoline at specific octane number and operating conditions depend mainly on the types of hydrocarbons in the feedstock. Paraffin-rich feedstock results in low yield of gasoline and requires high process severity. On the other hand, naphthenic feedstock is readily converted to aromatics and produces high yield of gasoline.

Moreover, the impurities in the different types of naphtha feedstock heavily affect the catalyst performance. For example, sulfur compounds are converted to hydrogen sulfide gas in the catalytic reforming conditions and build up in the recycle gas stream. Hydrogen sulfide poisons the platinum catalyst and reduces its activity especially for dehydrogenation and dehydrocyclization reactions.

In addition, organic nitrogen is usually converted to ammonia in the CNR units, and it decreases the acidic sites on the catalyst surface. Thus, the reactions controlled by acidic sites, such as isomerization and dehydrocyclization, are suppressed.

For these reasons, the impurities in the reformer feed are to be removed in a naphtha hydrotreatment (NHT) unit prior to being fed to the reformer.

Moreover, high acidity of the catalyst leads to high hydrocracking activity, which results in production of low-value light gases. In order to control the acidity of the catalyst at a certain range, the catalyst is usually presulfided. However, the presulfiding process is not adequate for CCR processes because of the limitation in the unit infrastructure. Therefore, novel catalysts have been developed to avoid the presulfiding step and to be able to use the modified catalyst in CCR processes.

In the following section, recent attempts in literature to modify the structure of reformer catalysts to overcome these issues are discussed.

6.7 RECENT ADVANCES IN NAPHTHA REFORMING CATALYSTS

As discussed above, a reformer catalyst is required to possess a bifunctional feature with interplay between acidic and metallic sites for the complex reaction network. Thus, at least one of these functions needs to be modified to achieve an improvement in the catalytic performance or bring solutions to the issues mentioned above. Below are examples from the last decade on such modifications of the catalyst, first focusing on acid function followed by those focusing on metal function.

Addition of a third metal on a bimetallic Pt–Re/Al_2O_3 catalyst results in changes in both the acidic and metallic functions of the naphtha reforming catalysts. Measurement of the evacuated amount of pyridine (base) as a function of temperature was performed on a set of a Ga–Pt–Re/Al_2O_3 catalyst with varying Ga amounts to estimate the acid strength and the distribution of Brønsted and Lewis sites. In general, the total acidity is decreased by Ga addition. This decrease was attributed to the amphoteric character of the gallium oxides deposited on the acidic sites of alumina. Another factor contributing to this decrease in total acidity was generated by the displacement of Cl atoms from Al sites to other sites by the addition of Ga. These results were also confirmed by means of Fourier transform infrared (FTIR) spectroscopy of adsorbed pyridine (Py). Bands in the region of 1612–1630 cm^{-1} were attributed to the pyridine coordinated to Lewis acid sites by aromatic ring, whereas those at 1451 and 1455 cm^{-1} correspond to pyridine adsorbed over Lewis acid sites. On the other hand, a band at 1540 cm^{-1} is assigned to pyridinium ion (PyH^+) formed on Brønsted acid sites. The IR spectra of Ga-containing samples after evacuation for 1 h at different temperatures (150°C, 250°C, and 350°C) show that adding Ga (0.1 or 0.3 wt.%) to the PtRe catalyst decreases approximately 16% of the total Lewis acidity, mostly the medium and strong Lewis acid sites. When these results were analyzed in parallel to the large decrease observed in the total acidity (Lewis and Brønsted) determined by pyridine TPD, it can be inferred that the impregnation of Ga decreases more of the Brønsted sites than the Lewis acid sites. Consequences of Ga addition on the catalytic performance were mainly attributed to this decrease in the acidic sites as well as the inhibition of the metal function. As a result of these effects, the total activity of the Pt–Re decreases for n-C_5 isomerization with a gain in its selectivity to i-C_5 isomers accompanied by a decrease in the selectivity to low-value light gases (C_1–C_3). Moreover, catalytic tests for n-C_7 reforming illustrated improvements in catalyst stability accompanied by an increase in selectivity to aromatics with a decrease in the formation of light gases and coke (Vicerich et al. 2011).

As in the case of Ga addition, the addition of Ge to the reforming catalysts influences the properties of the acid and metal functions. With respect to the acid function, the addition of Ge to the reforming catalyst resulted in a change in the acid strength distribution, increasing the amount of mild and weak acid sites (as germanium oxide is considered as a source of acid sites) and decreasing the amount of strong ones. The influence of these changes on the catalytic performance will be discussed once the effects on the metallic function are discussed below (Mazzieri et al. 2009a).

Indium is another metal that can be introduced to the Pt–Re/Al_2O_3 naphtha reforming catalyst as a third metal. Similar to the case with the addition of Ga and Ge, In has dual effect on the catalytic performance both on metal and acid functions. Focusing on the effect on the acid function here, the addition of In significantly reduces the number of acid sites (Benitez and Pieck 2010; Jahel et al. 2010), resulting in a decrease in the activity in acid-catalyzed reactions, such as *n*-alkane isomerization (Benitez et al. 2009).

The presence of Sn in the Pt/Al_2O_3 CNR catalyst formulation is found to reduce the rate of coke formation significantly as well as sintering of Pt particles. The synthesis and pretreatment methods, the nature of support, and the type of precursors affect the interaction of Sn with Pt and support, hence the catalytic performance. These factors may affect the accessibility to the metal active sites and the particle size. Furthermore, addition of Sn (0–0.9 wt.%) to the bimetallic Pt–Re/Al_2O_3–Cl catalyst influences the acidity of the catalyst as well as catalytic performance toward specific CNR reactions. As seen in Figure 6.13a, the addition of Sn to the Pt–Re/Al_2O_3–Cl decreases both the total acidity and distribution of acidic strength in terms of weak, mild, and strong acid sites. Among these acid sites, weak and strong acid sites are influenced the most by the Sn addition. The decrease in acidity by the addition of Sn was attributed to the displacement of Cl atoms from Al sites, where they generate acidity by polarizing the bonds of Al cation. Sn addition is assumed to neutralize 10 acid sites per Sn atom; in other words, Sn inhibits three Cl atoms on the whole. However, there is no significant change in the acidity at higher amounts of Sn addition possibly because of Sn cluster formation at higher contents resulting in low Sn dispersion throughout the chlorinated alumina surface. With respect to catalytic performance, addition of small amounts of Sn enhances the catalytic activity and selectivity toward isomerization reactions; however, it hinders the hydrocracking activity because of loss of acidic sites. Furthermore, the test reaction conducted with *n*-heptane revealed that addition of Sn decreases the benzene/*i*-C_7 molar ratio in the reformate, which decreases even almost three times at 0.1 wt.% of Sn addition (Figure 6.13b). Thus, the trimetallic catalyst with 0.1 wt.% of Sn content was found to demonstrate the best catalytic performance in terms of higher selectivity toward aromatics and *i*-C_7 isomers, but lower selectivity toward benzene (Mazzieri et al. 2005b).

Modifications on the acid function of a reforming catalyst can also be done by altering the alumina support by converting it into mixed metal oxides, such as the introduction of CeO_2. For instance, cerium nitrate was impregnated onto Al_2O_3 and calcined at 650°C to obtain Al_2O_3–CeO_2 mixed oxide supports. These mixed metal oxides were used as support for a reforming catalyst containing Pt and PtGe. Characterization of the samples revealed that when cerium oxide is present in the

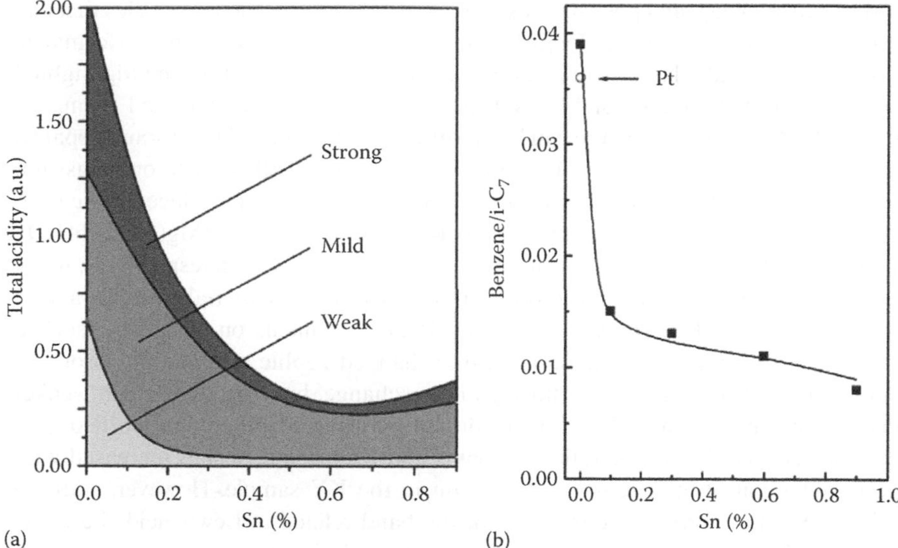

FIGURE 6.13 (a) Total acidity and acid strength distribution for Pt–Re–Sn/Al_2O_3-Cl catalysts obtained by temperature programmed deabsorption of pyridine. (b) Molar ratio of benzene/i-C_7 as a function of the Sn content. (Reprinted from *Catalysis Today* 107–108, Mazzieri, V. A. et al., Role of Sn in Pt–Re–Sn/Al_2O_3–Cl catalysts for naphtha reforming, 643–650, Copyright 2005, with permission from Elsevier.)

support, the chlorine content increases approximately by 25% in comparison with the alumina-supported catalyst without CeO_2. This increase in Cl content was attributed to the ability of CeO_2 to retain chlorine. The NH_3-TPD presented a shift to lower temperatures on the CeO_2-modified catalysts presenting weaker acid sites than those obtained on unmodified catalysts. Thus, the data revealed that the incorporation of CeO_2 onto the support resulted in a higher number of acid sites with lower strength than the catalysts with unmodified alumina support. However, the FTIR-pyridine adsorption spectra confirmed that the presence of CeO_2 does not modify the nature of acid sites, as Lewis acid sites were still present upon modification of the support with CeO_2. In these samples, CeO_2 also modifies the metal function because of its redox properties. Combination of these effects strongly modifies the catalytic performance for naphtha reforming. As a consequence, the bifunctionality of the catalyst is reduced, and the formation of isomerized olefins is increased for *n*-heptane dehydrocyclization with low aromatic yields and low deactivation rates (Medellín et al. 2010).

Aromatization of *n*-paraffins by dehydrocyclization reaction is one of the main reactions of the CNR reaction network. In order to maximize alkylbenzene production by this reaction, the catalysts should have low surface acidity to avoid cracking and isomerization (Jongpatiwut et al. 2003); its support should stabilize the well-dispersed metal nanoparticles (Jongpatiwut et al. 2003; Trakarnroek et al. 2006) with a pore system reducing the formation of large amounts of coke, possibly leading to catalyst deactivation (Iglesia et al. 1993; Jacobs et al. 1998). Thus, zeolites can offer opportunities to overcome these challenges because of their controllable acidity, morphology, and pore sizes as well as distribution of metal particles that can be dispersed over the surface. Jongpatiwuta et al. (2005) compared the performance of several non-acidic large-pore zeolites for *n*-octane aromatization and highlighted the potential of Pt–K/LTL and Pt–K/FAU with the highest activity and aromatics selectivity. The zeolite structure can be modified easily by introducing some dopants, such as magnesium, calcium, or barium. Fonseca et al. (2010) worked on potassium exchanged form of a commercial zeolite Y (Si/Al = 12.7) and introduced these dopants into its structure by impregnating with magnesium nitrate ($Mg(NO_3)_2{\cdot}6H_2O$), calcium nitrate ($Ca(NO_3)_2{\cdot}4H_2O$), and barium nitrate ($Ba(NO_3)_2$), respectively, using incipient wetness impregnation at room temperature. These materials were then used as supports for a platinum catalyst. Their IR measurements on pyridine adsorbed zeolite KY illustrated that the potassium exchanged zeolite still had the Brønsted and Lewis acid sites after going through ion exchange, but they were much weaker than the original zeolite HY. FTIR results on pyridine adsorbed on modified zeolites showed that the concentration of Lewis and Brønsted acid sites increased with the introduction of magnesium and calcium in the KY sample. However, addition of barium only improved the intensity of the band related to Lewis acid sites. The effects of these structural modifications on the catalytic performance were investigated by testing the performance of Pt catalyst prepared on these supports. Among the modifiers considered, although the barium provides the electron-enriched platinum species, as compared to others, calcium was the most efficient promoter for improving aromatization, most probably because of the role of acidic sites in favoring the bifunctional mechanism involving the ring formation on acidic sites (Fonseca et al. 2010).

In another study, the Pt/KL catalyst prepared by the CVD method was investigated for hexane aromatization. By the help of isotope labeling on *n*-hexane molecules, it was found that entry of hexane into wider lobes of the L-zeolite channels, in which Pt clusters are placed, controls the dehydrocyclization reaction on the Pt/KL catalyst. However, in the case of nonmicroporous-supported Pt catalysts, windows of L-zeolite channels make hexane molecules pass one by one to larger lopes (Azzam et al. 2010).

Aromatization of *n*-alkanes was also investigated for Pt promoted and mesoporous gallosilicates of MEL zeolites. Mesoporosity was introduced to gallium containing HZSM-11 samples by desilication. As acid sites are composed of silanol groups, Si–OH–Al, gallium species, such as Si–OH–Ga, and coordinatively unsaturated (Lewis) sites, total acidity was found to be increased slightly by desilication. Moreover, there is a small increase in Lewis acid sites and terminal silanol groups upon this process. The increase in the external surface area introduced by mesoporosity was indicated as the reason of this finding. Higher conversion and better selectivity to benzene over toluene were achieved for the mesoporous samples, which have the generation of mesopores inside the zeolite particles, because of the improved accessibility to the active extra-framework Ga species (Akhtar et al. 2012).

Zeolites with wide pores, such as H-beta (HBEA) and HY, were also studied in naphtha upgrading. Monometallic Pt and bimetallic Pt–Cr metals were loaded on these supports, and naphtha reforming reaction was performed using real feedstocks. When Cr is introduced to HBEA, high-temperature NH_3-TPD peaks shift to lower temperatures suggesting the interaction between Cr and HBEA acid sites. On the other hand, introduction of Pt as second metal increases both the strong and weak acidities of the sample. This was attributed to the fact that strong interaction of Pt and Cr may decrease the interaction between Cr and HBEA support leading to formation of additional acidic species like $[Cr(H_2O)_n]^{3+}$ and $[Cr(OH)_2]^{+}$. This interaction of Cr with the HBEA framework leads to the increase in medium acid sites and Pt dispersion. For the case of HY zeolite, there is no interaction of Pt and Cr with the zeolite framework. This special property of the Pt–Cr–HBE catalyst resulted in an increase in the RON of industrial feedstocks with negligible amount of benzene and other aromatics (Saxena et al. 2013).

Recently, mesoporous molecular sieves have also been utilized as supports for naphtha reforming catalysts. In particular, HMS is a hexagonal mesoporous silicate with a wormlike pore structure, and it has a simple preparation method. However, the main problem associated with it is the weak reactivity of surface silanol groups capable of forming only hydrogen bonds with foreign gaseous molecules or hardly ionizing in aqueous solutions. Among the additives used to improve acid properties of such molecular sieves, zirconia has both weak acid and base properties along with redox activities and high ion-exchange capacity. In these respects, Pt on zirconium-hexagonal mesoporous silica (Pt/Zr-HMS) was used as a reforming catalyst in *n*-heptane reforming. Results show that the amount of both Brønsted and Lewis acid sites increases linearly with increasing Zr amount, most of the sites being of the Lewis type. Moreover, the catalytic activity is also increased with decreasing Si/Zr amount (Peyrovi et al. 2014). Aluminum is also another additive used in HMS support in reforming catalysts. Hamoule et al. (2011) investigated Pt/Al–HMS catalysts with varying Al content. Similar to Zr, Al enhanced general acidic properties of this support. With the increasing Al content, the total number of acid sites, both Brønsted and Lewis, and their strengths are increased. Moreover, Brønsted to Lewis acid site ratio decreased due to the fact that tetrahedrally coordinated framework aluminum, a potential Brønsted acid site, decreases with increasing Al content, while octahedrally coordinated extra framework aluminum, a potential Lewis acid site, increases. *n*-Heptane reforming reaction on these catalysts revealed that there is a superior activity, high selectivity to aromatic and multibranched isomers, indicating the potential of such catalysts in reforming industry (Hamoule et al. 2011).

Al-HMS support was also utilized in preparation of bimetallic and trimetallic Pt–Sn and Pt–Sn–Re catalysts. For the case of bimetallic systems, Pt–Re has lower selectivity to aromatization than Pt–Sn because of the inhibition of strong acid sites by Sn addition. Similarly, selectivities to isomerization and aromatization are enhanced with addition of Sn while those of hydrocracking and hydrogenolysis decrease. These catalytic systems are found to be more active than their equivalent with Al_2O_3–Cl support because of the higher specific surface area, higher acidity, and porous structure of the materials. Besides, they are also superior in terms of higher aromatization and lower hydrogenolysis selectivity. Moreover, presence of combined microporosity and mesoporosity and smaller metal particles induced by the silanol groups in mesoporous structure was speculated to enhance catalytic properties (Peyrovi et al. 2012).

Carbon aerogels were also investigated as a new alternative to zeolite- or alumina-based supports in reforming reactions. Cr, Co, and W were doped to carbon aerogel, and catalysts were tested for *n*-hexane reforming. Since pure carbon is inactive, catalytic performance depends on the nature of metal phases formed. Results show that only cracking and aromatization reactions occur, and thus catalysts are monofunctional. Sintering of metal phases is prohibited by the carbon matrix even for the case of high-temperature treatments. Reduction of metal phases decreases the acidic character of the catalysts and improves the aromatization stability. Moreover, Ni and Co addition makes the catalysts completely selective to cracking (Maldonado-Hódar 2011). Platinum was also deposited on carbon aerogels, and the resulting materials were tested for *n*-hexane reforming. It was found that aerogels having mesoporous structure enhance the metal dispersion and sintering resistivity. For the case of microporous supports, the situation is just the opposite. Monofunctionality is also present in these catalysts indicated by the cracking, isomerization, and aromatization reactions. The activity of the catalysts is enhanced by Pt dispersion since carbon is inactive. The aromatization is favored by the ability of the supports for hydrogen spillover. Besides, increasing the Pt particle size favors the aromatization with an increase in temperature (Maldonado-Hódar 2012).

Modification of metal function can also significantly alter the reforming performance. Pt–Re/Al_2O_3 was the first patented bimetallic catalyst (Kluksdahl 1968). The use of Re along with Pt enabled better resistivity to coke formation, higher yields of high-quality products, and hydrogen. However, the downside of using Re is higher activity of hydrogenolysis reactions, producing low-quality gaseous products such as methane (Barbier 1986). Re was also reported to change dehydrogenation capacity to a lower extent than other metals, such as Ge and Sn; however, hydrogenolysis activity was found to be enhanced at the same time (Mazzieri et al. 2009b). It was also found that at low Re concentrations, such as 0.6 wt.%, where the Re/Pt weight ratio is 2, there is an increase in the production of aromatic products associated with a decrease in cracking yield. Nevertheless, further increase of Re leads to formation of lighter products, and thus it was concluded that 0.6 wt.% of Re was the best performing naphtha reforming catalyst (Viswanadham et al. 2008).

One of the disadvantages of the Pt–Re/Al_2O_3 catalysts is that they have high hydrocracking activity, which leads to a decrease in the reformate octane number by formation of light gases. Therefore, Pt–Re/Al_2O_3 catalysts have to be sulfided to suppress their high hydrocracking activity. Currently, most of the semiregenerative CNR units use sulfided Pt–Re/Al_2O_3 catalysts because of their highly stable catalytic activity that enables operation of the units at longer cycles. In addition to their superior stability, they offer higher resistance to catalyst deactivation providing operation at lower hydrogen partial pressures and higher process severities. During the presulfiding step, it is more favorable for sulfur to be strongly bonded to Re than to Pt and the surface of the metal particles, hence promoting the formation of Pt–ReS–Pt species. Since C–C bond cleavage requires higher groups of adjacent Pt atoms to be in close vicinity, the formation of Pt–ReS–Pt species by the presulfiding step dilutes or segregates the Pt groups and thus decreases the probability of the C–C bond cleavage reaction to occur. Although sulfided Pt–Re/Al_2O_3 catalysts offer aforementioned advantages in the semiregenerative CNR units, the presulfiding step requires a more complex infrastructure that is not compatible with the continuous CNR units. Therefore, studies have been conducted to develop new catalyst formulations based on Pt–Re/Al_2O_3 catalysts without presulfiding to be able to utilize them in continuous CNR units. In one of those studies, *n*-heptane and cyclopentane reforming reactions were performed over a monometallic Pt/Al_2O_3–Cl (0.3 wt.% Pt) catalyst, bimetallic sulfided Pt–Re/Al_2O_3–Cl (0.3 wt.% Pt, *x* wt.% Re, $x = 0.1$, 0.3, 0.9, and 2.0) catalysts, and trimetallic unsulfided Pt–Re–Sn/Al_2O_3–Cl (0.3 wt.% Pt, 0.3 wt.% Re, *y* wt.% Sn, $y = 0.1$, 0.3, 0.6, and 0.9) catalysts. Among the bimetallic sulfided Pt–Re/Al_2O_3–Cl catalysts studied, the 0.3Pt–0.3Re catalyst with 0.06 wt.% of S indicated the best catalytic activity and stability with a high *i*-heptanes/toluene ratio. On the other hand, among the trimetallic unsulfided Pt–Re–Sn/Al_2O_3–Cl catalysts studied, the 0.3Pt–0.3Re–0.6Sn catalyst showed similar catalytic activity and selectivity as sulfided 0.3Pt–0.3Re. However, it exhibited a lower hydrocracking activity and higher stability than the sulfided catalyst. As discussed before, addition of Sn reduces the total acidity of bimetallic Pt–Re/Al_2O_3–Cl catalysts, hence decreasing the hydrocracking activity and increasing the stability and catalytic selectivity toward isomerization reactions in return. Consequently, better catalytic activity and stability of the trimetallic unsulfided Pt–Re–Sn/Al_2O_3–Cl catalysts overcome the drawbacks associated with the conventional bimetallic sulfided Pt–Re/Al_2O_3–Cl catalysts such as presulfiding, and renders utilization of Pt–Re catalysts in the continuous CNR units as well (Mazzieri et al. 2005a). In a very similar study, the effect of Sn addition between 0.06 and 0.32 wt.% on the catalytic performance of bimetallic sulfided Pt–Re/Al_2O_3–Cl catalysts was investigated for reforming of *n*-octane. Accordingly, addition of Sn increased the selectivity toward *i*-paraffins and olefins, but decreased toward aromatics. Reduced hydrocracking activity in the presence of Sn resulted in a decreased yield of light gases, hence the higher liquid yield of the reformate. The catalysts with Sn content in the range of 0.14–0.32 wt.% exhibited the best catalytic performance in terms of selectivity toward *i*-parrafins and high liquid reformate yield (Elfghi and Amin 2015).

To complement these studies, the influence of Sn addition in different media (HCl and H_2O) on the performance of Pt–Re/Al_2O_3 catalysts for the *n*-C_7 reforming reaction was investigated. Results indicated a strong interaction between Sn and the active metal sites (Pt–Re) for the trimetallic Pt–Re–Sn/Al_2O_3–Cl catalysts synthesized in the presence of H_2O or HCl. The lower hydrocracking and dehydrogenating activity of the trimetallic catalysts as compared to the bimetallic counterparts correlated well with this pronounced interaction of Sn with the metal active sites. As discussed above, addition of Sn decreases the total acidity and Brønsted acidity. However, the total and Brønsted acidity of the catalysts prepared in HCl medium were found to be higher than those prepared in H_2O. This higher acidity of the catalysts prepared in HCl decreased with an increasing amount of Sn, which led to a lower yield of C_2–C_4 gases in the reformate because of the hindrance in the cracking activity with lower catalyst acidity. Furthermore, the yield of toluene reached a maximum value at 0.2 wt.% Sn (0.3 wt.% Sn nominal) content and further reduced with higher Sn content for both trimetallic catalysts series. This behavior was attributed to the fact that toluene was formed through dehydrocyclization reaction that is catalyzed by both acid and metal sites. However, acid-catalyzed cyclization is the limiting step and requires sites with lower acidic strength than those for hydrocracking. Therefore, the proportion of mild acid sites is increased with increasing Sn content, hence promoting dehydrocyclization and suppressing hydrocracking. Consistent with the previous studies, the addition of small amounts of Sn (≤0.2 wt.% or 0.3 wt.% nominal) can eliminate the presulfiding step for Pt–Re/Al_2O_3 catalysts (D'Ippolito et al. 2009b).

Moreover, most of the CNR reactions are controlled by metal sites that are usually provided by Pt atoms. The measurement of accessible active metal sites for the reactants and the metal dispersion throughout the alumina surface is crucial to correlate them with the catalytic activity. In accordance with this, a study was conducted to measure the metal dispersion in monometallic, bimetallic, and trimetallic naphtha reforming catalysts containing Pt, Re, and Sn on alumina by CO chemisorption. Results showed that CO chemisorption is an easy and reliable technique with high reproducibility to determine dispersion of metals in monometallic, bimetallic, and trimetallic naphtha reforming catalysts. Additionally, cyclohexane dehydrogenation, which is a "structure-sensitive" reaction controlled by metallic function, was used as a test reaction to elucidate the relation between the metal sites and catalytic activity. As illustrated in Figure 6.14,

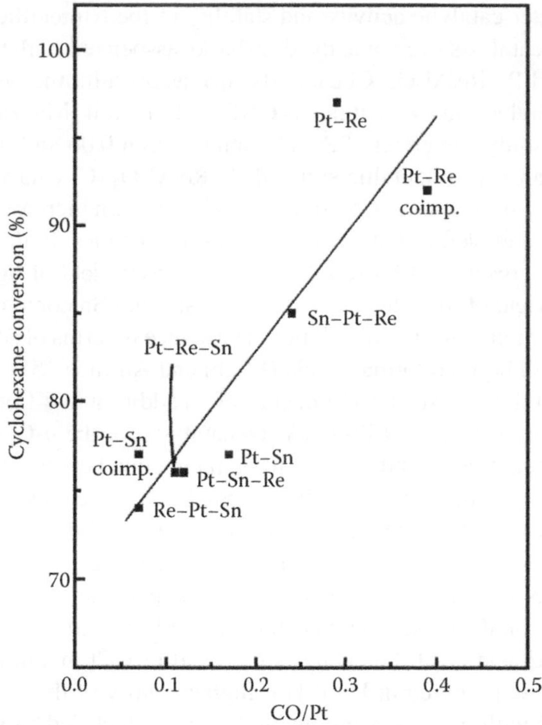

FIGURE 6.14 Correlation between catalytic activity during dehydrogenation of cyclohexane and CO/Pt ratio (regression coefficient, $R^2 = 0.934$). Coimp: catalysts prepared by coimpregnation of metal precursors. Suc: catalysts prepared by successive impregnations of metal precursors in the order indicated. Reaction conditions: temperature = 400°C, pressure = 0.1 MPa, H_2/hydrocarbon molar ratio = 30. (Reprinted from *Catalysis Today* 107–108, Pieck, C. L. et al., Metal dispersion and catalytic activity of trimetallic Pt–Re–Sn/Al_2O_3 naphtha reforming catalysts, 637–642, Copyright 2005, with permission from Elsevier.)

a linear correlation was established between the dehydrogenation activity of bimetallic and trimetallic catalysts and metal dispersion (or the number of accessible Pt atoms) measured by the CO chemisorption. Furthermore, the divergence from the linear behavior for some of the catalysts was attributed to the differences in electron density around Pt atoms induced by the presence of different metals (i.e., Sn and Re) (Pieck et al. 2005).

Addition of Sn to the Pt catalysts has been shown to create Pt_xSn alloys that increase the stability of the catalysts in return. However, different preparation, pretreatment, or posttreatment methods may provoke formation of different Pt_xSn alloy phases, which may or may not exhibit catalytic activity for certain reactions. For instance, there exist five different Pt_xSn alloy phases such as $PtSn_4$ orthorombic, $PtSn_2$ cubic, Pt_2Sn_3 hexagonal, PtSn hexagonal, and Pt_3Sn centered cubic phases. Among these alloy phases, catalytic activity was specifically attributed to Pt_3Sn and partly to PtSn phases. However, alloys with higher Sn content were considered to exhibit no catalytic activity for most of the reactions. In order to elucidate the formation and the catalytic behavior of different PtSn alloys, unsupported PtSn samples were prepared and exposed to different treatments such as pre-sintering, O_2 and H_2 treatments. After each treatment, their surface states and bulk structures were characterized, and catalytic activities were investigated in various test reactions. As illustrated in Figure 6.15, the mixture of Pt_3Sn and SnO_2 represented the final stabilized state formed after successive heating in air, and in H_2 eventually. Activity tests revealed that this mixed PtSn alloy presents no catalytic activity for "structure-sensitive" reactions such as dehydrogenation of cyclohexane. Nevertheless, it was active in "structure-insensitive" hydrogenation as well as in the dehydrogenation of cyclohexene (Paál et al. 2011).

Since catalytic activity in CNR reactions is controlled by a complex interplay of both metallic and acidic sites, various studies have been focused on modifying these sites by the addition of promoters. Nearly all of the commercial CNR catalysts accommodate chlorinated alumina in their formulations as the source of the acidic function. Some of the studies have been focused on application of different support materials in CNR catalysts. In this respect, zirconia- and alumina-supported Pt and Pt–Sn catalysts were tested in the dehydrocyclization of *n*-octane by using pure

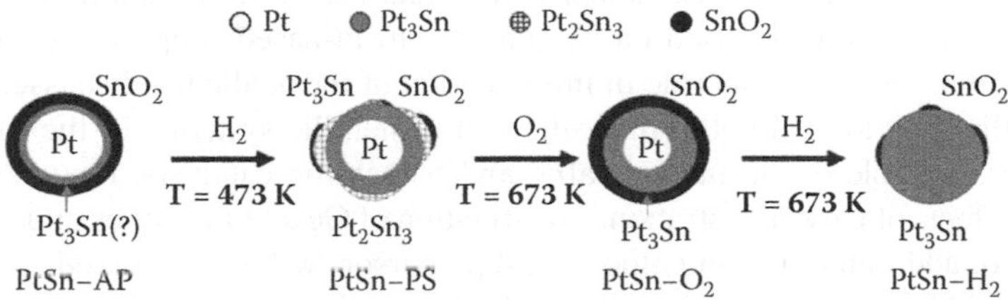

FIGURE 6.15 Representation of solid-state transformations after different pretreatments. The question mark after Pt_3Sn in PtSn–AP (AP: as prepared without pretreatment) means that it is not detected as a phase, but its formation may have started from dissolved Sn in Pt. (Reprinted from *Applied Catalysis A: General* 391 (1–2), Paál, Z. et al., Structural properties of an unsupported model Pt–Sn catalyst and its catalytic properties in cyclohexene transformation, 377–385, Copyright 2010, with permission from Elsevier.)

hydrogen (H_2) as well as water-vapor containing hydrogen (WVH_2). As anticipated, the Pt/Al_2O_3 and $Pt–Sn/Al_2O_3$ catalysts exhibited a bifunctional activity in H_2 atmosphere, resulting in the formation of a mixture of benzene and mono- and dialkylaromatics. However, they indicated a monofunctional activity in WVH_2 atmosphere and produced a mixture of *o*-xylene and ethylbenzene. The monofunctional activity of Pt/Al_2O_3 and $Pt–Sn/Al_2O_3$ catalysts in WVH_2 atmosphere was attributed to the loss of acidity caused by water vapor. Furthermore, the addition of Sn to these catalysts significantly decreased the catalyst deactivation rate by the virtue of well-known "ensemble effect" caused by the formation of PtSn alloys. On the other hand, zirconia-supported Pt and Pt–Sn catalysts showed a monofunctional activity toward the formation of *o*-xylene and ethylbenzene independent of the reaction atmosphere. In contrast to the alumina-supported catalysts, the zirconia-supported catalysts were found to be more stable in WVH_2 than in H_2 atmosphere. Moreover, the addition of Sn to the zirconia-supported catalysts significantly increased the catalytic activity but could not improve their resistance to deactivation in H_2 atmosphere because of the fact that PtSn alloys could not form (Hoang et al. 2007).

Similar to $Pt–Re/Al_2O_3$ catalysts, $Pt–Ir/Al_2O_3$ catalysts are presulfided in industrial operations to suppress their high hydrocracking activities. In order to eliminate the presulfiding step while retaining the high catalytic activity of the sulfided $Pt–Ir/Al_2O_3$ catalysts, the effect of addition of Sn to this catalyst system was also investigated for *n*-heptane dehydrocyclization. Accordingly, addition of Sn to the monometallic Pt/Al_2O_3 and bimetallic $Pt–Ir/Al_2O_3$ catalysts can substitute the presulfiding step with an increased stability and selectivity toward toluene. However, the deposition of Sn influences both the acidic and the metallic functions of the catalyst. Results also showed that the yield of toluene is the same for $Pt–Sn/Al_2O_3$ and $Pt–Ir–Sn/Al_2O_3$ catalysts; however, less Sn is required in the case of the trimetallic catalyst. The yield of toluene was found to be dependent on the amount of Lewis acid sites. As the amount of Lewis acid sites increases, the yield of toluene increases as well. Furthermore, the coke deposition resulted in a decrease in the amount of strong Lewis and Brønsted acid sites (Epron et al. 2005).

It is known that Ge, an inactive metal, promotes both stability and selectivity toward the desired reactions of the naphtha reforming reaction network, such as dehydrogenation of naphthenes, isomerization, and dehydrocyclization of paraffins. Numerous studies were focused on Ge addition to Pt-based supported catalysts in order to better understand its role in modification of the acidic (as discussed above) and metallic functions in both bimetallic and trimetallic systems. In the following paragraphs, the role of Ge in bimetallic and trimetallic catalysts, methods of Ge addition, effects of Ge concentration, investigation of Ge addition steps and their pH, and order of addition of Ge and other metal precursors will be discussed.

Effect of Ge content was studied for the bimetallic systems such as Ge–Pt on Al_2O_3 (Mariscal et al. 2007). Coimpregnation (CI) method was utilized to synthesize the catalysts with fixed 0.5 wt.% Pt loading on chlorinated Al_2O_3, whereas Ge amount was changed so that Pt/Ge atomic ratios of 0.3, 0.7, 1.4, 2.1, and 3.5 were obtained. Again, FTIR study of CO chemisorbed surfaces showed that Ge acts as electron acceptor decreasing the electron density of Pt (Santos et al. 2005). Thus, by decreasing electron density on Pt and generating charged electron-deficient atoms, adsorption capacity of Pt surface atoms was decreased. This finding was also in agreement with a decline in metallic activity in cyclohexane dehydrogenation. Increase in the Ge content also resulted in a decline in hydrogenolysis activity and the formation of aromatics, and a reduction in the production of gas and in deactivation by coke, thus improving catalyst stability. The maximum isoparaffin yield in *n*-heptane reforming tests was obtained for a Ge loading of 0.27 wt.% corresponding to a Pt/Ge atomic ratio of 0.7. Both electronic and geometric effects were suspected to be responsible for modification of metal function (Mariscal et al. 2007). Moreover, FTIR study of CO chemisorbed surfaces showed that Ge acts as an electron acceptor and decreases the electronic density of Pt. This finding was also in agreement with the decreased hydrogenation activity of Pt. The electronic effect was thought to be because of the neighborhood metal atoms and their stoichiometric alloys, as the neutral Pt was found to be much more active than electron-rich/deficient Pt. Besides, addition of Ge decreased the CH dehydrogenation activity more than Re, which was in agreement with the XPS result and indicated electronic effect. On the other hand, as a result of Ge addition, isomerization activity was found to be increased, whereas that of the aromatization and hydrogenolysis decreased. Moreover, enhanced stability of the trimetallic catalyst was present due to Ge addition (Santos et al. 2005).

Another study made use of organometallic grafting to prepare Ge–Pt/Al_2O_3 catalysts in amounts corresponding to 1/12, 1/8, 1/2, 1, or 2 monolayers to understand Pt–Ge interactions and the surface. EXAFS analysis showed that Ge atoms scattered on the surface of Pt with the 1/12 and 1/8 monolayer samples. Ge positioned selectively on the high-coordination sites for 1/12 and 1/8 monolayers; however, deposition is less selective for 1/2 monolayer sample. Since high coordination sites were active for benzene hydrogenation, this reaction was suppressed for 1/8 monolayer samples. In the case of larger monolayers, 1 or 2, more Ge was found in the neighborhood of Pt, and shorter bond length between Pt–Ge was observed. Thus, Pt_xGe_y solid solution might be formed as the Ge atoms were intercalated in the Pt subsurface layers or in the particles. These catalysts behaved similarly in hydrocarbon transformations as the non-Ge added Pt catalysts (Wootsch et al. 2006).

Catalytic reduction method was also utilized to prepare the Ge–Pt–Ir/Al_2O_3 catalyst in order to investigate the influence of the pretreatment method. It was found that Ge influences both the metal and acidic functions at different levels with respect to the deposited Ge content and on the activation treatment. The deposition of Ge occurs mainly on the Pt–Ir phase, which leads to an inhibition of the metallic activity. However, if the trimetallic catalyst is calcined before reduction, such kind of effect is not strongly present and in that case, a slight segregation of Ge occurs. On the other hand, the best catalytic performance was achieved when Ge content was as low as 0.1 wt.% and there was a direct reduction as an activation step (Samoila et al. 2007).

One of the most common trimetallic catalysts used in naphtha reforming is Pt–Re–Ge/Al_2O_3. Two types of active sites as Pt–Re ensembles and Re interacting with the support exist in a Pt–Re reforming catalyst (Grau and Parera 1991). Early studies in the last decade were dedicated to understand the effects of metal addition on the electronic environment at the surface. It was found that addition of Ge affects the Pt electronic state in a greater extent than Re (Santos et al. 2005). In this study, XPS analysis illuminated that there was a higher binding energy shift on Pt when Ge was introduced to the catalyst as shown in Figure 6.16.

Besides the work of Santos et al. (2005), several other studies investigate the effect of Ge content in both bimetallic and trimetallic systems. One of them made use of catalytic reduction method to prepare Pt–Re–Ge/Al_2O_3 and Pt–Ir–Ge/Al_2O_3 catalysts with varying Ge content. For Pt–Re–Ge/Al_2O_3, relative activity of the catalyst to the parent Re–Pt bimetallic one as a function of Ge content covers two distinct regions as shown in Figure 6.17. Between 0 and 0.3 wt.% loading, there is a 40% reduction in the activity, whereas there is a smaller slope for the high Ge-loaded region, from 0.3 to 1.4 wt.%. This result indicates that preferential deposition of Ge on Pt–Re ensembles is present at low Ge contents, and possible deposition of

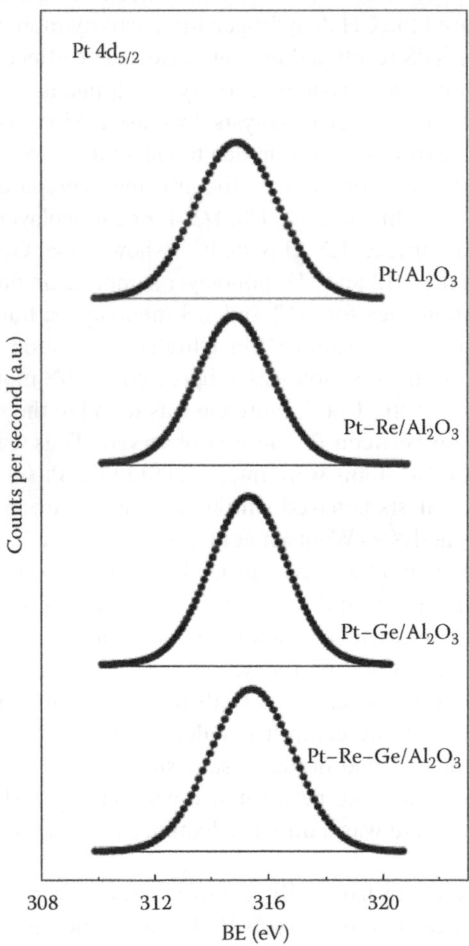

FIGURE 6.16 Pt $4d_{5/2}$ line profiles of catalysts pre-reduced under H_2 flow. (With kind permission from Springer Science+Business Media: *Catalysis Letters*, The effect of the addition of Re and Ge on the properties of Pt/Al_2O_3, 103 (3–4), 2005, 229–237, Santos, M. C. S. et al.)

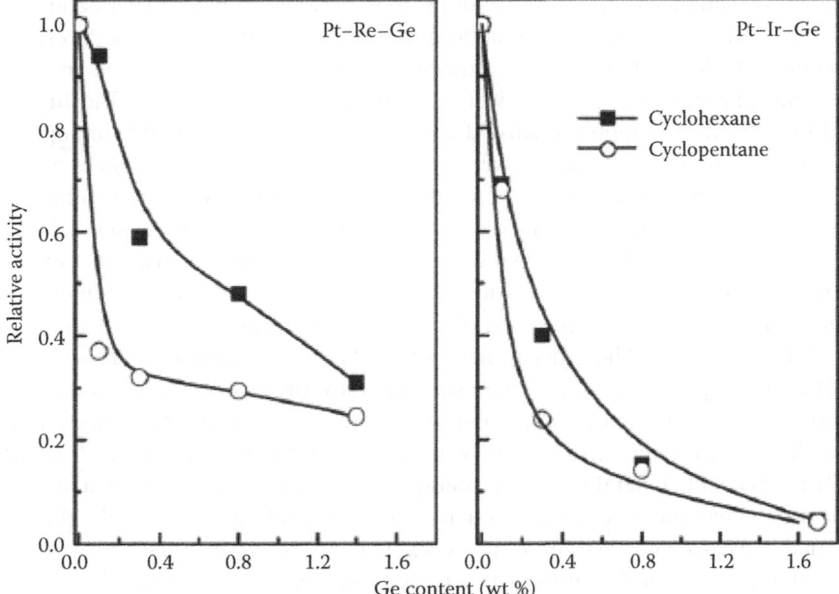

FIGURE 6.17 Relative activity for reactions of cyclohexane dehydrogenation and cyclopentane hydrogenolysis as a function of the Ge content. (Reprinted from *Applied Catalysis A: General* 319, Benitez, V. et al., Preparation of trimetallic Pt–Re–Ge/Al_2O_3 and Pt–Ir–Ge/Al_2O_3 naphtha reforming catalysts by surface redox reaction, 210–217, Copyright 2007, with permission from Elsevier.)

Ge to the support can be speculated at high loadings. In the case of the Pt–Ir–Ge catalysts, Ge preferentially deposits on Pt; hence, similar decay pattern of the cyclohexane dehydrogenation and cyclopentane hydrogenolysis activity values is obtained because only Pt is responsible for dehydrogenating activity and oppositely hydrogenolysis capacity is mainly supplied by Ir. Even a smaller amount of Ge is thought to be sufficient to inhibit the activity of Ir, and higher loadings of Ge would inhibit the metal function of the catalyst (Benitez et al. 2007). In a similar study, it was found that the addition of Ge in small quantities, such as 0.1 wt.%, causes the breaking of Pt–Re ensembles and modification of the electronic density of Pt in the Pt–Re–Ge/Al_2O_3 catalyst, resulting in a decline in activity for both hydrogenolysis and dehydrogenation (Mazzieri et al. 2009a).

In another study, the effect of Ge addition method (with 0.3 wt.%) to Pt–Re/Al_2O_3 and Pt–Ir/Al_2O_3 catalysts was investigated. For this purpose, the successive impregnation (SI) and CI methods followed by different catalytic reduction procedures were considered. For all of the cases, Ge addition decreases the activity of the metal function. Dehydrogenation and hydrogenolysis activities were inhibited by Ge addition; however, the effect is softer in the case of Pt–Re–Ge catalysts. This was thought to be because of the weaker interaction of Ge with the bimetallic Pt–Re particles. Lowest hydrogenolytic activity was observed for the catalysts prepared by successive impregnations. This method was found to be the most effective for producing stronger interactions of Ge clusters with the metal function. On the other hand, catalytic reduction (i.e., surface redox reaction) was found to be the best method, resulting in the most stable catalysts, with the lowest cracking with the highest toluene yield. These two impregnation methods were also utilized to alter the metallic function of Al_2O_3–SIRAL 40 supported Rh–Pd catalysts for MCH and decalin ring opening (D'Ippolito et al. 2015). Catalysts prepared by CI had smaller particle sizes as indicated by the higher H_2 chemisorption values. It was inferred that the formation of metal alloys is responsible for this difference as random deposition of second metal was previously reported (Del Angel et al. 2003). Because of the structure-sensitive nature of the hydrogenolysis reactions and the effect of the preparation method to the acidic function of the supports, the CI method yielded a more active catalyst than that of the SI method. The order of addition of the metal precursor is also found to be affecting the properties of the catalysts. In a study, the metal precursors are added in a different order using the SI method for the preparation of Pt–Re–Ge/Al_2O_3. For all of the cases, the interaction of Pt with Re is found to be much more than with Ge. If the last added metal is Ge, Pt interacts more with Re, and there is a decrease in electron depletion effect of Ge. When Re is added before other metals, the most distorted Pt electron band structure was obtained, indicating several electronic states of Pt. Moreover, when Pt is impregnated after other metals, it is more alloyed to the other metals in the solid. In contrast, the samples prepared by impregnating Pt before the other metals are more active in n-octane reforming tests than the others, indicating that Pt is mostly segregated from the other metals. Accordingly, the least activity among the catalysts is obtained when impregnating Ge before other metals because of stronger interactions among the metals. An intermediate behavior is obtained in the case of Re addition before other metals (Carvalho et al. 2012).

Another aspect is the pH of the Ge addition step in preparing the catalyst. This effect was investigated by using different impregnation media, such as H_2O, HCl, and NH_3 for the preparation of Pt–Re–Ge on γ-alumina. It was found that there is a weak metal–support interaction when ammonia is used as the catalyst retains less chlorine. In contrast, there is a strong metal–support interaction for the case of HCl. Correspondingly, neutral water resulted in in-between metal–support interactions (D'Ippolito et al. 2008). In a similar study, FTIR results for CO chemisorbed samples revealed that the trimetallic catalysts prepared with H_2O or HCl have a good interaction between Ge and Pt–Re metals. This interaction is found to be in a weaker level for the catalyst prepared with NH_3; thus, Ge is mainly in a segregated state. On the other hand, trimetallic catalysts prepared by using a neutral aqueous solution are the most active and toluene-selective in *n*-heptane reforming reaction tests, compared to those prepared using HCl or NH_3 solution (D'Ippolito et al. 2009a).

Several studies have been also carried out to elucidate the influence of In addition on the catalytic activity and selectivity of various CNR catalysts. For instance, the addition of In to the trimetallic Pt–Re–Ge/Al_2O_3 and Pt–Re–Sn/Al_2O_3 catalysts was reported to restrain both the acidic and metallic functions. The total acidity of the catalysts was reduced by the addition of In, which resulted in a decrease in the acid function-controlled reactions, such as hydrocracking and isomerization of *n*-heptane and *n*-pentane. Besides, a significant decrease in CO chemisorption capacity corroborated the fact that the metal active sites were neutralized by the addition of In.

As demonstrated in Figure 6.18, hydrogenolysis of cyclopentane and dehydrogenation of cyclohexane reactions were greatly hampered in the presence of In (Benitez et al. 2009).

Similarly, the addition of In to a bimetallic Pt–Re/Al_2O_3 catalyst yielded a significant decrease in acidity. Another significant effect of In addition was found to be the deactivation of the metallic function, which inhibited dehydrogenation and hydrogenolysis reactions. The total activity of the catalysts in the presence of In was reduced for n-C_5 isomerization reactions. However, the selectivity toward i-C_5 isomers was increased, and it was decreased toward low-value light gases. The results of activity test for the reaction of n-C_7 reforming indicated that the catalyst stability and the selectivity toward aromatics were increased, and the yield of light gases was decreased. Additionally, the deactivation of the catalyst due to coke formation was reduced in the presence of In (Benitez and Pieck 2010).

Another study was focused on elucidation of the effect of In on the catalytic activity and selectivity of trimetallic Pt–Sn–In/Al_2O_3–Cl catalysts. The results showed that In may react with the highly reactive alumina sites; thus it weakens the interaction between Sn and alumina. Temperature-programmed reduction studies revealed that In interacts strongly with Pt and also promotes reduction of Sn.

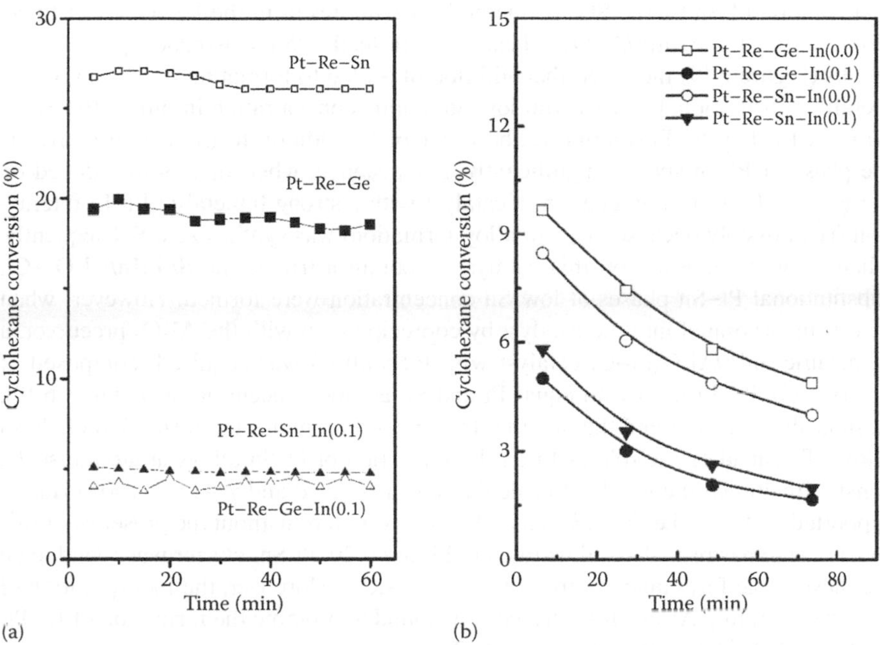

FIGURE 6.18 Effect of In addition to trimetallic Pt–Re–Ge/Al_2O_3 and Pt–Re–Sn/Al_2O_3 catalysts on the conversion of (a) cyclohexane and (b) cyclopentane as a function of time on stream. The numbers in parenthesis represent the weight percent of In. (Reprinted with permission from Benitez, V. M. et al., Modification of multimetallic naphtha-reforming catalysts by indium addition. *Industrial and Engineering Chemistry Research* 48 (2):671–676, Copyright 2009 American Chemical Society.)

Furthermore, addition of even a small amount of In (0.06 wt.%) yielded formation of a Pt_3Sn alloy. As the content of In was increased, formation of Pt_xSn alloys with almost equal Pt and Sn atomic concentrations was also increased. Moreover, the CO chemisorption capacity of these catalysts was decreased by In addition, provoking a modification of Pt electronic properties at higher In contents. The addition of In also decreased the acidity of the support in trimetallic catalysts. Consequently, In interacts with the metallic and acidic sites, which is reflected by the decrease in overall activity because of the blockage of Pt sites, and the decrease in hydrocracking selectivity but the increase in isomerization selectivity because of loss of acid sites (Jahel et al. 2010).

As discussed above, the addition of Sn to the Pt-based CNR catalysts enhances the catalyst stability through the formation of Pt_xSn alloys. Catalysts incorporating different metals have been studied to reveal their promoting effect on the formation of Pt_xSn alloys. In this context, the role of In doping on the formation of stable Pt_xSn alloys on γ-Al_2O_3 support was investigated. Aiming to explore the effect of In on Pt_xSn alloy formation, monometallic Pt-based, bimetallic SnPt-based, and trimetallic SnPtIn-based catalysts were prepared by "surface organometallic chemistry on metals (SOMC)" method with addition of In at different steps of synthesis. In contrast to the majority of the studies, data gathered in this study reveal the fact that addition of Sn by the SOMC method does not result in the formation of Pt_xSn alloys in bimetallic SnPt/Al_2O_3–Cl catalysts. Instead, ^{119}Sn Mössbauer spectroscopy results (Figure 6.19) indicated that addition of Sn led to formation of $Pt_xSn(O)$ oxo-metallic substitutional phases with low Sn atomic concentration in bimetallic SnPt/Al_2O_3–Cl catalysts. Furthermore, the order of In addition to the catalysts affects the phase of Pt_xSn species significantly. For instance, when In was introduced on the γ-Al_2O_3 by wet impregnation, a catalyst with a strong bimetallic Pt–In interaction (i.e., possibly because of Pt_xIn alloy formation) was synthesized. Subsequently, when Sn was deposited on this catalyst to obtain a trimetallic SnPtIn/Al_2O_3–Cl, substitutional Pt–Sn phases at low Sn concentration were formed. However, when In was incorporated into the catalyst by coprecipitation with the Al_2O_3 precursor, a trimetallic SnPt/Al_2O_3In–Cl catalyst with PtSn alloys was acquired, composed of a very high Sn^0/Pt ratio with equal Pt and Sn atomic concentrations (Figure 6.19). Results also corroborated the fact that In^{3+} species promotes the formation of Pt_xSn alloys. To elucidate the role of In on the formation of Pt_xSn alloys at atomic scale, density functional theory (DFT) calculations of isolated and γ-Al_2O_3 (100 surface)-supported Pt_xSn_{13-x} (i.e., $x = 13$, 7, and 0) clusters with or without the presence of In^{3+} species were performed. As illustrated in Figure 6.20, Pt_7Sn_6 cluster was found to be the most stable form among other Pt_xSn_{13-x} clusters. Moreover, the incorporation of In^{3+} species into γ-Al_2O_3 (100 surface) was found to promote the formation of In–Pt, In–Sn, and O–Sn bonds (Jahel et al. 2012).

In a similar study, ^{119}Sn Mössbauer spectroscopy was used to elucidate the effect of In addition on the trimetallic Pt/Al_2O_3SnIn–Cl catalyst framework with different In loading. The results showed that addition of even a small amount of In (0.06 wt.%) led to formation of Pt_xSn alloys that also increased with increasing In loading (Figure 6.21). Results illustrated that the overall conversion decreased because of the loss of active metal surface of Pt with alloy formation. Furthermore, the selectivity

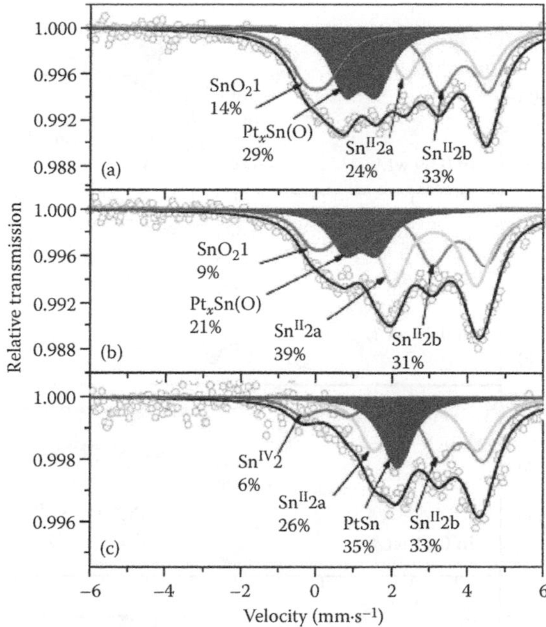

FIGURE 6.19 Sn Mössbauer spectra of reduced (a) SnPt/Al_2O_3–Cl, (b) SnPtIn/Al_2O_3–Cl, and (c) SnPt/Al_2O_3In–Cl catalysts obtained by Sn SOMC. (Reprinted with permission from Jahel, A. N. et al., Effect of indium doping of γ-alumina on the stabilization of PtSn alloyed clusters prepared by surface organostannic chemistry. *The Journal of Physical Chemistry C* 116 (18):10073–10083, Copyright 2012 American Chemical Society.)

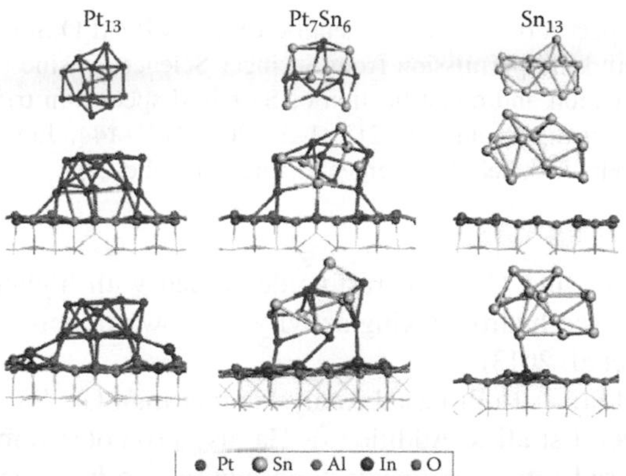

FIGURE 6.20 DFT structures of the optimized Pt_xSn_{13-x} systems: in gas phase (upper panel), Pt_xSn_{13-x}/γ-Al_2O_3 (100) without In (middle panel), and Pt_xSn_{13-x}/γ-Al_2O_3 with In^{3+} promotion in the support (lower panel). (Reprinted with permission from Jahel, A. N. et al., Effect of indium doping of γ-alumina on the stabilization of PtSn alloyed clusters prepared by surface organostannic chemistry. *The Journal of Physical Chemistry C* 116 (18):10073–10083, Copyright 2012 American Chemical Society.)

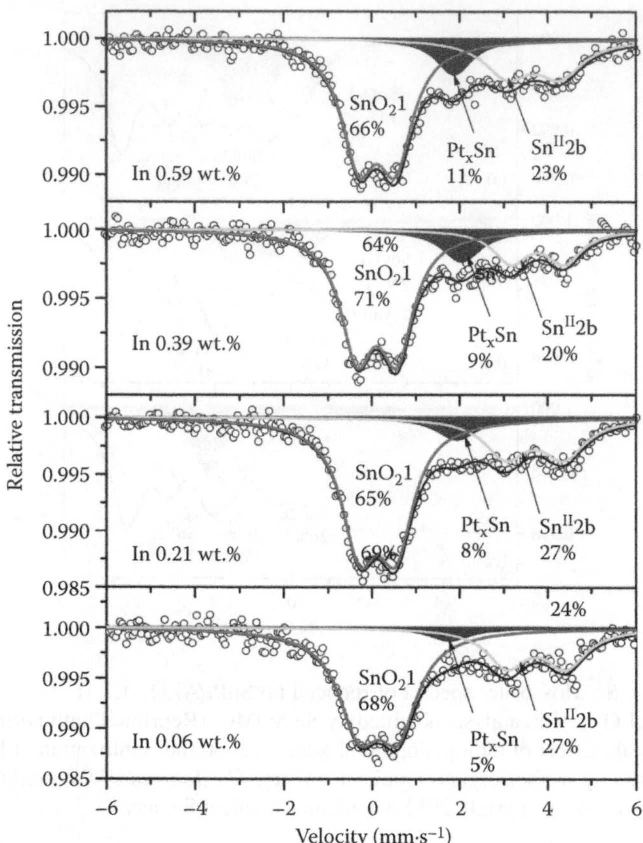

FIGURE 6.21 Sn spectra of reduced trimetallic catalysts Pt/Al_2O_3SnIn–Cl with different indium content. (With kind permission from Springer Science+Business Media: *Hyperfine Interactions*, Identification and quantification of Sn-based species in trimetallic Pt–Sn–In/Al_2O_3–Cl naphtha-reforming catalysts, 217 (1–3), 2013, 137–144, Jumas, J. C., Sougrati, M. T., Olivier-Fourcade, J., Jahel, A., Avenier, P., and Lacombe, S.)

of catalyst toward C_1 and C_3–C_4 paraffins decreased with higher indium loading as a result of decreased hydrocracking activity, but toward increased isomerization selectivity (Jumas et al. 2013).

Several other additives to Pt-based catalysts other than Ge, Sn, and In have been investigated in several studies. Addition of Ga also promotes some of the metallic properties of Pt-based reforming catalysts in addition to its effect on acidic function as discussed above. Catalyst stability and selectivity toward diminishing hydrogenolysis activity were found to be enhanced by the modification of the Pt/Al_2O_3 catalyst with Ga around 3.5 wt.% (Romero et al. 1990). Electronically deficient state of Pt clusters was also found to be present, as in the case of Ge addition, and associated with the electronic interaction between Pt and low-valance gallium ions. Transformations of *n*-hexane on Ga-introduced Pt/Al_2O_3 and Pt/SiO_2 catalysts also revealed that Ga increases the Pt dispersion on Al_2O_3, whereas there is a formation of PtGa alloy species for bimetallic SiO_2-supported catalysts (Gobölös et al. 2006).

Species of iron are widely used promoters that can chemically anchor preventing the aggregate formation of noble metals such as Pt. In a study, iron is one of the elements used to promote Pt supported on SiO_2 with varying loadings. In this case, there is a formation of electron-rich Pt because of the electronic interactions between Pt and Fe. The degree of electron density transfer between Fe and Pt reduces with increasing Fe loading. Small concentrations of Fe enhance the activity of Pt for the dehydrogenation of cyclohexane. This improvement in performance is associated with either an electronic effect or probably the presence of bimetallic Pt–Fe sites. On the other hand, large amounts of Fe ultimately cause the deactivation of Pt. Formation of iron oxide was suggested as the reason for this phenomenon (Siani et al. 2009). In a more recent study, effect of Fe in Pt/KL catalysts was studied, and it was reported that there is a formation of electron-rich Pt species when nonframework Fe is introduced; besides, high dispersion of Pt is obtained. On the other hand, when Al in the KL zeolite framework is partially changed with Fe, electron-deficient Pt species are observed with a decrease in Pt dispersion. Overall, results were promising with respect to the use of Fe as a promoter since Pt can also be tuned (Song et al. 2015).

As discussed above, germanium and tin are two of the most commonly investigated IVA Group additives for a Pt–Re/Al_2O_3 reformer catalyst. In a study, Pb, another IVA Group element, was also analyzed to reveal its effects on catalytic properties of the bimetallic pair. TPR studies indicated a close interaction of Pt with Pb, and the latter is anchored on the surface of the former. Besides, there is an implication of Pb deposition on strongest acid sites, which makes Pt oxides occupy the weaker acid sites. Decline in the metal function activity is also present as in the cases of Ge and Sn addition due to the interaction of Pb with Pt–Re. Hydrogenation activity is more affected than dehydrogenation, and thus there is improvement on the stability and selectivity of catalysts to C_7 isomers and aromatics in *n*-heptane reforming tests. A small amount of Pb, such as 0.1 wt.% Pb accompanied by Pt and Re loadings of 0.3 wt.%, is found to be sufficient to bring the stated promoter effects; see (Figure 6.22) (Sánchez et al. 2011).

Pb–Pt bimetallic catalysts were also investigated on $MgAl_2O_4$ support for *n*-butane dehydrogenation. Moreover, trimetallic catalysts such as Pt–Pb–In on the same support were also studied as a novel catalytic system. In contrast to the Pt–Re system, H_2 chemisorption studies indicated that there is no appreciable change in the Pt catalyst for low concentration of Pb, up to 0.25 wt.%. In the case of higher concentrations, such as 0.52 and 0.87 wt.%, there is a substantial decrease in H_2 uptakes indicating the presence of geometric and electronic effect of Pb atoms since Pt particle sizes do not change significantly. On the other hand, activation energy of cyclohexane dehydrogenation reaction increases with respect to the monometallic catalyst for low Pb-containing samples suggesting the electronic effects of Pb on Pt sites. For high concentrations of Pb, however, both rate and activation energy decrease due to the blocking effect of Pt surface atoms. Introduction of Pb at low concentrations also promotes the catalytic performance of *n*-butane dehydrogenation compared to monometallic Pt because of the electronic and dilution effects. These effects are also

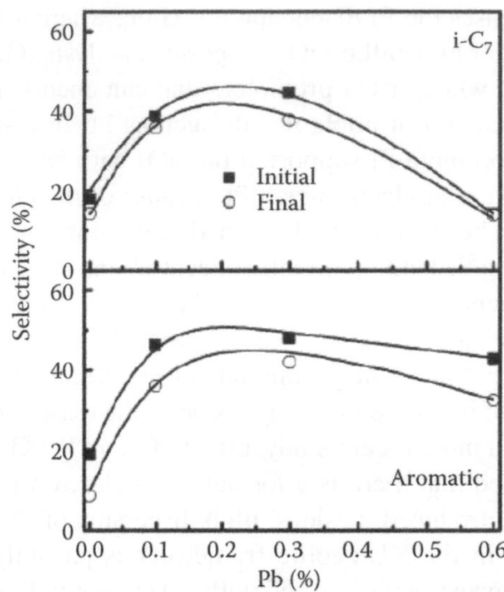

FIGURE 6.22 Selectivity to C_7 isomers and aromatics as a function of Pb content. (Sánchez, M. A. et al. Catalytic activity of Pt-Re-Pb/Al_2O_3 naphtha reforming catalysts. *Journal of Chemical Technology & Biotechnology* 86 (9). 1198–1204. 2011. Copyright Wiley-VCH Verlag GmbH & Co. KGaA. Reproduced with permission.)

present in a trimetallic Pt–Pb–In system in which impregnation sequence has a great effect on catalytic performance (Bocanegra et al. 2013).

Surface structure of a metallic component and its interaction with support have a significant effect on obtaining high-quality reformate with an improved octane number. Besides adding a third metal, the preparation method of the catalyst is another tool to modify the surface structure. A study made use of polyacrylic acid (PAA) as a template in preparation of Pt/Re/γ-alumina catalysts and obtained less coordinated surface Pt atoms. Up to 2.4% increase in reformate yield and 2% lower aromatic production than that of benchmark commercial catalysts were associated with the low coordination number. Besides, increase in the coke resistivity was also present in the studies with PAA template catalysts due to the same effect (Liu et al. 2010).

The effect of the preparation method was also present in a study in which sol–gel method was used to obtain Pt/In–Al_2O_3 catalysts. The formation of benzene was suppressed when In was doped using the sol–gel method instead of CI due to the change in Pt properties by the influence of the support (Rodríguez-González et al. 2007). Moreover, catalytic behavior of the Pt–Sn/Al_2O_3 catalysts in *n*-heptane reforming was also investigated for the comparison of the two preparation methods. The one-step sol–gel method resulted in enhancements in both coke resistivity and activity. NMR spectroscopy revealed that there was a shift to formation of coordinatively unsaturated aluminum (Al^v), which was associated with the formation of Al defects, and increase in support acidity when the sol–gel method was utilized. A stronger interaction with Pt and unsaturated Al^v sites led to a high dispersion of Pt particles, as revealed by TEM imaging and CO chemisorption measurements. Moreover, metal–support boundary sites, which are promoted by the acidity of the support, were suggested as the reason for the observed inhibition of benzene production (Ignacio et al. 2009).

Lastly, alternative catalysts to Pt-based systems were also investigated for several reforming reactions. Chromium supported on zirconia was one of the representatives of this class of catalysts. It was found to be active for dehydrogenation and ring closure reactions due to isolated Cr^{3+} ions (Raissi et al. 2010). Molybdenum and tungsten oxide-based bifunctional catalysts were also introduced in replacement of Pt (Al-Kandari et al. 2008; Katrib et al. 1996). By the help of controlled hydrogen treatment as a function of temperature, ensembles containing both Mo and MoO_2, supported on TiO_2, were obtained, and they were found to possess both acidic and metallic functions (Gulino et al. 1996; Jones et al. 1997). Moreover, Pd was also utilized in the form of Pd complexes, Pd-di(ethylthio)ethane, and Pd-di(ethylthio)propane, supported on MCM-41 to obtain effective catalysts for hydrocarbon reforming. These complexes were covalently anchored to the MCM-41 matrix, and remarkable catalytic properties were stated to be observed in *n*-hexane and *n*-heptane reforming reactions. Thermogravimetric analysis shows that these complexes were thermally stable up to the reaction temperature and recyclable for further utilization (Mishra and Kumar 2011). Ni was considered as a cheap alternative to Pt-based catalysts, especially as Ni phosphides. Ni_2P/Al_2O_3 catalysts have superior activity, stability, and selectivity over those of non-phosphide ones. Promotion of the activity of Ni_2P in dehydrogenation of cyclohexane and elimination of sintering are achieved by the ensemble effect of P in Ni_2P and small positive charge of Ni (Li et al. 2014).

6.8 CONCLUSIONS

CNR is a key process in producing high-octane gasoline components. Different CNR technologies such as semiregenerative, cyclic, and continuous catalytic reformers have been commercialized to boost the octane rating of gasoline. For these CNR technologies, the bifunctional reforming catalyst plays an indispensable role in achieving this goal. Their evolution from monometallic to trimetallic has provided enhanced catalytic performance and stability. Recent attempts in the last decade on the modification of reformer catalysts to overcome the issues related to catalytic performance and stability, specifically focusing on the modification of metal (i.e., addition of Re, Sn, In, Ge, Ga, etc.) and acid functions (i.e., CeO_2, Zr-HMS, etc.), have been discussed in detail. Studies covered illustrate that both the performance and stability of the catalysts can be improved tremendously. However, there is still so much to explore; research should be directed to achieve a fundamental level of understanding, and then to utilize this understanding to solve the issues regarding real-life applications.

REFERENCES

Aitani, A. M. 2004. Licensed Reforming Processes. In *Catalytic Naphtha Reforming, Revised and Expanded*: CRC Press, Boca Raton, FL.

Akhtar, M. N., AlYassir, N., Al-Khattaf, S., and Čejka, J. 2012. Aromatization of alkanes over Pt promoted conventional and mesoporous gallosilicates of MEL zeolite. *Catalysis Today* 179 (1):61–72.

Al-Kandari, H., Al-Kharafi, F., and Katrib, A. 2008. Isomerization reactions of n-hexane on partially reduced MoO_3/TiO_2. *Catalysis Communications* 9 (5):847–852.

Ali, S. A. 2004. Naphtha Hydrotreatment. In *Catalytic Naphtha Reforming, Revised and Expanded*: CRC Press, Boca Raton, FL.

Axens, Accessed on February 19, 2016. http://www.axens.net/document/703/octanizing-reformer-options/english.html

Azzam, K. G., Jacobs, G., Shafer, W. D., and Davis, B. H. 2010. Aromatization of hexane over Pt/KL catalyst: Role of intracrystalline diffusion on catalyst performance using isotope labeling. *Journal of Catalysis* 270 (2):242–248.

Barbier, J. 1986. Deactivation of reforming catalysts by coking—A review. *Applied Catalysis* 23 (2):225–243.

Benitez, V. M., and Pieck, C. L. 2010. Influence of indium content on the properties of Pt–Re/Al_2O_3 naphtha reforming catalysts. *Catalysis Letters* 136 (1–2):45–51.

Benitez, V., Boutzeloit, M., Mazzieri, V. A. et al. 2007. Preparation of trimetallic Pt–Re-Ge/Al_2O_3 and Pt–Ir–Ge/Al_2O_3 naphtha reforming catalysts by surface redox reaction. *Applied Catalysis A: General* 319:210–217.

Benitez, V. M., Vera, C. R., Rangel, M. C., Yori, J. C., Grau, J. M., and Pieck, C. L. 2009. Modification of multimetallic naphtha-reforming catalysts by indium addition. *Industrial and Engineering Chemistry Research* 48 (2):671–676.

Bocanegra, S. A., Scelza, O. A., and De Miguel, S. R. 2013. Behavior of $PtPb/MgAl_2O_4$ catalysts with different Pb contents and trimetallic PtPbIn catalysts in n-butane dehydrogenation. *Applied Catalysis A: General* 468:135–142.

Carvalho, L. S., Conceição, K. C. S., Mazzieri, V. A., Reyes, P., Pieck, C. L., and Rangel, M. D. C. 2012. Pt–Re–Ge/Al_2O_3 catalysts for n-octane reforming: Influence of the order of addition of the metal precursors. *Applied Catalysis A: General* 419–420:156–163.

D'Ippolito, S. A., Vera, C. R., Epron, F., Especel, C., Marécot, P., and Pieck, C. L. 2008. Naphtha reforming Pt–Re–Ge/γ-Al_2O_3 catalysts prepared by catalytic reduction. Influence of the pH of the Ge addition step. *Catalysis Today* 133–135 (1–4):13–19.

D'Ippolito, S. A., Vera, C. R., Epron, F. et al. 2009a. Catalytic properties of Pt–Re/Al_2O_3 naphtha-reforming catalysts modified by germanium introduced by redox reaction at different pH values. *Industrial and Engineering Chemistry Research* 48 (8):3771–3778.

D'Ippolito, S. A., Vera, C. R., Epron, F. et al. 2009b. Influence of tin addition by redox reaction in different media on the catalytic properties of Pt-Re/Al_2O_3 naphtha reforming catalysts. *Applied Catalysis A: General* 370 (1–2):34–41.

D'Ippolito, S. A., Especel, C., Epron, F., and Pieck, C. L. 2015. Selective ring opening of methylcyclohexane and decalin over Rh-Pd supported catalysts: Effect of the preparation method. *Fuel Processing Technology* 140:180–187.

Del Angel, G., Bonilla, A., Peña, Y., Navarrete, J., Fierro, J. L. G., and Acosta, D. R. 2003. Effect of lanthanum on the catalytic properties of PtSn/γ-Al_2O_3 bimetallic catalysts prepared by successive impregnation and controlled surface reaction. *Journal of Catalysis* 219 (1):63–73.

Elfghi, F. M., and Amin, N. A. S. 2015. Influence of tin content on the texture properties and catalytic performance of bi-metallic Pt–Re and tri-metallic Pt–Re–Sn catalyst for n-octane reforming. *Reaction Kinetics, Mechanisms and Catalysis* 114 (1):229–249.

Epron, F., Carnevillier, C., and Marécot, P. 2005. Catalytic properties in n-heptane reforming of Pt–Sn and Pt–Ir–Sn/Al_2O_3 catalysts prepared by surface redox reaction. *Applied Catalysis A: General* 295 (2):157–169.

Fonseca, J. D. S. L., Júnior, A. D. C. F., Grau, J. M., and Rangel, M. D. C. 2010. Ethylbenzene production over platinum catalysts supported on modified KY zeolites. *Applied Catalysis A: General* 386 (1–2):201–210.

Gobölös, S., Margitfalvi, J. L., Hegedus, M., and Ryndin, Y. A. 2006. Transformation of n-hexane on Al_2O_3 and SiO_2 supported Pt, Pt+Ga and Ir+Pt+Ga catalysts prepared by anchoring methods. *Reaction Kinetics and Catalysis Letters* 87 (2):313–324.

Grau, J. M., and Parera, J. M. 1991. Deactivation of Pt-Re/Al_2O_3 catalysts with different metallic charge. *Applied Catalysis* 70 (1):9–18.

Gulino, A., Parker, S., Jones, F. H., and Egdell, R. G. 1996. Influence of metal-metal bonds on electron spectra of MoO_2 and WO_2. *Journal of the Chemical Society—Faraday Transactions* 92 (12):2137–2141.

Hamoule, T., Peyrovi, M. H., Rashidzadeh, M., and Toosi, M. R. 2011. Catalytic reforming of n-heptane over Pt/Al-HMS catalysts. *Catalysis Communications* 16 (1):234–239.

Hoang, D. L., Farrage, S. A. F., Radnik, J. et al. 2007. A comparative study of zirconia and alumina supported Pt and Pt–Sn catalysts used for dehydrocyclization of n-octane. *Applied Catalysis A: General* 333 (1):67–77.

Honeywell UOP, Accessed on February 19, 2016. http://www.uop.com/reforming-ccr-platforming/

Iglesia, E., Soled, S. L., and Kramer, G. M. 1993. Isomerization of alkanes on sulfated zirconia: Promotion by Pt and by adamantyl hydride transfer species. *Journal of Catalysis* 144 (1):238–253.

Ignacio, C. A., Armando, V. Z., and Viveros, T. 2009. Influence of the synthesis method on the catalytic behavior of Pt and PtSn/Al_2O_3 reforming catalyst. *Energy and Fuels* 23 (8):3835–3841.

Jacobs, G., Padro, C. L., and Resasco, D. E. 1998. Comparative study of n-hexane aromatization on Pt/KL, Pt/Mg(Al)O, and Pt/SiO_2 catalysts: Clean and sulfur-containing feeds. *Journal of Catalysis* 179 (1):43–55.

Jahel, A., Avenier, P., Lacombe, S., Olivier-Fourcade, J., and Jumas, J.-C. 2010. Effect of indium in trimetallic Pt/Al_2O_3SnIn–Cl naphtha-reforming catalysts. *Journal of Catalysis* 272 (2):275–286.

Jahel, A. N., Moizan-Baslé, V., Chizallet, C. et al. 2012. Effect of indium doping of γ-alumina on the stabilization of PtSn alloyed clusters prepared by surface organostannic chemistry. *The Journal of Physical Chemistry C* 116 (18):10073–10083.

Jones, F. H., Egdell, R. G., Brown, A., and Wondre, F. R. 1997. Surface structure and spectroscopy of WO_2(012). *Surface Science* 374 (1–3):80–94.

Jongpatiwut, S., Sackamduang, P., Rirksomboon, T., Osuwan, S., and Resasco, D. E. 2003. n-Octane aromatization on a Pt/KL catalyst prepared by vapor-phase impregnation. *Journal of Catalysis* 218 (1):1–11.

Jongpatiwuta, S., Trakarnroek, S., Rirksomboon, T., Osuwan, S., and Resasco, D. E. 2005. N-octane aromatization on Pt-containing non-acidic large pore zeolite catalysts. *Catalysis Letters* 100 (1–2):7–15.

Jumas, J. C., Sougrati, M. T., Olivier-Fourcade, J., Jahel, A., Avenier, P., and Lacombe, S. 2013. Identification and quantification of Sn-based species in trimetallic Pt–Sn–In/Al_2O_3–Cl naphtha-reforming catalysts. *Hyperfine Interactions* 217 (1–3):137–144.

Katrib, A., Leflaive, P., Hilaire, L., and Maire, G. 1996. Molybdenum based catalysts. I. MoO_2 as the active species in the reforming of hydrocarbons. *Catalysis Letters* 38 (1–2):95–99.

Kluksdahl, H. E. 1968. Reforming a sulfur-free naphtha with a platinum–rhenium catalyst. Google Patents.

Li, J., Chai, Y., Liu, B. et al. 2014. The catalytic performance of Ni_2P/Al_2O_3 catalyst in comparison with Ni/Al_2O_3 catalyst in dehydrogenation of cyclohexane. *Applied Catalysis A: General* 469:434–441.

Liu, C., Zhu, Q., Wu, Z. et al. 2010. Increase of reformate yield by using polyacrylic acid as template in preparation of Pt/Re naphtha reforming catalysts. *Applied Catalysis A: General* 390 (1–2):19–25.

Maldonado-Hódar, F. J. 2011. Metal-doped carbon aerogels as catalysts for the aromatization of n-hexane. *Applied Catalysis A: General* 408 (1–2):156–162.

Maldonado-Hódar, F. J. 2012. Platinum supported on carbon aerogels as catalysts for the n-hexane aromatization. *Catalysis Communications* 17:89–94.

Mariscal, R., Fierro, J. L. G., Yori, J. C., Parera, J. M., and Grau, J. M. 2007. Evolution of the properties of PtGe/Al_2O_3 reforming catalysts with Ge content. *Applied Catalysis A: General* 327 (2):123–131.

Mazzieri, V. A., Grau, J. M., Vera, C. R., Yori, J. C., Parera, J. M., and Pieck, C. L. 2005a. Pt–Re–Sn/Al_2O_3 trimetallic catalysts for naphtha reforming processes without presulfiding step. *Applied Catalysis A: General* 296 (2):216–221.

Mazzieri, V. A., Grau, J. M., Vera, C. R., Yori, J. C., Parera, J. M., and Pieck, C. L. 2005b. Role of Sn in Pt–Re–Sn/Al_2O_3–Cl catalysts for naphtha reforming. *Catalysis Today* 107–108:643–650.

Mazzieri, V. A., Pieck, C. L., Vera, C. R., Yori, J. C., and Grau, J. M. 2009a. Effect of Ge content on the metal and acid properties of Pt–Re–Ge/Al_2O_3–Cl catalysts for naphtha reforming. *Applied Catalysis A: General* 353 (1):93–100.

Mazzieri, Vanina A., Grau, Javier M., Yori, Juan C., Vera, Carlos R., and Pieck, Carlos L. 2009b. Influence of additives on the Pt metal activity of naphtha reforming catalysts. *Applied Catalysis A: General* 354 (1–2):161–168.

Medellín, B., Gómez, R., and del Angel, G. 2010. Effect of Ge and CeO_2 on Pt–Ge/Al_2O_3–CeO_2 reforming catalysts. *Catalysis Today* 150 (3–4):368–372.

Mishra, G. S., and Kumar, A. 2011. Immobilized Pd complexes over MCM-41 as supported catalysts for effective reformation of hydrocarbons. *Catalysis Science and Technology* 1 (7):1224–1231.

Paál, Z., Wootsch, A., Teschner, D. et al. 2011. Structural properties of an unsupported model Pt-Sn catalyst and its catalytic properties in cyclohexene transformation. *Applied Catalysis A: General* 391 (1–2):377–385.

Peyrovi, M. H., Hamoule, T., Sabour, B., and Rashidzadeh, M. 2012. Synthesis, characterization and catalytic application of bi- and trimetallic Al-HMS supported catalysts in hydroconversion of n-heptane. *Journal of Industrial and Engineering Chemistry* 18 (3):986–992.

Peyrovi, Mohammad H., Parsafard, Nastaran, and Peyrovi, Parnian. 2014. Influence of zirconium addition in platinum–hexagonal mesoporous silica (Pt-HMS) catalysts for reforming of n-heptane. *Industrial & Engineering Chemistry Research* 53 (37):14253–14262.

Pieck, C. L., Vera, C. R., Parera, J. M. et al. 2005. Metal dispersion and catalytic activity of trimetallic Pt–Re–Sn/Al_2O_3 naphtha reforming catalysts. *Catalysis Today* 107–108:637–642.

Prestvik, R., Moljord, K., Grande, K., and Holmen, A. 2004. Compositional Analysis of Naphtha and Reformate. In *Catalytic Naphtha Reforming, Revised and Expanded*: CRC Press, Boca Raton, FL.

Rahimpour, M. R., Jafari, M., and Iranshahi, D. 2013. Progress in catalytic naphtha reforming process: A review. *Applied Energy* 109:79–93.

Raissi, S., Younes, M. K., Ghorbel, A., and Garin, F. 2010. Effect of sulphate groups on catalytic properties of chromium supported by zirconia in the n-hexane aromatization. *Journal of Sol-Gel Science and Technology* 53 (2):412–417.

Rodríguez-González, V., Gómez, R., Moscosa-Santillan, M., and Amouroux, J. 2007. Synthesis, characterization, and catalytic activity in the n-heptane conversion over Pt/In–Al_2O_3 sol–gel prepared catalysts. *Journal of Sol-Gel Science and Technology* 42 (2):165–171.

Romero, T., Arenas, B., Perozo, E. et al. 1990. A study of the platinum–gallium catalytic system. *Journal of Catalysis* 124 (1):281–285.

Samoila, P., Boutzeloit, M., Benitez, V. et al. 2007. Influence of the pretreatment method on the properties of trimetallic Pt–Ir–Ge/Al_2O_3 prepared by catalytic reduction. *Applied Catalysis A: General* 332 (1):37–45.

Sánchez, M. A., Mazzieri, V. A., Grau, J. M., Yori, J. C., and Pieck, C. L. 2011. Catalytic activity of Pt–Re–Pb/Al_2O_3 naphtha reforming catalysts. *Journal of Chemical Technology & Biotechnology* 86 (9):1198–1204.

Santos, M. C. S., Grau, J. M., Pieck, C. L. et al. 2005. The effect of the addition of Re and Ge on the properties of Pt/Al_2O_3. *Catalysis Letters* 103 (3–4):229–237.

Saxena, S. K., Viswanadham, N., and Garg, M. O. 2013. Cracking and isomerization functionalities of bi-metallic zeolites for naphtha value upgradation. *Fuel* 107:432–438.

Siani, A., Alexeev, O. S., Lafaye, G., and Amiridis, M. D. 2009. The effect of Fe on SiO_2-supported Pt catalysts: Structure, chemisorptive, and catalytic properties. *Journal of Catalysis* 266 (1):26–38.

Song, J., Ma, H., Tian, Z. et al. 2015. The effect of Fe on Pt particle states in Pt/KL catalysts. *Applied Catalysis A: General* 492:31–37.

Trakarnroek, S., Jongpatiwut, S., Rirksomboon, T., Osuwan, S., and Resasco, D. E. 2006. n-Octane aromatization over Pt/KL of varying morphology and channel lengths. *Applied Catalysis A: General* 313 (2):189–199.

Vicerich, María A., Especel, Catherine, Benitez, Viviana M., Epron, Florence, and Pieck, Carlos L. 2011. Influence of gallium on the properties of Pt–Re/Al_2O_3 naphtha reforming catalysts. *Applied Catalysis A: General* 407 (1–2):49–55.

Viswanadham, N., Kamble, R., Sharma, A., Kumar, M., and Saxena, A. K. 2008. Effect of Re on product yields and deactivation patterns of naphtha reforming catalyst. *Journal of Molecular Catalysis A: Chemical* 282 (1–2):74–79.

Wootsch, A., Paál, Z., Gyorffy, N. et al. 2006. Characterization and catalytic study of PtGe/Al_2O_3 catalysts prepared by organometallic grafting. *Journal of Catalysis* 238 (1):67–78.

7 Recent Developments in FCC Catalysts and Additives, Patent and Open Literature Survey*

Deniz Onay Atmaca

CONTENTS

7.1 FCC TECHNOLOGY IN PERSPECTIVE

In refineries, catalytic cracking processes are used for producing valuable lighter products such as liquefied petroleum gas (LPG), gasoline, naphtha, kerosene, and diesel fuels from crude oil. Initially, thermal cracking had been used for production

* The review for the historical overview of the FCC technology, process chemistry, and the developments of the FCC catalyst as well as FCC additives has been conducted by using five literature resources, namely *Handbook of Petroleum Processing* (Jones, D.S.J); *Fluid Catalytic Cracking Handbook* (Sadeghbeigi R.); *Petroleum Refining Processing* (Speight J.G. and Ozum B.); *Guide to Fluid Catalytic Cracking* (GRACE Davison); and *Catalytic Cracking, Catalysts, Chemistry and Kinetics* (Wojciechowksi B.W. and Corma A.) [1–5].

of lighter products. Following the invention of the catalytic cracking process by Eugene Houdry in 1936, which mainly replaced the thermal cracking process, the fluid catalytic cracking (FCC) process was commercialized by Standard Oil of New Jersey between the years 1939 and 1942 [1,2]. The main purpose of the FCC unit, where the catalyst aerated by gas is flowing like a liquid, is to produce high-octane gasoline and heating oil by cracking high-boiling point fractions known as *gas oil* [2]. Since the start-up of the first FCC unit, several different designs and configurations have been employed by different technology licensors such as ABB Lummus Global, Exxon Research and Engineering, Kellogg Brown & Root—KBR, Shell Oil Company, Stone & Webster Engineering Corporation (SWEC)/IFP, and UOP to meet the demand of the developments in terms of processing objectives, equipment types, operating modes, and catalyst types [1,2].

In the 1940s, the FCC units were taller with dense bed reactors made of carbon steel. Later, more compact units were designed including straight feed risers in the 1950s. Residue cracking had first been practiced in the 1960s. Major developments in the FCC technology took place after the invention of using zeolites at higher activity when compared to former catalysts made of amorphous alumina in the 1960s. The use of a catalyst at higher activities resulted in the feed riser to be the only conversion vessel by replacing the reactor dense beds. Since the coke on the zeolite decreased the activity and selectivity of the catalyst, the requirement of higher regenerator temperatures can be achieved by full CO burn instead of partial CO burn. For this reason, reactor internals are adapted for bearing higher temperatures. In the 1970s, regenerator modifications resulted in favorable coke burning kinetics, ultimately leading to minimized catalyst inventories. New designs of resid cracking units had been established in the 1980s including staged catalyst regeneration and several other modifications. Since light olefins are valuable products in the field of petrochemicals, FCC units to maximize propylene were commercialized in the 1990s. A catalytic pyrolysis process with more severe cracking conditions to obtain a wide range of petrochemicals was introduced in the 2000s [1].

Feed injection system, reactor riser, riser termination device, and vapor quench technology are the main constituents of the reaction section of the FCC units, which show differences in terms of equipment and configuration changing for each licensor. Feed is vaporized and a hot catalyst is quenched quickly in the feed injection system [1]. Then vaporized feed and the fluidized catalyst as dispersed in hydrocarbon vapors go to the reaction chamber. After the formation of cracked vapors in the reaction chamber, the cracked vapors flow to the cyclones at the top of the reaction chamber. Then the cracked vapors are fractionated into the products for the formation of light and heavy cracked gas oils, cracked gasoline, and cracked gases. Following the completion of the cracking reactions, the catalyst is stripped of from the hydrocarbons. For instance, the centrifugal force enables the catalyst to be separated from the vapors. After this stage, there is coke formation on the catalyst, which will be burned away in the regenerator while heating the catalyst. After the withdrawal of the catalyst from the bottom of the reactor, it is lifted into the regenerator for the coke to be removed by controlled burning. Then, a hot-regenerated catalyst goes to the feed riser with enough heat to vaporize the feed, which is fed by feed nozzles [3].

Process variables of the FCC units can be listed as feed quality, preheating of the feed, pressure of the feed, feedstock conversion, reactor temperature, recycle rate, space velocity (catalyst contact time or residence time), catalyst activity, catalyst/oil ratio, regenerator temperature, and air rate of the regenerator [3].

FCC feed, namely gas oil, is mainly a blend obtained from refinery units such as crude, vacuum, solvent deasphalting, and coker. Recently, heavier gas oils and residue are mainly processed in the FCC units [2]. There are three main types of FCC process applications, namely gasoil cracking, resid cracking, and cracking for production of petrochemicals. Gasoline, light cycle oil (LCO), and LPG are the target products in gas oil cracking produced from vacuum gas oils and in resid cracking produced from atmospheric resid, vacuum gas oil (VGO), and vacuum resid. Light olefins and aromatics are obtained mainly by cracking vacuum gas oils and added resins in cracking processes for petrochemicals [1].

Hydrocarbon types in the FCC feed are mainly paraffins, olefins, naphthenes, and aromatics. Typically 50–60 wt.% of the feed is paraffinic, and the corresponding content is easier to crack and yields mainly gasoline but with a lower octane number. Olefins, as a typical content of 5% in the FCC feed, are less desirable since undesired products such as coke and slurry are formed by the polymerization of olefins. Since gasoline content is richer in terms of aromatics due to the cracking of naphthenes leading to a higher octane number, presence of naphthenes in the FCC feed is preferred. Aromatic content in the FCC feed is not desired since aromatics are harder to crack together with excess fuel gas yield due to the involvement of side chain breaking. Moreover, aromatics may lead to coke formation on the catalyst. In addition to different types of hydrocarbons, FCC feed also includes impurities known as contaminants such as nitrogen, sulfur, nickel, vanadium, and sodium. Due to the environmental regulations, feed and product treatment is required to remove the sulfur, ultimately resulting in higher operational costs. Nitrogen in the feed is a temporary poison for the catalyst, since nitrogen in the coke is converted to elemental nitrogen in the regenerator, and due to the burning, the remaining nitrogen is converted to NO_x leaving the unit with the flue gas. Metals such as nickel, vanadium, and sodium in the feed permanently poison the FCC catalyst resulting in the decrease in activity and selectivity toward the desired products by promoting reactions such as dehydrogenation and condensation. Iron and lesser amounts of copper are the other impurities in the FCC feed [2].

The conversion of gas oil in the FCC unit results in products such as dry gas, LPG, gasoline, LCO, HCO, decanted oil, and coke. Dry gas components are H_2, methane, ethane, ethylene, and H_2S. Dry gas is the undesirable by-product of the FCC process. Nickel, vanadium accumulation on the catalyst, higher reactor/regenerator temperatures, higher amount of aromatics in the feed, and decrease in performance in feed nozzles are several causes of higher yields of dry gas. LPG is rich in terms of olefins, propylene, and butylene. Gasoline is generally the main target product of FCC units. Octane number, benzene, and sulfur content are the main quality features of the gasoline. LCO is another valuable product of the FCC units. It can be blended with heating oil or diesel fuel. Cetane number is one of the quality features of LCO. HCO is the product stream that boils between LCO and decants oil (DO). It can be used for transferring heat to fresh feed as a recycle, withdrawn as a product for further processing in hydrocracker units or blended with DO. DO, as being the heaviest product of the FCC unit with the lowest economical value, is also known as slurry oil, clarified oil, bottoms, or FCC residue. Coke is the deposit on the catalyst produced from secondary cracking reactions or polymerization reactions. Since coke is burned in the regenerator resulting in heat for providing heat of the reaction, it is a required by-product in the FCC units to a certain extent. There are four main types of coke in the FCC unit. Catalytic coke is formed as a by-product of the cracking reactions. Contaminant coke is the result of catalytic activity of the contaminants such as vanadium and nickel. Direct deposition of the feed on the catalyst results in feed residue coke. Hydrogen-rich coke from the reactor stripper leads to formation of catalyst circulation coke [2].

7.2 FCC PROCESS CHEMISTRY, CATALYSTS, AND ADDITIVES IN PERSPECTIVE

Before the establishment of the first FCC unit, acid activated clay was used mainly in fixed bed and moving bed reactor systems in the form of a pellet as a cracking catalyst. Later, clay catalysts in the form of sized powder were used in the FCC process. Mixtures of about 90% SiO_2 and 10% Al_2O_3 were used in 1944 in the FCC units. However, it was not found to be effective for the promotion of catalytic cracking reactions. Combinations of mixed oxides including combinations of SiO_2, MgO, ZrO, Al_2O_3, and B_2O_3 were also produced. Silica magnesia was the only one that had been commercialized. However, due to the operating difficulties in terms of regeneration, it was found that further improvements should be made. Starting from the 1940s to the 1960s, acid leached clays and amorphous silica–alumina as low alumina (~13% Al_2O_3) or high alumina (~25% Al_2O_3) were mainly used as the FCC catalyst. Semisynthetic catalysts with kaolin dispersed in silica–alumina gel were used in manufacturing FCC catalysts. Semisynthetic catalysts, used during 1958–1960, showed a better yield in terms of desired products; however, they showed lower catalytic activity. Since FCC catalysts are exposed to actions of CO_2, air, nitrogen compounds, steam, and sulfur during the operation, in addition to resistance to such chemical impacts, they should also be resistant to physical impacts and thermal shocks. The milestone in the FCC catalyst field was achieved with the discovery of using spray-dried zeolite in 1964 [1,3,4]. Use of crystalline aluminosilicates (zeolites) has replaced the use of amorphous silica alumina and other materials as the only component source of the FCC catalyst. Initial patents for the use of faujasite structure as an FCC catalyst were held by Exxon and Mobil. Use of ultrastable Y zeolite (USY) without rare earth improved gasoline octane rate. Later, it was also discovered that use of rare earth in the USY zeolite (REUSY) resulted in lower selectivity for coke. Quality and performance of the FCC catalysts in terms of retention properties were enhanced by silica-sol and alumina-sol binding technologies in 1973 and 1981 by Grace. Coke-selective deep bottom cracking matrix in 1986, nickel-resistant matrix in 1990, and gasoline selectivity enhancements in the catalysts in 1990 are several developments in terms of the matrix of FCC catalysts. Platinum CO combustion promoter in 1974 by Mobil, antimony nickel passivation in 1975 by Phillips, bismuth nickel passivation in 1987 by Chevron, ZSM-5 octane enhancement and increase in C3–C4 yield in 1986 by Mobil, and spinel SO_x additive in 1982 by ARCO are several developments in terms of additives in the FCC process [3,4].

As a result of the historical developments in the field of FCC catalysts, major components of the catalysts are zeolite and matrix, including active matrix, binder, and filler [2]. Synthesis of FCC catalysts is either by "incorporation" or "in situ" processes. The incorporation process involves making the zeolite and the matrix separately followed by being brought together with the binder, whereas zeolite growth takes place within preformed microspheres in the in situ process [2]. Preparation of the cracking catalysts mainly involves the following basic steps: synthesis, ion exchange and activation of the zeolite, synthesis of the matrix gel, combination of the matrix gel with the zeolite, washing and exchanging the wet catalyst, and finally drying and calcination of the catalyst [5]. Concentration and temperature of the initial sodium silicate solution, acid amount introduced for gelation, aging time of the gel, and conditions for aluminum salt added to the corresponding gel are several synthesis parameters that have an influence on the final performance of the catalyst [3].

Zeolites, also known as molecular sieve, have a well-defined lattice structure where the basic structure is made of silica and alumina tetrahedron with oxygen atoms on each corner [2]. Zeolites with three-dimensional structures should have high stability (>870°C) to both heat and steam, high acidity to have the required activity, and larger pores (~0.75 nm) [1]. Acidity on the surface of the zeolite is initially generated by replacement of sodium ions by protons generally via ammonia during the synthesis of the zeolite [5]. Rare earth materials such as lanthanum and cerium are used to enhance the strength of Brønsted acid sites where the catalytic activity is provided. Rare earth exchange also increases the hydrothermal stability of the FCC catalysts. Synthetic versions of natural zeolites known as faujasites are mainly used for the preparation of the catalyst. X, Y, and ZSM-5 are the most common types of zeolites for FCC catalysts. Y zeolites have a higher silica alumina ratio and higher hydrothermal stability when compared with X zeolite. ZSM-5 mainly increases olefin yields and improves the octane number of the gasoline. Unit cell size (UCS), rare earth level, and sodium content are the most common parameters that define zeolite behavior [2].

A matrix includes components of the FCC catalyst other than the zeolite, and an active matrix includes components of the catalyst other than the zeolite with catalytic activity. The main source of the active matrix is mostly amorphous alumina. Since zeolite has small pores, larger molecules cannot diffuse for further cracking. With the help of the active matrix with larger pores, it becomes possible to crack the larger molecules. Therefore, primary cracking sites are provided by the active matrix. The active matrix also acts as a protective component for the zeolite from impurities such as vanadium and basic nitrogen by trapping those molecules to a certain extent [2].

Kaoline [$Al_2(OH)_2$, Si_2O_5], which is the most common type of clay, is used as a filler and enables the balance in terms of the catalytic activity by diluting the catalyst activity. Binder is the component that holds the zeolite, matrix, and filler together. Filler and binder mainly provide the physical catalyst properties such as attrition resistance, density, and particle size distribution (PSD) together with the benefits in terms of enabling required heat transfer and fluidizing medium [2].

Important characterization properties of fresh FCC catalysts are PSD, surface area, and sodium and rare earth content in wt.%. PSD shows the fluidization properties of the catalyst, and surface area is one of the indicators for catalyst activity. Sodium content is crucial since deactivation of the zeolite is observed due to the presence of sodium. Catalytic activity and hydrothermal stability of the catalyst are attained by the rare earth content. The catalyst circulating within the FCC unit is called the equilibrium catalyst (e-cat). Samples withdrawn as e-cat are taken by refineries and sent to catalyst manufacturers regularly for analysis to have the results related with catalytic, physical, and chemical properties, which ultimately give valuable information on the conditions of the FCC unit [2].

Additives are used for improving the performance of FCC units. Change of the product yield can also be achieved by the catalyst or feed additives. There are environmental limits such as amount of coke burned in the regenerator, emission of particulates, SO_x, NO_x, and sulfur in the gasoline. FCC additives are also used for reduction of the pollutant gases. CO promoter as an additive enables the acceleration of CO combustion leading to uniform coke burning; SO_x additives are used for reduction of SO_x emissions that are formed naturally by the conversion of sulfur in the coke to SO_2 and SO_3 [2].

Cracking reactions include formations of olefins and smaller paraffins from paraffins, smaller olefins formed from olefins, side chain scission of aromatics, and formation of olefins and smaller ring compounds from naphthenes. In the FCC units, thermal cracking via free radicals is minimized mainly by process modifications in feed, riser sections, and proper catalyst selections. Catalytic cracking reactions go through carbenium ions in association with Brønsted acid sites on the catalyst [1]. Carbenium ions are positive-charged carbon ions that are formed when feed is vaporized at the contact with the catalyst. Main reactions of carbenium ions are cracking reactions (beta scission as splitting of C–C bond two carbons away from the positive charge carbon atom), isomerization reactions (which results in a higher octane number of gasoline or a lower cloud point of diesel), and hydrogen transfer reactions (bimolecular hydride transfer where the reactants are "olefin and olefin" or "olefin and naphthenes," ultimately resulting in the increase in gasoline yield but with a lower octane number). Isomerization reactions occurring in FCC units are bond shift of olefins, iso-olefin formation from normal olefins, iso-paraffin formation from normal paraffins, and cyclopentane formation from cyclohexane. Dehydrogenation, coking, cycloaromatization, trans-alkylation, cyclization of olefins to naphthenes, and dealkylation are the other types of reactions observed in FCC units. Dehydrogenation reactions are formed due to the catalyst contaminants such as vanadium and nickel within the feed.

7.3 ANALYSIS BASED ON PATENT APPLICANTS AND COUNTRIES OF THE APPLICANTS

In this section, FCC patents in terms of catalysts and additive-based developments between the years 2008 and 2014 will be reviewed. Throughout the study, 577 FCC targeted patents are examined, of which 279 are mainly related with "FCC catalyst and additives" and the remaining are mostly describing "process"-based technologies. The total numbers for "FCC catalyst and additives" and "process"-based developments indicate that developments in "FCC catalyst and additives" are as high as developments in the "FCC process." The statistical analysis for this chapter was based on the final available issue date considering that the latest revised version covers the most updated technologies relevant to the development.

Figure 7.1 shows the annual distribution of FCC patents in terms of two main targets, namely "catalyst and additive"-based developments and process-based developments. There is no clear trend in terms of increase or decrease in the total count by years. However, the distribution in terms of numbers "in 2010 and before 2010" and "after 2010" shows that the number of patents for FCC catalysts and additives is increasing "after 2010."

Analysis of the FCC catalyst and additive patents in terms of inventors (stated also as applicants) shows that catalyst manufacturers are the governing category with 49% of the total count as can be seen in Figure 7.2. This result indicates that FCC catalysts and additives are a core topic mainly for manufacturers. Refineries are ranked second having 20% of the total number of FCC catalyst and additive–based patents. The remaining inventor categories are for other individuals without a collaborator and universities/research institutes/R&D companies without collaborators having 12% and 9% of the total patents, respectively.

Patent applicants in collaboration only consist of 10% of the total count showing that FCC catalysts and additives are mainly investigated by individual research entities. Collaborated patents constitute 10% of the total count, 8% of which are due to the cooperation of refineries with universities/research institutes/R&D companies.

Figure 7.3 shows the distribution of FCC patents and catalysts in terms of countries. The United States has the highest number of FCC catalysts and patents. Since gasoline is the governing fuel in the United States, this result is consistent together

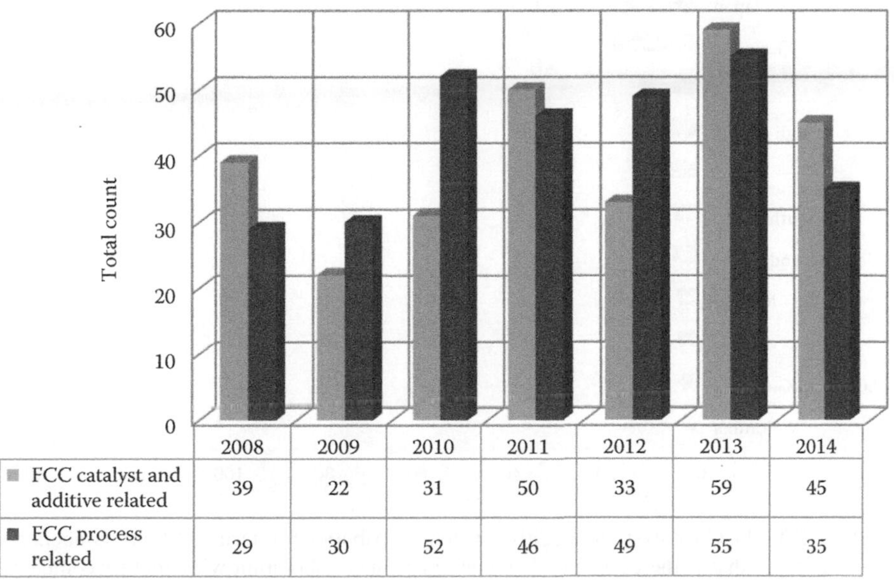

	2008	2009	2010	2011	2012	2013	2014
■ FCC catalyst and additive related	39	22	31	50	33	59	45
■ FCC process related	29	30	52	46	49	55	35

FIGURE 7.1 Annual distribution of FCC patents between the years 2008 and 2014.

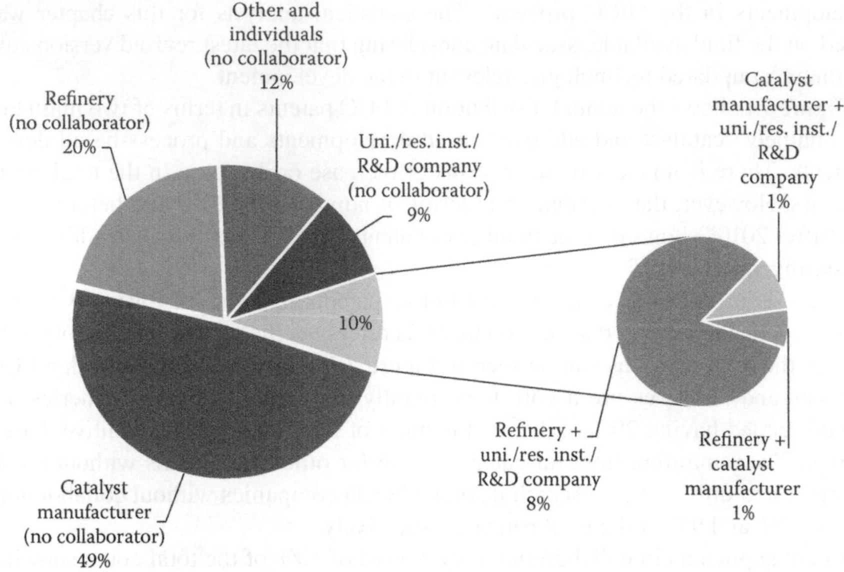

FIGURE 7.2 Patent distributions of FCC catalyst and additives based on applicants.

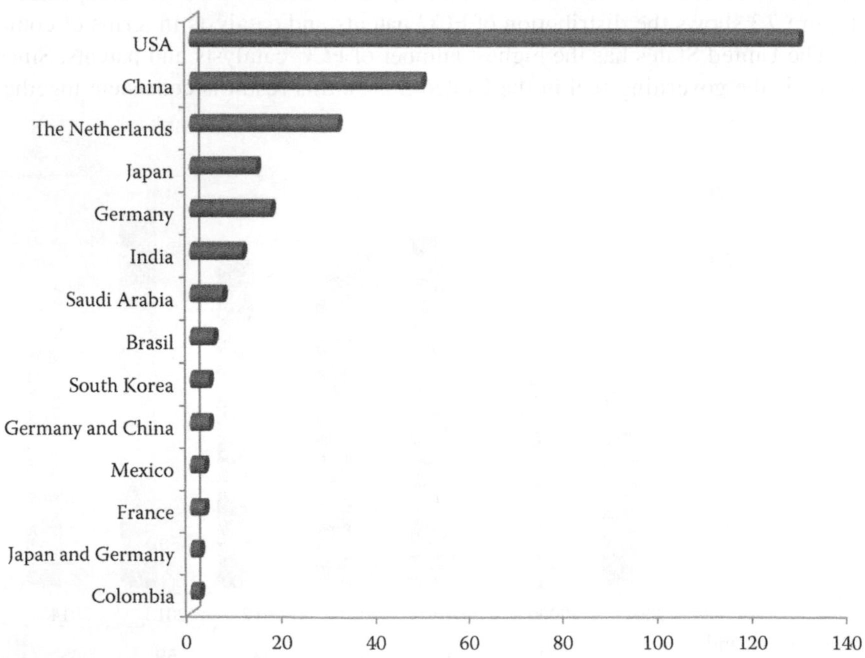

FIGURE 7.3 FCC catalyst and additives patent distribution* in terms of total numbers by countries (note that if the corresponding country is in collaboration with another company, their collaborative studies is taken into account while making the percentage calculations). *First 14 countries are shown in the figure.

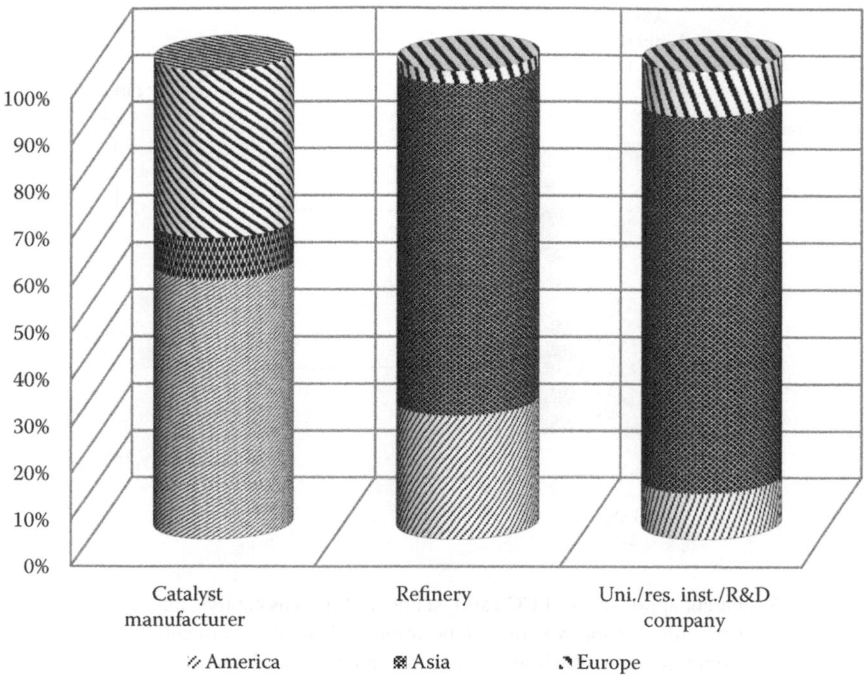

FIGURE 7.4 Distribution of the patent applicant categories for FCC catalyst and additives.

with the fact that gasoline is the target product of a conventional FCC process. Other than the countries shown in Figure 7.3, there are also several other countries where there is a patent application in the general main topic of "FCC catalyst and additives" in Australia, Italy, Switzerland, United Kingdom, and Venezuela.

As seen in Figure 7.4, catalyst manufacturers who have a patent application in the general topic of "FCC catalyst and additive" are mainly located in America and Europe, whereas patent applicants of both refineries and universities/research institutes/R&D companies are predominantly of Asian origin.

7.3.1 Involvement of Catalyst Manufacturers

FCC catalyst and additive patent applicants as catalyst manufacturers mainly consist of companies such as BASF, Grace, Albemarle, UOP, and Rive Technology as can be seen in Figure 7.5. BASF has the most number of patents in the catalyst manufacturer patent group by including several of its companies named as BASF Se, BASF Corporation, BASF Catalysts LLC, and BASF (China) Company.

7.3.2 Involvement of Refineries

Refineries that have patents in the major topic of FCC catalysts and additive are given in Figure 7.6. There are several refineries with one or two patents in a group

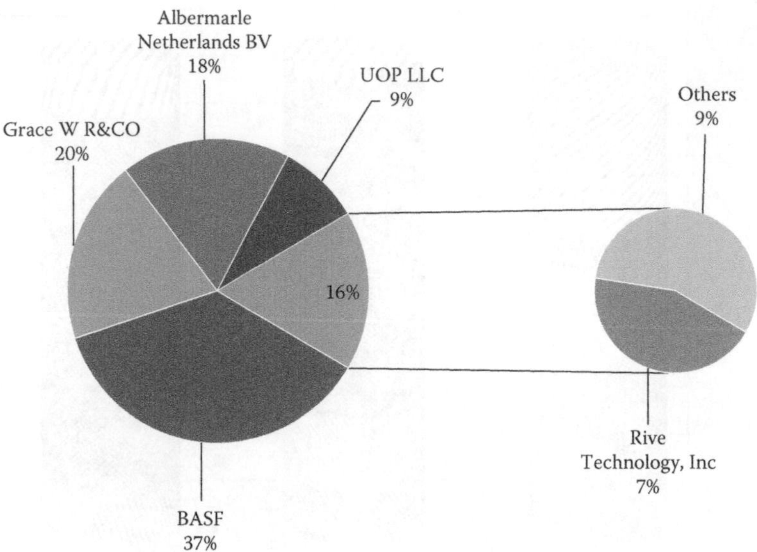

FIGURE 7.5 Patent applicants of FCC catalyst and additives as catalyst manufacturers (note that if the corresponding company is in collaboration with another company, their collaborative studies is taken into account while making the percentage calculations).

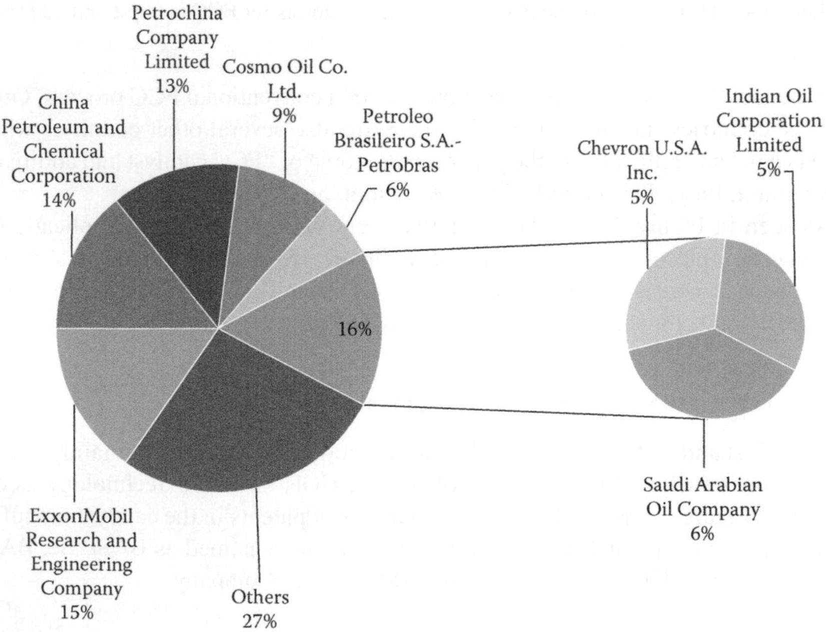

FIGURE 7.6 Patent applicants of FCC catalyst and additives as refineries (note that if the corresponding company is in collaboration with another company, their collaborative studies is taken into account while making the percentage calculations).

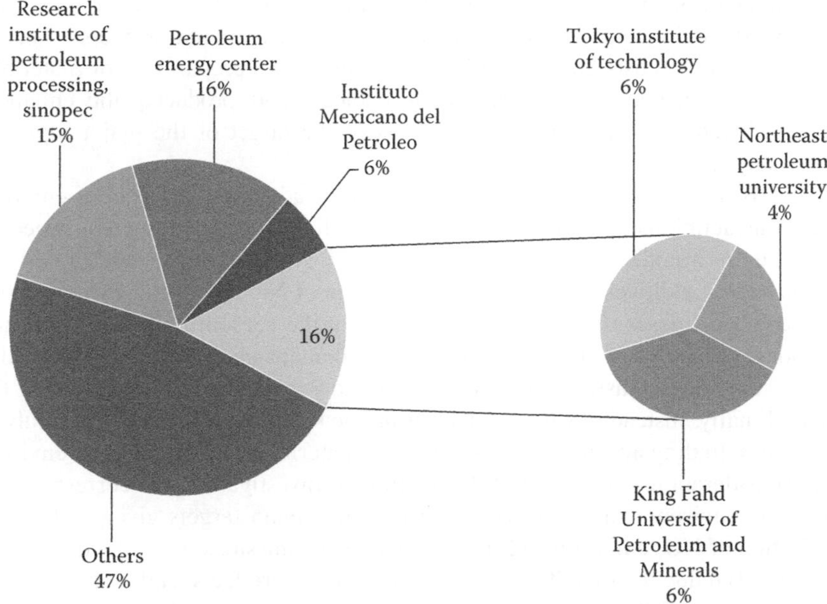

FIGURE 7.7 Patent applicants of FCC catalyst and additives as universities/research institutes/R&D companies (note that if the corresponding company is in collaboration with another company, their collaborative studies is taken into account while making the percentage calculations).

called "others." This group has the major proportion of 27% of the total count of patents of refineries. This is followed by ExxonMobil (15% of the total count), China Petroleum & Chemical Corporation (14% of the total count), Petrochina Company Limited (13% of the total count), Cosmo Oil (9% of the total count), and Petrobras (6% of the total count). The remaining 16% of the total count is divided almost equally among Saudi Arabian Oil Company, Chevron, and Indian Oil Corporation.

7.3.3 Involvement of Universities/Research Institutes/R&D Companies

As can be seen in Figure 7.7, there are several patent applicants in the category of universities/research institutes/R&D companies that are gathered as "others." This group named as "others" constitutes 47% of the total count for the corresponding category. The remaining percentage comes from Research Institute of Petroleum Processing, Petroleum Energy Center, Instituto Mexicano Del Petroleo, King Fahd University of Petroleum and Minerals, Tokyo Institute of Technology, and Northeast Petroleum University.

7.4 ANALYSIS BASED ON RESEARCH AREA

Patents of FCC catalysts and additives have been categorized mostly in terms of clearly stated topics and targets in order to make a rough but detailed technical classification throughout the investigated patents. Mostly stated major topics for FCC catalysts and patents are named as follows: "environmental aspects," "stability enhancement," "cost-effective, efficient, or alternative preparation of material," "product yield/selectivity," "feedstock-based technology," and "product quality enhancement." Each topic has a breakdown in terms of the target of the major topic for further analysis.

Synthesis developments of FCC catalysts and/or additives in terms of environmental considerations are grouped under the heading of "environmental aspects." Due to the environmental regulations in terms of gas emissions, patents of FCC catalysts and/or additives have a target of reduction of NO_x, SO_x, and CO emissions from the exhaust gases of the FCC units. With similar reasoning of environmental regulations, reduction of sulfur and/or nitrogen content of the FCC products ultimately results in decreasing SO_x and NO_x emissions from the exhaust gases of the vehicles. Finally, instead of directly depositing the spent/equilibrium FCC catalysts or FCC fines, finding alternative ways to reuse a material is also taken as an environmental consideration. As a result of this method of investigation, the corresponding heading of "environmental aspects" consists of five main targets given as follows: (1) reduction of NO_x emissions; (2) reduction of SO_x emissions; (3) reduction of CO emissions; (4) reduction of sulfur and/or nitrogen in the products; and (5) reuse of the spent/equilibrium FCC catalyst, the materials in the catalysts, or FCC fines [6–110].

In the FCC units, maintaining the stability of the FCC catalyst throughout the process is another crucial issue where the corresponding patents are gathered together with the topic name of "stability enhancement" [35,41,50,55,59,95,111–180]. Several of those patents have the purpose of achieving attrition/wear-resistant and more stable catalysts. Preventing formation or deposition of excess amount of coke is also another way to provide more stable FCC process. Since heavier feeds for FCC units contain higher metals such as vanadium, nickel, and iron [173], metal contamination of the catalyst has also been interrelated with the stability of the process. This approach for the target breakdown for the "stability enhancement" topic leads to three main categories named as "production of attrition/wear-resistant catalyst and/or more stable catalyst," "less coke formation/deposition," and "metal passivator/trap."

According to the modification approach in terms of the synthesis procedure, several patents of FCC catalysts and additives are grouped under a more general heading named as "cost-effective, efficient, or alternative preparation of material" [14,20,21,33,35,38,95,125,130,131,133,134,138,140–161,164–167,181–213]. There are patents where the catalyst content is modified in terms of rare earth or silica and alumina. Rare earth free FCC catalysts are aimed in order to cover the severe cut of export of rare earth metals from China, which supplies the majority of rare earth metals [142]. When the Si/Al ratio of the catalyst is low, the number of acid sites and acid density is high, which results in quicker hydrogen transfer and more coke yield [144]. Patents for synthesis of FCC catalysts and additives have also focused on pore structure, mostly for achieving mesoporous structures to enable more accessible sites for the feeds and the products [164]. To overcome the difficulties during catalyst preparation processes, patents describing the solid content of the sprayed catalyst slurry are grouped with a target heading of "catalyst with high solid content"

[202]. The FCC catalysts should be fluidized in the unit; therefore, there is a requirement of small particle–sized material production [152]. Those patents are grouped as "particle size targeted modification." Patents related with adjustment of the crystallinity of the catalysts are under the heading of "crystallinity targeted modification." Remaining patents related with the description of preparation methods are for decreasing the cost of synthesis, eliminating the complex steps from the synthesis procedure, or finding alternative ways for catalyst production. These patents are gathered under the heading of "cost reduction and/or less complex process and/or an alternative catalyst preparation."

Another major topic are patents related "product yield/selectivity" [62,96,111, 113–117,130,131,133,137,138,142,143,146,155,162,169,170,203,209,212,214–259]. Subheadings that are investigated as the targets of the corresponding major topic are "improvement of activity and/or selectivity," "fewer dry gas by-products," "increasing propylene and light olefins and lightweight yield," and "improved bottoms conversion."

Several patents in terms of the catalyst and/or additive production suitable for cracking of alternative feedstocks or heavier feeds resulted in a major heading named as "feedstock-based technologies" [62,113,139,155,181,194,219,246,247,260–279]. Names of the target headings of that major heading are "cracking of alternative feedstock (biomass/bio-oil/pyrolysis oil/plastics)" and "use of feedstock containing hydrocarbons of higher molecular weight."

Patents that describe catalysts and/or additives for the properties of the FCC products are investigated under the major heading of "product quality," where the target breakdown headings are given in terms of the octane number of gasoline and aromatic content of LCO, named as "increasing octane number of gasoline" and "LCO with low aromatic content," respectively [219,223,237,251–253,280–282].

Target breakdown for each topic will be given in the following sections for deeper understanding of the developments in the FCC catalyst and additive research area. Target classification has been implemented based on the most distinctive features of the corresponding patents. The investigation of the classification shows that the research area of the FCC catalyst and additive patents can be interrelated. For instance, preparation of a material as a mesoporous structure results in enhanced transportation of bigger molecules leading to less coke accumulation within the corresponding material. Therefore, mesoporous structure can be correlated with the enhanced stability of the catalyst. Considering the intersection between the topics, an effort has been put in order to make a classification in terms of topics and corresponding targets of the patents (total count of 279) investigated. The summary of the classified topics and targets in the investigated patents of FCC catalysts and additives is given in Table 7.1.

The distribution of FCC catalyst and additive patents related to mostly stated major topics within the patents is given in Figure 7.8. Since FCC is one of the well-established refinery processes, major developments are conducted in terms of improving "environmental aspects" (30% of the total count). This is followed by "stability enhancement" issues (21% of the total count), techniques for "cost-effective, efficient, or alternative preparation of material" (20% of the total count), and developments in terms of "product yield/selectivity" (20% of the total count),

TABLE 7.1
Referenced Patent Numbers Related to the Target Topics for the Patents of FCC Catalysts and Additives

Topic	Target	Referenced Patents
Environmental aspects	Reduction of NO_x emissions	[6–49]
	Reduction of SO_x emissions	[19,41,48–60]
	Reuse of spent/equilibrium FCC catalyst, the materials in the catalysts, or FCC fines	[61–90]
	Reduction of sulfur and/or nitrogen in the products	[41,55,91–102]
	Reduction of CO emissions	[18,75,103,104,107–110]
Stability enhancement	Production of attrition/wear-resistant catalyst and/or more stable catalyst	[35,50,55,111–138]
	Less coke formation/deposition	[139–167]
	Metal passivator/trap	[41,59,95,129,168–180]
Cost-effective, efficient, or alternative preparation of material	Achieving mesoporous structure and/or pore structure targeted modification	[125,130,133,140,141,145, 147–151,153,154, 156–161,164–167,181–187]
	Cost reduction and/or less complex process and/or with an alternative catalyst preparation	[14,21,33,35,38,95,131,134, 187–197]
	Catalyst with high solid content	[133,138,198–203]
	Crystallinity targeted modification	[20,204–208]
	Rare earth free FCC catalyst	[141,143,209,210]
	Si/Al ratio targeted modification	[144,147,211,212]
	Particle size targeted modification	[152,190,213]
Product yield/selectivity	Increasing propylene and light olefins and lightweight yield	[62,96,111,115,117,137,146, 169,170,209,214–245]
	Improvement of activity and/or selectivity	[113,114,116,117,130,133, 138,142,143,146,162, 203,212,243,246–258]
	Fewer dry gas by-products	[146,155,162,236,255,257]
	Improved bottoms conversion	[131,259]
Feedstock-based technology	Cracking of alternative feedstock (biomass/bio-oil/pyrolysis oil/plastics)	[62,139,181,260–276]
	Use of feedstock containing hydrocarbons of higher molecular weight	[113,155,194,219,246,247, 277–279]
Product quality	Increasing the octane number of gasoline	[219,223,251–253,280]
	LCO with low aromatic content	[237,281,282]

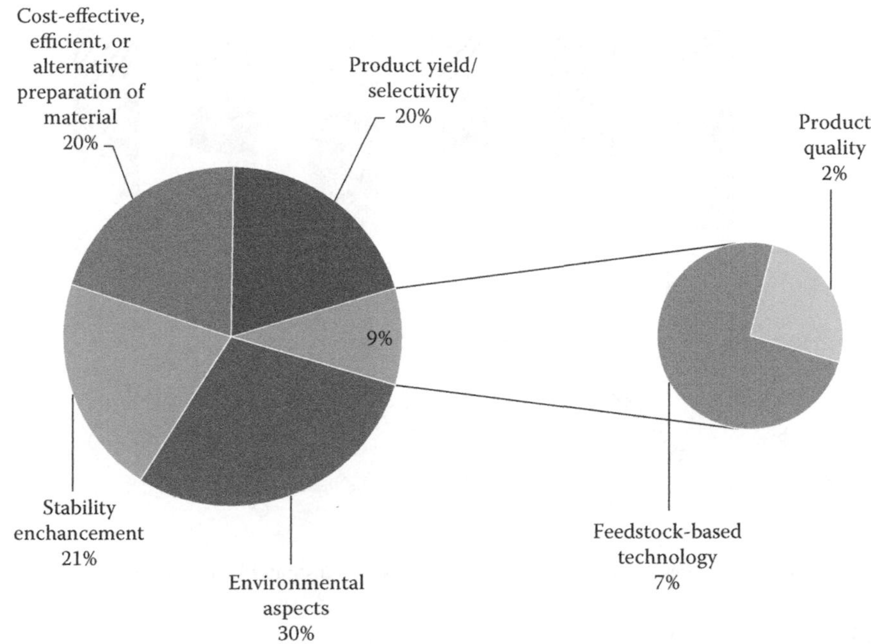

FIGURE 7.8 Patent distributions of FCC catalysts and additives based on mostly stated topics.

respectively. The remaining patents include feedstock-based technologies (7% of the total count) and product quality related topics (2% of the total count).

Analysis of topics together with applicants is given in Figure 7.9 for FCC catalysts and additives. Catalyst manufacturers mainly focused on developments related with "environmental aspects," "cost-effective, efficient, or alternative preparation of materials," and "stability enhancement." As expected, the main concern of refineries in terms of the classified topics for FCC catalysts and additives are the developments that will ensure and increase profitability. Relevant topics, which enable refineries to accomplish the objective of profitability, are "product yield/selectivity," "product quality," and "feedstock-based technologies." Studies of universities/research institutes/R&D companies are in line with the topics that are the main scope of refineries, namely "product quality," "product yield/selectivity," and "feedstock-based technologies."

Topic analysis with respect to the continents of the applicants given in Figure 7.10 shows the distribution of the studies in terms of the worldwide focus points. The focus of American studies in terms of FCC catalysts and additives is on the developments of "feedstock-based technologies" and "stability enhancement." Asia, where the main applicants are refineries and universities/research institutes/R&D companies, has been conducting the studies for the developments of "product quality" and "product yield/selectivity" issues in terms of FCC catalysts and additives. Studies in Europe do not cover the governing percentage in any of the main topics. However, to

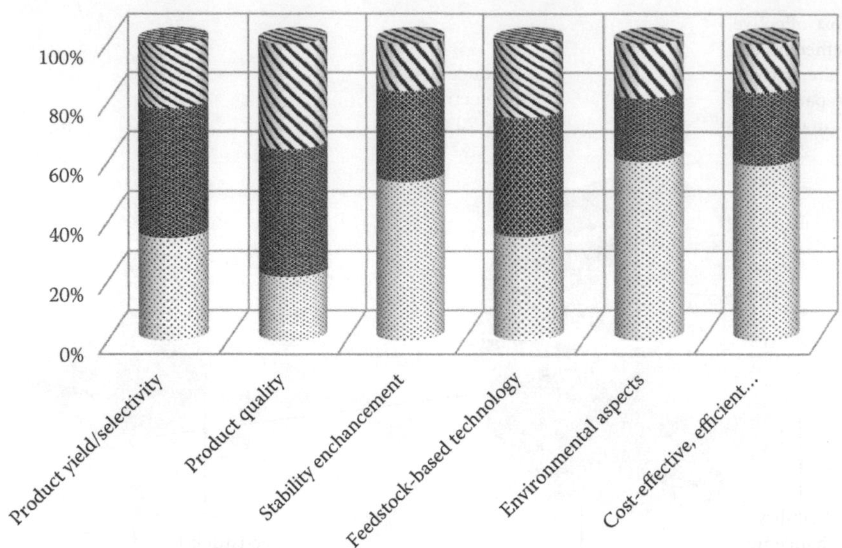

FIGURE 7.9 Topic analyses with respect to applicants.

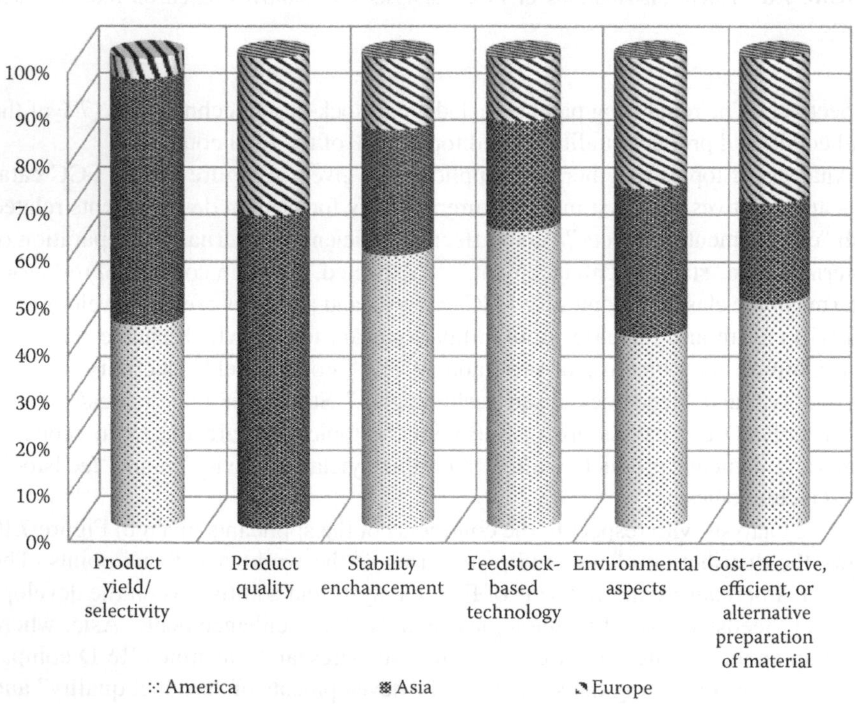

FIGURE 7.10 Topic analyses with respect to continents of the applicants.

name specifically, the main studies of Europe can be classified as developments in terms of "environmental aspects" and "cost-effective, efficient, or alternative preparation of material."

7.4.1 Developments in Terms of Environmental Aspects

Developments in environmental aspects have been divided into five main target subtopics, namely "reduction of NO_x emissions," "reuse of spent equilibrium FCC catalyst, the materials in the catalysts, or FCC fines," "reduction of SO_x emissions," "reduction of sulfur and/or nitrogen in the products," and "reduction of CO emissions" as given in Figure 7.11. Developments in the corresponding main topic are mostly related to reduction of detrimental gaseous product emissions, where the major proportion is for the reduction of NO_x emissions (39% of the total count of environmental aspects). Reduction of air pollutants in the atmosphere can be prevented either by employing additives to the system or by modifying the catalyst. Twenty-six percent of the developments for environmental aspects include the reuse of spent or equilibrium FCC catalysts, the materials within the catalyst formulations, or FCC fines. The subtopic named as "reduction of sulfur and/or nitrogen in the products" can also be related to developments in terms of "product quality." Since, as an ultimate target, reduction of sulfur and/or nitrogen in the FCC products results in the decrease in emissions of SO_x and NO_x, the corresponding subheading, having

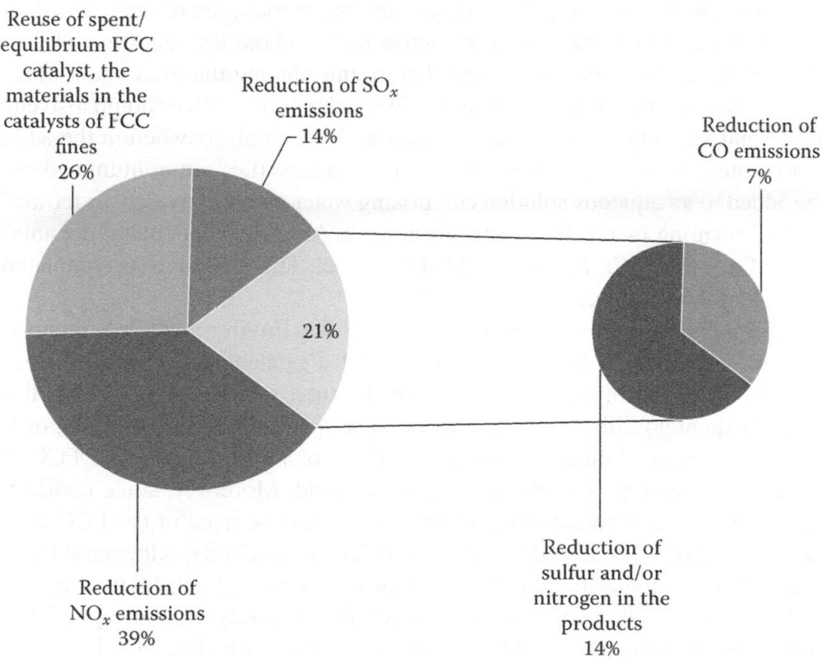

FIGURE 7.11 Target breakdowns of "environmental aspects"–related patents for FCC catalyst and additive.

14% of the total count, has been categorized in the main topic of environmental aspects.

BASF, Grace, Albemarle, and ExxonMobil are the prominent companies that have a patent in the subtopic of reduction of NO_x emissions. NO_x reduction can be achieved by a metal additive of FeSbCu system, which can be added to the FCC unit by an additive loader or can be preblended with a fresh FCC catalyst that is introduced to the system. Another approach is to use copper-promoted eight-ring small-pore molecular sieve at chabazite (CHA) structure, which includes an alkaline earth component such as barium. Compositions of anionic clay, which are hydrotalcite-like materials with a crystal structure, and rare earth metal hydroxyl carbonate can also be used as a NO_x reduction material. Silicoaluminophosphate (SAPO-34) together with Cu content, acetaldehyde, or propylene as a reductant in the catalyst system, BEA-type framework structure with Cu and/or Fe as nonframework elements, and copper containing Levyne molecular sieves are also included within different types of materials that are used as additives to the FCC process for NO_x reduction [6–49].

Patents related with FCC catalysts and additive developments in terms of SO_x reduction are issued by Albemarle, BASF, Instituto Mexicano Del Petroleo, Intercat, and Beijing Sj Environmental Protection and New Material Co., Ltd. Similar to NO_x reduction materials, anionic clays consisting of positively charged layers of specific combinations of divalent and trivalent metal hydroxides between which there are anions and water molecules are also used for SO_x reduction. For SO_x reduction additives, there are also patents describing the preparation of materials that are convenient for in situ formation of the active sites without the requirement of peptization, aging, and/or calcination steps before introducing the final material to the FCC process. Preparation technique of the SO_x reduction additive should also ensure hydrothermal stability of the zeolite-containing FCC catalysts wherein the additive is incorporated. In order to achieve this purpose, a base (i.e., ammonium hydroxide) can be added to an aqueous solution comprising water-soluble trivalent or tetravalent salts of gel-forming metals in order to form a gel. After that step, metal dopants can be Ti, Zr, Ce, La, Al, Cr, P, and Fe added to the gel. The procedure is completed by calcination of the doped gel [19,41,48–60].

Asian companies such as Yueyang Dingge Yuntian Environment Prot Technology Co Ltd and Kitakyushu Foundation for the Advancement of Industry Science & Technology have the highest percentage of the total number of patent applicants for reuse of spent/equilibrium FCC catalysts, the materials in the catalysts, or FCC fines. Recoveries of aluminum oxide, vanadium, or nickel from waste FCC catalysts are the major topics in the corresponding field. Moreover, since equilibrium catalysts have a lower vanadium content and can still be used in the FCC process as a start-up catalyst because they retain sufficient accessibility, Albemarle has conducted a study for upgrading an FCC equilibrium catalyst (e-cat) by treating it with an acidic solution to obtain an acid-treated equilibrium catalyst. This step is followed by contact with an aqueous solution or suspension of a divalent metal compound. One of the investigations of Albemarle is about process description for recovering rare earth metals from a waste FCC equilibrium catalyst, which has a zeolitic structure [61–90].

Saudi Arabian oil Company, King Fahd University of Petroleum and Minerals, Grace, and Albemarle are the main applicants that have a patent for reduction of sulfur in the gasoline. Metal vanadate compound supported on the exterior surface of the pore structure of the zeolite, zinc, and rare earth element as cations that have been exchanged on Y-type zeolite are several different types of materials that are used for reduction of sulfur in the FCC gasoline [41,55,91–102].

Albemarle, BASF, Bharat Petroleum Corporation Limited, and Sud Chemie India have patents for reduction of CO emissions by FCC catalysts and additive developments. The material used for this purpose is rhodium-supported anionic clay. Hydrotalcite-like compound as basic support, a lanthanide series element, a transition metal element of group 1b and 2b, and a precious metal, is a type of composition for the corresponding CO reduction material. Three main components, namely acidic oxide support, precious metal for CO oxidation, and the metals or metal oxides for reduction of NO_x, can also be used as an additive during CO reduction. Precious metal for the purpose of CO reduction is distributed in the central interior of the particulate additive, and the remaining metals and metal oxides are distributed in the particulate additive as a shell around the corresponding precious metal [18,75,103,104,107–110].

7.4.2 Developments in Terms of Stability Enhancements

Since maintaining the stability of the catalyst within the FCC reactor has high importance, there are several studies related to enhancement of catalyst stability. Subtopics are categorized into three main fields as can be seen in Figure 7.12,

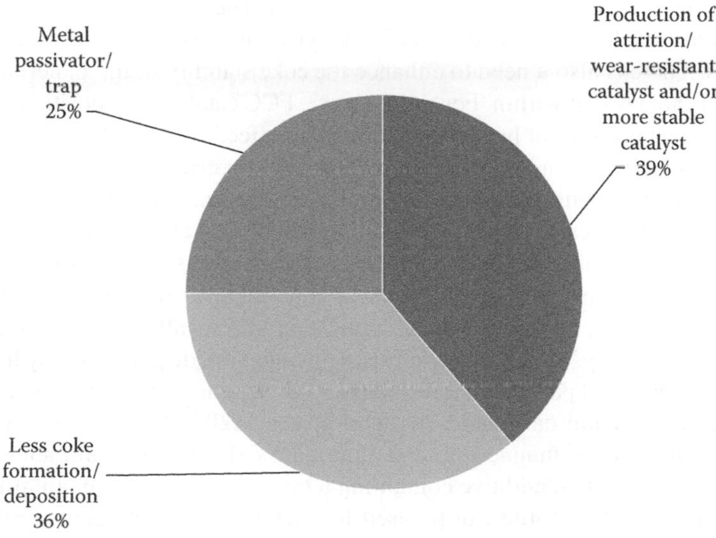

FIGURE 7.12 Target breakdown of "stability enhancement"–related patents for FCC catalyst and additive.

namely "production of attrition/wear-resistant catalyst" (39%), "less coke formation/deposition" (36%), and "metal passivator/trap" developments (25%).

BASF, Petrochina Company Limited, Grace W R & CO, Albemarle, China Petroleum & Chemical Corporation, and several other companies have a patent application that describes methods for production of attrition wear-resistant catalysts to ensure higher stability. Magnesium oxide, rare earth oxide, sodium oxide, and a molecular sieve may compose the type of material where rare earth ions are located in sodalite cages in order to provide a catalyst at higher stability. Another procedure described in the patents includes the steps starting from formation of a slurry of inorganic components, clay as binder, and matrix materials; followed by slurry milling and cooling the milled slurry; and finally spray drying and calcining and/or washing the final material. Aluminum phosphate can be used for binding zeolite and clay in FCC catalyst composition as a hardening agent. Including alumina of less than 10% in the microsphere formulation and obtaining the total alumina composition of less than 30% can also lead to the production of additives with 40–80% ZSM-5 having a good attrition resistance. Another approach can consist of at least a portion of the microsphere with particles of higher density, low surface area unreactive species, and microspheres at more than 30% of ZSM-5 to have higher attrition resistance [35,50,55,111–138].

Since formation of a mesoporous structure within the materials enhances the resistance to coke deposition, Rive Technology is the main applicant for the corresponding topic due to the involvement of mesoporous material production technology within the patents owned. For instance, acid washing of initial zeolite with low silicon-to-aluminum ratio and formation of at least one mesopore with the zeolite is the approach of Rive Technology with regard the issue. ExxonMobil, China Petroleum & Chemical Corporation, and Grace are the other prominent applicants within the field. For instance, adjustment of the particle size of the gibbsite to a smaller value results in the reduction of coke yield at constant conversion and bottom yields. There is also a need to enhance the coke stability of the material in case rare earth is not present within. For this purpose, FCC catalyst compositions of silica source and magnesium salt have been found to be effective [139–167].

There is an increase in the use of heavier and sourer crudes, which contain more organic compounds with vanadium and nickel. Those fractions of nickel and vanadium deposit on the catalyst surface resulting in the decrease in catalytic activity. Therefore, there is a need for developments of FCC catalysts and additives known as "metal passivator/traps." Albemarle, BASF, Ecopetrol S.A., Indian Oil Corporation, and Grace have more than one patent application specifically for developments in "metal passivator/trap." Materials comprising manganese dispersed on hydrotalcite; a calcined spray-dried particle that includes kaolin; magnesium oxide or magnesium hydroxide and calcium carbonate; pyrophosphates M2P207 (M = Ba or Ca) supported on a mixture of magnesium and aluminum oxide (magnesium aluminate in the spinel phase); and an additive containing a rare earth component, alumina, clay, colloidal silica, and a zeolite can be used for metal passivation agents in the FCC units. There are also catalyst formulations specifically for passivating the detrimental effect of vanadium and nickel in the feedstock. Discrete particles of rare earth (i.e., lanthanum) carbonate compound dispersed in matrix formed by alumina have been found to be effective for cracking vanadium-containing hydrocarbons. Sodium is another type of contaminant that causes deactivation of FCC catalysts. Sodium can be present either within the feedstock or in raw materials during the zeolite manufacturing process. Therefore, during zeolite synthesis, zeolite undergoes several exchange processes to lower the sodium content by ammonium, rare earths, and other cations [41,59,95,129,168–180].

7.4.3 Developments in Terms of Cost-Effective, Efficient, or Alternative Preparation of Material

Developments that describe the modifications in the manufacturing process of the FCC catalysts and additives are classified under the heading of "cost-effective, efficient, or alternative preparation of material." There are certainly concrete targets underlying the corresponding category of patents. Several different preparation techniques lead to differences in the characterization properties of the materials. Developments related with the pore structure of the materials are grouped under the heading of "achieving mesoporous structure and/or pore targeted modification." There are also subtopics named as "crystallinity targeted modifications" and "particles size targeted modifications." "Catalyst with high solid content," "rare earth free FCC catalyst," and "Si/Al targeted modifications" are related with the content of the material. Several patents, grouped as "cost reduction and/or less complex process and/or alternative catalyst preparation," aim to reduce cost in the manufacturing process of the material. Those findings can also lead to reduction of the complexity of the manufacturing processes for the preparation of the material. The distribution of the corresponding subheading of the category can be seen in Figure 7.13, where the pore structure targeted developments take the lead with 42% of the total number of stability enhancement developments.

Rive Technology, ExxonMobil, and BASF are known to be the prominent applicants for the subtopic category of "material preparation for achieving mesoporous structure and/or general modifications resulting in changes of the pore structure." A material in a mesoporous structure can be initiated by treating the zeolite in a dehydroxylation environment at about 500–1200°C, followed by contacting with a pH-controlling agent comprising a basic surfactant. Another possible approach includes the addition of rare earth elements to a fully crystalline nonmesoporous zeolite with long-range crystallinity followed by acid washing and formation of a plurality of mesopores, which is subjected to hydrothermal regeneration. The technique for synthesis of mesoporous alumina including aluminum, rare earth element, and phosphorous begins with the formation of a reaction mixture together with a mesopore templating agent in a liquid medium, followed by separation of mesoporous alumina and separation of a portion of the templating agent by solvent extraction, finalized with calcination methods that are sufficient to remove any residual mesopore templating agent that result in condensation reactions with hydroxyl groups to liberate H_2O. There are also procedures for preparation of zeolitic materials where the porosity is given directly without the need for more postsynthetic treatments for removing structure directing agents from the crystallized framework [125,130,133,140,141,145,147–151,153,154,156–161,164–167,181–187].

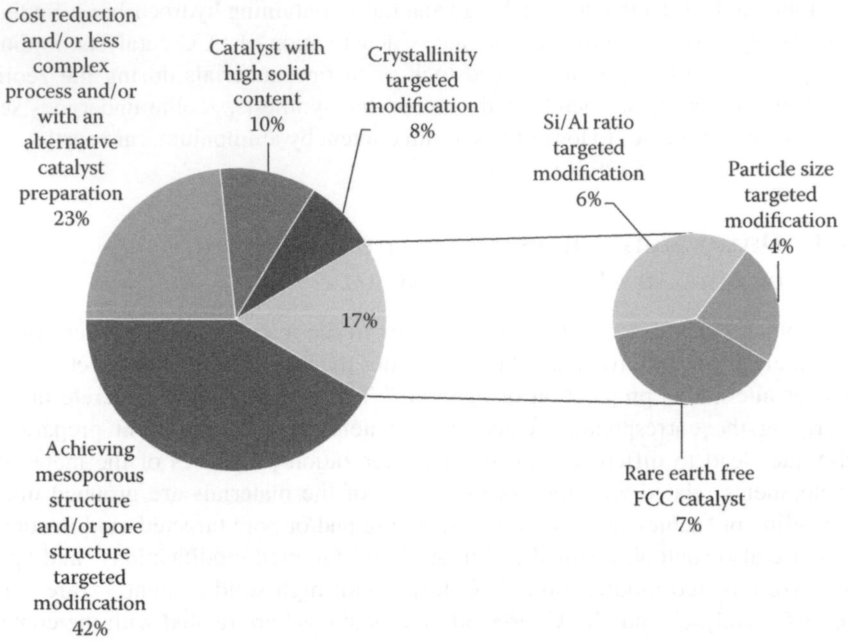

FIGURE 7.13 Target breakdown of "cost-effective, efficient, or alternative preparation of material"-related patents for FCC catalyst and additive.

The patents that describe the preparation methods in a more general manner with a less clear-cut target topic are grouped under the heading "alternative catalyst preparation" together with "cost reduction and/or less complex catalyst process" descriptions for the material synthesis. When compared to other applicants, BASF has a large number of patents in the corresponding general topic. Catalyst preparation techniques involve prevention of dealumination of the zeolite and the consequent loss of metal-containing active centers within the zeolite. Process descriptions for preparing different types of zeolites include Beta (BEA) framework in nanoporous structure with 3D 12-membered ring pore/channel system, Levyne (LEV-type) framework with heptadecahedral cavities at large micropore volume containing only 8-membered ring pore openings, and Chabazite (CHA) and Erionite (ERI) frameworks with small pores. There are several techniques that describe processes without using expensive tetraalkylammonium salts and other organic compounds together with eliminating the removal step of organotemplates that are encapsulated within the zeolitic frameworks. Use of ammonium carbonate solution instead of ammonium sulfate, ammonium nitrate, and ammonium chloride for exchanging sodium ions in zeolites comprising sodium ions and rare earth metal ions for ammonium ions has been found to result in minimization of the amount of rare earth metal ions that are leached out. Another advantage of using ammonium carbonate for ion exchange in zeolites is said to be lowering the amount of discharge of the salt due to recycling of excess ammonium carbonate. Since sodium results in loss of crystallinity of the zeolite, there are also efforts to decrease the sodium content of synthetic zeolites made of sodium silicate by adding yttrium compound to the zeolite [14,21,33,35,38,95,131,134,188–197].

Particularly, Petrochina Company Limited has patents in the topic of catalyst preparation with "high solid content." During the semisynthetic process of preparation of an FCC catalyst, alumina sol as binder, boehmite, clay, inorganic acid, and other raw materials are formed as a slurry, which is spray-molded followed by post-treatment steps. Before spray molding, colloid solid content is low resulting in nonuniform mixing of the active component of the molecular sieve and matrix components. Addition of inorganic acid after clay and aluminum-containing binder and before pseudoboehmite together with the addition of phosphorous-containing dispersant have been found to increase the solid content. Preparation method specifically for alumina particles that forms aqueous dispersions with relatively high solid content is also described within the contents of the corresponding list of patent topics [133,138,198–203].

BASF, Chevron, and UOP are some of the applicants with patents in the subtopic of "crystallinity targeted modification." Formation of CHA framework structure where the crystal size is greater than 1 μm is determined by SEM analysis. The corresponding material has been found to have phase-pure chabazite without impurities by a time- and cost-saving preparation procedure. A high molar ratio of Si/Al is also another outcome of the aforementioned material. Another preparation technique is to obtain ZSM-5 with small crystals, which can be used for processes in need of higher external surface area. Uniform distribution of the zeolitic crystallites for amorphous silica can be achieved by filling the pores with a sodium hydroxide solution, which is used as a crystallization agent [20,204–208].

Presence of rare earth metals in FCC catalysts enhances the catalytic activity and hydrothermal stability. Since gasoline demand increases together with the need for processing heavy feeds, the amount of rare earth in the formulation of FCC catalysts has increased. The supply of rare earth metals to the market has recently been stopped by China, which is known as the main supplier of rare earth metals (95%). This leads to the increase in the cost of the catalyst. In order to overcome such a difficulty, the industry has conducted research for reduction of rare earth metal content in the FCC catalysts while achieving desired yields. Synthesis of rare earth free FCC catalysts includes use of magnesium oxide. Non-rare earth (NRE) exchanged Y zeolites having a lower overall conversion and lower cracking efficiency have been found to be more effective for increasing conversion of propylene in case they are prepared by magnesium exchange together with ZSM-5. Therefore, the topic of rare earth free catalyst has an interrelation point with the target topic of increasing propylene yield in the FCC processes within the contents of the examined patents [142,143,209,210].

The topic of "Si/Al targeted modifications" has a strong relation with the topic of "crystallinity targeted modifications." Leaching with acid can be employed during the synthesis of molecular sieve (i.e., SSZ-81) to have a higher silica-to-alumina ratio (SAR), namely Si/Al greater than 10. Increasing the alumina-to-silica ratio at the surface of ultrastable zeolite Y (USY) includes heating the ammonium exchanged USY, adding the ammonium exchanged USY bath through hydrothermal conditions, and recycling the sodium content. The resulting catalyst has been found to enhance activity and selectivity, and reduce problems such as nickel and vanadium poisoning, lack of temperature structural stability in the presence of sodium, and coke formation and deposition [144,147,211,212].

ExxonMobil and Sud-Chemie have been found to possess patents for modification in terms of "particle size." During heating in the FCC unit, gibbsite structure has been found to show different phase trajectories for different precursors at different particle sizes. A zeolite cracking component in a gibbsite matrix with a median particle size of less than 400 nm has been found to reduce formation of coke showing that small particle size is a necessity for the process. Since spray drying is a technique for production of larger quantities of FCC catalysts at the desired particle size, it is not suitable for high throughput experimentations (HTEs) where there is a requirement of smaller quantities of catalysts (50 g) with controlled particle size. Therefore, there are also developments for obtaining the controlled particle size of smaller quantity FCC catalysts within the content of the patents [152,190,213].

7.4.4 Developments in Terms of Product Yield/Selectivity

More than half of the patents in the main category of product yield/selectivity are for increasing propylene/light olefin/lightweight yield (56%) in terms of product outlets as seen in Figure 7.14. Thirty-four percent of the corresponding category consist of patents investigating "improvement of activity and/or selectivity" of the FCC catalyst in a more general aspect. Formation of fewer dry gas by-products (8%)

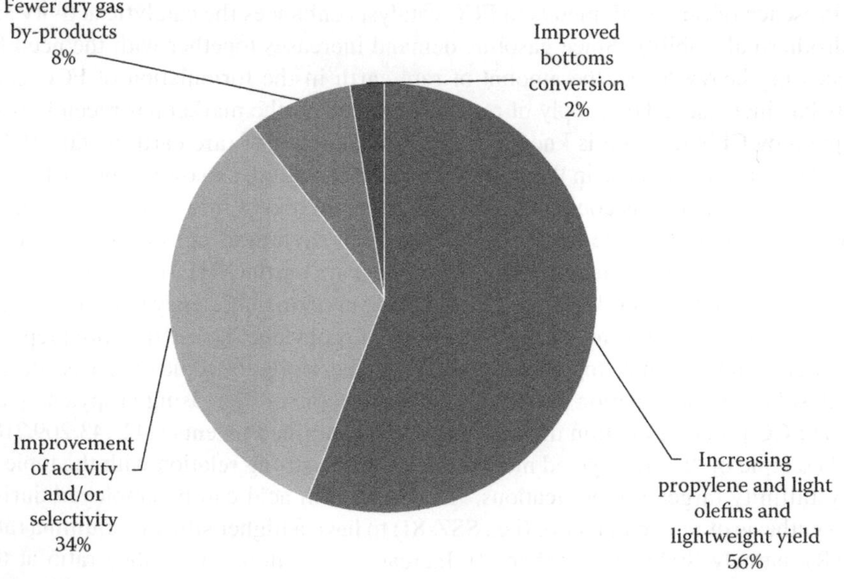

FIGURE 7.14 Target breakdown of "product yield/selectivity"–related patents for FCC catalyst and additive.

and increasing bottoms conversion (2%) are the remaining topics within the investigated patents.

UOP, Petrobras, Grace, BASF, Indian Oil Corporation, China Petroleum & Chemical Corporation, Albemarle, and several other companies have patents for increasing propylene, light olefins, and lightweight yield. Since ethylene and propylene are known to be the building block materials for production of plastics and packaging materials, the demand for light olefins increases. Due to the increasing demand, a well-known production technique, steam cracking, for propylene failed to satisfy the required amounts. FCC process together with the catalyst developments enables the production of propylene as the desired product by cracking larger hydrocarbons. Catalyst mixture consisting of Y zeolite as a large-pore zeolite, ZSM-5 as a molecular sieve, and metals deposited are mainly used in the production of propylene, light olefins, and lightweight products. ZSM-5 can also be used as an additive for the corresponding target. ZSM-5 can be combined by phosphorous and iron oxide. Using magnesium substitution to Y zeolite together with ZSM-5 has been found to increase the propylene yield in the desired manner. Medium-pore zeolite ZSM-5 can be stabilized by phosphate salts at a pH of 7–9 to be used for increasing propylene yield together with silica-rich binder to end up as an FCC additive with an enhanced stability under hydrothermal conditions [62,96,111,115,117,137,146,169,170,209,214–245].

Improvement of activity and/or selectivity is a more general group name that includes catalyst preparation methods to enhance activity and or selectivity toward gasoline formation. There are several companies, including Petroleum Energy Center, Petrochina Company Limited, Grace, and BASF, with patents grouped under the corresponding general heading. Defined components of the FCC catalyst to improve activity or selectivity are diverse, which include, for instance, clay, crystal grains, crystalline aluminosilicate, silica binder, and a binding agent. Details of the synthesis for ultrastable rare earth type molecular sieve where the rare earth ions are precisely localized are given within the content of the patents. Rare earth free FCC catalyst compositions improving catalyst activity include combination of magnesium salt and ultralow soda (amount of Na_2O in the catalyst). Preparation of boehmite alumina to be used for active matrix in the FCC catalyst includes heating in the presence of alkali metal hydroxide at 100°C under steam in an autoclave. A catalyst with a substantially inert core (mullite or alpha alumina) and active catalytic shell (in situ crystallized zeolite and matrix) around the core has been found to be both attrition-resistant and at increased activity. Modification of the FCC catalyst with phosphorous is also another technique, which includes the reduction of the sodium content of the formed catalyst, a treatment with phosphate solution, a second ammonium exchange for further reduction of the sodium content, and finally a second treatment with phosphate [113,114,116,117,130,133,138,142,143,146,162,203,212,243,246–258].

Dry gas fraction of the FCC process is the product that does not liquefy when compressed at ambient temperature, and it is defined to be the undesired gas by-products (hydrogen, methane, ethane, and ethylene). The main components are crystalline aluminosilicate, clay mineral, and silica binder. Combinations of H-ZSM-5 molecular sieve, USY molecular sieve, natural clay, and binder have also been used for the synthesis of the FCC catalyst for the purpose of less dry gas by-products. All these materials are mixed and pulped, followed by standing the pulped mixture for

1–4 h, continuously pulping for 10–30 min, drying naturally, drying in an oven at 100–120°C, heating in a muffle furnace at 50–650°C, roasting, and naturally cooling to obtain an FCC catalyst that is used for low yield of dry gas. In another patent, preparation of polymorph kaolin microspheres and in situ crystallization of cracking catalyst have been described by using ultrafine raw kaolin and spinel kaolin slurry as the basic raw material. As a conclusion, the final FCC catalyst product has been found to consist of many other attributions that are required for the FCC process, including less dry gas yield formation, mesoporous structure, ability to crack heavy feeds, and high lightweight yield [146,155,162,236,255,257].

For the improvement of bottoms conversion, inorganic metal oxide particles and alumina obtained from aluminum sulfate with a binder to form a particulate inorganic metal oxide together with the faujasite zeolite can be combined in the synthesis procedure. The corresponding patent has defined the synthesis steps to be economical and the catalyst to have an improved bottoms conversion together with decreased coke production and also good attrition properties. Another approach includes the catalyst comprising faujasite zeolite and a siliceous metal oxide, both with a negative surface charge at a pH of 7, together with precipitated alumina. Finally, the catalyst has a high matrix surface area, which is one of the key factors that enables improved bottoms conversion [131,259].

7.4.5 Developments in Terms of Feedstock-Based Technologies

Feedstock-based technologies are highly gathered around the subtopic of "cracking of alternative feedstock (biomass/bio oil/pyrolysis oil/plastics)" as can be seen in Figure 7.15 with 77% of the total value of the corresponding topic. Other than the

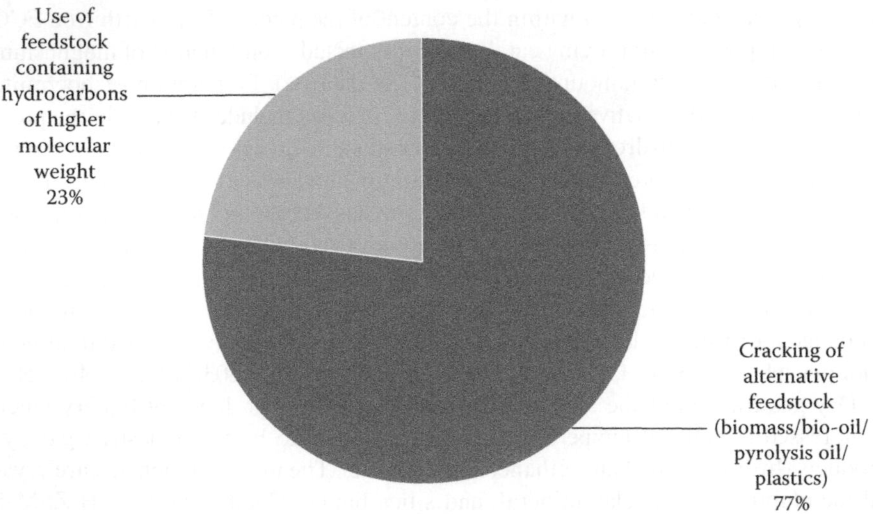

FIGURE 7.15 Target breakdowns of "feedstock-based technologies"–related patents for FCC catalyst and additive.

alternative feedstock, the other main subtopic classification that aims at the existing common FCC feedstock in terms of higher molecular weight has been determined to be 23% of the total number of patents for feedstock-based technologies.

The governing portion of "feedstock-based technologies" category includes patents of cracking alternative feedstocks such as biomass, pyrolysis oil, and plastic. Main applicants for this category are Kior and several other companies such as UOP, ExxonMobil, and Grace. Thermal pyrolysis process for converting biomass to bio-oil or gas can be achieved by using fluidized-bed reactors due to the requirement of fast heating rates, short reaction times, and a rapid quench of the liquid reaction products. Liquid transportation fuels can be obtained by the conversion of crude bio-oil in the modified refinery FCC processes. Hybrid silica–alumina catalysts with a controlled Lewis acidity and a controlled porosity can be applied to be used as a biomass conversion catalyst. Silica and alumina precursors are mixed to form slurry. Zeolite is phosphorus promoted and then mixed with silica–alumina slurry together with clay. The final slurry is shaped and then calcined to become the corresponding catalyst. Another approach includes the preparation of a biomass catalytic cracking catalyst starting with modification of phyllosilicate by leaching with acidic solutions. This step is followed by contact with suspension of metal ions, mixing the modified phyllosilicate with inorganic materials and finally shaping the final product to achieve fluidizable microspheres. UZM-39 aluminosilicate zeolite can be used for catalytic pyrolysis of biomass after the catalyst is finely ground to be used in the FCC unit. For treatment of hydrocarbon feedstocks with a biorenewable feed, a high zeolite-to-matrix ratio catalyst combination including Y-type zeolite, a matrix, and rare earth metal oxide can also be used [62,139,181,260–276].

Due to the depletion of lighter feedstocks, the demand for FCC catalysts that can achieve conversion of heavy oils to the desired products at higher yields has increased. Localization of rare earth ions in sodalite cages by magnesium modification of USY molecular sieve has been found to have a high heavy oil conversion capacity since rare earth ions are not lost during reverse exchange process. Preparation of Y-type molecular sieve including hydrothermal crystallization with synthetic fluid, which contains clay, oriented agent, silicon source, sodium hydroxide, and water, resulted in abundant mesopores in the catalyst with the centralized distribution. The prepared material also favors the macromolecular reactant (heavy feed components) to approach the active catalyst centers and then diffusion of the desired product molecules from the pores. Ultimately, the corresponding Y-type molecular sieve has been found to be a suitable catalyst carrier for treating heavy feedstocks [113,155,194,219,246,247,277–279].

7.4.6 Developments in Terms of Product Quality

Patents of FCC catalysts and additives in terms of increasing product quality have been grouped into two main categories as can be seen in Figure 7.16, with 67% of them for increasing the octane number of gasoline and 33% for production of LCO with low aromatic content.

Branched chain hydrocarbons, alkyl groups, and olefins are the required components for a higher octane number in gasoline, whereas straight-chain hydrocarbons have a low octane number. Attempts to increase the octane number of gasoline may

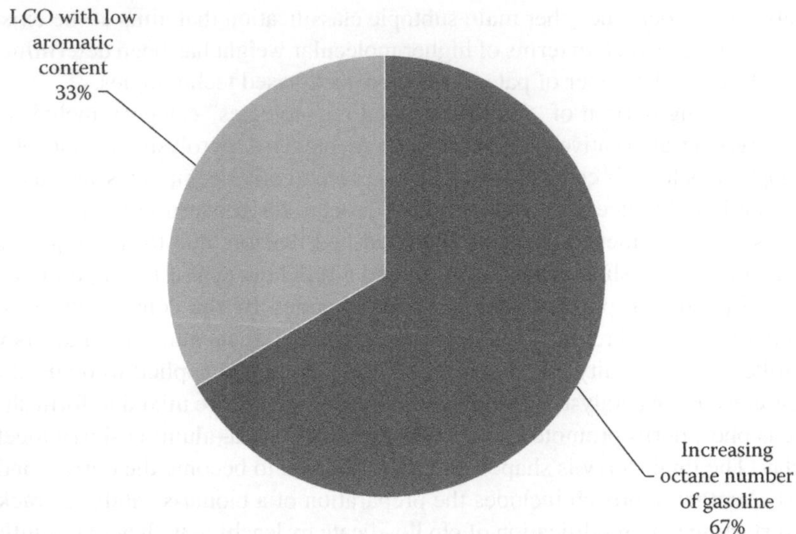

FIGURE 7.16 Target breakdown of "product quality"–related patents for FCC catalyst and additive.

lead to low yields of gasoline or tar deposits on the catalyst surface. There are several patents to overcome these problems during the increase of the octane number. Crystalline aluminosilicate, sodium, potassium, and rare earth content together with a xenon adsorption have been proposed for achieving the corresponding purpose. There are also studies for preparing a prosperous FCC catalyst in terms of nickel tolerance together with high-octane gasoline formations with a yield of a large amount of light olefins by using acidic catalysts with large-pore rare earth faujasite zeolite, pentasil zeolite, and pseudoboehmite with resid cracking component and lanthanum oxide and/or aluminum oxide impregnation [219,223,251–253,280].

LCO is the FCC product that is known to be the basis for fuel oil. It can also be converted to diesel by an additional hydrotreatment process. Since diesel is to have a high cetane number, LCO should consist of straight-chain hydrocarbons in order to be blended for the feed that will be converted to diesel fuel. Therefore, branched chain hydrocarbons, olefins, and aromatics should be eliminated from LCO in order for LCO to have a higher cetane number. Instead of conventional acidic FCC catalyst compositions, using the FCC catalyst with an intermediate and/or small-pore zeolite and preferably free of large-pore zeolite together with the basic material has been found to result in LCO with lower aromatic content [237,281,282].

7.5 FUTURE PROSPECTS IN FCC TECHNOLOGY

Since its establishment in the 1940s, FCC has become quite a flexible conversion process for refineries in terms of meeting the demand of product versatility under the strict requirements of heavier feedstocks and environmental regulations. In addition to process-based advances such as improvements of feed and catalyst injection systems, elimination of long dilute-phase residence time downstream of the riser, and developments of mechanical reliability of FCC reactor–regenerator components, improvements in the research area of FCC catalysts and additives have played an important role for the success of FCC processes [2].

Initiating from the discovery of using zeolites in the FCC catalyst formulations in the 1960s, there are several advances in the field of FCC catalysts and additives in terms of zeolite quality, properties of the binder and active matrix, use of ZSM-5, reduction of sulfur in gasoline, and reduction of NO_x and SO_x emissions. Quality of the zeolite has been improved by greater SAR to achieve higher stability, more olefin yield, improvement of the octane number of gasoline, and better selectivity. Crystallinity improvement of the zeolite is accomplished by synthesis of more uniform crystals. Acid site distribution of the zeolites together with choice of the active matrix is required to achieve the desired catalyst activity. Attrition and wear resistance of the catalyst are mainly dependent on the improvements in the properties of the binder. There are also efforts for the improvements of spent catalyst distribution and minimization of the disposal of the equilibrium catalyst for the operational performance as well as meeting the demand of environmental regulations [2].

Developments in compatible numbers are noticed throughout the patent investigation study based on the classified main patent topics between the years 2008 and 2014 as can be seen in Figure 7.17. The main research area for FCC catalysts and

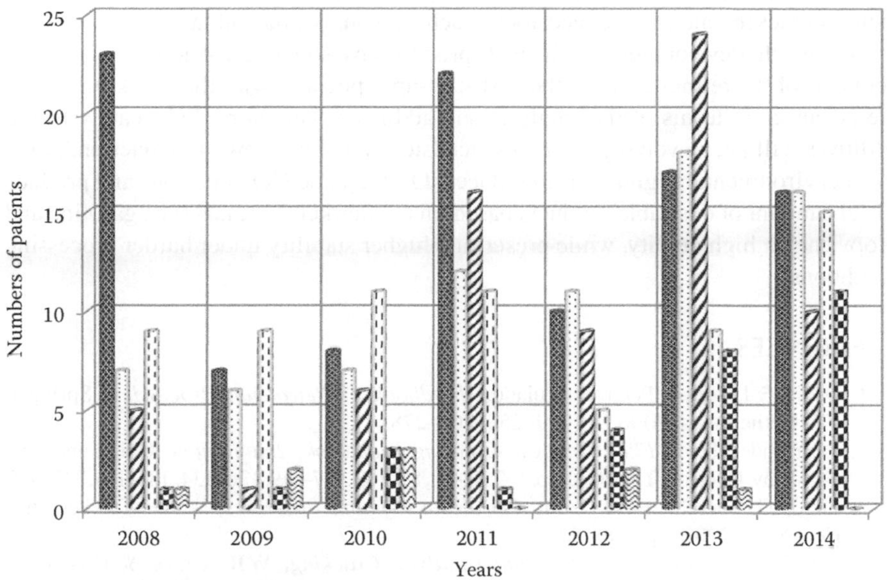

FIGURE 7.17 Annual distributions of the investigated patents in terms of the topic classification.

additives has been found to be environmental considerations such as reduction of NO_x, SO_x, and CO emissions, sulfur in the gasoline, and finding alternative applications for reuse of spent/equilibrium FCC catalysts or the materials in the catalyst or FCC fines for decreasing the disposal of the corresponding component.

In terms of finding a cost-effective, more efficient, and alternative catalyst, preparation methods are also developed in higher numbers. Especially patents related to production of mesoporous FCC catalysts have been found to be higher in number within the corresponding topic. Mesoporous structure of the zeolites used in the FCC catalyst has made it possible for the heavier molecules to be cracked while the desired products can rapidly flow out of the zeolite.

Increasing share can be seen to be related to the advances in terms of stability enhancements and product yield/selectivity to achieve a higher amount of desired products with more stable catalysts for the FCC units to become more profitable. As the demand for propylene increases in the petrochemical industry, developments are achieved to shift the production of FCC from gasoline to higher yields of lighter olefins by using modified catalyst formulations including ZSM-5.

The main focus of stability enhancements is to synthesize wear/attrition-resistant catalysts with less coke formation and deposition, and use additives as metal passivators/traps to endure feed contaminants such as Ni, Va, or Na.

Despite having less number of patents in terms of feedstock-based technologies, developments steadily increase in the corresponding fields mainly due to the requirement of cracking alternative feedstocks such as biomass, bio-oil, and pyrolysis oil.

As a result, developments in the FCC process have shown great adaptability to the demands of the refineries since the first start-up especially with the breakthrough of the advances in terms of the catalysts and additives. Therefore, FCC catalysts and additives will keep evolving to process feedstocks that are new or heavier under the strict environmental regulations to reduce NO_x, SO_x, and CO emissions and produce a high amount of desirable products based on the market demands (i.e., gasoline and propylene) at high quality, while preserving higher stability under harder processing conditions.

REFERENCES

1. David S.J. Jones, Peter R. Pujado, *Handbook of Petroleum Processing*, Springer, The Netherlands, 2006, pp. 239–250, 266–278.
2. Reza Sadeghbeigi, *Fluid Catalytic Cracking Handbook, 2nd Edition*, Gulf Publishing Company, Houston, TX, 2000, pp. 1–2, 41–55, 85–102, 117–119, 126–134, 182–199, 332–335.
3. James G. Speight, Baki Özüm, *Petroleum Refining Processes*, Marcel Dekker, Inc., New York, 2002, pp. 400–420.
4. GRACE Davison, *Guide to Fluid Catalytic Cracking*, W.R. Grace & Co.-Conn., Baltimore, MA, 1993, pp. 11–12.
5. Bohdan W. Wojciechowksi, Avelino Corma, *Catalytic Cracking, Catalysts, Chemistry and Kinetics*, Marcel Dekker, Inc., New York, 1986, p. 65.
6. Martin Evans, Xunhua Mo, Raymond Paul Fletcher, Process of removing NO_x from flue gas, EP 2817083 A1, filed February 22, 2013, and issued December 31, 2014.
7. Natalia Trukhan, Ulrich Mueller, Ivor Bull, Process for the direct synthesis of Cu containing zeolites having CHA structure, US 8715618 B2, filed May 19, 2009, and issued May 6, 2014.

8. Jaya L. Monahan, Patrick Burk, Michael J. Breen, Barbara Slawski, Makoto Nagata, Yasuyuki Banno, Eunseok Kim, 8-ring small pore molecular sieve as high temperature SCR catalyst, WO 2014062949 A1, filed October 17, 2013, and issued April 24, 2014.
9. Jaya L. Monahan, Patrick Burk, Michael J. Breen, Barbara Slawski, Makato Nagata, Yasuyuki Banno, Eunseok Kim, Mixed metal 8-ring small pore molecular sieve catalyst compositions, catalytic articles, systems and methods, WO 2014062944 A1, filed October 17, 2013, and issued April 24, 2014.
10. Jaya L. Monahan, Patrick Burk, Makato Nagata, Yasuyuki Banno, Eunseok Kim, 8-ring small pore molecular sieve with promoter to improve low temperature performance, WO 2014062952 A1, filed October 17, 2013, and issued April 24, 2014.
11. Ajit B. Dandekar, Richard L. Eckes, Richard F. Socha, S. Beau Waldrup, Jason M. Mcmullan, Catalytic reduction of NO_x with high activity catalysts with NH_3 reductant WO 2014025529 A1, filed July 23, 2013, and issued February 13, 2014.
12. Ajit B. Dandekar, Richard L Eckes, Richard F. Socha, S. Beau Waldrup, Jason M. Mcmullan, Catalytic reduction of NO_x with high activity catalysts with acetaldehyde reductant, WO 2014025531 A1, filed July 23, 2013, and issued February 13, 2014.
13. Ajit B. Dandekar, Richard F. Socha, Richard L. Eckes, S. Beau Waldrup, Catalytic reduction of NO_x with high activity catalysts with propylene reductant, US 8815195 B2, filed July 10, 2013, and issued August 26, 2014.
14. Ivor Bull, Ulrich Müller, Cu containing silicoaluminophosphate (Cu-SAPO-34) EP 2687284 A1, filed June 2, 2010, and issued January 22, 2014.
15. Mathias Feyen, Ulrich Müller, Roger Ruetz, Thomas Bein, Karin Möller, Cha type zeolitic materials and methods for their preparation using cycloalkylammonium compounds, WO 2013182974 A1, filed June 3, 2013, and issued December 12, 2013.
16. Stacey Ian Zones, Preparation of molecular sieve ssz-23, WO 2013130240 A1, filed February 8, 2013, and issued September 6, 2013.
17. Stefan Maurer, Michael Wycisk, Julia Petry, Stephan Deuerlein, Weiping Zhang, Chuan Shi, Hihara Takashi, Iron- and copper-containing zeolite beta from organotemplate-free synthesis and use thereof in the selective catalytic reduction of NO_x, WO 2013118063 A1, filed February 6, 2013, and issued August 15, 2013.
18. Chiranjeevi Thota, Dattatraya Tammannashastri Gokak, P. S. Viswanathan, Multifunctional catalyst additive composition and process of preparation thereof, US 20130130888 A1, filed May 7, 2012, and issued May 23, 2013.
19. Martin Evans, Raymond Fletcher, Mehdi Allahverdi, Guido Aru, Paul Diddams, Xunhua Mo, William Reagan, Shanthakumar Sithambaram, Cracking catalysts, additives, methods of making them and using them, US 8444941 B2, filed May 25, 2011, and issued May 21, 2013.
20. Bilge Yilmaz, Ulrich Müller, Feng-Shou Xiao, Takashi Tatsumi, Vos Dirk De, Xinhe Bao, Weiping Zhang, Hermann Gies, Hiroyuki Imai, Bart Tijsebaert, Limin Ren, Chengguan Yang, Process for the production of a core/shell zeolitic material having a CHA framework structure, WO 2013038372 A2, filed September 14, 2012, and issued March 21, 2013.
21. Bilge Yilmaz, Ulrich Müller, Faruk Özkirim, Meike Pfaff, Mike Haas, Feng-Shou Xiao, Process for the organotemplate-free synthetic production of a zeolitic material using recycled mother liquor, US 20130064758 A1, filed September 7, 2012, and issued March 14, 2013.
22. Ahmad Moini, Saeed Alerasool, Subramanian Prasad, Molecular sieve precursors and synthesis of molecular sieves, WO 2013028958 A1, filed August 24, 2012, and issued February 28, 2013.
23. Chen Yanguang, Han Hongjing, Lu Jia, Song Hua, Chang Zhigang, Zhang Mei, Method for decreasing FCC regenerative process NO_x by $DeNO_x$ additive, CN 102888242 B, filed October 20, 2012, and issued January 23, 2013.

24. Chen Yanguang, Han Hongjing, Lu Jia, Song Hua, Chen Ying, Li Jie, Method for reducing NO_x emission in FCC (fluid catalytic cracking) regeneration process, CN 102824849 B, filed August 3, 2012, and issued December 19, 2012.
25. Qi Wenyi,Yuan Qiang, Qin Ruyi, Wang Longyan, Huang Xinlong, Hao Daijun, Liu Shufang, Wang Hongbin, Wang Shaofeng, Hu Yanfang, Zhao Zhigang, Catalyst for removing NO in smoke and preparation method thereof, CN 102631933 B, filed February 14, 2011, and issued August 15, 2012.
26. Ivor Bull, Ulrich Müller, Copper containing ZSM-34, OFF and/or ERI zeolitic material for selective reduction of NO_x, US20120014865 B2, filed July 15, 2011, and issued January 19, 2012.
27. Martin Evans, Raymond Fletcher, Mehdi Allahverdi, Guido Aru, Paul Diddams, Xunhua Mo, William Reagan, Shanthakumar Sithambaram, Cracking catalysts, additives, methods of making them and using them, US 20110308620 B2, filed May 25, 2011, and issued December 22, 2011.
28. Hong-Xin Li, William E. Cormier, Bjorn Moden, Novel metal-containing zeolite beta for NO_x reduction and methods of making the same, US 20110286914 A1, filed May 18, 2011, and issued November 24, 2011.
29. David Matheson Stockwell, FCC additive for partial and full burn NO_x control, EP2380950 A1, filed November 17, 2006, and issued October 26, 2011.
30. George Yaluris, John Allen Rudesill, NO_x reduction compositions for use in FCC processes, sg173330), filed March 24, 2006, and issued August 29, 2011.
31. Tilman Beutel, Martin Dieterle, Ivor Bull, Ulrich Müller, Ahmad Moini, Michael Breen, Barbara Slawski, Saeed Alerasool, Process for preparation of copper containing molecular sieves with the CHA structure, catalysts, systems and methods, US 20110165052 B2, filed December 16, 2010, and issued July 7, 2011.
32. Tilman Beutel, Martin Dieterle, Ulrich Müller, Ivor Bull, Ahmad Moini, Michael Breen, Barbara Slawski, Saeed Alerasool, Wenyong Lin, Xinsheng Liu, Process of direct copper exchange into Na+-form of chabazite molecular sieve, and catalysts, systems and methods, US 20110165051 B2, filed December 16, 2010, and issued July 7, 2011.
33. Ivor Bull, Ulrich Müller, Process for the preparation of zeolites having CHA structure, WO 2011064186 A1, filed November 23, 2010, and issued June 3, 2011.
34. Michael S. Ziebarth, Meenakshi Sundaram Krishnamoorthy, Roger J. Lussier, Compositions and processes for reducing NO_x emissions during fluid catalytic cracking, sg169976 A, filed March 24, 2006, and issued April 29, 2011.
35. Ivor Bull, Ulrich Müller, Bilge Yilmaz, Copper containing levyne molecular sieve for selective reduction of NO_x, WO 2011045252 A1, filed October 11, 2010, and issued April 21, 2011.
36. Ren Xiaoguang, Song Yongji, Ren Chao, Hexaaluminate metal oxide catalyst, preparation methods and application thereof, CN 102008954 A, filed October 14, 2010, and issued April 13, 2011.
37. Michael S. Ziebarth, Meenakshi Sundaram Krishnamoorthy, Roger J. Lussier, Compositions and methods for reducing NO_x emissions during catalytic cracking with fluidised catalyst, ru0002408655 C2, filed April 15, 2005, and issued January 10, 2011.
38. Ivor Bull, Ulrich Muller, Process for the direct synthesis of cu containing silicoaluminophosphate (cu-sapo-34), EP2269733 A1, filed June 2, 2010, and issued January 5, 2011.
39. George Yaluris, Michael Ziebarth, Xinjin Zhao, Ferrierite compositions for reducing NO_x emissions during fluid catalytic cracking, US 20100276337 A1, filed July 9, 2010, and issued November 4, 2010.

40. George Yaluris, Michael Scott Ziebarth, Xinjin Zhao Fluidizing cracking catalyst and method for reducing NO_x emissions during fluid catalytic cracking, CN 101503632 B, filed November 4, 2004, and issued August 12, 2009.
41. Dennis Stamires, Paul O'connor, Process for the preparation of an aluminum phosphate-containing catalyst composition, US 20090149317 A1, filed November 7, 2006, and issued June 11, 2009.
42. Song Haitao, Tian Huiping, Jiang Wenbin, Zhu Yuxia, Zheng Xueguo, Tang Liwen, Chen Beiyan, Da Zhijian, Zhang Jiushun, Long Jun Composition for reducing discharge of NO_x in FCC stack gas, CN 101311248 B, filed May 24, 2007, and issued November 26, 2008.
43. David Matheson Stockwell, Process for regenerating FCC catalysts with a reduced amount of NO_x emissions in the flue gas, EP 1969091 B1, filed November 17, 2006, and issued September 17, 2008.
44. George Yaluris, John Allen Rudesill, Meenakshi Sundaram Krishnamoorthy, Timothy J. Dougan, Method for controlling NO_x emissions in the FCCU, US 20080213150 A1, filed February 23, 2006, and issued September 4, 2008.
45. George Yaluris, Michael Ziebarth, Xinjin Zhao, Compositions and processes for reducing NO_x emissions during fluid catalytic cracking, za2006/08953, filed October 30, 2006, and issued April 30, 2008.
46. Dennis Stamires, Paul O'Connor, William Jones, Oxidic metal composition, its preparation and use as catalyst composition, EP 1896171 A1, filed June 2, 2006, and issued March 12, 2008.
47. John Allen Rudesill, George Yaluris, Meenaksh Sundaram Krishnamoorthy, Timothy J. Dougan, Method for controlling NO_x emissions in the FCCU, mxMX/a/2007/011531, filed September 19, 2007, and issued February 22, 2008.
48. William Jones, Dennis Stamires, Paul O'Connor, Michael F. Brady, Catalyst composition comprising anionic clay and rare earth metals, its preparation and use in FCC, US 20080039314 B2, filed December 6, 2004, and issued February 14, 2008.
49. William Jones, Dennis Stamires, Paul O'Connor, Michael F. Brady, Process for the preparation of an additive-containing anionic clay, US 20080032883 B2, filed April 26, 2005, and issued February 7, 2008.
50. Zhang Weihang, Li LIying, Fan Li, Liu Zhenyi, Method for preparing microsphere sulfur transfer agent with high activity and high wear resistance, CN 103920455 A, filed January 14, 2013, and issued July 16, 2014.
51. Jaime Sánchez Valente, Roberto Quintana Solórzano, Lázaro Moisés García Moreno, Rodolfo Juventino Mora Vallejo, Francisco Javier Hernández Beltrán, Multimetallic anionic clays and derived products for SO_x removal in the fluid catalytic cracking process, US 20120067778 A1, filed September 16, 2010, and issued March 22, 2012.
52. Graaf Elbert Jan De, Jorge Alberto Gonzalez, Julie Ann Francis, Maria Margaret Ludvig, Additive-containing anionic clays for reducing SO_x emissions from an FCC regenerator and process for making them, US 20100111797 A1, filed March 20, 2008, and issued May 6, 2010.
53. Dennis Stamires, Paul O'Connor, Hydrocarbon conversion process using a catalyst composition comprising aluminum and a divalent metal, US 20090305872 A1, filed November 19, 2005, and issued December 10, 2009.
54. David M. Stockwell, Catalyst for NO_x and/or SO_x control, US 20080227625 A1, filed April 25, 2008, and issued September 18, 2008.
55. Dennis Stamires, Paul O'Connor, William Jones, Erik Laheij, Metal-doped mixed metal oxide, its preparation and use as catalyst composition, CN 101238198 A, filed June 2, 2006, and issued August 6, 2008.

56. Dennis Stamires, Paul O'Connor, Erik Jeroen Laheij, Michael F. Brady, Julie A. Francis, Maria M. Ludvig, Use of anionic clay in an FCC process, CN 101208410 A, filed June 30, 2006, and issued June 25, 2008.
57. Dennis Stamires, Paul O'Connor, William Jones, Oxidic metal composition, its preparation and use as catalyst composition, EP 1896172 A1, filed June 2, 2006, and issued March 12, 2008.
58. William Jones, Dennis Stamires, Paul O'Connor, Michael Brady, Process for the Preparation of a Metal-Containing Composition, US 20080039313 A1, filed April 26, 2005, and issued February 14, 2008.
59. Luis Francisco Pedraza Archila, Luis Cedeño Caero, Bifunctional additive with lanthanum and niobium for the FCC process, mxPA/a/2006/003397, filed March 27, 2006, and issued February 8, 2008.
60. William Jones, Dennis Stamires, Paul O'Connor, Michael Brady, Process for the preparation of an additive-containing anionic clay, US 20080032884 A1, filed April 26, 2005, and issued February 7, 2008.
61. Amutha Rani Devaraj, Hai Xiang Lee, Diego Alfonso Martinez Velandia, Nikolaos Vlasopoulos, Binder composition, US 20140290535 A1, filed August 8, 2011, and issued October 2, 2014.
62. Ravichander Narayanaswamy, Krishna Kumar Ramamurthy, P. S. Sreenivasan, Conversion of plastics to olefin and aromatic products, US 8895790 B2, filed February 12, 2013, and issued November 25, 2014.
63. Xiaolin David Yang, Pascaline Harrison Tran, Lawrence Shore Pollutant emission control sorbents and methods of manufacture and use US 8728974 B2, filed August 3, 2012, and issued May 20, 2014.
64. Albert A. Vierheilig, Methods of recovering rare earth elements, US 8216532 B1, filed June 17, 2011, and issued July 10, 2012.
65. Roger G. Etter, System and method for introducing an additive into a coking process to improve quality and yields of coker products, US 8888991 B2, filed February 12, 2013, and issued November 18, 2014.
66. Liao Le, Wang Xiaoyun, Method for recycling Ni from waste FCC (Fluid Catalytic Cracking) catalyst, CN 103343232 A, filed July 11, 2013, and issued October 9, 2013.
67. Liao Le, Wang Xiaoyun, Method for recovering vanadium from waste FCC catalyst, CN 103332741 A, filed July 11, 2013, and issued October 2, 2013.
68. Liao Le, Wang Xiaoyun, Method for recovering aluminum oxide from waste FCC catalyst, CN 103332715 A, filed July 11, 2013, and issued October 2, 2013.
69. Manish Agarwal, V. Chidambaram, P. S. Choudhury, A. V. Khathikeyani, Prabhu K. Mohan, S. Rajagopal, Biswanath Sarkar, Balaiah Swamy, Value added spent fluid catalytic cracking catalyst composition and a process for preparation thereof, US 20130168290 A1, filed July 4, 2011, and issued July 4, 2013.
70. Yu-Lung Sun, Ming-Zhe Tsai, Yung-Hao Liu, Method for recovering rare earth compounds, vanadium and nickel, US 8986425 B2, filed November 28, 2011, and issued April 18, 2013.
71. Albert A. Vierheilig, Methods of re-using a spent FCC Catalyst, US 8614159 B2, filed August 13, 2012, and issued December 20, 2012.
72. Augusto R. Quinones, Process to improve formulations of hydrocarbon conversion catalysts through removal and modification of detrimental particles and reuse of modified fractions, WO 2012129065 A3, filed March 16, 2012, and issued December 6, 2012.
73. Albert A. Vierheilig Methods of recovering rare earth elements WO2012174454, filed June 15, 2012, and issued December 20, 2012.
74. Xingtao Gao, William Todd Owens, Process for metal recovery from catalyst waste, US20120156116 A1, filed December 15, 2010, and issued June 21, 2012.

75. Chiranjeevi Thota, Dattatraya Tammannashastri Gokak, Ravikumar Voolapalli, Nettam Venkateshwaralu Choudary, Mohammad Amir Siddiqui, Rajan Bosco, Raghunath Prasad Mehrotra, A carbon monoxide combustion catalyst and a process of preparation thereof, US20120157293 A1, filed December 30, 2009, and issued June 21, 2012.
76. Alexander Kehrmann, Method for recovering lanthanum from zeolites containing lanthanum, US 20120087849 A1, filed October 5, 2011, and issued April 12, 2012.
77. Zhang Junji, Wu Xiujuan, Wu Bin, Gao Hong, Method for producing white carbon black with spent catalyst of FCC (fluid catalytic cracker) as raw material, CN 102275935 A, filed May 7, 2011, and issued December 14, 2011.
78. Ban Bong Chan, Jin Woo Hyun, Ji Hye, Mee So Lim, Chang Gil Paik Producing method of a fire extinguishing agent by alkali fritting and mechano-chemistry technologies, KR1020110113407, filed April 9, 2010, and issued October 17, 2011.
79. Yao Hua, Zheng Shuqin, Tan Licheng, Yang Guangjun, Zhen Yuhong, Tan Fuping, Method for synthesizing porous microsphere material containing NaY zeolite by waste FCC (fluid catalytic cracking) catalyst, CN 102125872 B, filed January 17, 2011, and issued July 20, 2011.
80. Kaoru Fujimoto, Xiaohong Li, Method for catalytically cracking waste plastics and apparatus for catalytically cracking waste plastics, US 20110166397 B2, filed March 18, 2011, and issued July 7, 2011.
81. Biswanath Saha, Sonal Maheshwari, Paramasivam Senthivel, Nettem Venkateswarlu Choudary, Crumb rubber modified bitumen (crmb) compositions and process thereof, WO 2011074003 A2, filed December 14, 2010, and issued June 23, 2011.
82. Xingtao Gao, James Fu, Catalyst composition for reducing gasoline sulfur content in catalytic cracking process, US 20110139684 B2, filed February 23, 2011, and issued June 16, 2011.
83. Liu Xinmei, Liang Haining, Li Liang, Yang Tingting, Yan Zifeng, Method for synthesizing superfine Y-type molecular sieve, CN 101891221 A, filed July 22, 2010, and issued November 24, 2010.
84. Xiaolin D. Yang, David M. Stockwell, Pascaline H. Tran, Lawrence Shore, Methods of manufacturing mercury sorbents and removing mercury from a gas stream, US 20100266468 B2, filed March 31, 2010, and issued October 21, 2010.
85. Naoki Fujiwara, Method for decomposing dinitrogen monoxide, US 20100196238 A1, filed July 26, 2007, and issued August 5, 2010.
86. Kaoru Fujimoto, Xiaohong Li, Method for catalytically cracking waste plastics and apparatus for catalytically cracking waste plastics, JP 2010013657 A, filed August 31, 2009, and issued January 21, 2010.
87. Paul O'Connor, Erik Jeroen Laheij, Dennis Stamires, Oscar Rene Chamberlain Pravia, Rodolfo Eugenio Roncolatto, Yiu Lau Lam, Process for upgrading an FCC equilibrium catalyst, US 20080210599 A1, filed January 6, 2006, and issued September 4, 2008.
88. Liu Xinmei, Zhang Xingong, Yan Zifeng, Method for reliving FCC dead catalyst CN 101219396 B, filed January 28, 2008, and issued July 16, 2008.
89. Naoki Fujiwara, Method for decomposing dinitrogen monoxide, WO 2008020535 A1, filed July 26, 2007, and issued February 21, 2008.
90. Paul O'Connor, Erik Jeroen Laheij, Dennis Stamires, Oscar Rene Chamberlain Pravia, Rodolfo Eugenio Roncolatto, Yiu Lau Lam, Process for upgrading an FCC equilibrium catalyst, CN 101102840 A, filed January 6, 2006, and issued January 9, 2008.
91. Abdennour Bourane, Omer Refa Koseoglu, Musaed Salem Al-Ghrami, Christopher F. Dean, Mohammed Abdul Bari Siddiqui, Shakeel Ahmed, Clay additive for reduction of sulfur in catalytically cracked gasoline, US 8623199 B2, filed July 30, 2009, and issued January 7, 2014.

92. Christopher F. Dean, Musaed Salem Musaed Al-Ghrami Al-Ghamdi, Khurshid K. Alam, Mohammed Abdul Bari Siddiqui, Shakeel Ahmed, Catalyst additive for reduction of sulfur in catalytically cracked gasoline, US 8409428 B2, filed June 28, 2006, and issued April 2, 2013.
93. Hu, Ruizhong, Gasoline sulfur reduction catalyst for fluid catalytic cracking process, EP 2 604 340 A1, filed February 24, 2006, and issued June 19, 2013.
94. Dattatraya Tammannashastri Gokak, Chiranjeevi Thota, Pragya RAI, N. Jose, P. S. Viswanathan, Sulphur reduction catalyst additive composition in fluid catalytic cracking and method of preparation thereof, US 20130081980 A1, filed September 30, 2011, and issued April 4, 2013.
95. Zhang Feng, Yang Hua, Tang Wei, Multifunctional desulphurization auxiliary agent of distillate of FCC device and preparation method thereof, CN 102229813 B, filed May 23, 2011, and issued November 2, 2011.
96. Zheng Jinyu, Zhou Jihong, Luo Yibin, Shu Xingtian, Cracking catalyst, CN 102166528 A, filed February 26, 2010, and issued August 31, 2011.
97. Abdennour Bourane, Omer Refa Koseoglu, Musaed Salem Al-Ghrami, Christopher F. Dean, Mohammed Abdul Bari Siddiqui, Shakeel Ahmed, Metallic clay based FCC gasoline sulfur reduction additive compositions, US 20110120912 B2, filed July 30, 2009, and issued May 26, 2011.
98. Terry Glynn Roberie, Ranjit Kumar, Michael S. Ziebarth, Wu-Cheng Cheng, Xinjin Zhao, Nazeer Bhore, Additive for reducing gasoline sulfur in fluid catalytic cracking, EP 2325283 A1, filed September 18, 2000, and issued May 25, 2011.
99. William Jay Turner, Ronald Lee Cordle, David J. Zalewski, Jeffrey A. Sexton Gasoline sulfur reduction in FCCU cracking, US 7763164 B1, filed May 3, 2007, and issued July 27, 2010.
100. Richard Franklin Wormsbecher, Ruizhong Hu, Gasoline sulfur reduction catalyst for fluid catalytic cracking process, US 20100133145 B2, filed January 28, 2008, and issued June 3, 2010.
101. Akira Iino, Toshio Itoh, Shinji Akashi, Fluid catalytic cracking catalyst having desulfurizing functions, process for production of the same, and process for production of low-sulfur catalytically cracked gasoline with the catalyst, US 20090230023 A1, filed June 25, 2007, and issued September 17, 2009.
102. Lu Shanxiang, Ouyang Danxia, Chen Hui, Wang Su, Xu Feng, Yu Jiyong, Catalytic cracking auxiliary agent for reducing sulfur content in gasoline and preparation method thereof, CN 101148601 A, filed September 10, 2007, and issued March 26, 2008.
103. Darrell Ray Rainer, Jorge Alberto Gonzalez, FCC CO Oxidation Promoters, US 20090050529 A1, filed June 9, 2008, and issued February 26, 2009.
104. Darrell Ray Rainer, Julie Ann Francis, Jorge Alberto Gonzalez, Lin Luo, Low NO_x CO oxidation promoters, US 20090050528 A1, filed June 9, 2008, and issued February 26, 2009.
105. Allan J. Jacobson, Pradeep Samarasekere, Methods and systems for recovering rare earth elements WO 2014144463 A1, filed March 14, 2014, and issued September 18, 2014.
106. Albert A. Vierheilig, Sulfur oxide removing additives and methods for partial oxidation conditions, CN 101631844 A, filed October 30, 2007, and issued January 20, 2010.
107. David Matheson Stockwell, Lower temperature CO promoters for FCC with low NO_x, CN 101351259 A, filed November 28, 2006, and issued January 21, 2009.
108. Jorge Alberto Gonzalez, Darrell Ray Rainer, FCC co oxidation promoters, WO 2008148683 A1, filed May 28, 2008, and issued December 11, 2008.
109. Julie Ann Francis, Lin Luo, Darrell Ray Rainer, Fluid Catalytic Cracking Additive, US 20080287284 A1, filed March 8, 2006, and issued November 20, 2008.

110. Marius Vaarkamp, David Matheson Stockwell, CO oxidation promoters for use in FCC processes, EP 1879982 B1, filed March 3, 2006, and issued January 23, 2008.
111. Gopal Ravichandran, Praveen Kumar Chinthala, Tejas Doshi, Arun Kumar, Amit Gohel, Sukumar Mandal, Asit Kumar Das, Srikanta Dinda, Amit Kumar Parekh, FCC catalyst additive and a method for its preparation, WO 2014207756 A8, filed January 28, 2014, and issued December 31, 2014.
112. Michael Sigman, Mitchell Willis, Kenneth Folmar, Method of producing fcc catalysts with reduced attrition rates, WO2014204720, filed June 10, 2014, and issued December 24, 2014.
113. Xionghou Gao, Haitao Zhang, Hongchang Duan, Chaowei Liu, Di Li, Xueli Li, Zhengguo Tan, Yunfeng Zheng, Xiaoliang Huang, Jinjun Cai, Chenxi Zhang, Magnesium modified ultra-stable rare earth y-type molecular sieve and preparation method therefor, EP2792408 A1, filed April 13, 2012, and issued October 22, 2014.
114. Chandrashekhar Pandurang Kelkar, Qi Fu, Phosphorus modified cracking catalysts with enhanced activity and hydrothermal stability, WO 2013067481 A1, filed November 5, 2012, and issued May 10, 2013.
115. Xingtao Gao, David Hamilton Harris, Zsm-5 additive activity enhancement by improved zeolite and phosphorus interaction, US 20140206526 A1, filed January 23, 2014, and issued July 24, 2014.
116. Jun Long, Yuxia Zhu, Yibin Luo, Jinghui Deng, Jinyu Zheng, Fei Ren, Xue Yang, Ying Ouyang, Catalytic cracking catalyst having a rare earth-containing Y zeolite and a preparation process thereof, US 20140080697 A1, filed June 27, 2013, and issued March 20, 2014.
117. Gan Jun, FCC (fluid cracking catalyst) catalyst containing silicon binder and preparation method thereof, CN 103447070 A, filed June 1, 2012, and issued December 18, 2013.
118. Dennis Stamires, Paul O'Connor, Erik Jeroen Laheij, Charles Vadovic, FCC catalyst, its preparation and use, US 20130203586 A1, filed March 14, 2013, and issued August 8, 2013.
119. Yun-Feng Chang, Lian Du, Method of catalyst making for superior attrition performance, US 20130098804 A1, filed October 24, 2011, and issued April 25, 2013.
120. Dong X. Li, Attrition selective particles, US 20130048541 A1, filed August 28, 2012, and issued February 28, 2013.
121. Shu Yuying, Wormsbecher Richard F, Cheng Wu-Cheng, Process for making improved zeolite catalysts from peptized aluminas US 20130005565 A1, filed March 1, 2011, and issued January 23, 2013.
122. Chen Beiyan, Zhu Yuxia, Jiang Wenbin, Shen Ningyuan, Tian Huiping, Huang Zhiqing, Song Haitao, Luo Yibin, Ouyang Ying, Inorganic binder containing phosphorus and aluminum compounds CN 102847547 A, filed June 30, 2011, and issued January 2, 2013.
123. Yuying Shu, Richard F. Wormsbecher, Wu-Cheng Cheng, Michael D. Wallace, Process for making improved catalysts from clay-derived zeolites, WO 2011115746 A1, filed March 1, 2011, and issued December 27, 2012.
124. Xionghou Gao, Dong JI, Haitao Zhang, Hongchang Duan, Di Li, Zhengguo Tan, Yi Su, Zhicheng Tang, Yi Wang, Yanqing Ma, Yanbo Sun, Double-component modified molecular sieve with improved hydrothermal stability and production method thereof, CA 2779312 A1, filed December 1, 2009, and issued May 5, 2011.
125. Folmar Kenneth, Willis Mitchell, Thermochemical structuring of matrix components for FCC catalysts, US2 0120228194 B2, filed March 8, 2011, and issued September 13, 2012.
126. Chang Yunfeng, Du Li'an, Preparing method of wearing-resistant catalyst, CN 102631945 A, filed December 23, 2011, and issued August 15, 2012.

127. Qi Fu, Walter Babbitt, Mark Schmalfeld, Chandrashekhar P. Kelkar, Rare earth-containing attrition resistant vanadium trap for catalytic cracking catalyst, WO2012096744, filed December 12, 2011, and issued July 19, 2012.
128. Scott Michael Babitz, Catalyst, a process for its preparation, and its use, US 20110263412 A1, filed June 29, 2011, and issued October 27, 2011.
129. Qi Fu, Walter Babbitt, Mark Schmalfeld, Chandrashekhar P. Kelkar, Rare earth-containing attrition resistant vanadium trap for catalytic cracking catalyst, US 20110132808 B2, filed January 12, 2011, and issued June 9, 2011.
130. David M. Stockwell, Stephen H. Brown, Ji-Yong Ryu, Structurally enhanced cracking catalysts, US 20110000821 B2, filed June 12, 2008, and issued January 6, 2011.
131. Ranjit Kumar, Aluminum sulfate bound catalysts, US 20100264066 A1, filed June 11, 2007, and issued October 21, 2010.
132. Ranjit Kumar, Kenneth Bryden, Process for preparing high attrition resistant inorganic compositions and compositions prepared therefrom, EP 2242572 A2, filed December 18, 2008, and issued October 27, 2010.
133. Duan Hongchang, Gao Xionghou, Ji Dong, Liu Tao, Liu Yinding, Ma Yanqing, Tan Zhengguo, Wang Yi, Zhang Aiqun, Zhang Haitao, Preparation method of fluidizing, catalyzing and cracking catalyst with high solid content, CN 101829592 B, filed March 10, 2009, and issued September 15, 2010.
134. Kazunori Honda, Chizu Inaki, Hirofumi Ito, Atsushi Okita, Koji Oyama, Alkaline earth metal compound-containing zeolite catalyst, preparation method and regeneration method thereof, and method for producing lower hydrocarbon, US 20100168492 A1, filed August 29, 2007, and issued July 1, 2010.
135. Jan Hendrik Hilgers, Der Zon Monique Van, Process for the preparation of an FCC catalyst, EP 2086675 B1, filed November 7, 2007, and issued August 12, 2009.
136. Ji Dong, Zhang Haitao, Zhang Zhongdong, Gao Xionghou, Wang Yi, Liu Tao, Su Yi, Sun Shuhong, Modified method for improving hydrothermal stability of molecular sieve in FCC catalyst with high efficiency, CN 101537365 B, filed March 19, 2008, and issued September 23, 2009.
137. Gary M. Smith, Barry K. Speronello, ZSM-5 additive, US 20090023971 B2, filed September 23, 2008, and issued January 22, 2009.
138. Liu Conghua, Tan Zhengguo, Gao Xionghou, Zheng Shuqin, Teng Qiuxia, Ding Wei, Lu Tong, Wang Yi, Liu Tao, Method for improving solid content of catalytic cracking catalyst slurry, CN 101134906 B, filed August 30, 2006, and issued March 5, 2008.
139. David C. Dayton, Maruthi Sreekanth Pavani, John R. Carpenter III, Holle Matthew Von, Catalyst compositions and use thereof in catalytic biomass pyrolysis, WO 2014089131 A1, filed December 4, 2013, and issued June 12, 2014.
140. Wenyih Frank Lai, Nicholas S. Rollman, Synthesis of zsm-5 crystals with improved morphology, US 20140162867 A1, filed October 25, 2013, and issued June 12, 2014.
141. Javier Garcia-Martinez, Mesoporous zeolite catalyst supports, US 20140128246 A1, filed November 8, 2013, and issued May 8, 2014.
143. Yuying Shu, Wu-Cheng Cheng, Richard F. Wormsbecher, Kevin J. Sutovich, Ranjit Kumar, Michael S. Ziebarth, Magnesium stabilized ultra-low soda cracking catalysts, WO 2014018484 A1, filed July 23, 2013, and issued January 30, 2014.
143. Ranjit Kumar, Wu-Cheng Cheng, Kevin J. Sutovich, Michael S. Ziebarth, Yuying Shu, High matrix surface area catalytic cracking catalyst stabilized with magnesium and silica, US 20140021098 A1, filed July 23, 2013, and issued January 23, 2014.
144. Jun Long, Fei Ren, Yuxia Zhu, Yibin Luo, Jiasong Yan, Xue Yang, Huiping Tian, Li Zhuang, Beiyan Chen, Minggang Li, Ying Ouyang, Xiangtian Shu, Catalyst containing a modified y-type zeolite and a preparation process thereof, US 20140005032 A1, filed June 27, 2013, and issued January 2, 2014.

145. Garcia-Martinez Javier Mesostructured zeolitic materials and methods of making and using the same, US 20130299389 B2, filed July 17, 2013, and issued November 14, 2013.
146. Liu Haiyan, Ding Jiajia, Bao Xiaojun, Fan Yu, Shi Gang, Productive propylene fluidized catalytic cracking (FCC) catalyst and preparation method thereof CN 103357429 A, filed March 28, 2012, and issued October 23, 2013.
147. Javier Garcia-Martinez, Marvin M. Johnson, Ioulia Valla, Introduction of mesoporosity in low Si/Al zeolites, US 8486369 B2, filed January 18, 2010, and issued July 16, 2013.
148. Ernest Senderov, Mohammad Ibrahim Qureshi, Introduction of mesoporosity into zeolite materials with sequential acid, surfactant, and base treatment, US 20130183231 A1, filed January 14, 2013, and issued July 18, 2013.
149. Kunhao Li, Javier Garcia-Martinez, Michael G. Beaver, Introduction of mesoporosity into low silica zeolites, WO 2013106816 A1, filed January 14, 2013, and issued July 18, 2013.
150. Javier Garcia-Martinez, Introduction of mesoporosity into inorganic materials in the presence of a non-ionic surfactant, US 20130183229 A1, filed January 14, 2013, and issued July 18, 2013.
151. Sulaiman S. Al-Khattaf, Rabindran Jermy Balasamy, Mohammed Abdul Bari Siddiqui, Abdullah M. Aitani, Mian Rahat Saeed, Method of forming a hydrocarbon cracking catalyst, US 20130165315 A1, filed December 27, 2011, and issued June 27, 2013.
152. John Scott Buchanan, William A. Wachter, Kun Wang, Kathryn L. Peretti, Daniel Mark Giaquinta, Hongyi Hou, Gibbsite catalytic cracking catalyst, US 20130131419 A1, filed November 22, 2011, and issued May 23, 2013.
153. Jason Wu, Extra mesoporous y zeolite, US 20130115162 A1, filed December 12, 2012, and issued in May 9, 2013.
154. Delphine Minoux, Nadiya Danilina, Process for preparing a mesoporized catalyst, catalyst thus obtained and use thereof in a catalytic process, WO 2013060705 A2, filed October 24, 2012, and issued May 2, 2013.
155. Wang Baihua, Wang Baihua Zhang Yongming, He Zhigui, In-situ crystallization cracking catalyst prepared by polymorph kaolin microspheres and preparation method of in-situ crystallization cracking catalyst, CN 102764670 A, filed July 31, 2012, and issued November 7, 2012.
156. David H. Olson, Dehydroxylation pretreatment of inorganic materials in mesopore introduction process, US 20120275993 A1, filed April 27, 2012, and issued November 1, 2012.
157. Javier Garcia Martinez, Ernest Senderov, Richard Hinchey, Mesoporous framework-modified zeolites, US 20120258852 A1, filed April 5, 2012, and issued October 11, 2012.
158. Kun Wang, Robert C. Lemon, John S. Buchanan, Christine E. Kliewer, Wieslaw J. Roth, Aggregates of small particles of synthetic faujasite zeolite, US 20120227584 B2, filed March 7, 2011, and issued September 13, 2012.
159. Kun Wang, Robert C. Lemon, Mesoporous aluminas stabilized with rare earth and phosphorous, US20120088654 B2, filed October 8, 2010, and issued April 12, 2012.
160. Ernest Senderov, Javier Garcia Martinez, Marvin Johnson, Compositions and methods for making stabilized mesoporous materials, US 20110171121 A1, filed January 7, 2011, and issued July 14, 2011.
161. Javier Garcia-Martinez, Lawrence B. Dight, Barry K. Speronello, Methods for enhancing the mesoporosity of zeolite-containing materials, US2 0110118107 B2, filed October 19, 2010, and issued May 19, 2011.
162. Youhao Xu, Shouye Cui, Jun Long, Jianhong Gong, Zhijian Da, Jiushun Zhang, Yuxia Zhu, Yibin Luo, Jinlian Tang, Catalytic cracking catalyst having a higher selectivity, processing method and use thereof, JP 2011005488 A, filed June 25, 2010, and issued January 13, 2011.

163. Zhiping Tan, Fluid catalytic cracking catalyst with low coke yield and method for making the same, US 20100298118 A1, filed May 21, 2009, and issued November 25, 2010.
164. Lawrence B. Dight, Javier Garcia-Martinez, Ioulia Valla, Marvin M. Johnson, Compositions and methods for improving the hydrothermal stability of mesostructured zeolites by rare earth ion exchange US 20100190632 B2, filed January 8, 2010, and issued July 29, 2010.
165. William A. Wachter, FCC catalyst, US20090082193 A1, filed October 7, 2008, and issued March 26, 2009.
166. Jozef Cornelis Groen, Jacob Adriaan Moulijn, Javier Perez-Ramirez, Mesoporous mordenite, preparation and use thereof, WO 2008147190 A1, filed May 29, 2008, and issued December 4, 2008.
167. Javier Garcia-Martinez, Methods for making mesostructured zeolitic material, US 20080138274 A1, filed March 13, 2007, and issued June 12, 2008.
168. Bram W. Hoffer, David M. Stockwell, Improved metal passivator/trap for FCC processes, WO 2013077836 A1, filed November 21, 2011, and issued May 30, 2013.
169. Robert Mehlberg, Erick D. Gamas-Castellanos, Chad R. Huovie, Resid catalytic cracker and catalyst for increased propylene yield, WO 2013074191 A1, filed September 11, 2012, and issued May 23, 2013.
170. Arumugam Velayutham Karthikeyani, Mohan Prabhu Kuvettu, Biswanath Sarkar, Pankaj Kumar Kasliwal, Balaiah Swamy, Ganga Shankar Mishra, Kamlesh Gupta, Santanam Rajagopal, Ravinder Kumar Malhotra, Catalyst composition for fluid catalytic cracking, process for preparing the same and use thereof, WO 2014016764 A1, filed July 23, 2013, and issued January 30, 2014.
171. Mohan Prabhu Kuvettu, Manish Agarwal, A.V. Karthikeyani, Balaiah Swamy, Biswanath Sarkar, Bhanu Patel Mitra, S. Rajagopal, Metal passivator additive and process for preparing, WO 2011154973 A1, filed June 1, 2011, and issued April 10, 2013.
172. Mitchell James Willis, Kenneth Warren Folmar, Improved heavy metals trapping co-catalyst for FCC processes, EP 2482973 B1, filed September 29, 2010, and issued August 8, 2012.
173. Julie Ann Francis, Charles Vadovic, Additives for removal of metals poisonous to catalysts during fluidized catalytic cracking of hydrocarbons, US 20110042271 B2, filed November 1, 2010, and issued February 24, 2011.
174. Hernández Cesar Augusto Vergel, Rubiano Luis Oswaldo Almanza, Marin Luis Javier Hoyos, Vandium traps for catalytic cracking processes and preparation thereof, US 20110152071 B2, filed December 6, 2007, and issued June 23, 2011.
175. Luis Oswaldo Almanza Rubiano, Luis Javier Hoyos Marin, Cesar Vergel Hernández, Method for producing vanadium traps by means of impregnation and resulting vanadium trap, US 20110143932 B2, filed November 30, 2007, and issued June 16, 2011.
176. Philip S Deitz, Wilson Suarez, Ranjit Kumar, Richard Franklin Wormsbecher, Rare earth carbonate compositions for metals tolerance in cracking catalysts, EP 2280777 A1, filed January 8, 2009, and issued February 9, 2011.
177. Albert A. Vierheilig, Additives for metal contaminant removal, US 20100025297 B2, filed November 25, 2008, and issued February 4, 2010.
178. Dennis Stamires, Paul O'Connor, William Jones, Oxidic metal composition, its preparation and use as catalyst composition, US 20090269266 A1, filed June 2, 2006, and issued October 29, 2009.
179. Albert A. Vierheilig, Mixed metal oxide additives, US 20080202984 A1, filed January 28, 2008, and issued August 28, 2008.
180. Julie Ann Francis, Charles Vadovic, Additives for removal of metals poisonous to catalysts during fluidized catalytic cracking of hydrocarbons, WO 2008003091 A8, filed June 29, 2007, and issued January 3, 2008.

181. George W. Huber, Yu-Ting Cheng, Zhuopeng Wang, Wei Fan, Method for converting a hydrocarbonaceous material to a fluid hydrocarbon product comprising p-xylene, US 20130324772 A1, filed September 27, 2012, and issued December 5, 2013.
182. Herrera Héctor Armendáriz, Castillo María de Lourdes Alejandra Guzmán, Beltrán Francisco Javier Hernández, Romo Patricia Pérez, Valente Jaime Sánchez, Fripiat José Marie Maurice Julien, Process for altering the physico-chemical properties of faujasite y-type zeolites, EP 2570386 B1, filed May 13, 2011, and issued July 4, 2013.
183. Stefan Maurer, Hanpeng JIN, Jeff Yang, Ulrich Muller, Organotemplate-free synthetic process for the production of a zeolitic material of the cha-type structure, CA 2855572 A1, filed November 9, 2012, and issued May 16, 2013.
184. Bilge Yilmaz, Ulrich Müller, Meike Pfaff, Feng-Shou Xiao, Bin Xie, Halyan Zhang, Organotemplate-free synthetic process for the production of a zeolitic material of the lev-type structure, WO 2011157839 A1, filed June 17, 2011, and issued April 24, 2013.
185. Wenyih Frank Lai, Robert E. Kay, Jason Wu, Kun Wang, Robert C. Lemon, Stabilized aggregates of small crystallites of zeolite Y, US 8882993 B2, filed March 6, 2012, and issued January 31, 2013.
186. Bilge Yilmaz, Ulrich Müller, Meike Pfaff, Hermann Gies, Feng-Shou Xiao, Takashi Tatsumi, Xinhe Bao, Weiping Zhang, Vos Dirk De, Hiroyuki Imai, Bin Xie, Haiyan Zhang, Alkali-free synthesis of zeolitic materials of the LEV-type structure, US 20110312486 A1, filed June 17, 2011, and issued December 22, 2011.
187. Bilge Yilmaz, Ulrich Müller, Moreno Bibiana Andrea Betancur, Hermann Gies, Feng-Shou Xiao, Takashi Tatsumi, Xinhe Bao, Weiping Zhang, Vos Dirk De, Meike Pfaff, Bin Xie, Haiyan Zhang, Zeolitic materials of the LEV-type structure and methods for their production, US 20110313226 A1, filed June 17, 2011, and issued December 22, 2011.
188. Kirsten Spannhoff, Florina Corina Patcas, Kerem Bay, Manuela Gaab, Ekkehard Schwab, Michael Hesse, Catalyst and process for the conversion of oxygenates to olefins, US 20140005457 A1, filed June 26, 2013, and issued January 2, 2014.
189. Kirsten Spannhoff, Florina Corina Patcas, Ekkehard Schwab, Alexander Weck, Kerem Bay, Catalyst coating and process for the conversion of oxygenates to olefins, US 20140005456 A1, filed June 26, 2013, and issued January 2, 2014.
190. Jürgen Ladebeck, Jürgen Koy, Stephen Wellach, Josef Schönlinner, Götz Burgfels, Mechanochemical production of zeolites, US 20130266507 A1, filed November 28, 2011, and issued October 10, 2013.
191. Hermann Luyken, William Todd Owens, Process for ion exchange on zeolites, WO 2013072808 A1, filed November 7, 2012, and issued May 23, 2013.
192. XIAO Feng-Shou, Xie Bin, Mueller Ulrich, Yilmaz Bilge, Organotemplate-free synthetic process for the production of a zeolitic material, US 20130123096 A1, filed December 19, 2012, and issued May 16, 2013.
193. Bilge Yilmaz, Ulrich Berens, Vijay Narayanan Swaminathan, Ulrich Müller, Gabriele Iffland, Laszlo Szarvas, Synthesis of zeolitic materials using N,N-dimethyl organotemplates, US 20130059723 A1, filed September 6, 2012, and issued March 7, 2013.
194. Wu-Cheng Cheng, Yuying Shu, Richard F. Wormsbecher, Sodium tolerant zeolite catalysts and processes for making the same, WO2012071368, filed November 22, 2011, and issued May 31, 2012.
195. Ivor Bull, Ulrich Müller, Process for the preparation of zeolites having B-CHA structure, US 20110142755 B2, filed November 19, 2010, and issued June 16, 2011.
196. Zhou Lingping, Zhang Weilin, Xu Mingde, Li Zheng, Tian Huiping, Zhu Yuxia, Yan Jiasong, Method for reducing ammonium and nitrogen consumption in FCC (fluid catalytic cracking) catalyst production process, CN 102078820 B, filed November 27, 2009, and issued June 1, 2011.

197. Lance L. Jacobsen, Brian S. Konrad, David A. Lesch, Beckay J. Mezza, James G. Vassilakis, Cynthia R. Berinti-Vondrasek, A process for preparing molecular sieve beads, WO 2010036252 A1, filed September 25, 2008, and issued April 1, 2010.
198. Gan Jun, FCC (fluid cracking catalyst) catalyst containing silicon binder with high solid content and preparation method thereof, CN 103447071 A, filed June 1, 2012, and issued December 18, 2013.
199. Demetrius Michos, Alumina particles and methods of making the same, WO 2008147519 A1, filed September 12, 2012, and issued February 21, 2013.
200. Yun-Feng Chang, High solids catalyst formulation and spry drying, US 20110108462 A1, filed November 10, 2009, and issued May 12, 2011.
201. Xionghou Gao, Haitao Zhang, Zhengguo Tan, Hongchang Duan, Dong JI, Yinding Liu, Yi Wang,Yanqing Ma, Tao Liu, Aiqun Zhang, Preparation method for increasing solid content of fcc catalyst, WO 2010102430 A1, filed May 7, 2009, and issued September 16, 2010.
202. Lıu Conghua, Tan Zhengguo, Ding Weı, Zheng Shuqin, Pang Xinmei, Sun Shuhong, Wang Dong, Teng Qıuxıa,, Lu Tong, Method for increasing solid content of catalytıic cracking catalyst slurry, JP 2008055416 B2, filed August 13, 2007, and issued March 13, 2008.
203. Conghua Liu, Zhengguo Tan, Wei Ding, Shuqin Zheng, Xinmei Pang, Shuhong Sun, Dong Wang, Qiuxia Teng, Tong Lu, Method to raise the solid content of catalytic cracking catalyst slurry, US 20080058197 B2, filed August 28, 2007, and issued March 6, 2008.
204. Deng-Yang Jan, Jaime G. Moscoso, Methods for forming zeolites from homogeneous amorphous silica alumina, WO 2013036359 A1, filed August 14, 2012, and issued March 14, 2013.
205. Gu Jianfeng, Cui Louwei, Wang Xinxing, He Guanwei, Preparation method of NaY molecular sieve, CN 102701232 B, filed June 8, 2012, and issued October 3, 2012.
206. Bull Ivor, Mueller Ulrich, Process for the preparation of zeolites having CHA structure, US 8883119 B2, filed November 23, 2010, and issued October 3, 2012.
207. Andres M. Quesada Perez, Gerardo Vitale-Rojas, Offretite aluminosilicate composition and preparation and use of same, US 20120132567 A1, filed February 7, 2012, and issued May 31, 2012.
208. Allen W. Burton, Method for making mfi-type molecular sieves, US 20110117007 A1, filed November 13, 2009, and issued May 19, 2011.
209. Maria Margaret Ludvig, Erja Paivi Helena Rautiainen, Albert Carel Pouwels, Modified y-zeolite/zsm-5 catalyst for increased propylene production, WO 2014096267 A1, filed December 19, 2013, and issued June 26, 2014.
210. Kevin J. Sutovich, Wu-Cheng Cheng, Ranjit Kumar, Michael S. Ziebarth, Yuying Shu, Silica sol bound catalytic cracking catalyst stabilized with magnesium, US 20140021097 A1, filed July 23, 2013, and issued January 23, 2014.
211. Stacey I. Zones, Anna Jackowski, Method for preparing molecular SSZ-81, US20110318262 B2, filed June 16, 2011, and issued December 29, 2011.
212. Richard Franklin Wormsbecher, Wu-Cheng Cheng, Michael Wallace, Wilson Suarez, Yuying Shu, Novel ultra-stable zeolite y and method for manufacturing the same, US 20110224067 A1, filed December 17, 2009, and issued September 15, 2011.
213. William A. Wachter, Jeffrey T. Elks, Brenda A. Raich, Theodore E. Datz, Mary T. Van Nostrand, Gordon F. Stuntz, David O. Marler, Nicholas Rollman, Drying device for producing small quantities of controlled particle size catalysts which are appropriate for use in fluidized bed operations such as fluid catalytic cracking, US 20110152062 A1, filed December 1, 2010, and issued June 23, 2011.
214. Christopher P. Nicholas, Christian D. Freet, Bussche Kurt M. Vanden, Todd M. Kruse, Process for fluid catalytic cracking oligomerate, WO 2014074856 A1, filed November 8, 2013, and issued May 15, 2014.

215. David H. Harris, Novel catalyst to increase propylene yields from a fluid catalytic cracking unit, US 20130066131 A1, filed September 13, 2011, and issued March 14, 2013.
216. Musaed Salem Al-Ghrami, Cemal Ercan, Catalyst for enhanced propylene in fluidized catalytic cracking, WO 2013177388 A1, filed May 23, 2013, and issued November 28, 2013.
217. Chad R. Huovie, Process for producing at least one of ethene, propene, and gasoline, WO 2013133932 A1, filed February 11, 2013, and issued September 12, 2013.
218. Raquel Bastiani, Lam Yiu Lau, Rezende Pinho Andrea De, Rosana Wasserman, Espirito Santo Ivanilda Barboza Do, Additives for the maximization of light olefins in fluid catalytic cracking units, and process, WO 2013126974 A1, filed March 2, 2012, and issued September 6, 2013.
219. A. V. Karthikeyani, B. Sarkar, V. Chidambaram, B. Swamy, P. K. Kasliwal, G. S. Mishra, K. M. Prabhu, A process for enhancing nickel tolerance of heavy hydrocarbon cracking catalysts, WO 2013054174 A1, filed October 11, 2012, and issued April 18, 2013.
220. Ravichandran Gopal, Kumar Chinthala Praveen, Doshi Tejas, Kumar D. Arun, Gohel Amit, Mandal Sukumar, Kumar Das Asit, Dinda Srikanta, Parekh Amitkumar, Fcc catalyst additive and preparation method thereof, WO 2013011517 A1, filed September 2, 2011, and issued January 24, 2013.
221. Yuying Shu, Richard F. Wormsbecher, Wu-Cheng Cheng, High light olefins FCC catalyst compositions, US 8845882 B2, filed September 17, 2012, and issued May 29, 2013.
222. Wang Dingbo, Ba Haipeng, Guo Jinghang, Ma Zhiyuan, Catalytic cracking assistant for propylene yield increase, CN 102872902 A, filed July 12, 2011, and issued January 16, 2013.
223. Jun Long, Wenbin Jiang, Mingde Xu, Huiping Tian, Yibin Luo, Xingtian Shu, Jishun Zhang, Beiyan Chen, Haitao Song, Catalyst and a method for cracking hydrocarbons, US 20120292230 A1, filed August 2, 2012, and issued November 22, 2012.
224. Wu-Cheng Cheng, Ranjit Kumar, Meenakshi Sundaram Krishnamoorthy, Michael Scott Ziebarth, Philip S. Deitz, Pentasil zeolite catalyst for light olefins in fluidized catalytic units CN102553633 A, filed March 23, 2006, and issued July 11, 2012.
225. Lawrence L. Upson, Lazlo T. Nemeth, Catalyst compositions for improved fluid catalytic cracking (FCC) processes targeting propylene production, US20120149550 B2, filed April 23, 2009, and issued June 14, 2012.
226. Christopher P. Nicholas, Deng-Yang Jan, Jaime G. Moscoso, Process for catalytic cracking of hydrocarbons using UZM-35, US 20110308998 B2, filed June 21, 2010, and issued December 22, 2011.
227. Almeida Marlon Brando Bezerra De, Alexandre De Figueiredo Costa, Lam Yiu Lau, Sergio Augusto Santos Rodrigues, Marcelo Andre Torem, Additive with multiple system of zeolites and method of preparation, US 20110207984 A1, filed September 18, 2009, and issued August 25, 2011.
228. Kenji Akagishi, Hiroyuki Yano, Ryusuke Miyazaki, Zeolite-containing catalyst and method for producing the same, and method for producing propylene, US 20110092757 A1, filed June 25, 2009, and issued April 21, 2011.
229. Dae Hyun Choo, Hong Chan Kim, Suk Joon Kim, Ji Min Kim, Tae Jin Kim, Sun Choi, Seung Hoon Oh, Yong Seung Kim, Deuk Soo Park, Young Ki Park, Chul Wee Lee, Hee Young Kim, Won Choon Choi, Na Young Kang, Bu Sub Song, Catalyst for catalytic cracking of hydrocarbon, which is used in production of light olefin and production method thereof, US 20110039688 B2, filed February 5, 2009, and issued February 17, 2011.
230. Dae Hyun Choo, Hong Chan Kim, Suk Joon Kim, Ji Min Kim, Tae Jin Kim, Sun Choi,Seung Hoon Oh, Yong Seung Kim,Deuk Soo Park, Young Ki Park, Chul Wee Lee, Hee Young Kim, Won Choon Choi, Na Young Kang, Bu Sub Song, Catalyst for

catalytic cracking of hydrocarbon, which is used in production of light olefin and production method thereof, EP 2248582 A4, filed February 5, 2009, and issued November 10, 2010.

231. William J. Reagan, Lawrence L. Upson, FCC catalysts for producing light olefins, mxMX/a/2009/000937, filed January 23, 2009, and issued September 8, 2010.
232. William J. Reagan, Lawrence L. Upson, FCC catalyst for producing light olefin, CN 101786013 A, filed January 23, 2009, and issued July 28, 2010.
233. Hayim Abrevaya, Ben A. Wilson, Stephen T. Wilson, Suheil F. Abdo, Nanocrystalline silicalite for catalytic naphtha cracking, WO 2010053482 A1, filed November 6, 2008, and issued May 14, 2010.
234. Gavin P. Towler, Hayim Abrevaya, Mixture of catalysts for cracking naphtha to olefins, US 20100105974 B2, filed October 24, 2008, and issued April 29, 2010.
235. Yoshikazu Takamatsu, Kouji Nomura, Method for producing ethylene and propylene US 20100063340 B2, filed September 14, 2006, and issued March 11, 2010.
236. Jean-Pierre Dath, Walter Vermeiren, André Noiret, Catalytic Cracking, US 20140051901 A9, filed December 14, 2006, and issued August 27, 2009.
237. Graaf Elbert Arjan De, Leendert Arie Gerritsen, Fcc process employing basic cracking compositions, WO 2009087576 A2, filed January 7, 2009, and issued July 16, 2009.
238. Pan Huihua, Luo Yibin, Wang Dianzhong, Mu Xuhong, Jiang Wenbin, Shu Xingtian, Method for preparing ZSM-5 zeolite by in situ crystallization, CN 101462740 B, filed December 20, 2007, and issued June 24, 2009.
239. William J. Reagan, Lawrence L. Upson, Fcc catalyst for light olefin production, US 20090124842 A1, filed July 12, 2006, and issued in May 14, 2009.
240. Claudia Maria De Lacerda Alvarenga Baptista, Raquel Bastiani, Lam Yiu Lau, Catalytic system and additive for maximisation of light olefins in fluid catalytic cracking units in operations of low severity, EP 2055760 A1, filed October 29, 2008, and issued May 6, 2009.
241. Sun Choi, Deuk Soo Park, Suk Joon Kim, Ahn Seop Choi, Hee Young Kim, Yong Ki Park, Chul Wee Lee, Won Choon Choi, Sang Yun Han, Jeong Ri Kim, Solid acid catalyst for producing light olefins and process using the same, US 20090112035 B2, filed November 5, 2008, and issued April 30, 2009.
242. Alexandre De Figueiredo Costa, Almeida Marlon Brando Bezerra De, Lam Yiu Lau, Eliane Bernadete Castro Mattos, Additive for maximizing light olefins in FCC and process for preparation thereof, US 20090099006 A1, filed October 10, 2008, and issued April 16, 2009.
243. Jun Long, Zhonghong Qiu, Youbao Lu, Jiushun Zhang, Zhijian Da, Huiping Tian, Yuxia Zhu, Wanhong Zhang, Zhenbo Wang, Cracking catalyst and a process for preparing the same, US 20080293561 A1, filed July 14, 2005, and issued November 27, 2008.
244. Wu Cheng Cheng, Meenakshi Sundaram Krishnamoorthy, Ranjit Kumar, Xinjin Zhao, Michael Scott Ziebarth, Catalyst for light olefins and LPG in fluidized catalytic units, US 20080093263 A1, filed November 3, 2005, and issued April 24, 2008.
245. Lam Yiu Lau, Raquel Bastiani, Claudia Maria de Lacerda Alvarenga Baptista, Additive to maximize GLP and propene suitable for use in low-severity operations of a fluid catalytic cracking unit and its preparatory process, US 20080015105 B2, filed March 6, 2007, and issued January 17, 2008.
246. Xionghou Gao, Haitao Zhang, Zhengguo Tan, Di Li, Dong JI, Hongchang Duan, Chenxi Zhang, Ultra-stable rare earth Y-type molecular sieve and preparation method therefor, EP2792408 A1, filed April 13, 2012, and issued October 22, 2014.
247. Srikanta DINDA, Praveen Kumar CHINTHALA, Amit GOHEL, Ashwani Yadav, Sukumar Mandal, Gopal Ravichandran, ASIT Kumar DAS, Process and composition of catalyst/additive for reducing fuel gas yield in fluid catalytic cracking (FCC) process, WO 2013005225 A1, filed September 29, 2011, and issued January 10, 2013.

248. Xionghou Gao, Dong Ji, Haitao Zhang, Hongchang Duan, Di Li, Zhengguo Tan, Yi Su, Chenxi Zhang, Yi Wang, Yanqing Ma, Yanbo Sun, Modified molecular sieve characterized by improved sodium-contamination-resisting activity and preparation method thereof, US 20120213695 B2, filed December 1, 2009, and issued August 23, 2012.
249. Jianxin Jason Wu, William A. Wachter, Colin L. Beswick, Edward Thomas Habib, JR., Terry G. Roberie, Ruizhong Hu, Low small mesoporous peak cracking catalyst and method of using, US20110220549 B2, filed February 17, 2011, and issued September 15, 2011.
250. Ranjit Kumar, Process for making boehmite alumina, and methods for making catalysts using the same, WO 2011075286 A1, filed November 22, 2010, and issued June 23, 2011.
251. Katsuya Watanabe, Catalytic cracking catalyst, process for producing the same, and method of catalytic cracking of hydrocarbon oil, JP 2010247146 A, filed March 9, 2010, and issued November 4, 2010.
252. Katsuya Watanabe, Catalytic cracking catalyst, process for producing the same, and method of catalytic cracking of hydrocarbon oil, US 20100236983 A1, filed March 2, 2007, and issued September 23, 2010.
253. Katsuya Watanabe, Cracking catalyst, process for preparation thereof, and process for catalytic cracking of hydrocarbon oil, CN 101405078 B, filed March 2, 2007, and issued April 8, 2009.
254. Tadashi Shibuya, Catalyst for catalytic cracking hydrocarbon oil and method for catalytic cracking hydrocarbon oil by using the same, JP 2008200579 B2, filed February 19, 2007, and issued September 4, 2008.
255. Tadashi Shibuya, Catalytic cracking of hydrocarbon oil and catalytic cracking method of hydrocarbon oil using it, JP 2008173530 B2, filed July 31, 2008, and issued January 16, 2007.
256. Giuseppe Bellussi, Roberto Millini, Caterina Rizzo, Daniele Colombo Cracking process and enhanced catalysts for said process, US 20110163006 A1, filed May 28, 2009, and issued July 7, 2011.
257. Tadashi Shibuya, Catalytic cracking of hydrocarbon oil and catalytic cracking method of hydrocarbon oil using it, JP 2008173581 B2, filed January 19, 2007, and issued July 31, 2008.
258. Junya Hamani, Tadashi Shibuya, Catalytic cracking of hydrocarbon oil and catalytic cracking method of hydrocarbon oil using it, JP 2008104925 B2, filed October 24, 2006, and issued May 8, 2008.
259. Wu-Cheng Cheng, Kevin John Sutovich, Ruizhong Hu, Ranjit Kumar, Xinjin Zhao, Catalytic cracking catalyst compositions having improved bottoms conversion, US 20110036754 B2, filed November 21, 2007, and issued February 17, 2011.
260. Gopal Juttu, Kelsey Shogren, Bruce Adkins, Phosphorous promotion of zeolite-containing catalysts, US 20140275588 A1, filed March 14, 2014, and issued September 18, 2014.
261. Richard Wormsbecher, Kevin Sutovich, Process of cracking biofeeds using high zeolite to matrix surface area catalysts, WO2010068255 C2, filed December 8, 2009, and issued July 10, 2014.
262. John Scott Buchanan, Halou Oumar-Mahamat, Wayne Richard Kliewer, Catalytic cracking process for biofeeds, US 20140163285 A1, filed November 8, 2013, and issued June 12, 2014.
263. Johannes Antonius Hogendoorn, Sasha Reinier Aldegonda Kersten, Process for the catalytic cracking of pyrolysis oils, EP 2325281 A1, filed November 24, 2009, and issued May 25, 2011.
264. Christopher P. Nicholas, Edwin P. Boldingh, Catalytic pyrolysis using UZM-44 aluminosilicate zeolite, US 8609911 B1, filed March 11, 2013, and issued December 17, 2013.

265. Christine M. Henry, Stephen Schuylen, Jerry J. Springs, Dennis Stamires, Zongchao Zhang, Ling Zhou, Hybrid silica and alumina as catalyst matrix and/or binder in biomass conversion catalysts and bio-oil upgrading, US 20140007493 A1, filed July 2, 2013, and issued January 9, 2014.
266. Christopher P. Nicholas, Edwin P. Boldingh, Catalytic pyrolysis using UZM-39 aluminosilicate zeolite, US 8853477 B2, filed November 8, 2013, and issued October 7, 2014.
267. Dennis Stamires, Catalyst compositions for use in a two-stage reactor assembly unit for the thermolysis and catalytic conversion of biomass, US 20130261355 A1, filed March 7, 2013, and issued October 3, 2013.
268. Dennis Stamires, Michael Brady, Mesoporous zeolite-containing catalysts for the thermoconversion of biomass and for upgrading bio-oils, WO 2013123299 A1, filed February 15, 2013, and issued August 22, 2013.
269. Dennis Stamires, Michael Brady, Catalyst composition with increased bulk active site accessibility for the catalytic thermoconversion of biomass to liquid fuels and chemicals and for upgrading bio-oils, WO 2013123297 A2, filed February 15, 2013, and issued August 22, 2013.
270. Paul O'Connor, Optimized catalyst for biomass pyrolysis, WO 2013098195 A1, filed December 20, 2012, and issued July 4, 2013.
271. Andrew Holt, Kirsty Bonner, Ioulia Zimozdra, A process and catalyst system for cracking pyrolysis derived organic molecules, WO 2013087480 A9, filed December 5, 2012, and issued December 19, 2013.
272. Bruce Adkins, Robert Bartek, Michael Brady, John Hackskaylo, Dennis Stamires, Improved catalyst for thermocatalytic conversion of biomass to liquid fuels and chemicals, WO 2012142490 A1, filed April 13, 2012, and issued October 18, 2012.
273. Robert Bartek, Michael Brady, Dennis Stamires, Phyllosilicate-based compositions and methods of making the same for catalytic pyrolysis of biomass WO 2012122245 A1, filed March 7, 2012, and issued September 13, 2012.
274. Richard Wormsbecher, Kevin Sutovich, Process of cracking biofeeds using high zeolite to matrix surface area catalysts, EP 2373291 A1, filed December 8, 2009, and issued September 15, 2011.
275. Robert Bartek, Michael Brady, Dennis Stamires, Controlled activity pyrolysis catalysts, WO 2010124069 A2, filed April 22, 2010, and issued October 28, 2010.
276. Paul O'Connor, Steve Yanik, Robert Bartek, Biomass conversion using solid base catalyst, WO 2010065872 A1, filed December 4, 2009, and issued June 10, 2010.
277. Kong Lingjiang, Wang Weijia, He Mingyuan, Luo Yibin, Mu Xuhong, Method for preparing Y type molecular sieves, CN 101676207 A, filed September 19, 2008, and issued March 24, 2010.
278. Walter Vermeiren, Sébastien Decker, Marc Bories, Jean-Pierre Dath, Catalytic composition and catalytic cracking method in a fluidised bed using such a composition, EP 1892041 A1, filed December 11, 2006, and issued February 27, 2008.
279. David H. Harris, Mingting Xu, David Stockwell, Rostam J. Madon, FCC catalysts for feeds containing nickel and vanadium, KR 1020090120525 B1, filed October 9, 2002, and issued November 24, 2009.
280. Katsuya Watanabe, Cracking catalyst, process for preparation thereof, and process for catalytic cracking of hydrocarbon oil, EP2008711 A4, filed March 2, 2007, and issued December 31, 2008.
281. Paul O'Connor, King Yen Yung, Avelino Corma, Graaf Elbert Jan De, Erja Paeivi Helena Rautiainen, Novel cracking catalytic compositions, US 20120222991 A1, filed December 22, 2006, and issued September 6, 2012.
282. Graaf Elbert Arjan De, Danielle Alofs-Gimpel, King Yen Yung, Raymond Paul Fletcher, Avelino Corma Canos, Highly acidic catalyst for use in fluid catalytic cracking, US 20100298126 A1, filed October 3, 2008, and issued November 25, 2010.

8 Hydrogen Production by Catalytic Hydrocarbon Steam Reforming

Peter Broadhurst, Jumal Shah, and Raimon Perea Marin

CONTENTS

8.1 INTRODUCTION

The refining industry is currently the biggest consumer of on-purpose hydrogen using 48% of total hydrogen production. Hydrogen demand in refineries is increasing with continued growth forecast. The main drivers for this demand are the decreasing quality of crude oil, which requires more hydrogen to process, and increasingly stringent regulations of transport fuels, which grow the need of hydroprocessing, e.g., production of transport diesel. A recent study forecasts a sixfold increase in hydrogen demand from refineries, rising from 2.23 million Nm^3/day in 2015 to 13.39 million Nm^3/day in 2035.[1] In hydroprocessing technology, there are several units consuming hydrogen, as listed in Table 8.1.

There are two main sources of hydrogen in the refinery: the catalytic reformer and on-purpose hydrogen production via hydrocarbon steam reforming. The catalytic reformer is normally the main source of hydrogen for the refinery. However, market demand for gasoline, largely produced in the catalytic reformer, is dropping in favor of diesel, which is mainly produced in the hydrocracker. Also new environmental regulations are due to bring down the permitted benzene concentration in gasoline, further reducing the hydrogen available from the catalytic reformer. These trends increase the need for a continuous and reliable source of hydrogen in the refinery that is not dependent on the operation and throughput of another linked unit and is leading to a growth in on-purpose hydrogen production via hydrocarbon steam reforming.

TABLE 8.1
Hydrogen Consumers on a Refinery

Process	H_2 Consumption Nm^3 per bbl
VGO hydrotreating	21.0–31.5
Distillate hydrotreating	7.9–21.0
Naphtha hydrotreating	5.3–13.1
Shell's conventional hydrocracker	23.7–31.5
UOP mild hydrocracking	17.1
ExxonMobil MPHC	13.9–34.1
LC-FINING	17.2–49.9
$H\text{-}OIL_{RC}$	31.5
Microcat-RC	43.4–52.6
Veba Combi Cracker	22.3–46.0
Aromatics saturation	5.3–13.1
Isomerization	1.3–3.9

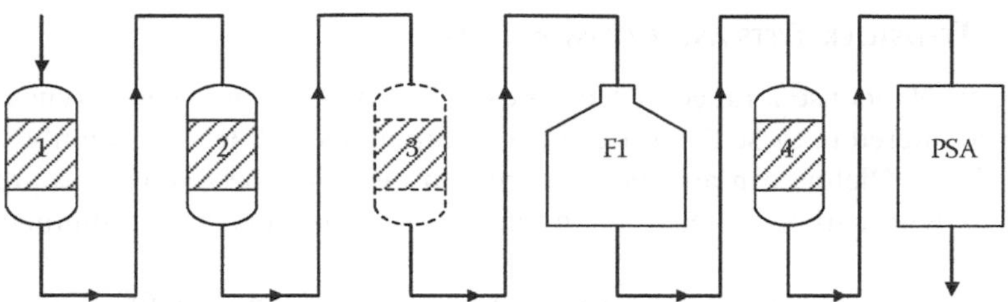

FIGURE 8.1 Modern hydrogen plant design. 1, hydrodesulphurization; 2, de-sulphurizer; 3, pre-reformer (optional); F1, steam reformer; 4, high temperature shift.

Steam reforming for hydrogen production is a well-established process and is illustrated in Figure 8.1. The heart of the plant is a steam methane reformer (SMR) that extracts hydrogen from a mixture of hydrocarbons and water. Prior to the SMR, the feed is purified to remove catalyst poisons. A pre-reformer may also be employed upstream of the SMR to convert higher hydrocarbons to methane. The gas exiting the SMR is a mixture of carbon oxides, hydrogen, and methane known as synthesis gas (syngas). The heat contained in the syngas is recovered by raising steam in a process gas boiler. Following temperature conditioning, the syngas is fed to the water gas shift (WGS) section. The shift converters decrease the carbon monoxide (CO) levels and further increase the hydrogen concentration, and are followed by hydrogen purification. Older hydrogen plants have a two-stage WGS section (high temperature shift [HTS] and low temperature shift [LTS]) followed by purification using CO_2 removal and methanation. These plants produced hydrogen with 95%–98% purity. In 1976, pressure swing absorption (PSA) technology was introduced as an alternative means of purifying hydrogen, replacing the need for the LTS, CO_2 removal, and methanator. Modern hydrogen plants therefore have a single-stage WGS section followed by a PSA. The PSA allows hydrogen purity to be increased to >99% and consequently is the most common flowsheet arrangement at present.

In recent years, there has been a greater focus on improving the hydrogen plant efficiency as a result of increasing natural gas prices. This has encouraged consideration of medium temperature shift (MTS) in place of the HTS. An MTS provides a more desirable equilibrium position and removes the limit on steam-to-gas ratio resulting from HTS operational constraints. The improved CO slip requires less feed for the same production rate. However, the change affects other parts of the process, e.g., the calorific value of the PSA purge gas is reduced, and consequently, an increase in makeup fuel gas is needed.

8.2 FEEDSTOCK PURIFICATION

Many catalysts used in modern steam methane reforming-based hydrogen plants are susceptible to poisons that are present in the hydrocarbon feedstock. To maximize the operating efficiency and the lives of expensive downstream catalysts, effective feedstock purification is essential.

8.2.1 Feedstock Types and Common Poisons

Hydrogen plants use a range of feedstocks. The two catalyst poison types usually encountered in these feeds are sulfur compounds and, less commonly, chloride species. Olefins can also be encountered and must be removed to prevent problems with carbon formation. Common feeds and poisons are summarized in Table 8.2.

Off gases are by-product gases from various processes in the refinery. Catalytic reformer off gases will normally contain approximately 1 ppmv of S and 2–3 ppmv of Cl. Other poisons are removed upstream to protect the catalytic reformer catalyst. Coker off gas tends to have high levels of S as well as olefins, dienes, arsenic, and metals. Generally, it cannot be used as a direct feed, but it is occasionally considered after pretreating. Fluid catalytic cracker (FCC) off gas and refinery fuel gas (RFG) usually contain significant levels of olefins and high levels of other poisons, and this generally precludes their direct use as a feed.

In natural gas feeds, S is the prevalent poison; it is generally found at levels between 1 and 30 ppmv depending on the gas source and pretreatment. Gases that are shipped as liquefied natural gas (LNG) or pipelined long distances are low in sulfur due to deliberate desulfurization. In some locations, the gas has a stenching agent added near the point of use to aid in leak detection. This stenching agent is typically 2–3 ppmv of a sulfur compound such as tetrahydrothiophene (THT) or tBuSH (TBM). Other than this, the sulfur compounds will be a combination of hydrogen sulfide, mercaptans, organic sulfides, and disulfides. Thiophenes are rarely observed.

Liquefied petroleum gas (LPG) poisons vary depending on the source. If the LPG derives from a refinery source where it has been processed through one or more catalytic units, it tends to have relatively low levels of poisons. Chloride tends to be associated with storage of LPG in brine caverns. Some streams contain low levels

TABLE 8.2
Types of Poisons in Hydrocarbon Feeds

Poison Type	Forms	Off Gases	Natural Gas	LPG	Naphtha
Sulfur	H_2S; COS; mercaptans (RSH); organic disulfides (R_2S_2); organic sulfides (R_2S); thiophenes	Yes	Yes	Yes	Yes
Chloride	HCl; organic Cl (RCl)	Yes	Rare	Unusual	Rare
Olefins	Various $\left(R'_2C=CR'_2\right)$	Yes	No	Yes	Yes
Arsenic and metals	AsH_3; AsR_3	Yes	Yes	Yes	Unusual
	Ni/V/Pb compounds	No	No	No	Unusual
Mercury	Hg; R_2Hg	Yes	Yes	Yes	Yes

Note: R = alkyl, aryl; R′ = H, alkyl, aryl.

of AsH_3 and COS. Olefin content can be single figure percentages, and sulfur levels can be as high as 50 ppmv.

Most naphtha feeds come from hydrotreated sources and so contain very low poisons, typically 1–2 ppmv sulfur and little else. Naphtha from other sources, for example, if derived directly from distillation, can have significant levels of sulfur, metals, and olefins.

The levels to which poisons should be removed to protect the downstream catalysts are shown in Table 8.3.

A pre-reformer, if present, is highly susceptible to traces of sulfur, chlorine, arsine, and metals, which might escape the purification section. In the absence of a pre-reformer, the steam reformer does not act as an absolute trap for the very small traces of sulfur and chlorine that slip through the purification section. These compounds are kept mobile by the high operating temperature and increasing H_2 partial pressure in the steam reformer and HTS. If the slip from the purification section exceeds tolerable levels, the performance of the SMR catalyst will be impacted as greater levels of sulfur and chlorine remain on the surface of the active Ni metal, but traces of sulfur and chlorine will continue to pass through the SMR. These traces are effectively captured by either the MTS or LTS. Both the MTS and LTS catalysts suffer steady poisoning, which causes a progressive loss of activity at the top of the bed. Severe sulfur poisoning events have a larger impact and can completely deactivate the MTS or LTS catalyst. Chlorine is also a virulent poison for these catalysts as it promotes rapid sintering of the active metal crystallites; thus, chlorine levels should be reduced to less than 5 ppbv (Figure 8.2).

There is good evidence that mercury passes through the hydrogen plant flowsheet.[2] Mercury is highly toxic and may accumulate in colder parts of the process presenting a safety issue during turnarounds. Mercury can also cause catastrophic

TABLE 8.3
Allowable Levels of Poisons for Each Unit

Poison	Duty	Maximum	Preferred
Sulfur	Pre-reforming	0.1 ppmv[a]	<50 ppbv[a]
	Steam reforming	0.1 ppmv[a]	<50 ppbv[a]
	Medium temperature shift	0.1 ppmv	
	Low temperature shift	0.1 ppmv	
Chloride	Pre-reforming	0.1 ppmv[a]	<50 ppbv[a]
	Steam reforming	0.1 ppmv[a]	<50 ppbv[a]
	Medium temperature shift	5 ppbv	<5 ppbv
	Low temperature shift	5 ppbv	<5 ppbv
Olefins	Pre-reforming	<2 vol%	<1 vol%
	Steam reforming	<2 vol%	<1 vol%
Arsenic and metals	Pre-reforming	5 ppbv	<5 ppbv
	Steam reforming	5 ppbv	<5 ppbv
Mercury	Nonspecific	5 ppbv	<1 ppbv

[a] Lower limits for forcing conditions are 10 ppbv maximum, <10 ppbv preferred.

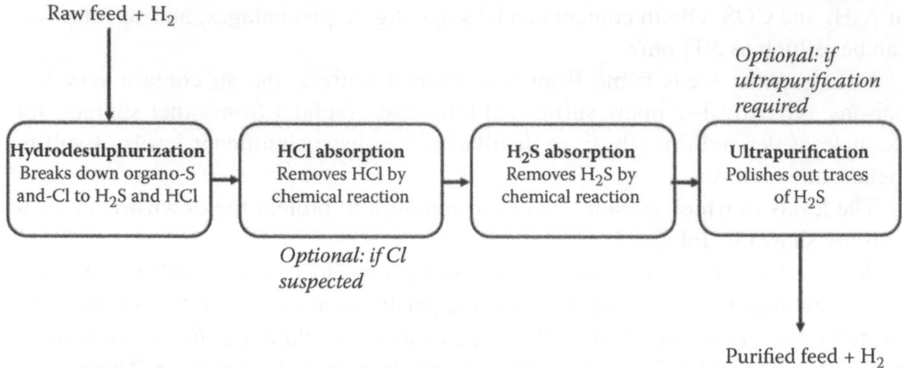

FIGURE 8.2 Block diagram of purification stages.

failures in cryogenic equipment or poisoning of process catalysts, particularly those based on Pd and Pt. There are three general approaches to Hg removal from hydrocarbon feeds:

- A two-stage process reacting H_2 with organo-Hg compounds liberating Hg followed by absorption
- Absorption over sulfided activated carbon
- Absorption over metal sulfide

8.2.2 Feedstock Purification in the Flowsheet

Johnson Matthey's PURASPEC$_{JM}$™* absorbents use the last approach to remove Hg from feed streams. The technology uses a fixed-bed solution for mercury removal. The absorbents contain an active mixed metal sulfide finely dispersed through an inert support looking up the Hg as HgS. These systems operate effectively in temperatures ranging from 0°C to 90°C and pressures up to 200 bar.

Figure 8.3 shows the potential stages in a typical feedstock purification flowsheet. In the hydrodesulfurization (HDS) stage, organosulfur is converted catalytically to hydrogen sulfide, and if present, organochloride is also converted to hydrogen chloride. This conversion requires hydrogen, present in the feed or recirculated product from the back end of the plant. Where necessary, an alkali promoted aluminum oxide-based absorbent is installed next to remove hydrogen chloride. The third stage is removal of hydrogen sulfide using absorbents based on zinc oxide, which reacts to form zinc sulfide. The degree of desulfurization achieved after hydrogen sulfide removal is typically at a level of 30–50 ppbv with end of run usually defined as 0.1 ppmv sulfur slip. In some cases, the downstream catalysts, such as the pre-reforming catalyst, are more susceptible to poisoning even at a low level of sulfur. In these cases, an ultrapurification stage can be added to remove sulfur components to below 10 ppbv.

* PURASPEC is a trademark of the Johnson Matthey group of companies.

8.2.3 Hydrodesulfurization

The HDS step is catalytic hydrogenolysis to convert organosulfur and organochloride species into hydrogen sulfide and hydrogen chloride, respectively, by the cleavage of C–S (or C–Cl) bonds by the action of hydrogen. Typical reactions, shown in Equations 8.1 to 8.5, are relatively exothermic; however, feeds very rarely contain more than 100 ppmv organic sulfur, and a temperature rise is typically not detected. The reaction is carried out over a cobalt–molybdenum (CoMo) or nickel–molybdenum (NiMo) catalyst on a promoted alumina support formed into shaped extrudates (e.g., $\text{KATALCO}_{\text{JM}}$™ 41-6T or $\text{KATALCO}_{\text{JM}}$ 61-1T*). Operation is between 300°C and 400°C and at a pressure in the range 25–45 bara. Long service lives are common, often in excess of 10 years. The reaction requires H_2, typically 2 mol% with natural gas feedstocks and up to 25 mol% with naphtha feedstocks. The H_2 not only drives hydrogenation but also inhibits hydrocarbon cracking through the front end of the flowsheet.

$$C_2H_5SH + H_2 \rightarrow H_2S + C_2H_6 \tag{8.1}$$

$$CH_3SSCH_3 + 3H_2 \rightarrow 2H_2S + 2CH_4 \tag{8.2}$$

$$CH_3SCH_3 + 2H_2 \rightarrow H_2S + 2CH_4 \tag{8.3}$$

$$C_4H_4S \text{ (thiophene)} + 4H_2 \rightarrow H_2S + C_4H_{10} \tag{8.4}$$

$$C_2H_5Cl + H_2 \rightarrow HCl + C_2H_6 \tag{8.5}$$

Mercaptans are relatively easy to convert, followed by disulfides and then sulfides, whereas thiophenic compounds are very difficult to convert. If the duty becomes more onerous (e.g., increased throughput, higher organosulfur, more difficult organosulfur, or lower operating temperature), continued operation may be possible if a more active catalyst, such as $\text{KATALCO}_{\text{JM}}$ 61-2F, is retrofitted into the existing converter. HDS catalysts are more active when the CoMo or NiMo phases are sulfided. Adequate sulfidation is usually achieved from the sulfur in the feed.

Side reactions can also occur over HDS catalysts in normal operation as shown in Equations 8.6 to 8.9. Some of these consume hydrogen, and inlet H_2 levels will therefore require appropriate compensation.

$$H_2 + CO_2 \rightleftharpoons CO + H_2O \tag{8.6}$$

$$H_2S + CO_2 \rightleftharpoons COS + H_2O \tag{8.7}$$

$$R_2C = CR_2 + H_2 \rightleftharpoons R_2HC\text{-}CHR_2 \text{ (R = H, alkyl)} \tag{8.8}$$

$$O_2 + 2H_2 \rightleftharpoons 2H_2O \tag{8.9}$$

---

* KATALCO is a trademark of the Johnson Matthey group of companies.

There is also a low background level of carbon formation due to cracking reactions. The reverse WGS, COS formation/hydrolysis, and also COS hydrogenation reactions (Equation 8.10) reach equilibrium over HDS converters, which can lead to increased COS at the bed exit when the feed contains higher levels of CO_2 and total sulfur. Where COS may be a problem, additional control measures may be needed to protect downstream catalysts. These include injection of a small amount of steam to move the equilibrium position that may be combined with specialized downstream COS removal products.

$$H_2 + COS \rightleftharpoons CO + H_2S \tag{8.10}$$

Although the presence of olefins or O_2 is uncommon, the hydrogenation reactions are strongly exothermic. A NiMo catalyst is preferred where olefin hydrogenation is needed as it is more active than the CoMo alternative. To process high olefin levels, a recirculation system is usually employed around the HDS reactor.[3] Alternatively, a tube cooled converter[4] or a high activity catalyst capable of operating at a lower inlet temperature can be used.[5]

Undesirable side reactions such as methanation, hydrocracking, and carbon formation can occur, but these are rare and associated with less usual feeds (e.g., high CO_x, high H_2, LPG/naphtha, and olefinic feeds) and at abnormal operating conditions (e.g., low flow, very low sulfur, and high temperatures). Where methanation or hydrocracking is a concern, presulfiding the HDS catalyst hugely decreases the potential for problems.

8.2.4 Chloride Removal

If chlorides are present, hydrogen chloride removal follows the HDS. This is carried out using an alkali promoted high surface area alumina absorbent operating at the same temperature and pressure as the upstream HDS catalyst. For a high chloride capacity, the absorbent must combine an optimized porosity, density, and alkali level. An example is $KATALCO_{JM}$ 59-3, which uses sodium hydroxycarbonate and sodium aluminate phases to react with hydrogen chloride (Equations 8.11 and 8.12). The reaction between hydrogen chloride and alkali is kinetically and thermodynamically favored. This ensures removal to very low levels (<5 ppbv) even with shallow absorbent beds operating at high space velocity.

$$Na_3(OH)(CO_3) + 3HCl \rightarrow 3NaCl + CO_2 + 2H_2O \tag{8.11}$$

$$2NaAlO_2 + 2HCl \rightarrow 2NaCl + Al_2O_3 + H_2O \tag{8.12}$$

Hydrogen chloride is removed before the zinc oxide–based hydrogen sulfide removal absorbents, as hydrogen chloride will react with the zinc oxide to form $ZnCl_2$, which consumes capacity for hydrogen sulfide and sublimes, travelling to downstream processes.

8.2.5 Sulfur Absorption

In the next stage, hydrogen sulfide is removed by reaction with a zinc oxide–based absorbent, which locks up the sulfur as zinc sulfide (Equation 8.13). This is usually at the same operating conditions as the upstream HDS catalyst.

$$ZnO + H_2S \rightleftharpoons ZnS + H_2O \tag{8.13}$$

Products that contain alumina or an alumina containing binder are also usually able to remove low levels of COS exiting the HDS catalyst by hydrolysis and reaction of the resulting hydrogen sulfide (Equations 8.7 and 8.13).

The performance of a zinc oxide–based absorbent in this duty is a balance among porosity, surface area, and the corresponding density. A higher density provides higher theoretical absorption capacity, but the associated lower porosity leads to lower hydrogen sulfide removal reaction rate and a broader reaction zone within the bed. Keeping a high porosity for a given density is particularly important to avoid excessive pore blocking after the zinc oxide is converted to zinc sulfide, which causes lattice expansion. As a result of this process, the achieved sulfur pickup is less than the theoretical (i.e., complete conversion of zinc oxide).

Suppliers take different approaches that affect the final product performance. Some use high zinc oxide levels with a binder, whereas others use very high zinc oxide levels without a binder. Forming may use granulation or extrusion. Johnson Matthey uses a binder and granulation method to make a product range with different densities and surface areas to optimize product selection for a particular duty. This includes $PURASPEC_{JM}$ 2020, which has relatively high porosity but low density to optimize performance for low-temperature duties (<250°C).

8.2.6 Sulfur Ultrapurification

Ultrapurification or sulfur polishing uses high copper content materials to capture trace hydrogen sulfide/COS/organosulfur slip from the upstream beds to lower the sulfur level to below 10 ppbv. As the product is usually located in the same converter(s) as the sulfur removal absorbent, operation is at the same temperature and pressure. In the active form, these products comprise copper, zinc oxide, and alumina (e.g., $PURASPEC_{JM}$ 2084). The sulfur polishing mechanism is thought to be a surface adsorption (chemisorption) on metallic copper sites with reaction limited to the copper surface. The sulfur capacity in ultrapurification duties is only a few wt.%, mainly due to the very low levels of sulfur slip that define the end-of-run conditions. The products are generally supplied and loaded in the oxide form and require in situ reduction prior to use due to the high copper content, similar to LTS catalysts (see Section 8.4.10).

8.2.7 Combined Function Purification

Products are available that combine sulfur removal functions in a single product; however, there are some constraints on their operating envelopes. The products have a lower level HDS activity than conventional HDS catalysts but combine this with hydrogen sulfide removal capacity comparable to zinc oxide–based sulfur removal absorbents. In some cases, they also offer sulfur removal down to ultrapurification levels.[6–9] Formulations are based on sulfur removal absorbents. The high zinc oxide content provides the absorption capacity and is combined with low level catalytic additives or promoters to provide other functionality. An example of this product type is KATALCO$_{JM}$ 33-1. This is used for natural gas and lighter off gas feeds with organosulfur levels up to 20 ppmv, depending on speciation, and total sulfur up to 30 ppmv. During normal operation, the feed must contain H_2 for the HDS reaction and to inhibit cracking. The potential advantages of the product are that

- New plant designs require fewer converters, saving CAPEX.
- Total installed catalyst/absorbent volume may be less.
- No minimum feed sulfur level is required for sulfiding.
- Controlled reduction during commissioning is not required unlike copper-based ultrapurification products.
- Handling is simplified as only one product type is involved.
- Enhanced protection against COS slip as any hydrogen sulfide formed through hydrogenation is captured immediately eliminating subsequent reaction of hydrogen sulfide with CO_2 to form/reform COS.

8.3 STEAM METHANE REFORMING

After poisons have been removed from the feed stream, the steam methane reforming section converts hydrocarbons and steam to syngas (a mixture of carbon oxide, hydrogen, and methane).

8.3.1 Thermodynamics of the Steam Reforming Reaction

The reaction steam reforming reaction (Equation 8.14) is endothermic and heat has to be supplied to the system.

$$CH_4 + H_2O \rightleftharpoons CO + 3\,H_2 \tag{8.14}$$

The steam–methane reforming reaction is a reversible reaction favored by high temperatures (endothermic reaction) and low pressures (volumetric expansion). In the same fashion, heavy saturated hydrocarbons react following the general reaction in Equation 8.15.

$$C_nH_{2n+2} + nH_2O \rightarrow nCO + (2n+1)\,H_2 \tag{8.15}$$

The WGS reaction also runs alongside the steam reforming reaction under these conditions.

$$CO + H_2O \rightleftharpoons CO_2 + H_2 \tag{8.16}$$

The steam reforming of alkanes is an equimolar relationship between C and H_2O. A more desirable equilibrium position is, however, achieved by increasing the steam-to-carbon ratio (S:C). The actual operating S:C is determined by the need to avoid carbon formation over the catalyst and the requirements of the downstream HTS, if present. SMRs normally operate with a S:C molar ratio greater than 2.0.

8.3.2 Kinetics of the Steam Methane Reforming Reaction

The kinetics of the steam methane reforming reaction have been widely studied, and there are several rate equations in the literature.[10] It is widely understood that the dissociation of the H–C bond is the rate determining step. The high temperature at which the reaction takes place induces fast reaction rates, and the reaction dominantly happens on the catalyst surface.[11] Thus, the catalyst activity is strongly related to the geometric surface area (GSA):

$$d_{CH4}/dt \propto GSA \times k \times P_{CH4} \times P_{H2O}^{-\alpha} \times Keq \tag{8.17}$$

where P_{CH4} and P_{H2O} are the hydrocarbon and steam partial pressure, respectively, α is an exponential factor that varies between 0 and 1 depending on the catalyst and reaction conditions, k represents the kinetic constant at temperature, and Keq stands for the equilibrium constant.[10]

8.3.3 Carbon Formation during Steam Methane Reforming

The mechanism of carbon deposition on catalysts depends on the nature of the hydrocarbons present and the operating temperature. Methane-rich feeds may form carbon on and within the catalyst structure by cracking. Heavy feeds, such as naphtha, have a greater potential for carbon deposition. Carbon formation occurs via four mechanisms[11]: hydrocarbon cracking/polymerization (Equations 8.18 and 8.19), CO disproportionation (Equation 8.20), and CO reduction (Equation 8.21).

$$C_nH_{2n+2} \rightarrow nC + (n+1)\,H_2 \tag{8.18}$$

$$C_nH_{2n+2} \rightarrow \text{olefins} + \text{aromatics} + \text{radicals} \rightarrow \text{polymers} \tag{8.19}$$

$$2\,CO \rightleftharpoons C + CO_2 \tag{8.20}$$

$$CO + H_2 \rightleftharpoons C + H_2O \tag{8.21}$$

Heavier hydrocarbons thermally crack at lower temperatures than methane. Less hydrogen is generated per mole of carbon, so the reverse reaction is minimal and alternative reaction paths are available from the various hydrocarbons in a heavier feed. This generates organic molecules and fragments that polymerize within the pore structure of the catalyst and degrades to a polymer-derived carbon (Equation 8.19). This leads to the formation of pyrolytic carbon, polymeric carbon, and whisker

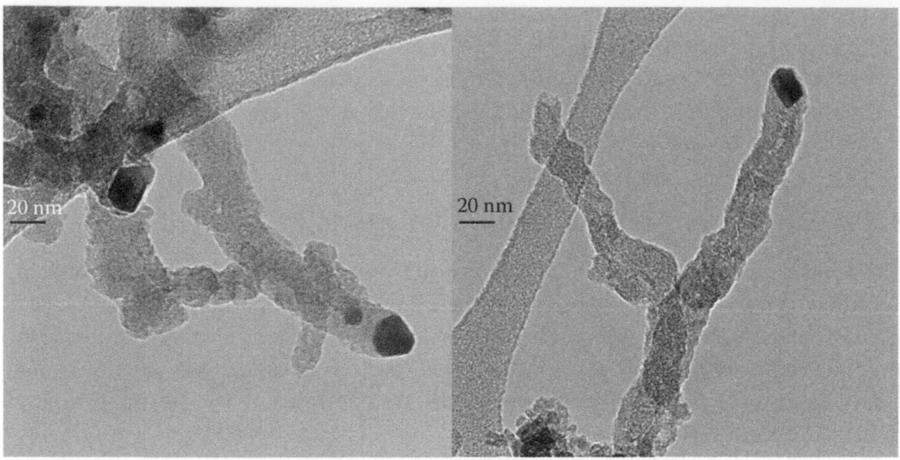

FIGURE 8.3 Transmission electron microscopy images of whisker carbon evolving from Ni crystallites.

carbon fiber (see Figure 8.3). The rates of formation are favored by the presence of heavy hydrocarbons, low steam-to-carbon ratio, and high temperatures. As the catalyst ages and loses activity, the potential for carbon formation increases, and carbon formation may occur even though the process conditions do not change.

8.3.4 Pre-Reforming

When a pre-reformer is included in a hydrogen flowsheet, it is situated prior to the SMR. Pre-reformers are occasionally retrofitted into older hydrogen plants to allow an increase in plant throughput. They are also becoming more common in modern hydrogen plants to either provide feedstock flexibility or reduce export steam. Within a pre-reformer, the C_{2+} species are reformed and the methane–steam reaction comes to equilibrium.[12]

Nickel is the active phase in the pre-reforming catalyst, which is normally supplied in a pre-reduced and passivated form. This allows handling and loading in air, as well as making start-up easier and quicker. The nickel content is relatively high, e.g., $KATALCO_{JM}$ CRG-LHR contains 47 wt.% with the balance made up of alumina and promoters. The reforming reaction is diffusion limited; hence, the catalyst is designed to have the optimum balance among the surface area, pore size, and pore volume. A large surface area is achieved by using small pellets, but this also has an adverse impact on the pressure drop. Thus, shaped catalysts with high voidage, such as $KATALCO_{JM}$ CRG-LHCR, are available.

The lifetime of pre-reforming catalysts is typically determined by one of three deactivation mechanisms: poisoning, thermal sintering, and carbon laydown. The operating feed has a significant impact on the catalyst life in terms of poison content, reaction temperature profile, and carbon forming potential.

For light feeds, such as natural gas, the steam reforming of methane dominates, and an endotherm is observed along the bed (Figure 8.4a). For heavier hydrocarbons,

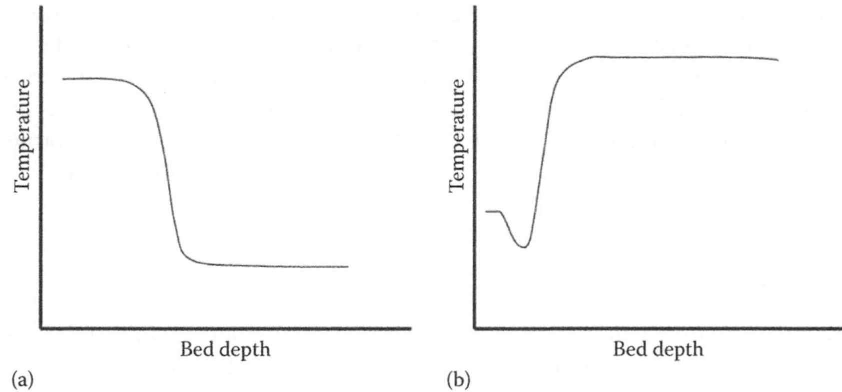

FIGURE 8.4 Pre-reformer temperature profile with natural gas (a) and naphtha (b) feeds.

such as LPG or naphtha, the endothermic steam reforming reaction dominates at the top of the bed, but, as the level of carbon oxides and H_2 in the process gas builds up, the exothermic methanation reaction (Equation 8.22) dominates, leading to a temperature profile with a net exotherm (Figure 8.4b).

$$CO + 3\ H_2 \rightleftharpoons CH_4 + H_2O \tag{8.22}$$

8.3.5 Steam Methane Reforming Technology

The SMR is usually a tubular furnace with a catalyst packed into the tubes and fuel burned on the furnace side to provide the heat needed to drive the reaction. The design of the feed, fuel, combustion air, and effluent headers are very important since poor design can lead to mal-distribution on either the process or flue gas side. Similarly, the packing characteristics of the catalyst and the ability to load the catalyst uniformly are also important for ensuring uniform distribution of the process gas.

SMR tubes are designed to work at relatively high pressures and temperatures, up to 45 barg and 920°C for the process gas. The flue gas temperature may reach 1100°C inside the radiant box, which is kept at a slight negative pressure as a safety measure. Tubes are normally made by centrifugal casting of nickel–chromium microalloy steel. The number of tubes in a reformer will depend upon the throughput; tube count varies from 3 tubes in the smallest reformers up to 900 tubes in the largest reformers. SMRs found in refinery hydrogen plants normally have between 50 and 250 tubes.

The heat of reaction is supplied by combustion of fuel in the burners and transferred to the tubes by radiation. The burners can be placed either on the top of the radiant box (top-fired reformers) or on the sides (terrace wall or side-fired reformers). In modern hydrogen plants, the fuel to burners is mostly provided by PSA off-gas with a makeup fuel, usually natural gas or refinery off gas, as the balance to meet the required calorific duty. Top-fired reformers have a peak heat flux at around 30%–40% down the tube. Side-fired and terrace wall reformers have a more homogeneous heat flux distribution along the tube. While top-fired reformers with fewer burners are easier to balance, they concentrate a large amount of heat in the top section of the tube, and this may increase the risk of carbon formation. Side-fired and terrace wall reformers can be more difficult to balance due to the large number of burners. In addition to these well-known reformer furnaces, another steam reforming concept is called a "can" reformer, usually used for small hydrogen demands. These reformers have between 6 and 14 tubes, and are fired from the bottom with either cocurrent or countercurrent flow of process gas and flue gas.

8.3.6 Steam Methane Reforming Catalysts

There are several transition metals that can catalyze the steam methane reforming reaction, with the optimum being nickel. Platinum group metals (pgms) are also very active for the reaction, but the economics and their sensitivity to poisons discourage the commercialization of pgm-based catalysts for standard duties. Nickel is dispersed on a ceramic support that contributes to the catalyst physical integrity and contains promoters to reduce potential for carbon formation. Common supports are alpha alumina or calcium and magnesium aluminates. These are pressed into complex shapes and fired at very high temperatures. Nickel precursors are impregnated onto the formed supports before calcination to fix the active phase onto the support. Nickel is active in the reduced phase; hence, catalyst reduction is required before it is put into operation. The selection of support affects the ease of catalyst reduction.

One of the major concerns during the life of steam methane reforming catalyst is deactivation via carbon formation. Thus, catalysts should be designed to minimize the rate of carbon deposition and maximize the rate of carbon gasification. Thermal cracking reactions are promoted by surface Lewis acid sites, whereas the carbon gasification is promoted by the alkali sites and is proportional to the surface coverage of OH^-. It may therefore be necessary to increase the alkalinity of the catalyst, especially for the inlet section of the SMR tube where the risk of carbon formation is greater. Besides calcium and magnesium, potassium is the preferred promoter to increase the catalyst alkalinity, reduce the formation of carbon, and induce carbon gasification. For heavy feeds, potassium is effective due to its mobility on the catalyst surface. Complex phases such as kalsilite ($K_2O{\cdot}Al_2O_3{\cdot}SiO_2$) convert slowly under reaction conditions to release potassium, which catalyzes the carbon gasification. The range of Johnson Matthey KATALCO$_{JM}$ catalysts with different levels of potassium promotion is given in Table 8.4.[13]

Potash promoted catalysts are designed to operate in the upper (inlet) section of the tubes, where there is the greatest carbon forming potential. The rate of potash release is controlled to avoid carbon formation over the catalyst life. The lower (exit) section of the reformer tubes operates at the highest temperatures, but at this point in the process, sufficient hydrocarbons have been converted and hydrogen formed to suppress carbon formation. Thus, a non-alkalized catalyst, such as KATALCO$_{JM}$ 57-4Q, is typically used in the bottom 50%–60% of the reformer tubes. The catalyst in the upper section of the reformer tube is generally determined by the carbon formation potential of the feed. A mildly alkalized catalyst such as KATALCO$_{JM}$ 25-4Q will generally provide sufficient protection for a lighter feed such as natural gas. Heavier feeds, such as LPG and naphthas, have the greatest potential for carbon

TABLE 8.4
KATALCO$_{JM}$ Steam Reforming Catalyst

K_2O (wt.%)	Series	Feedstock/Carbon Protection Requirement
0	KATALCO$_{JM}$ 23-4 or 57 series	Light feed/low C protection
1.5–2.5	KATALCO$_{JM}$ 25-4 series	↓
4–5	KATALCO$_{JM}$ 47 series	↓
6–7	KATALCO$_{JM}$ 46-3	Heavy feed/high C protection

Source: Carlsson M. 2015. Carbon Formation in Steam Reforming and the Effect of K-Promotion. *Johnson Matthey Technology Review*, Royston, UK.

formation and may necessitate a third layer of a highly alkalized catalyst, such as KATALCO$_{JM}$ 46-3Q, in the top 20%–30% of the tube.

The catalyst shape is of paramount importance. Raschig rings were used during the early days. In the 1980s, more complex shapes were developed to improve catalyst support properties such as GSA, heat transfer, and pressure drop while maintaining adequate strength to withstand typical operational events. This led to the development of the cylindrical four-hole shape that was introduced in the KATALCO$_{JM}$ 23-, 25-, 46-, and 57-series in the late 1980s. In the early twenty-first century, a further development was introduced with the QUADRALOBE™ shape. This geometry further increased the GSA, i.e., the catalyst activity, strength, and robustness, while maintaining the same pressure drop. Like its predecessors, it is designed to break

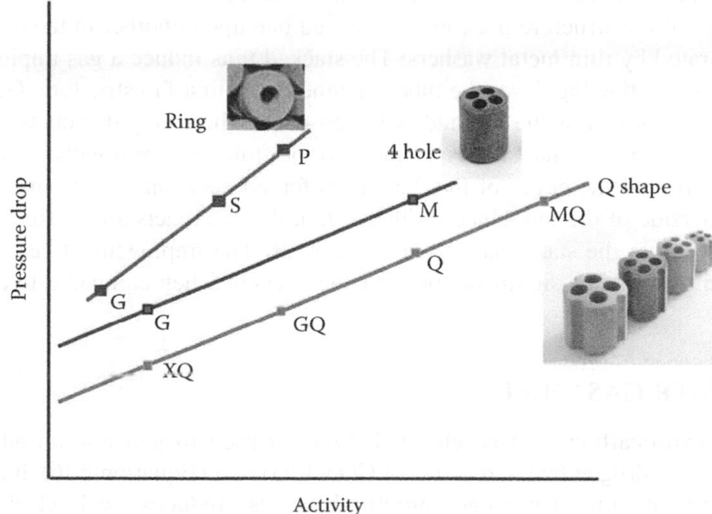

FIGURE 8.5 Comparison of steam methane reforming catalyst shapes. Q, standard quadralobe size; X, extra-large; G, large (giant); M, small (mini); S, short; P, very short.

in large fragments to minimize pressure drop growth in operation. Figure 8.5 shows the evolution of the shapes and compares the activity and pressure drop. Each shape is manufactured in different sizes to alter the pressure drop and activity to allow for performance optimization.

8.3.7 SMR Operation

The first task after a new catalyst has been installed is the in situ reduction using hydrogen-rich gas or natural gas. Once activated, the catalyst can quickly be brought to design operating conditions, where inlet temperature range is between 450°C and 600°C and process gas at outlet can be as high as 920°C. Methane slip is an easy measurement and is therefore widely used to evaluate the performance of the reformer.

The steam-to-carbon ratio depends on the nature of the feed, process design, and catalyst selection. Light feeds such as natural gas or pre-reformed feeds require a relatively low steam-to-carbon ratio, whereas heavier feeds such as LPG and naphtha will require a larger value. Steam-to-carbon ratios can range from 2.3 up to 4.0 and sometimes higher. Some plants may have a strong drive to reduce steam-to-carbon ratios to a minimum; however, this means that a small upset may place the plant transiently within the carbon formation region and/or HTS catalyst overreduction region.

8.3.8 Structured Reforming Catalyst

CATACEL$_{JM}$ SSR^{TM}* technology is a new and innovative approach to steam methane reforming catalysis. It is a high surface area structure formed from thin metal foil coated with nickel. This type of technology has been very effective in other sectors such as emission control and combustion technology.[14]

Modules of the structure packing are stacked one upon another in the reforming tube, separated by thin metal washers. The stacked fans induce a gas impingement mechanism: gas flowing down the tube encounters the first fan structure. Gas flows through corrugations of the fan and is forced out of the triangular ducts, impinging directly on the internal surface of the reformer tube, where it gathers heat. Gas then flows around the edges of the fan and is forced back into the triangular duct on the underside of the fan. Once inside the fan, the gas reacts and is free to move to the next fan in the stack and repeat the process. The impingement heat transfer mechanism results in a significant performance benefit when compared to classical pellets.[14]

8.4 WATER GAS SHIFT

After the hydrocarbon feed is reformed, WGS is used to generate an additional 10%–15% of hydrogen by the reaction of CO with steam (Equation 8.16). It not only increases the amount of hydrogen produced but also reduces the level of CO for removal by either methanation or the PSA unit (Figure 8.6).

* CATACEL and SSR are trademarks of the Johnson Matthey group of companies.

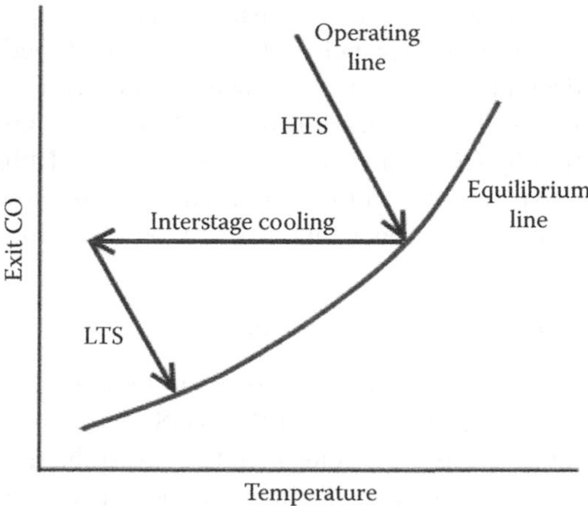

FIGURE 8.6 Two-stage WGS.

8.4.1 Thermodynamic Equilibrium

The equilibrium constant, Kp, for the WGS reaction is defined as follows:

$$\mathrm{Kp} = \frac{P[\mathrm{CO_2}] \cdot P[\mathrm{H_2}]}{P[\mathrm{CO}] \cdot P[\mathrm{H_2O}]} \tag{8.23}$$

The reaction is moderately exothermic and reversible so excess steam and low temperatures give preferable equilibrium conversion. If operated in an adiabatic bed, the reaction is equilibrium limited. Pressure has no influence on the equilibrium position, and the operating pressure is normally dictated by the upstream reformer.

8.4.2 WGS Reaction Kinetics

The reaction kinetics for the WGS reaction can be described by the following equation:

$$\frac{\mathrm{dCO}}{\mathrm{dt}} \propto \mathrm{activity} \times \mathrm{GSA} \times \exp^{[-\Delta E/(R.T)]} \times \frac{P[\mathrm{CO}]}{P[\mathrm{H_2O}]} \times (\mathrm{Kp}' - \mathrm{Kp}) \tag{8.24}$$

where activity is the activity of the catalyst, GSA is the geometric surface area of the pellets, ΔE is the activation energy for the WGS reaction, *R* is the universal gas constant, *T* is the absolute temperature, and *P* is the partial pressure of the said component.

The rate of reaction is effectively reliant on the rate constant, i.e., a feature of the catalyst, the operating temperature, and the driving force (how close the reaction is to equilibrium).

Despite being widely studied, there still remains some uncertainty of the reaction mechanism. The two main mechanisms that are considered are the associative and redox mechanisms. The associative mechanism describes the process of reactants adsorbing onto the catalyst surface before forming an intermediate, such as formates, followed by decomposition to the products and desorption from the surface. The regenerative redox mechanism suggests the catalyst surface undergoes oxidation by water and reduction by CO in a cyclic process. Work by Kubsh and Dumesic[15] supports the regenerative mechanism for iron-based catalysts. However, there is less certainty of the mechanism for copper catalysts and low temperatures.[16]

8.4.3 WGS Reaction in Hydrogen Plants

Older hydrogen processes use a scrubbing process to remove CO_2 and methanation to purify the hydrogen. The performance of the WGS section in these flowsheets is particularly important as hydrogen is consumed to convert the residual CO to methane, which is inert in most processes. To increase the conversion of CO, the WGS reaction is carried out in two stages with intermediate cooling to maximize CO conversion (Figure 8.7).

The bulk of the conversion is carried out across the HTS, where the exit CO concentration is typically in the range of 2%–4% and close to equilibrium. These units operate at relatively high inlet temperatures, 300–400°C. Following the HTS, interstage cooling of the process gas is achieved by an external gas–gas heat exchanger. The CO level can then be reduced to around 0.1%–0.3% by the copper-based LTS catalyst, which operates with inlet temperatures of 190–210°C. PSA units have now replaced the need for the LTS, CO_2 removal, and methanator; thus, most modern flowsheets usually have a single-stage WGS unit.

The minimum operating steam-to-gas ratio for a hydrogen plant can be limited by the conditions required for the iron-based HTS catalyst; thus, copper-based MTS is an alternative flowsheet option for improved efficiency if the export steam is not valued.

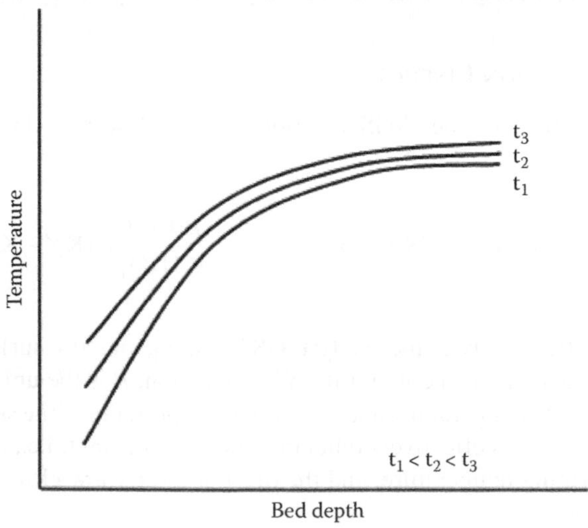

FIGURE 8.7 Change in HTS temperature profile over time.

8.4.4 High Temperature Shift

Conventional HTS catalysts consisting of iron and chromium oxides have been used for more than 100 years[17] with feeds ranging from 3% to 75% CO. Current HTS catalysts are composed of iron oxide promoted with chromia and copper oxide. Chromium oxide reduces the rate of sintering of the active iron crystallites and increases the pellet strength allowing greater robustness to process upsets. The copper oxide enhances the catalyst activity and helps avoid iron overreduction.

Catalyst strength is also affected by the pellet size that can influence the catalyst life. Large pellets can withstand bad operation better than smaller ones and reduce pressure drop, but smaller pellets are beneficial for the effects of pore diffusion. The pellet size is a balance between the required catalyst volume and the consequent pressure drop over the unit. Pore diffusion can become significant above 370°C. Thus, an effective HTS catalyst needs large pores for rapid diffusion and small pores for high surface area.

8.4.5 HTS Catalyst Reduction and Start-Up

The KATALCO$_{JM}$ 71-5 catalyst is manufactured and supplied in the hematite phase, Fe_2O_3, but the active phase is magnetite, Fe_3O_4, and reduction is required. This reduction normally occurs at the same time as the reduction of the reforming catalyst as the reduction envelopes for both overlap.

As part of the start-up process, the system must be purged and heated before steam can be introduced. If heating with nitrogen is prolonged, e.g., due to reformer refractory dry out requirements, then it is possible to dehydrate the HTS catalyst. Due to the high surface area of the fresh HTS catalyst, the consequent effect of this is an observable rehydration exotherm on its first exposure to steam. This effect is not observed on subsequent start-up due to the change in catalyst structure once the catalyst has been reduced.

Steam must be present during the reduction process to moderate the HTS catalyst reaction and avoid catalyst overreduction to metallic iron (see Section 8.4.8). The equilibrium is controlled by the ratios of H_2O/H_2 and the CO_2/CO.

$$3\,Fe_2O_3 + H_2 \rightleftharpoons 2\,Fe_3O_4 + H_2O \quad \Delta H = -6.0\ \text{kJ mol}^{-1} \tag{8.25}$$

$$3\,Fe_2O_3 + CO \rightleftharpoons 2\,Fe_3O_4 + CO_2 \quad \Delta H = -47.3\ \text{kJ mol}^{-1} \tag{8.26}$$

8.4.6 HTS Catalyst Operation

An HTS catalyst charge is easy to operate once it has been brought online. Over time, the activity will fall slowly as the iron crystallites sinter and the active surface area falls. To minimize this sintering effect, the catalyst should be operated at the lowest possible temperature consistent with minimum CO slip. It is usually necessary to adjust the inlet temperature periodically to compensate for the gradual decline in activity. Figure 8.7 shows typical HTS catalyst temperature profiles in which the slope of the temperature profile relaxes as the catalyst ages.

8.4.7 Condensation or Boiler Leaks

Wetting of the HTS catalyst due to condensing steam or process gas boiler leaks can cause a number of problems such as reduction in catalyst strength, agglomeration of the HTS catalyst, thermal shock, from catastrophic catalyst failure from wetting and then rapid evaporation of water within the pores of the catalyst, contamination of the catalyst with boiler solids blocking the active sites, increased pressure drop, and mal-distribution issues.

KATALCO$_{JM}$ 71-6 is specifically manufactured to have a higher in-service strength and can better handle upset conditions. Protective hold-down materials such as SHIFTSHIELD provide added protection against impingement of liquid water directly onto the catalyst and provide a high surface area and voidage to capture fouling solids.

8.4.8 Low Steam-to-Gas Ratio

The HTS catalyst can limit the minimum allowable steam-to-gas ratio for plant operation. At low steam levels, the syngas can overreduce the HTS catalyst to iron monoxide or metal iron (Equations 8.27 and 8.28).

$$Fe_3O_4 + H_2 \rightleftharpoons 3\ FeO + H_2O \tag{8.27}$$

$$FeO + H_2 \rightleftharpoons Fe + H_2O \tag{8.28}$$

The changes in crystallite structure weaken the catalyst. The metallic iron reacts with CO to form iron carbide that promotes side reactions, such as the Fischer–Tropsch synthesis, which consume hydrogen and form undesired hydrocarbons. Another issue is CO disproportionation, which causes carbon lay down on the catalyst, fouling the active sites and weakening the catalyst. Overreduction is less of a problem with modern copper promoted HTS catalysts. The copper increases the activity notably, reducing the CO level rapidly to a safer level both at the surface of the catalyst pellets and within the top section of the catalyst bed.

8.4.9 Low and Medium Temperature Shift

Unlike the iron-based HTS catalyst, LTS and MTS catalysts are composed of a mixture of copper oxide, zinc oxide, and alumina. Copper provides higher WGS activity than magnetite, but it is prone to sintering and must operate at lower temperatures than those used in the HTS.

The LTS is operated in the temperature range of 190–250°C and reduces the CO level to 0.1–0.3 mol%. An MTS usually operates over at 210–350°C and enables the CO slip to be reduced to 0.5–1.5 mol%.

An effective catalyst needs a formulation that provides a balance between thermal stability and activity. Catalysts need good copper crystallite dispersion for high activity and good thermal stability, and have high resistance to poisons. The dispersion of the zinc oxide and alumina enhances the catalyst strength and prevents copper crystallites from sintering. The zinc oxide also acts as a sulfur sink extending the catalyst life.

The shift reaction rate on commercial copper/zinc catalysts is on the borderline of diffusion limits. Enhanced performances in terms of activity may be achieved using smaller pellets, but practical limits are set by the balance between activity gained and physical properties, such as strength and pressure drop. The poisoning reactions with hydrogen chloride and hydrogen sulfide are more strongly diffusion-limited than the shift reaction; hence, smaller pellets are more effective for poison removal.

8.4.10 LTS and MTS Catalyst Reduction and Start-Up

The LTS and MTS catalyst is normally supplied in the oxidized form and needs to be converted to the active metal phase before operation. The reduction is carried out using a small controlled concentration of hydrogen in an inert carrier gas, such as nitrogen or natural gas. If not controlled, the potential reduction exotherm is very large, and the temperature rise could damage the catalyst, compromise the vessel mechanical integrity, and potentially lead to cracking reactions over an oxidic catalyst if a natural gas carrier is used.[18,19]

8.4.11 LTS and MTS Catalyst Operation

Copper/zinc catalysts should be operated with an inlet temperature as low as possible in order to achieve a minimum CO slip, but the minimum allowable inlet temperature is limited by the dew point of the gas feed. Initially, the CO is converted to its equilibrium level in the top of the bed, and most of the bed is unused. As the catalyst at the top of the bed starts to deactivate, normally through poisoning, the reaction profile moves further down the bed. An example of a typical exotherm movement through an LTS bed as a function of life is given in Figure 8.8. The exotherm across

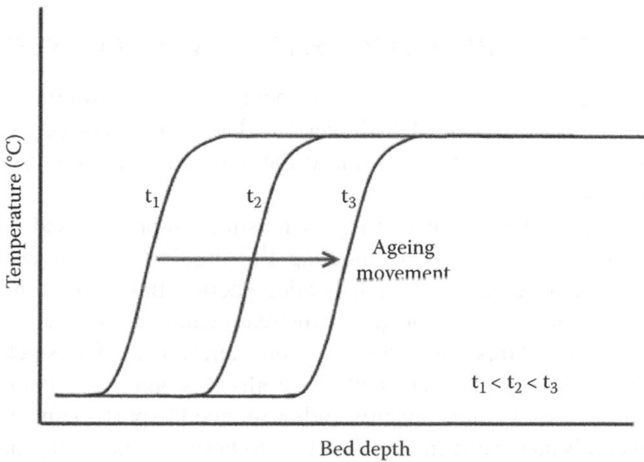

FIGURE 8.8 Change in LTS temperature profile over time.

an MTS bed is much higher as there is more CO converted. Thermal sintering is also more pronounced as it operates at a higher temperature compared to LTS catalysts. Once the reaction front has reached the bottom of the bed, change-out should be imminent as the CO slip will start to rise to unacceptable levels within a relatively short time.

8.4.12 Effect of Water

Condensation on LTS catalysts occurs during normal operation of the plant when the conditions at the inlet of the LTS converter are too close to the dew point.[20] Gross wetting can occur, for example, if there is a leak in an upstream heat exchanger or from faulty quench systems. The main effect of water on LTS catalysts is a reduction in catalyst strength, but it can also have a significant impact on the catalyst activity. Poisons such as chlorides trapped at the top of the catalyst bed can be washed down the catalyst bed, resulting in poisoning of the remaining active catalyst beneath the previously poisoned zone. Water may also contain solids and contaminants, which plate out on the catalyst surfaces covering the active sites and can cause increased pressure drop.

8.4.13 Methanol Formation

Several hundred parts per million of methanol and lesser quantities of other oxygenates are normally produced over the WGS section. This may have an impact on the quality of process condensate and be constrained by environmental legislation. Methanol increases with high temperatures, high inlet CO levels, low steam-to-carbon ratios, and low space velocities. Methanol formed across LTS and MTS catalysts decreases rapidly in the first few months of catalyst operation. Alkali promoted LTS and MTS catalysts, e.g., $KATALCO_{JM}$ 83-3X, can be used to reduce methanol levels to less than 10% compared with that of conventional catalysts.[21]

8.5 SUMMARY OF THE HYDROGEN PLANTS IN REFINERIES

Requirements for hydrogen in refineries are increasing and will continue to do so in the coming years. The pool of hydrogen available from indirect sources (e.g., catalytic reformers) has declined and raised the demand from dedicated hydrogen manufacturing units.

Hydrogen plant designs are becoming increasingly more focused on feedstock flexibility, improved efficiency, and meeting the requirements of environmental legislation. The performance of the purification section has become more vital in ensuring long-term performance of the downstream catalyst units. The inclusion of pre-reformers in new plants has increasing consideration for feedstock flexibility. Retrofits of pre-reformers in older plants have also become more common giving feed flexibility, increased plant capacity, and/or reduced export steam. Steam methane reforming catalysts have seen modifications to both the chemistry and the shape of the catalyst to provide optimal conversion with minimal pressure drop. Advances in coating technology have also started to yield structured steam methane reforming catalyst, such as $CATACEL_{JM}$ SSR. This has allowed a step change improvement in the GSA and heat transfer properties of the catalyst with a tailored pressure drop. The technology potentially could lead to a reduction in the cost and size of reformers for a specified duty. In the WGS section, more operators are using low methanol catalyst variants to meet environmental legislation, and an increasing number of plants are using MTS catalysts to provide low steam export flowsheets.

REFERENCES

1. Atkins L. 2015. Global Crude Oil, Refining and Fuels Outlook to 2035. *Global Outlook*. Hart Energy Research and Consulting.
2. Broadhurst P. V., Carnell P. J. H., and Foster A. 2007. Removing Mercury from Gas Feeds to Syngas Plants. *Proceedings of Nitrogen & Syngas 2007.* Bahrain.
3. Broadhurst P. V., and Hinton G. C. 2011. Challenges in Using Refinery Fuel Gas Feedstocks for Steam methane reforming Based Hydrogen Plants. Proceedings of 14th International Topical Conference on Refinery Processing. Session 114: Hydrogen Management. *AIChE 2011 Spring Meeting.* Chicago.
4. Musich N., Natarajan R. S., and Klein H. 2008. Process and System for Reducing the Olefin Content of a Hydrocarbon Feed Gas and Production of a Hydrogen-Enriched Gas Therefrom. *US Patent Application 20080237090.*
5. Davis R. A., Macleod N., and Wilson G. E. 2009. Process for Hydrogenating Olefins. *World Patent WO 2009/123909 A2.*
6. McKidd A. W., Macleod N., and Broadhurst P. V. 2010. Pure and Simple. *Hydrocarbon Engineering.* Vol 15. No. 3. 75.
7. Broadhurst P. V., Macleod N. and McKidd A. W. 2010. Introducing $KATALCO_{JM}$ 33-1—A Tri-functional Purification Solution for Sulphur Removal from Synthesis Gas Plant Feedstocks. *Proceedings of Nitrogen & Syngas 2010.* Bahrain.
8. Broadhurst P. V., Macleod N., and McKidd A. W. 2011. Improved Flexibility for Feedstock Purification using $KATALCO_{JM}$ 33-1. *Proceedings of International Methanol Technology Operators Forum 2011.* London. 125.
9. Kohler R., Stenseng M., Damand F. A., Mortensen H. J., Kustova M., and Hammershoi B. S. 2010. Development of New Topsoe Feed Purification Products. *Proceedings of Nitrogen & Syngas 2010.* Bahrain, 83.
10. Twigg M. V. 1996. *Catalyst Handbook.* Second edition. Manson Publishing. London.
11. Broadhurst P. V., French S., Lynch F.E., and Nijemeisland M. 2011. The Science of Catalysis—The Chemistry within Your Catalysts. Part 2—Steam Reforming. *Nitrogen & Syngas Journal.* Vol 311. 159.
12. Lloyd L. 2011. *Handbook of Industrial Catalysts.* Springer, Bath, UK.
13. Carlsson M. 2015. Carbon Formation in Steam Reforming and the Effect of K-Promotion. *Johnson Matthey Technology Review.* Royston, UK.
14. Whittenberger W., and Farnell P. W. 2015. Foil Supported Catalysts Deliver High Performance in Steam methane reformers. Nitrogen & Syngas 28th International Conference and Exhibition. Istanbul.
15. Kubsh J. E., and Dumesic J. A. 1982. In Situ Gravimetric Studies of the Regenerative Mechanism for Water-Gas Shift Over Magnetite. *AIChE Journal.* Vol 28. 793–800.
16. Anderson R., Broadhurst P. V., Cairns D., Lynch F.E, and Park C. 2011. The Science of Catalysis—The Chemistry within Your Catalysts. Part 3—Water Gas Shift and Methanation. *Nitrogen & Syngas Journal.* Vol 312. 48.
17. Bosch C., and Wild W. 1914. Hydrogen Production. *Canadian Patent No. 153379.*
18. Richardson J., Wagner J., Drucker R., and Rajesh H. 1997. Understanding Hydrocarbon Reactions during LTS Catalyst Reductions. *Ammonia Technical Manual.* Vol 37. 300.

19. Wishman M., and Reiter T. 2013. Safe Reduction of LK-853 FENCE at High Pressure. *Ammonia Technical Manual*. 147.
20. Appleton S., and Lynch F. E. 2003. Water and Your Shift Converter—Hero or Villain. 48th AIChE Safety in Ammonia Plants and Related Facilities Symposium. Orlando, USA.
21. Noelker K., and Pach J. 2013. Methanol Emission from Ammonia Plants and Its Reduction. 58th AIChE Safety in Ammonia Plants and Related Facilities Symposium. Germany.

9 Recent Developments in Hydrogen Production Catalysts—Patent and Open Literature Survey*

Ayşegül Bayat

CONTENTS

* The review for the historical overview of the hydrogen production technology, process chemistry, and the developments of the catalysts has been performed using six sources: *Petroleum Refining Process* (Speight, J.G. and Ozum, B.), *The Desulfurization of Heavy Oils and Residua* (Speight, J.G.), *Hydrogen in the Refinery: an Overview* (Navarre, J.), *Large-scale Hydrogen Production* (Rostrup-Nielsen, J.R. and Rostrup-Nielsen, T.), *The Hydrogen Plant Flowsheet* (van Uffelen, R.), and *The Fundamentals of Steam Reforming—Chemistry and Catalysis* (Broadhurst, P.) [1–6].

9.1 H_2 PRODUCTION TECHNOLOGY IN PERSPECTIVE

As the refineries began to process heavier feedstocks, hydrogen consumption of processes such as hydrotreating and hydrocracker has accelerated accordingly. In addition, the production capacities of catalytic hydrocracking, catalytic hydrotreating, and catalytic hydrorefining in refineries are increasing as well and require hydrogen in substantial quantities. Therefore, hydrogen production has gained significant importance in order to respond to the increasing hydrogen demand [1].

In earlier stages, hydrogen was used in refineries in naphtha hydrotreating processes in order to pretreat the feed stream. As the environmental regulations became stricter, the treatment processes also improved in order to treat heavier feed streams. Hydrogen was supplied to hydroprocessing units via a catalytic reforming process in which hydrogen gas is produced due to a dehydrogenation reaction. The hydrogen gas is a mixture of a high amount of hydrogen and a smaller amount of light ends such as methane, ethane, and propane that are separated from hydrogen in subsequent processes. However, as the refineries and processed feedstocks improved between the 1960s and the 1990s, the catalytic reforming process was not sufficient to compensate for the increasing hydrogen demand. In addition, the composition of crude oil determines the hydrogen consumption in a refinery [1]. It is known that for light crude oil with paraffinic nature, the hydrogen requirement is lower for subsequent hydrotreatment processes as compared to heavier crude oil [2]. Therefore, other sources of hydrogen were also considered such as the by-products of the coking, visbreaker, and catalytic cracking units. Even though these processes can be considered as hydrogen sources, most of the necessary hydrogen is generated by the steam-methane reforming process [1].

Hydrogen is generated from different pathways in a refinery in which the values of purity and pressure of hydrogen differ. The process that yields the highest hydrogen purity with 95–99.9+% and pressure between 14 and 32 kg/cm^{2}g is the steam reforming process. In this process, synthesis gas (syngas), which is a mixture of hydrogen and carbon monoxide, is produced and then purified to obtain high purity hydrogen. This process is followed by catalytic reforming that yields hydrogen with 70–90% purity. FCC off-gas provides hydrogen with the lowest hydrogen purity, which is 15–30% [3].

In general, steam and methane are used as a reactant in the steam reforming process to produce hydrogen as the main product and carbon monoxide. Apart from methane, or to generalize natural gas, this process has the flexibility and compatibility of processing other feed types such as refinery-off gas, LPG, or naphtha. Natural gas is the most widely used feedstock for steam reforming processes and contains methane and ethane of over 90% with a smaller amount of propane and light hydrocarbons [1,4].

Carbon dioxide, nitrogen, sulfur, and other impurities are also present in natural gas in trace amounts. These impurities are removed by treatment with zinc oxide and hydrogenation prior to the steam reforming process in order to avoid poisoning of the reforming catalyst. Light refinery gas, which is a by-product in refineries, contains a high amount of hydrogen together with olefins, propane, and heavier hydrocarbons. Since olefins cause coking problems, a promoted reformer catalyst may be used to prevent coke formation. Liquid feedstocks like naphtha are mostly considered as a backup feed. The sulfur in these feeds is in the form of mercaptans, thiophenes, or heavier compounds and is very stable. Therefore, zinc oxide is not sufficient, and a hydrogenation unit is necessary to remove these compounds. The reforming process is highly endothermic so that a high amount of heat is required in order for the reactions to proceed. Therefore, the reformer furnaces are one of the main types of equipment for this process [1,4].

Syngas is generated by oxygen-based technology (gasification/partial oxidation), which operates without a catalyst at high temperatures (>1200°C); by steam-based technology (SMR), which operates catalytically at lower temperatures (<950°C); or by using hybrid technology (autothermal reforming), which also requires a catalyst and intermediate temperatures (<1100°C). The steam reforming process has some advantages over the other technologies in terms of less severe operating conditions, lower cost of steam compared to oxygen and flexibility of feed such as gas and liquids that are lighter than FBP 170°C, and a higher H_2/CO ratio of the produced syngas [5].

The effluent gas stream contains carbon monoxide, carbon dioxide, hydrogen, and excess steam produced in a steam reformer, and then passes through a reactor called a shift converter for water-gas shift (WGS) reaction. In WGS reaction, carbon monoxide produced by steam reforming reaction is reacted with steam to generate more hydrogen. Monitoring and analysis of catalyst performance are vital for these catalytic processes. Hydrogen produced from these reactions may reach a purity level of above 70%, and hydrogen purity may be increased over 99% via separation in high-performance adsorption units, e.g., pressure swing adsorption (PSA) [2,4].

9.2 H_2 PRODUCTION PROCESS CHEMISTRY AND CATALYSTS IN PERSPECTIVE

Steam methane reforming has been used to produce hydrogen over several decades and is a continuous catalytic process. In steam reforming, the major reaction is methane and steam reforming to produce a hydrogen and carbon monoxide mixture called syngas:

$$CH_4 + H_2O \rightarrow CO + 3H_2O \quad \Delta H = +\,206\ \text{kJ/mol}$$

where ΔH is the heat of the reaction. Higher boiling hydrocarbons are also used to produce hydrogen:

$$C_3H_8 + 3H_2O \rightarrow 3CO + 7H_2$$

The general form of this reaction for steam reforming of higher boiling hydrocarbons may be shown as follows:

$$C_nH_m + nH_2O \rightarrow nCO + \left(n + \frac{m}{2}\right)H_2$$

The steam-methane reforming reaction occurs generally at 815°C and at 28.12 kg/cm^2. Generally, a nickel-based catalyst is used for this reaction, and the catalyst is packed into the tubes of the reformer furnace. Apart from steam reforming reaction, the cracking reactions of hydrocarbon feedstock and the reaction of carbon with steam also occur at this high temperature:

$$CH_4 \rightarrow 2H_2 + C$$

$$C + H_2O \rightarrow CO + H_2$$

As the steam reforming reaction of hydrocarbon proceeds to produce hydrogen and carbon monoxide, carbon is formed over the catalyst as well. When natural gas or light hydrocarbon is used as feedstock, carbon is converted into carbon monoxide and hydrogen by reaction with steam and removed from the catalyst very fast. For the heavier hydrocarbon feedstock, this reaction is not fast enough so that carbon is not removed effectively and accumulates on the catalyst. Therefore, the catalyst is to be either regenerated or replaced. For heavier feedstocks, accumulation of carbon can be reduced or prevented by using a catalyst that is promoted with alkali compounds such as potassium that could promote reaction of carbon with steam. In the case of using LPG occasionally as the feedstock, the steam-methane reforming catalyst can be used with a higher steam/carbon ratio to prevent coke formation [1].

For the heavier hydrocarbon feedstocks, such as naphtha, steam reforming reaction occurs generally at a temperature of 675–815°C and at 21.1 kg/cm^2. In this case, a pre-reforming reactor that operates at lower temperature with an adiabatic catalyst bed may be used as a pretreatment of the feed stream prior to steam reforming to reduce coke formation on the catalyst of the steam reformer [1].

Carbon monoxide in syngas formed by steam reforming reaction is further reacted with steam to produce more hydrogen. This reaction is called the WGS reaction:

$$CO + H_2O \rightarrow CO_2 + H_2 \quad \Delta H = -41 \text{ kJ/mol}$$

Carbon monoxide and carbon dioxide are converted back to methane and removed from the gas mixture as follows:

$$CO + 3H_2O \rightarrow CH_4 + H_2O$$

$$CO_2 + 4H_2 \rightarrow CH_4 + 2H_2O$$

Temperature, pressure, and steam/hydrocarbon ratio are important parameters for the steam reforming process. The steam reforming reaction is an equilibrium and endothermic reaction, so that high temperature favors the conversion. In contrast, WGS reaction is exothermic and favored by low temperatures. This reaction occurs at 315–370°C over an iron oxide catalyst and is limited by the catalyst activity at these low temperatures [1].

As mentioned briefly, coke formation occurs in the reformer, and high molecular weight hydrocarbons with poor hydrogen content such as naphtha are more prone to coking compared to methane. Coke may cover the active sites and block the pores of the catalysts; in fact, coke may damage the catalysts mechanically by forming nanorods. As a consequence, due to inefficient utilization of the catalyst, hotspots within the catalyst bed may occur and lead to sintering of catalyst particles, which reduce the catalyst activity even further [4].

After manufacturing, the steam reforming catalysts are usually in the nickel oxide phase; therefore, the catalyst has to be reduced into metallic nickel form with hydrogen after loading to the reactor [2]. In general, reforming catalysts are composed of ceramic supports coated with nickel and in the form of rings in cylindrical geometry with holes [4]. Active metal for these catalysts is mostly nickel due to robust nature, good activity, and tolerance to poisons. In some specific applications, precious metals can also be considered as the active metal due to their very high activity; however, they are rather expensive and sensitive to poisons [6].

In principle, a steam methane reforming catalyst contains approximately 15–25 wt.% nickel oxide supported on a mineral carrier [2]. Support material of the catalyst is required to be stable at high temperatures and in the presence of steam so that the catalyst would show no reactivity with steam under high steam partial pressures. The catalyst supports mostly have non-acidic nature to prevent hydrocarbon cracking that leads to coke formation. There are different types of support materials that can be considered for a steam reforming catalyst. α-Alumina is a tough, highly inert ceramic support material with little surface–Ni^{2+} interaction. Calcium aluminate is also a robust, ceramic material with moderate surface–Ni^{2+} interaction and high surface area. Magnesium aluminate is a strong and stable support material with basic nature and very high surface–Ni^{2+} interaction. Depending on the strength of surface–Ni^{2+} interaction, magnesium alumina requires higher reduction temperatures of 700–750°C followed by calcium alumina of 500–550°C and then by α-alumina of 400–500°C [6]. In the case of heavier feedstocks such as LPG or naphtha, less active catalysts can be used such as nickel-containing alkaline carriers or nickel-based magnesium oxide without alkaline metals. Thereby, rapid decomposition of hydrocarbons is reduced, and the formed carbon-based components can react with steam [2].

Catalyst activity, stability, types of promoters, and pressure drop are important parameters for the catalyst design. Activity of the catalyst is related to the metal dispersion, so that for high activity, small particles of metals must be dispersed over the support material. The shape and size of the catalyst affect the activity as well. Activity is proportional to the geometric surface area, i.e., pellet surface area per unit volume. It is known that a higher geometric surface area is observed for smaller pellets; on the other hand, higher pressure drop may also be seen for these catalysts. Stability is referred to as the integrity of the overall formulation of the catalyst in terms of chemical structure and composition of support, active metals, and promoters. During processes, metal particles sinter and agglomerate that can be minimized due to higher metal–support interaction. Promoters are used in order to improve the performance of the catalyst. Potash is the most generally used promoter for steam reforming catalysts. It decreases surface acidity to prevent coke formation and reacts with steam to form hydroxide species. These hydroxide species react with carbon to generate carbon dioxide and hydrogen to promote removal of carbon from the catalyst. Rare earth species, for example lanthanum, may also be used as a promoter to block sintering of nickel particles by forming refractory oxide at the surface. Pressure drop is related with void fraction within the bed and equivalent particle diameter. Catalyst particles to smaller size lead to higher pressure drop, while for larger particles, pressure drop is lower. Pressure drop buildup is also related to the strength of the catalyst and breakage characteristics of the shapes since smaller catalysts or smaller fragments tend to increase pressure drop buildup. Very dense packing of the catalyst leads to low voidage within the catalyst bed and causes high pressure drop. On the contrary, poor packing leads to too much void within the bed and causes formation of hot spots due to formation of areas with lower activity [6].

Depending on the type of feedstock, different types of components can act as a poison for the steam reforming catalyst leading to deactivation. Chloride or metals are the expected poisons in the case of using catalytic reformer off-gas or naphtha as feed, respectively. Sulfur is the most common poison and reduces the catalytic activity even in very low levels; the limit for the sulfur level should be lower than 0.1 ppmv [6]. However, the inactive nickel sulfide particles can be converted into active nickel particles again by increasing the reaction temperature so that this activity loss can be compensated for to a certain extent. For the feedstocks heavier than naphtha, strontium-based aluminum and calcium oxide catalysts can be used in order to resist coke formation and sulfur poisoning [2].

9.3 ANALYSIS BASED ON PATENT APPLICANTS AND COUNTRIES OF THE APPLICANTS

This chapter investigates H_2 production patents in terms of catalyst developments between the years of 2008 and 2014. In this chapter, the total number of investigated patents related to "H_2 production through steam reforming" is 385; among them, 104 are related to "catalysts for H_2 production through steam reforming;" the rest concern process developments for H_2 production through steam reforming. These numbers indicate that most of the studies related to H_2 production focused on proses-based developments rather than catalyst-based developments.

For the upcoming patent analyses and also for the year-based analyses, the latest issue date for each patent has been considered in order to cover the most recent update related to the given technological development, with similar approach as in the previous chapters. Annual distribution of H_2 production patents between the years of 2008 and 2014 is given in Figure 9.1 in terms of catalyst-related and process-related developments. Development of process-related technologies has gained most of the attention up to 2013. The number of catalyst-related studies has increased throughout the years and reached the number of the process-related studies in 2014. For the following analyses, the patents related to "catalysts for H_2 production through steam reforming" are the main focus point within Chapter 9.

Patent distributions of H_2 catalysts that are categorized with respect to inventors are given in Figure 9.2. The major part of the studies has been conducted by

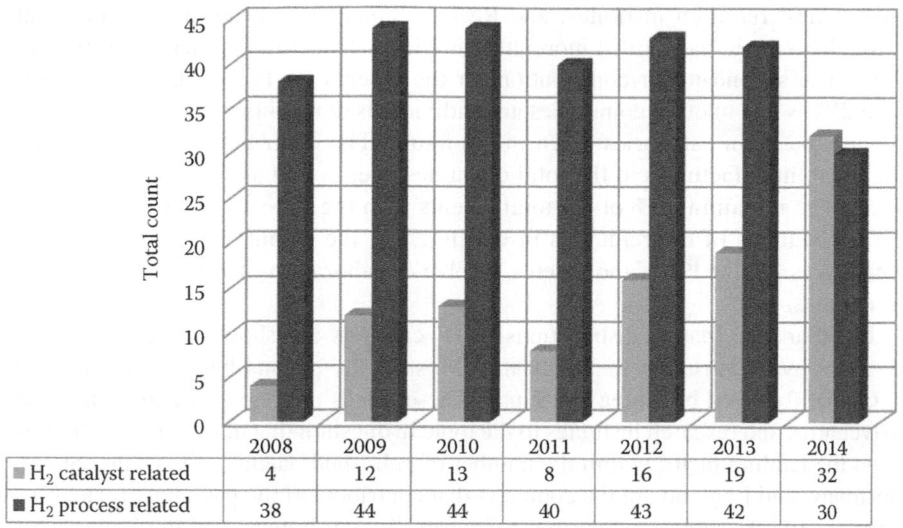

	2008	2009	2010	2011	2012	2013	2014
H_2 catalyst related	4	12	13	8	16	19	32
H_2 process related	38	44	44	40	43	42	30

FIGURE 9.1 Annual distribution of H_2 production patents between the years of 2008 and 2014.

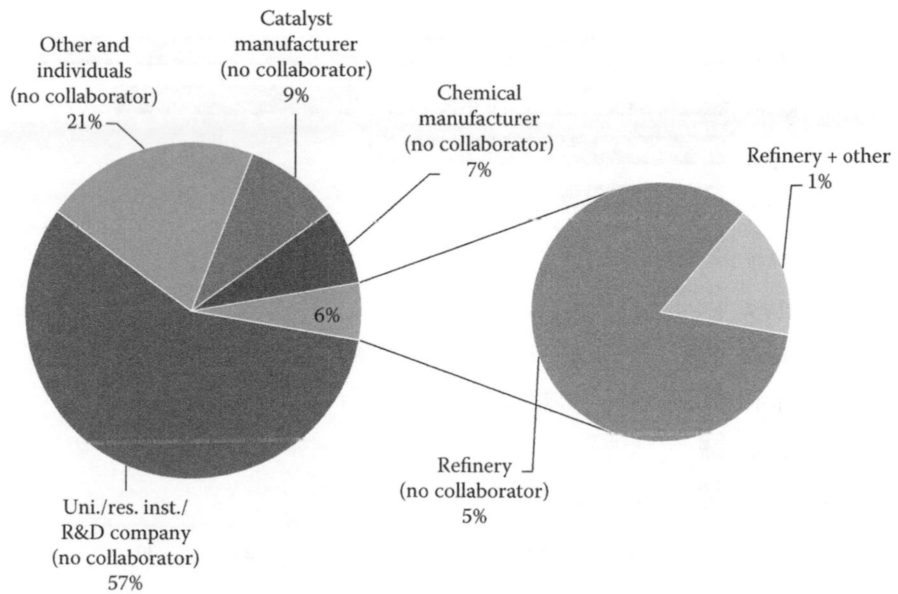

FIGURE 9.2 Patent distributions of H_2 catalysts based on applicants.

universities, research institutes, and R&D companies covering 57% of the total count. This topic has gained more attention from academic and technical institutions. The second major contribution for the patents for H_2 production catalysts with 21% is due to other companies and individuals that contain applicants that fall outside the main categories given in the figure. The contribution of catalyst and chemical manufacturers to the total count has been found as 9% and 7%, respectively. The remaining 6% of the total patents with the lowest contribution contain the applications of the refineries in which 5% of the patents belong to refineries themselves, while 1% of the patents involve collaboration of refineries and other companies.

In Figure 9.3, patent distributions of H_2 catalysts are shown in terms of total numbers by countries of the applicants. Most of the patents have been published in China, followed by Korea and Japan. These trends are due to the interest of the universities and research institutes to the topic in question that allows these countries to be the leading factor within the number of published patents. The United States, Germany, and England are the countries that contribute to the patent of H_2 catalysts significantly. As shown in Figure 9.3, Taiwan, Russia, France, and the Netherlands also own patents of H_2 catalysts.

According to Figure 9.4, refineries and universities, research institutes, and R&D companies located in Asia play a dominant role in the patents of H_2 catalysts, while European and American companies of the catalyst and chemical manufacturers are effective in the patent publications. The patents from European refineries, universities, research institutes, and R&D companies were not detected.

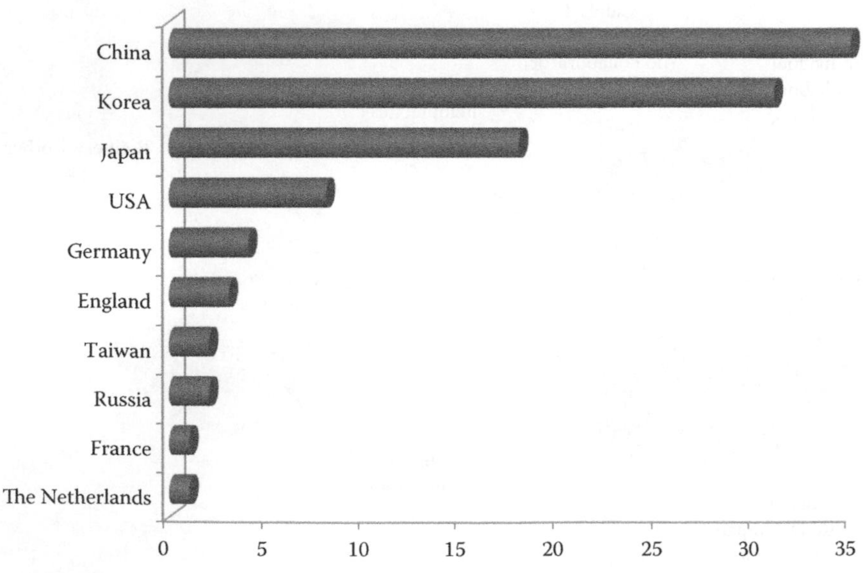

FIGURE 9.3 Patent distributions of H_2 catalysts in terms of total numbers by countries (note that if the corresponding country is in collaboration with another company, their collaborative studies are taken into account for the percentage calculations).

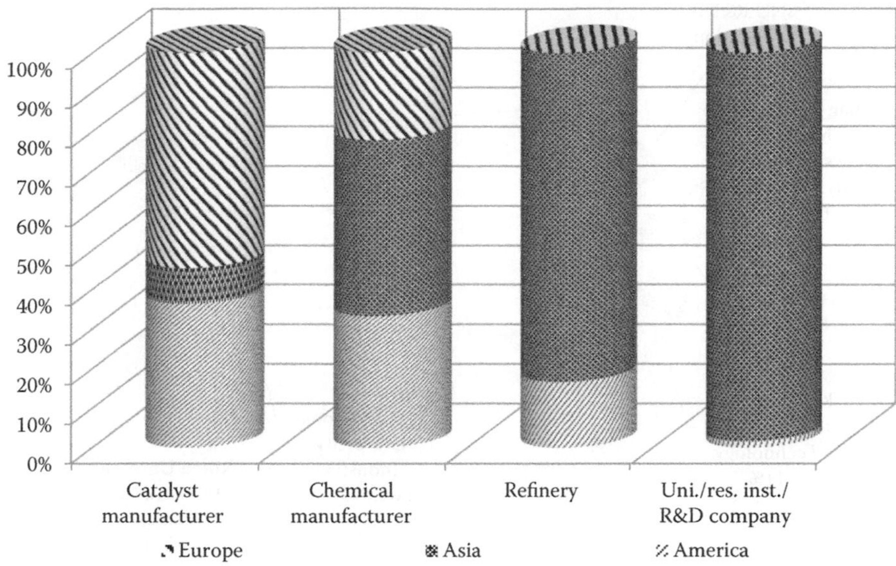

FIGURE 9.4 Patent distributions of H_2 catalysts in terms of total numbers by continents.

9.3.1 Involvement of Catalyst Manufacturers

Catalyst manufacturers that have published the patents related to H_2 production catalysts are mainly Sued Chemie, Johnson Matthey, and BASF as given in Figure 9.5. Sued Chemie has published more patents as compared to other companies and consists of 60% of the total count within the classification among the catalyst manufacturers.

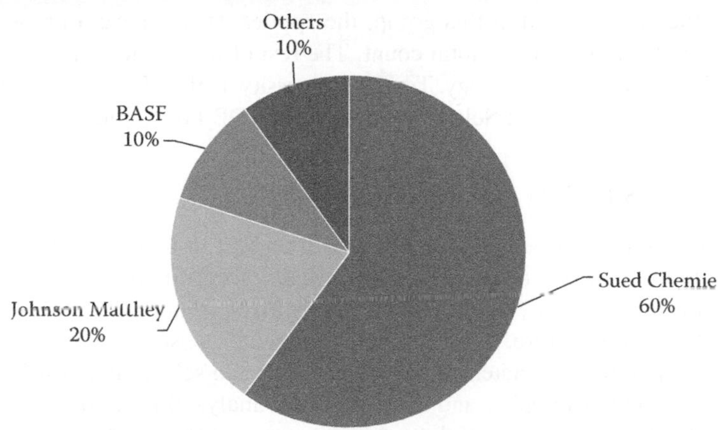

FIGURE 9.5 Patent applicants of H_2 catalysts as catalyst manufacturers (note that if the corresponding company is in collaboration with another company, their collaborative studies are taken into account for the percentage calculations).

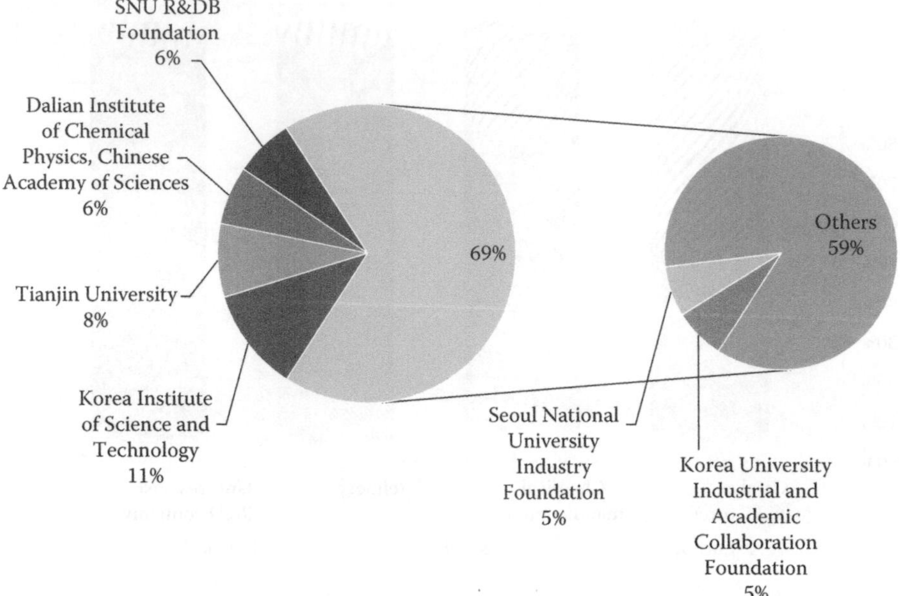

FIGURE 9.6 Patent applicants of H_2 catalyst with respect to universities/research institutes/R&D companies (note that if the corresponding company is in collaboration with another company, their collaborative studies are taken into account for the percentage calculations).

9.3.2 Involvement of Universities/Research Institutes/R&D Companies

In Figure 9.6, patent distributions of H_2 catalysts with respect to universities/research institutes/R&D companies are shown. The applicants that contribute to the total count in less than 5% are grouped into a category that constitutes 69% of the total number of the patents. Within this group, the applicants that have a proportion of 4% and below form 59% of the total count. The rest of the portion belong to Korea Institute of Science and Technology, Tianjin University, Dalian Institute of Chemical Physics, Chinese Academy of Sciences, and SNU R&DB Foundation.

9.4 ANALYSIS BASED ON RESEARCH AREA

Patents of H_2 catalysts have been classified with respect to main topics and related targeted subtopics in order to classify the patents accordingly. The main topics involve "reforming process," "active metal," "additive/promoter," "catalyst support material," "catalyst structure," "stability enhancement," "cost-effective, efficient, or alternative preparation of material," and "product yield/selectivity." Each topic is divided into targeted subtopics, and the patents are analyzed in terms of these subtopics. During the investigation of the patents, it was seen that the targeted subtopics for H_2 catalysts are related to other subtopics.

The processes that are studied in the patents are firstly grouped in terms of the type of feed streams used under the heading of "reforming process." Many different types of feed streams such as methane, natural gas, ethanol, dimethyl ether (DME), LPG, methanol, liquefied natural gas, and biomass type feeds are considered for the steam reforming processes. In addition, in some of the patents, carbon dioxide or carbon dioxide and oxygen are introduced into the feed stream for steam reforming of methane and natural gas. Therefore, the patents are classified with the target names of "SR: methane," "SR: natural gas," "SR: ethanol," "SR: dimethyl ether," "SR: LPG," "SR: methanol," "SR: liquefied natural gas," "SR: bio-oil/biomass/other hydrocarbons," "Steam+CO_2_R: methane," "Steam+CO_2+O_2_R: methane," "Steam+CO_2+O_2_R: natural gas," and "Steam+CO_2_R: natural gas" [7–105].

The referenced patent numbers related to the target topics are summarized in Table 9.1. The catalyst activity is related to the type of the active metal. Therefore, the catalysts that are studied in the patents are classified accordingly with the heading of "active metal." The target names of this topic are defined as "Ni," "transition metal," "noble metal," and "rare earth metal." Nickel is the most commonly used active metal for the steam reforming catalysts [106]. In addition, noble metals are also used due to higher sulfur resistance [107], and transition metals are known to have higher catalytic activity and stability [44].

Patents are also classified according to the topic of "additive/promoter." Addition of different promoter and additive metals into the catalyst structure improves resistance against coke formation, sintering, and sulfur poisoning [41]. The related target headings are defined as "alkali metal," "alkaline earth metal," "rare earth metal," "noble metal," and "transition metal."

There are different types of support materials used for the catalyst; therefore, the patents are also classified according to the type of "catalyst support material." The patents are classified with the target names of "alumina-type," "calcium aluminate/magnesium aluminate," "composite oxide," "nanosized-dispersed zinc aluminate," "transition metal oxide," "other," "rare earth oxide," and "zeolite/mesoporous alumina/powdery aerosil alumina/foam silicon carbide/silica alumina-SBA-15."

The patents are classified with respect to the "catalyst structure" since the structure of the catalyst affects the behavior of the catalyst. The target topics in this section is classified as "highly dispersed metal particles," "mesoporous/macroporous/bimodal pore structure," "optimized catalyst shape/low density," "increased specific surface area/pore properties," "reduced pressure drop," "increased mass transfer properties," "increased heat transfer properties," and "multifunctional catalyst: catalytic active site–absorption active site." For example, high dispersion of metal particles increases catalyst activity [41]. Mesoporous, macroporous, bimodal pore structure of the catalyst improves mass transfer properties within the catalyst structure, which enables processing larger molecules and provides resistance to coke deposition [18,35,41]. By optimizing the catalyst shape, pressure drop within the reactor can be reduced to obtain more stable pressure profile [107].

In steam reforming process, the stability of the catalyst is an important parameter. Therefore, the catalysts studies in the patents are grouped under the topic of "stability enhancement" and divided into subtopics as "conversion of coke to CO/CO_2," "improved catalyst life/service life," "improved mechanical resistance," "improved thermal stability," "more stable catalyst," "operation at low temperature/

TABLE 9.1
Referenced Patent Numbers Related to the Target Topics

Topic	Target	Referenced Patents
Active metal	Ni	[7–15,20–24,28,30,31,33–35, 37–39,41–43,45,46,48,51,53–56, 58–62,65–75,77–84,92,96–98, 100–103,105,106,108]
	Transition metal	[16,32,34,40,44,48,62,82,85,91,107]
	Noble metal	[7,16,19,25–27,36,47,49,62,82,83,88, 95,104,107]
	Rare earth metal	[21,32,38,44,48,56,82,83,107]
Additive/promoter	Alkali metal	[20,21,47,53,69,70,72,93]
	Alkaline earth metal	[8,21,24,29–31,33–35,37,44,48,53,54, 59,61–63,65,68,71,96,97,99,105,108]
	Rare earth metal	[7,12,13,20,22,26,35,38,41,43,46,53, 57,71,86,92,98–100,105]
	Noble metal	[15,29,33,38,59,69,97]
	Transition metal	[19,20,27,61,72,76,88,95,99,103]
Catalyst support	Alumina-type	[8,10,13,14,16,20–23,26,28,29,33–36, 50,53,55,56,58–62,71,74–81,83,86, 88,89,92,94,102,107,108]
	Calcium aluminate/magnesium aluminate	[11,12,15,22,33,37,38,43,54,56,70,106]
	Composite oxide	[9,24,27,30–32,34,45,51,52,69,74,87, 91,98,100]
	Transition metal oxide	[16,18,25,35,36,44,47,49,51,52,63, 65–68,70,83,85,94,105]
	Other	[9,17,19,43,44,46,48,57,58,64,66,72, 73,96,97,99]
	Rare earth oxide/nanosized-dispersed zinc aluminate	[18,49,50,52,60,67,85,90]
	Zeolite/mesoporous alumina/powdery aerosil alumina/foam silicon carbide/silica alumina-SBA-15	[39,41,78,82,84,94,95,104]
Catalyst structure	Highly dispersed metal particles	[24,30,31,36,41,55,80]
	Mesoporous/macroporous/bimodal pore structure	[15–18,35,41,74,75,77–79,81,93]
	Optimized catalyst shape/low density	[28,104,106–110]
	Increased specific surface area/pore properties	[18,36,55,73,77,89,102,106–110]
	Reduced pressure drop Increased mass transfer properties	[18,22,28,43,55,58,87,101–104, 106–110]

(*Continued*)

TABLE 9.1 (CONTINUED)
Referenced Patent Numbers Related to the Target Topics

Topic	Target	Referenced Patents
	Increased heat transfer properties	[18,28,43,55,109]
	Multifunctional catalyst: catalytic active site–absorption active site	[11,101]
Stability enhancement	Conversion of coke to CO/CO_2	[93]
	Improved catalyst life/service life	[26,34,51,69,71,110]
	Improved mechanical resistance	[100,101]
	Improved thermal stability	[19,23,25,55,62,72,99]
	More stable catalyst	[3,7,8,10,11,17,23–28,30,34,38,40,45, 51,52,61,63,64,69,71–73,79,80,87, 93,95–97,99–101]
	Operation at low temperature/wide temperature range/high temperature/high pressure	[9,10,14,15,17,19,25,27,32,40,47,49, 50,52,64,71,85,86,94,95,97,101,105]
	Resistance to coke formation	[8,9,20–22,27,33,35,37,38,41, 44–46,48–50,52,56–58,61,63–66, 68–70,92,93,96–99,101,103,105]
	Resistance to sintering	[8,41,48,54,87,98]
	Resistance to sulfur poisoning	[16,27,36,66,94]
Cost-effective, efficient, or alternative preparation of material	Core-shell, spray coating, hydrothermal, slurry, sol-gel method, coprecipitation, injection molding	[39,42,43,51,55,71,79,81,106]
	Physical mixture of two or more catalysts	[49,52,61,88]
	Support material by surfactant/ template	[18,74,77,78,81]
	Cost reduction/less complex process/utilization of waste catalyst/reduced environmental pollution	[9,10,12–14,40,42,44–46,58,60,64,66, 69,76,84]
Product yield/ selectivity	Improvement of activity/selectivity	[8–11,13–16,18,20,21,24,32, 36–38,40–42,44–46,48,51,53,54,56, 58–62,64,67,69–71,73,74,85, 89–91,94,95,98,99,101,102,105,110]
	Increased H_2 yield	[11,18,27,32,39,47,49–52,63,64,67,71, 76,90]
	Reduction of CO_2 yield by conversion/absorption	[11,34,42,99,105]

wide temperature range/high temperature/high pressure," "resistance to coke formation," "resistance to sintering," and "resistance to sulfur poisoning."

"Cost-effective, efficient, or alternative preparation of material" is emphasized in some of the patents so that the related patents are investigated in terms of the catalyst preparation methods such as "core-shell, spray coating, hydrothermal, slurry, sol-gel method, coprecipitation, injection molding," "physical mixture of two or more catalysts," "support material by surfactant/template," and "cost reduction/less complex process/utilization of waste catalyst/reduced environmental pollution."

The patents that describe the improvement in the "product yield/selectivity" of the catalysts are also investigated with respect to the target topics of "improvement of activity/selectivity," "increased H_2 yield," and "reduction of CO_2 yield by conversion/absorption."

In Figure 9.7, patent distributions of H_2 catalysts based on mostly stated topics are given except for "reforming process," "active metal," and "catalyst support material" since these topics are common for all patents that have been investigated. The differences in the patents emerge from whether or not an additive/promoter has been used, whether an improvement in stability, product yield, or catalyst structure has been explained, or the details of alternative preparation method have been focused on. As seen in Figure 9.7, most of the studies are based on "stability enhancement" of the H_2 catalysts with 29% of the total count, which is followed by improvements in "product yield/selectivity" with 25% of the total count. The studies that involve "additive/promoter" addition to the catalyst structure are also an important subject with a portion of 21%. Some studies have been conducted in terms of "cost-effective, efficient, or alternative preparation of material" and "catalyst structure" with 13% and 12%, respectively.

A topic analysis with respect to the applicants is given in Figure 9.8 for the patents of H_2 catalysts. Refineries focus on mostly "additive/promoter" addition to the catalyst structure, "cost-effective, efficient, or alternative preparation of material," and "product yield/selectivity." The main target areas of the catalyst manufacturers

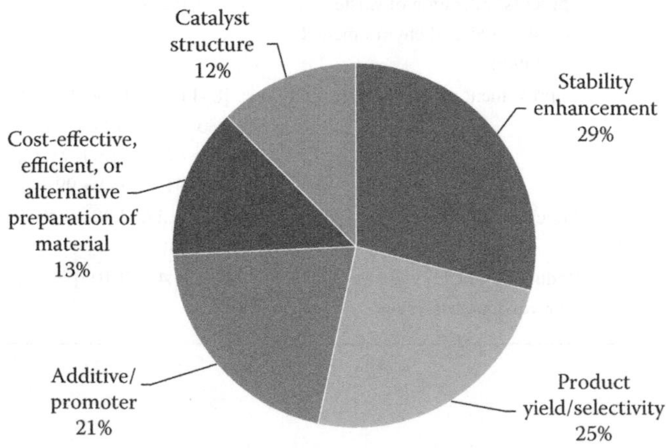

FIGURE 9.7 Patent distributions of H_2 catalysts based on mostly stated topics.

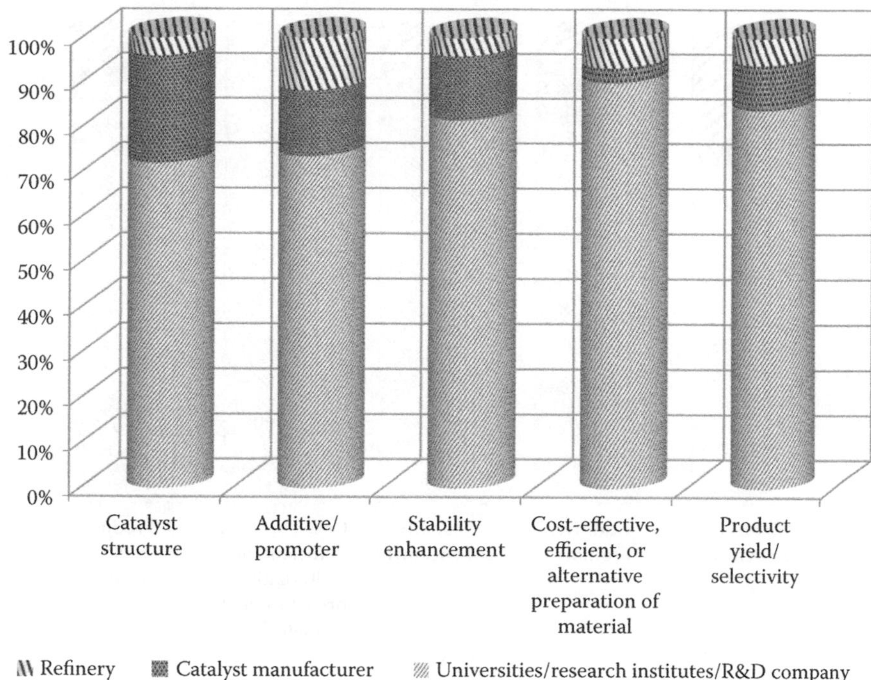

FIGURE 9.8 Topic analyses with respect to the applicants.

are improvement in the "catalyst structure," "additive/promoter" addition to the catalyst structure, and "stability enhancement" of the catalysts. On the other hand, universities/research institutes/R&D companies are interested in "cost-effective, efficient, or alternative preparation of material," "product yield/selectivity," and "stability enhancement" of the catalysts.

Topic analysis with respect to continents of the applicants is given in Figure 9.9. This figure represents that Asia is the major contributor for the distribution of the topics according to point of interest, which is also consistent with the previous analyses such that more extensive studies have been found among Asian applicants. The Asian applicants that include mostly universities/research institutes/R&D companies focus on "cost-effective, efficient, or alternative preparation of material," "product yield/selectivity," and "stability enhancement" of the catalysts. European applicants concern mostly for "catalyst structure" of the H_2 catalysts. The main study areas for the American applicants are found to be "additive/promoter" and "stability enhancement" of the catalysts, while for the topic of "cost-effective, efficient, or alternative preparation of material," no study has been found.

9.4.1 Developments in Terms of Reforming Process

In Figure 9.10, the target breakdown of "reforming process" for the patents of H_2 catalysts is investigated with respect to the type of main feed stream and studied

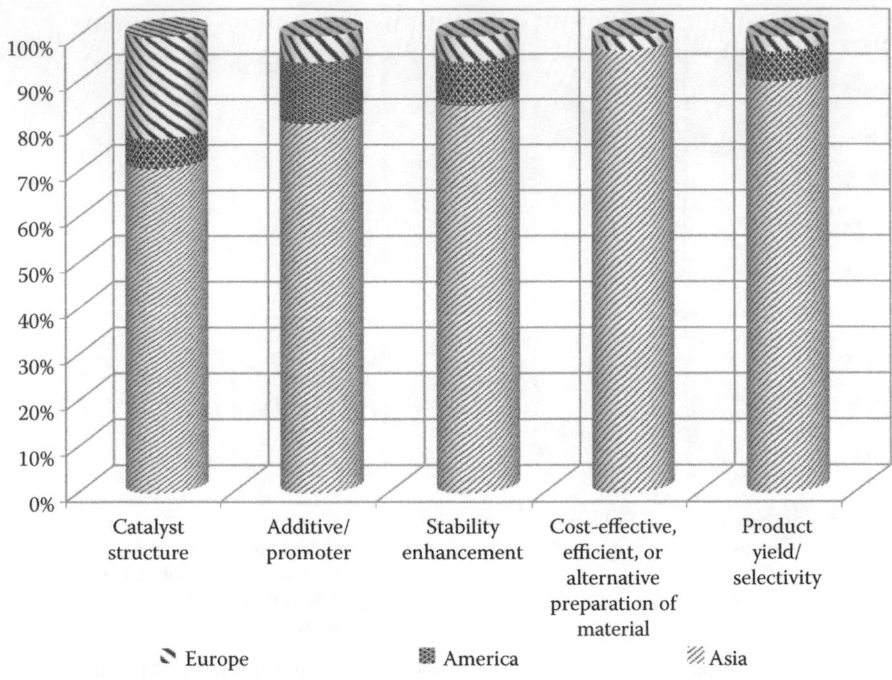

FIGURE 9.9 Topic analyses with respect to the continents of the applicants.

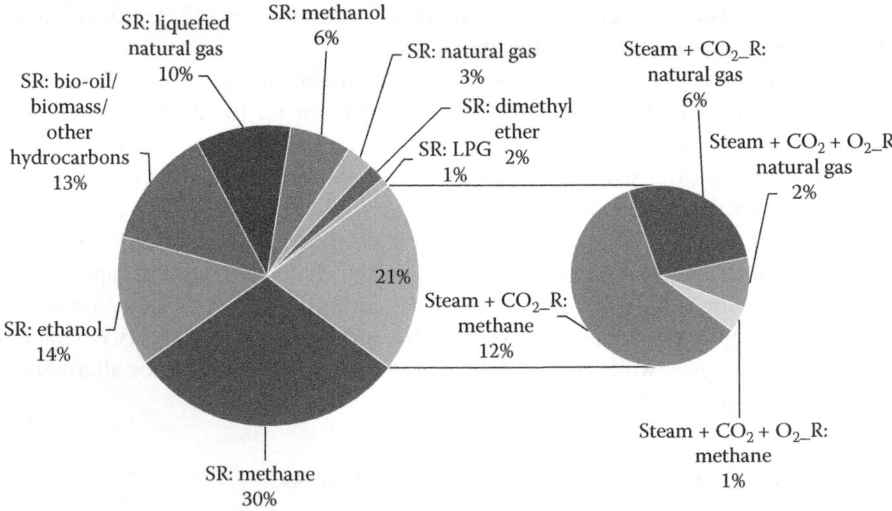

FIGURE 9.10 Target breakdown of "reforming process" for the patents of H_2 catalysts.

type of reforming process. In the given figure, steam reforming process is denoted as "SR;" steam–CO_2 reforming and steam–CO_2–O_2 reforming processes are denoted as "Steam+CO_2_R" and "Steam+CO_2+O_2_R," respectively. According to Figure 9.10, methane steam reforming process to produce hydrogen is found to be the main research area with 30% of the total count [7–38]. Steam reforming of ethanol, bio-oil, biomass, and liquefied natural gas has also gained much attention [17,21,39–84]. Researchers have also investigated different combinations of feed streams such that steam–carbon dioxide reforming of methane to produce hydrogen is an important research area with 12% of the total count [24,30,31,38,53–74]. Alternative feed streams are also studied such as methanol, DME, natural gas, and LPG [22,76,85–95].

Utilization of CO_2 by addition into feed stream of steam reforming has gained interest due to the fact that CO_2 is a major air pollutant. Therefore, conversion of CO_2 into valuable chemicals is a favored subject in environmental and economical point of view. In addition, carbon deposition over the catalyst is highly favorable for steam reforming and carbon dioxide reforming processes. Combined reforming processes with steam and carbon dioxide can overcome carbon deposition problem, as well as the composition of the produced syngas can be adjusted by changing feed gas composition [96–102].

In some processes, oxygen may be added to the feed stream for partial oxidation of methane. In such a case, since partial oxidation of methane is an exothermic reaction, this reaction may provide heat for steam reforming reactions, which are endothermic. In this way, the whole process may become self-heating and the energy consumption may be reduced. The mole ratio of the produced syngas can be adjusted to be between 1.5 and 2.0 by varying the composition of the feed gas stream. In addition, steam and oxygen present in the feed stream can react with carbon that is formed during the process so that coke deposition over the catalyst may be reduced to prolong catalyst life [103–105].

9.4.2 Developments in Terms of Type of Active Metals Used

Target breakdown of "active metals" used for H_2 catalysts are presented in Figure 9.11. For the steam reforming process to produce H_2, nickel is known to be the most widely used active metal for the steam reforming catalysts. Within the investigated patents, it is confirmed that nickel is the most used metal as the active component with 54% of the total count. Transition metals and noble metals have also gained much interest as the active metal with 26% and 13% of the total count, respectively. Some of the studies have been conducted by rare earth metals as the active metals with 7%.

Companies like Sued Chemie, Johnson Matthey, and China Petroleum & Chemical Corporation involve nickel in their studies [12,28,37,38,106,108]. It is seen in the studies that nickel is the most widely selected active metal for steam reforming catalysts to produce hydrogen [7–15,20–24,28,30,31,33–35,37–39,41–43,45,46,48,51,53–56,58–62,65–75,77–84,92,96–98,100–103,105,106,108].

Applicants such as Sued Chemie, Kyoto University, JX Nippon Oil & Energy Corporation, and SK Innovation used noble metals as active metals [19,27,36,88]. Noble metals such as platinum and palladium may also be used as the active metal

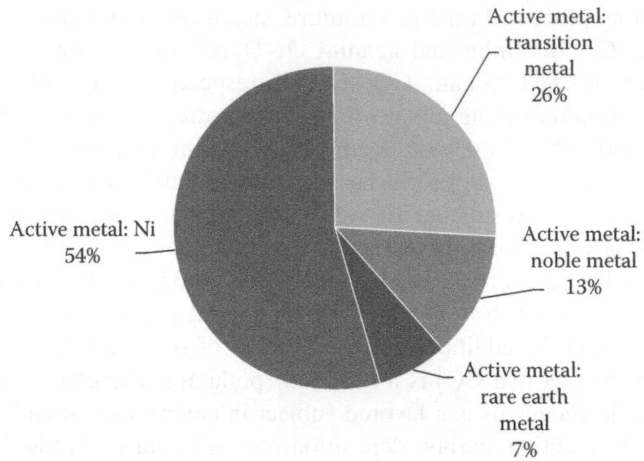

FIGURE 9.11 Target breakdown of "active meals" used for H_2 catalysts.

for steam reforming catalysts and are known to be highly resistant to sulfur poisoning and agglomeration [7,16,19,25–27,36,47,49,62,82,83,88,95,104,107].

Apart from nickel, transition metals such as iron, zinc, zirconium, and copper may also be used as active metals. It is stated in many studies that noble metals are expensive and economically unfavorable. Therefore, utilization of transition metals alternative to nickel and noble metals has gained importance. In some studies, these transition metals are reported to show higher activity and stability than nickel-based catalysts [16,32,34,40,44,48,62,82,85,91,107]. Universities such as Dalian Institute of Chemical Physics, Chinese Academy of Sciences, and Tianjin University have published some studies related to utilization of these metals. Rare earth metals are also known for high coke resistance capability; therefore, these metals are also used as active metals in some studies [21,32,38,44,48,56,82,83,107].

9.4.3 Developments in Terms of Type of Additive/Promoter Used

In Figure 9.12, the distribution of different types of metals for additive/promoter purposes is shown by investigating the patents that include an additive/promoter within the structure of the catalysts. Alkaline earth metals are preferred mostly as additive/promoter metals with 37%, followed by rare earth metals with 28%. Alkali, transition, and noble metals are also used as additive/promoter metals with similar percentages.

Additives and promoters are utilized in order to improve resistance to carbon formation and metal dispersion that in turn improves activity and selectivity. Alkaline earth metals, such as magnesium oxide and calcium oxide, may be involved in catalyst structure [8,21,24,29–31,33–35,37,44,48,53,54,59,61–63,65,68,71,96,97,99,105,108]. In a study of University of Science & Technology of China, magnesium oxide is used as a promoter with 5–15 wt.%, and it is reported that MgO as a promoter reduces coke formation and improves catalyst life. It is stated in some studies that reduction of magnesia promoted catalysts and maintaining a reduced state is difficult. In

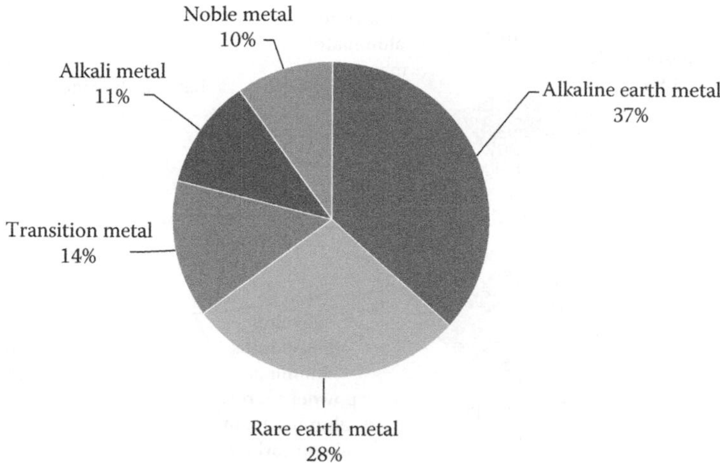

FIGURE 9.12 Target breakdown of "additive/promoter" used for H_2 catalysts.

addition, it is known that calcium oxides are also good promoters against coke formation; however, catalysts with rich calcium content are not favorable since they can damage the structure of the catalyst due to easy hydration capability [71].

Rare earth metals are also used for the purpose of improving resistance to coke formation [7,12,13,20,22,26,35,38,41,43,46,53,57,71,86,92,98–100,105]. In a study of Sued Chemie, nickel impregnated on calcium aluminate support material contains rare earth metals as a promoter for the production of syngas with a low H_2/CO ratio of less than 2.3. In an exemplar catalyst, lanthanum nitrate is impregnated on the support with 6.3 wt.% of lanthanum content; then, nickel is impregnated with 8.2 wt.% of nickel content. The proposed catalysts show a high activity index in whether or not sulfur is present in the feed stream. Lanthanum promoted catalysts present higher activity than that of titanium promoted catalysts [38].

In many studies, alkali metals are used as promoters [20,21,47,53,69,70,72,93]. However, alkali metals such as potassium may break down from the catalyst structure into base formation due to high temperature and water vapor partial pressure. This may cause corrosion in the equipment due to precipitation and also may cause deactivation of the catalyst because it covers the active sites of the surface. Therefore, it is important to synthesize the catalyst without the release of alkali metal components [37].

Transition metals like iron, titanium, and manganese may be considered to improve thermal stability, resistance to sulfur poisoning, and coke formation of the catalysts [19,20,27,61,72,76,88,95,99,103]. Apart from carbon resistance and thermal stability, the activity of the catalyst may be improved by addition of noble metals such as Pt, Pd, and Rh as promoters [15,29,33,38,59,69,97].

9.4.4 Developments in Terms of Type of Catalyst Support Material Used

The distribution of the type of the support materials of the catalysts is shown in Figure 9.13. According to the given figure, most of the support materials are of alumina

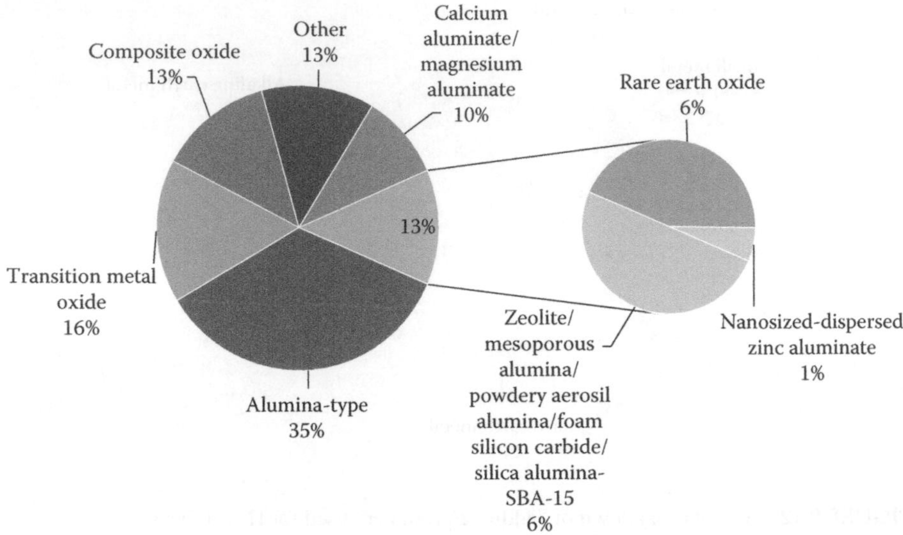

FIGURE 9.13 Target breakdown of "catalyst support material" used for H_2 catalysts.

type with 35% of the total count, which includes γ-alumina, θ-alumina, α-alumina, η-alumina, and hexaaluminate [8,10,13,14,16,20–23,26,28,29,33–36,50,53,55,56,58–62, 71,74–81,83,86,88,89,92,94,102,107,108]. Transition metal oxides are also studied within the patents with 16% that include oxides of zinc, zirconium, and titanium [16,18,25,35,36,44,47,49,51,52,63,65–68,70,83,85,94,105]. The group of composite oxides is created for the support materials that include more than one metal oxide group such as zirconium oxide–cerium oxide with 13% of the total count [9,24,27,30–32,34,45,51,52,69,74,87,91,98,100]. The support material in the form of clay type, mineral type, hydrotalcite type, perovskite type, carbon-based, and ceramic carrier are classified as "other" with 13% of the total count [9,17,19,43,44,46,48,57,58,64, 66,72,73,96,97,99]. Support materials like calcium aluminate, magnesium aluminate, mesoporous materials, and rare earth oxide are also investigated within the patents. Support materials like calcium aluminate and magnesium aluminate are 10% of the total count [11,12,15,22,33,37,38,43,54,56,70,106]. The support materials like zeolite, mesoporous alumina, powdery aerosil alumina, foam silicon carbide, and silica alumina-SBA-15 are collected in a group that constitutes 6% of the total count [39,41,78,82,84,94,95,104]. Rare earth oxides and nanosized-dispersed zinc aluminate support materials are also used with 6% and 1% of the total count, respectively [18,49,50,52,60,67,85,90].

9.4.5 Developments in Terms of Catalyst Structure

In Figure 9.14, the patents are investigated in terms of developments within the catalyst structure of H_2 catalysts. According to the figure, most of the studies are performed in order to improve the textural properties of the catalysts by synthesizing

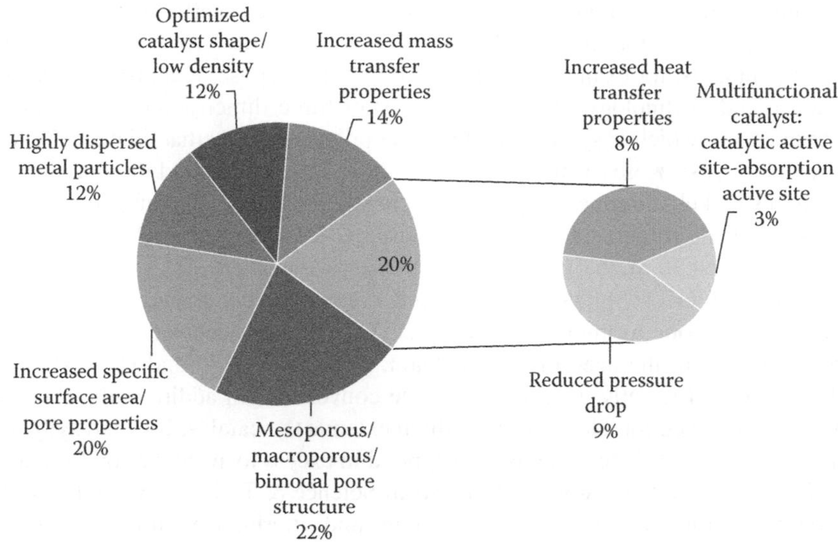

FIGURE 9.14 Target breakdown of "catalyst structure"–related patents for H_2 catalysts.

the catalysts with mesoporous, macroporous, or bimodal pore structure to increase mass transfer properties. In some patents, optimized catalyst shape is studied in order to reduce pressure drop within the reactor, to decrease flow resistance ensuring undisturbed flow pattern, and to improve heat transfer properties. In some patents, multifunctional catalyst structures are studied that include a catalytic active site for reforming process and an absorption active site for absorption of CO_2 by-product.

Mesoporous materials especially mesoporous molecular sieves and macroporous and bimodal materials have gained importance due to their large surface area, pore size, and controllable pore volumes [15–18,35,41,74,75,77–79,81,93]. According to the study of Tianjin University, pure SBA-15 mesoporous molecular sieve, which has one-dimensional ordered pore structure, high surface area, thick pore walls, adjustable pore size within 5–30 nm range, and high thermal stability, is used as support material. As nickel is loaded on SBA-15, the growth of nickel particles is suppressed due to the confinement effect within the pores of SBA-15 that prevents sintering of nickel particles [41].

The catalyst with calcium aluminate support and nickel as the active metal is used in a study of Air Products and Chemicals. The catalyst has a bimodal pore structure having pores smaller than 50 nm to be 30.4% of the total porosity. The pores with large diameters provide easy flow of the reactant and product through the catalyst pellet, while the pores with smaller diameters increase the surface area for the uniform distribution of the active metals [93].

The catalyst has an alumina support with macroporous and bimodal pore structure in the study of Chengdu Organic Chemicals. Pores greater than 300 nm constitute 65% of the total pore volume. Due to the bimodal pore structure of the catalyst, the catalyst can operate at a low water/carbon ratio (H_2O/CH_4) of 2.5 without formation of carbon [35].

In some studies, different methods are used to increase specific surface area and the pore properties of the catalysts [18,36,55,73,77,89,102,106–110]. Different surfactant and templates, for example, can be used for this purpose. Kunming University of Science and Technology proposes a catalyst of three-dimensional, macroporous Ce–Zr material, which is synthesized by using poly(methyl methacrylate) (PMMA) as a template [18]. Korea University Research & Business Foundation prepares a mesoporous nickel–alumina composite by using an anionic surfactant [77].

Preserving the dispersion of active metal particles is important for stable catalytic activity [24,30,31,36,41,55,80]. According to the study of Japan Petroleum Exploration, MgO or MgO/CaO as promoter converts the active metal, cobalt, into composite oxide to improve dispersion. Therefore, the deposition of carbon is suppressed without the need of excess steam, and syngas (CO and H_2 mixture) is produced more efficiently with high methane conversion. In addition, the catalytic activity is conserved for a long period, which extends the catalyst life, and syngas is produced at low cost since steam is not supplied in excess to avoid carbon formation [31]. National Institute of Advanced Industrial Science & Technology synthesized a catalyst containing alumina, zirconia, cerium, and ytterbium with coprecipitation method and reported that the catalyst is highly resistant to sintering [87].

Promotion of mass transfer properties and reduction of pressure drop are interconnected with optimization of the catalyst shape [18,22,28,43,55,58,87,101–104,106–110]. The subject of optimization of the catalyst shape is studied with different methods by applicants like Johnson Matthey and Sued Chemie [28,106–108]. For the catalyst with conventional shapes, an increase in pressure drop due to packing of small pellets is a common problem with steam reforming processes, and the reaction yield decreases as a consequence. Therefore, in order to overcome this problem, researchers have focused on optimizing the catalyst shape. These optimized catalyst shapes can be found in the form of pellets with multiple holes and curved exteriors. Thereby, the effective geometric surface area of the catalyst is increased, and as the catalyst is loaded in a reactor, the flow pattern is kept undisturbed by enabling alternative pathways due to presence of additional pores. This, in turn, reduces the mass transfer resistances and pressure drop within the reactor as well. Heat transfer properties can be increased by optimizing the catalyst shape as well [18,28,43,55,109]. In a study by Johnson Matthey, the catalytic material containing α-alumina support is molded into a specific shape with tetralobes and sprayed by a slurry solution that contains active metals to obtain the final coated egg-shell-type coated catalyst [107]. It is highlighted in a study of Sued Chemie that multiple holes within the pellet may decrease the mechanical strength of the catalyst. The proposed catalyst is in a honeycomb shape with lateral pressure resistance of at least 700 N and prepared by injection molding of the catalyst carrier of calcium aluminate and impregnation of active material nickel [106].

In some of the studies, a novel, multifunctional catalyst with a catalytic active site and an absorption active site is proposed [11,101]. East China University of Science and Technology synthesized a catalytic site that is composed of calcium aluminate support and nickel as the active metal. The absorption active site is provided by the presence of calcium oxide to absorb CO_2 for a product stream with high H_2 purity.

The proposed catalyst preserves its catalytic activity and absorption capacity after 50 absorption–regeneration cycles [11].

9.4.6 Developments in Terms of Stability Enhancements

In Figure 9.15, the patents are classified according to the objective of stability enhancement of the H_2 catalysts. The major concern for the catalysts of steam reforming process to produce hydrogen is formation of coke over the catalyst, which leads to deactivation of the catalyst. Therefore, the development of novel catalysts with resistance to coke formation is an important focus point among research studies. Within the subject classification, resistance to coke formation takes place with 31% followed by development of a more stable catalyst target with 26%. There are some patents in which the catalysts are developed to operate in different operation conditions, such as low temperature, wide temperature range, high temperature, and high pressure. Improvement in thermal stability, catalyst life and mechanical resistance, resistance to sintering and sulfur poisoning, and conversion of coke to CO/CO_2 are the subjects that are concerned within the patents.

During steam reforming process, formation of coke is highly favorable in process conditions. The formed coke deposits on the surface of the catalyst covering the active sites, which in turn causes deactivation of the catalyst. Further deposition of coke also plugs the pores of the catalyst and damages the structure of the catalyst. Due to large accumulation of coke, the contact of gaseous streams with the catalyst may decrease so that the reforming reaction rate decreases as well. The formation

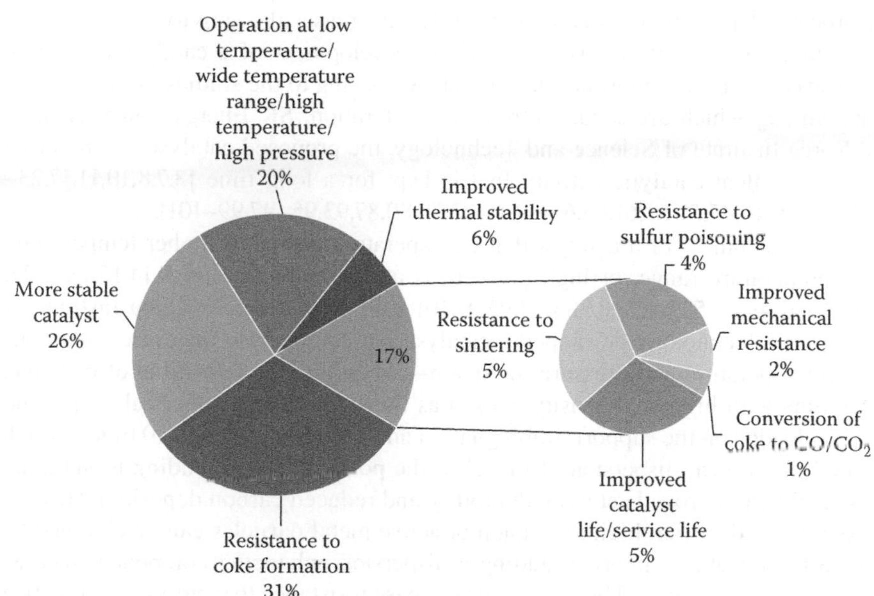

FIGURE 9.15 Target breakdown of "stability enhancement"–related patents for H_2 catalysts.

of carbonaceous materials may be reduced and prevented by using excess amount of steam in the feed; however, this method is not preferable due to high energy costs. Therefore, the most efficient pathway to overcome this problem is synthesizing the catalyst to be highly resistant to coke formation. Researchers propose novel catalysts in which promoters are added during synthesis of the catalysts that provide very high dispersion of the active metals to increase resistance to coke formation, or noble metals may be used as active metals [8,9,20–22,27,33,35,37,38,41,44–46,48–50,52,56–58,61,63–66,68–70,92,93,96–99,101,103,105].

In a study by Korea Institute of Science and Technology, the zirconia support is doped with yttrium and a metal of lanthanum and Ce, or an alkaline earth metal, Ca and Mg. It is stated that yttrium is bonded to the crystal lattice of zirconia to deform the lattice and enables transfer of an oxygen ion. As carbon is deposited on the nickel surface of the catalyst, it reacts with the released oxygen ion from the support and is converted into CO or CO_2 and separated from the nickel surface. Yttrium is used in the range of 5–10 mol% relative to zirconia for the optimum mobility of oxygen ion. The lanthanum and the alkaline earth metal are used in the range of 1.5–3 wt.% relative to zirconia for increasing activity and stability of the catalyst together with yttrium [105].

Another approach to reduce coke deposition is the addition of potassium during the synthesis of the catalyst. Potassium additive enables gasification of coke to avoid deactivation due to pore plugging of the catalyst. In a study of Air Products and Chemicals, 0.4–5 wt.% of potassium is added on the calcium aluminate support with 1–20 wt.% of nickel as the active metal [93].

Stability and life of the catalyst are related to the resistance of the catalysts against carbon formation. Since coke formation is a major problem in steam reforming process, deposition of coke on the catalyst decreases the activity of the catalyst and reduces its stability as well. Therefore, development of a catalyst with more stable structure is an important focus point. According to the studies of many applicants, among which are Japan Petroleum Exploration, SK Energy, Sued Chemie, and Korea Institute of Science and Technology, the proposed catalysts are reported to have excellent catalytic activity that is kept for a long time [3,7,8,10,11,17,23–28,30,34,38,40,45,51,52,61,63,64,69,71–73,79,80,87,93,95–97,99–101].

The development of a catalyst that can operate at lower or higher temperature, wide temperature range, or high pressure is also investigated [9,10,14,15,17,19,25,27,32,40,47,49,50,52,64,71,85,86,94,95,97,101,105]. For instance, Korea Institute of Science and Technology developed a catalyst that can operate in severe conditions of high temperature and pressure for steam–carbon dioxide reforming of methane. η-Alumina with high acid density is used as the support with spherical shape, and nickel is loaded on the support. Impregnation and drying is repeated 10 times, which resulted in uniform dispersion of nickel in the pores and firm binding to alumina. This results in improved catalytic durability and reduced carbon deposition [101].

As indicated earlier, high dispersion of active metal particles can be obtained by using a more stable support or adding a dispersion-enhancing component such as magnesium and calcium. These methods increase resistance to sintering of the active metals [8,41,48,54,87,98]. Agglomeration of the active metals can also be inhibited by using metals with high melting temperature such as noble metals Pt and Pd, as discussed in the study of Sued Chemie and Air Liquide [36].

For the conventional nickel-based catalysts, nickel is affected by the presence of sulfur even in trace amounts of ~1 ppm, and the catalyst is deactivated by sulfur poisoning. Therefore, sulfur present in the gaseous streams has to be removed by hydrodesulfurization process [16,27,36,66,94]. One other solution is to synthesize a sulfur-tolerant catalyst. In a joint study of Sued Chemie and Air Liquide, it is stated that the sulfur tolerance of active metals is determined with respect to their affinity to metal sulfide formation. Platinum, palladium, and iridium have low affinity to sulfur as compared to nickel and ruthenium. In their study, platinum, ruthenium, and iridium are used as active metals and impregnated on commercially available calcium aluminate/aluminate. 1 wt.% of Rh, Pt, and Ir is loaded on the support, and their performances are compared. Rh-containing catalyst shows higher catalytic activity as compared to Pt- and Ir-containing catalysts using the feed stream without contamination with sulfur or oxygen. On the other hand, after aging the catalyst for 100 h, the activity of Ir-containing catalyst is higher as compared to that of Rh- and Pt-containing catalysts with the feed stream containing sulfur [36]. In some studies, the catalysts that are developed are proposed to have improved thermal stability [19,23,25,55,62,72,99], to have reduced deactivation [97,99], and to have improved mechanical resistance [100,101].

9.4.7 Developments in Terms of Cost-Effective, Efficient, or Alternative Preparation of Material

The distribution of developments in the catalysts in terms of the subjects within "cost-effective, efficient, or alternative preparation of material" classification is shown in Figure 9.16. 51% of the patents state that the proposed catalysts are improved with

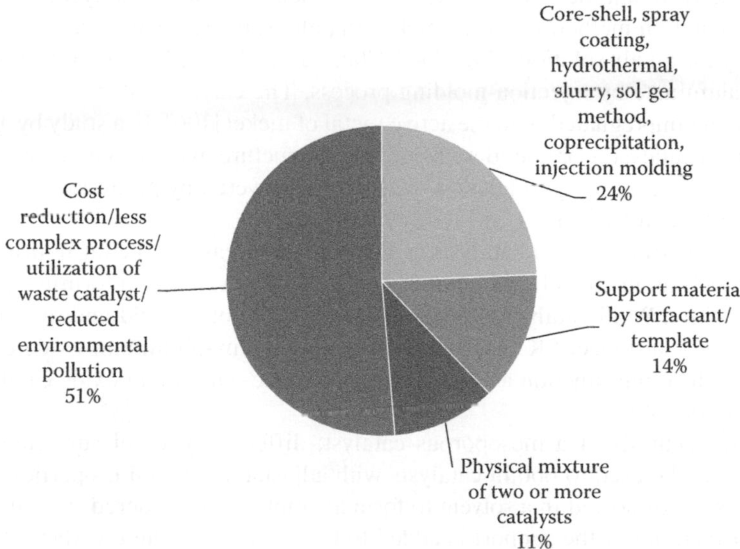

FIGURE 9.16 Target breakdown of "cost-effective, efficient, or alternative preparation of material"–related patents for H_2 catalysts.

respect to cost reduction, less complex process, utilization of waste catalyst, and reduced environmental pollution. Different types of synthesis procedures are also followed for the developed catalysts, which constitute 24% of the patents. In some patents, the support materials are synthesized by using a surfactant material that contributes 14% of this classification and followed by investigation of physical mixture of two or more catalysts with 11%.

Many of the researchers report that the proposed method of catalyst preparation is simple and efficient with low cost [9,10,12–14,40,42,44–46,58,60,64,66,69,76,84]. China Petroleum & Chemical Corporation published a patent in which a waste catalyst is utilized to prepare a novel catalyst with low cost. The waste catalyst is recycled by preprocessing, coprecipitation–wet mixing, and steam curing after-treatment method to obtain a NiO/CaO–Al_2O_3 type catalyst. The study reports that 95% of the waste catalyst is recovered with this method with lowered production cost within 10–20%. In addition, the physical and chemical properties, activity, and stability of the proposed catalyst are found to be similar to those of the catalyst prepared with conventional methods. Another benefit is that environmental pollution that is caused by the waste catalysts may be reduced by utilization of the waste catalyst [12].

Researchers propose different catalyst preparation methods in order to increase structural properties, stability, and activity of the catalyst [39,42,43,51,55,71,79,81,106]. In a study of Seoul National University Industry Foundation, mesoporous alumina support is synthesized by hydrothermal method with a cationic surfactant; then nickel is impregnated on the mesoporous alumina support [81]. SNU R&DB Foundation synthesized a mesoporous nickel–alumina–zirconia xerogel catalyst by sol-gel method in which an epoxide-based compound is added to a solution of alumina, zirconium, and nickel precursors followed by hydration, gelation, drying, and calcining [39]. In another study, a mesoporous nickel–alumina catalyst is prepared by coprecipitation method in which nickel and alumina precursors are precipitated by adding an alkali solution [79]. Sued Chemie synthesized a catalyst carrier of calcium aluminate by injection-molding process. The carrier is then coated with a washcoat and impregnated with the active metal of nickel [106]. In a study by Tianjin University, a catalyst is prepared with core-shell structure, which consists of magnesium aluminate core in hydrotalcite-type structure covered by perovskite-type shell containing La and Ni [43].

Physical mixture of two catalysts is considered in case of steam reforming of alcohols such as methanol and ethanol [49,52,61,88]. A steam reforming catalyst is mixed with a WGS catalyst in order to increase hydrogen yield and reduce coke formation. For instance, SK Innovation used a platinum–aluminum oxide catalyst for steam reforming function and a copper–zinc oxide–aluminum oxide catalyst for WGS function [88].

For the synthesis of a mesoporous catalyst, different types of surfactants and templates can be used to obtain catalysts with adjustable textural properties. These surfactants are dissolved in a solvent to form a template with ordered structure. The precursor material of the support is added to the surfactant solution where the precursor is located on the template. Upon calcination, the template is burned off to give the support material with ordered structure. Kunming University of Science and Technology used PMMA as a template [18]; SNU R&DB foundation used pluronic-based or tetronic-based block copolymer including F108, F98, F88, P123, P105, and P104 as a surfactant [74,78]. Korea University Research & Business Foundation and Seoul National University Industry Foundation used an anionic and a cationic surfactant, respectively [77,81].

9.4.8 Developments in Terms of Product Yield/Selectivity

In Figure 9.17, the patents are investigated in terms of product yield/selectivity. As seen in the figure, improvement of activity or selectivity of the catalyst is the major focus subject with 71% [8–11,13–16,18,20,21,24,32,36–38,40–42,44–46,48,51,53,54,56, 58–62,64,67,69–71,73,74,85,89–91,94,95,98,99,101,102,105,110]; 22% of the patents report an increase in H_2 yield [11,18,27,32,39,47,49–52,63,64,67,71,76,90], while the remaining patents report reduction of CO_2 yield [11,34,42,99,105].

In the study by Sued Chemie and Air Liquide, the activities of the synthesized catalysts are compared in terms of activity coefficient values. When comparing the activities of 4 wt.% of Ir impregnated on pure monoclinic zirconia support and calcium aluminate/alumina support, the activity coefficient of the catalyst with zirconia support is found to be higher than that of the catalyst with calcium aluminate/alumina support [36]. Sued Chemie also prepared a catalyst of strontium containing calcium aluminate with nickel as the active metal. The catalyst showed higher catalytic activity for a long period of time than the conventional nickel–alumina or nickel–calcium aluminate catalysts [37]. The catalyst proposed by Jiangsu University of Science and Technology is composed of nickel and magnesium oxide in which the nickel content varies from 6.9% to 43.4%. All of the catalysts are reported to have high catalytic activity with methane conversion of 95–98% and carbon dioxide conversion of 52–54% under the conditions of 875°C and H_2O/CH_4 of 1 [14].

Guangzhou Institute of Energy Conversion, Chinese Academy of Sciences, proposed a catalyst that contains γ-Al_2O_3 as support with Cu as the active metal and

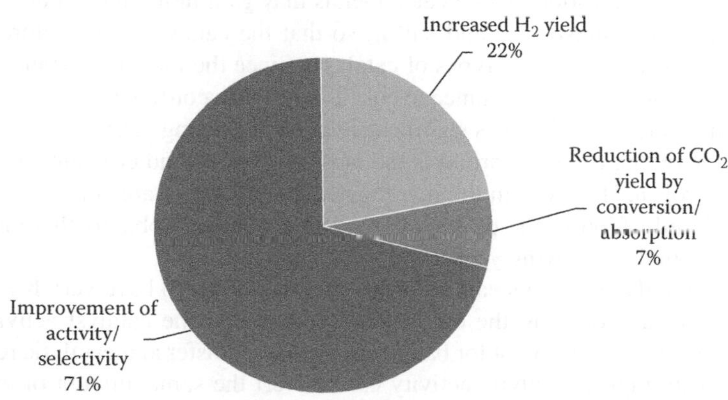

FIGURE 9.17 Target breakdown of "product yield/selectivity"–related patents for H_2 catalysts.

Fe as the promoter. The catalyst is used for steam reforming of DME and converts DME completely with H_2 yield above 93% [76]. For steam reforming of methane, Kunming University of Science and Technology prepared a catalyst with Ce–ZrO_2 support containing iron as the active metal. The conversion of methane is higher than 85%, and the H_2 to CO mole ratio in the produced synthesis gas is found to be about 2 [32].

Zhejiang University synthesized a composite catalyst that has a catalytic function and an absorbing function due to presence of CaO. The absorbing function is used for absorption of carbon dioxide. For the catalytic function, alumina is used as support and nickel and zirconia precursors are used as active metals. The synthesized catalysts provide methane conversion of at least 90% and product stream containing hydrogen between 93% and 95%, carbon monoxide between 0.77% and 1.5%, and carbon dioxide between 0.03% and 1.05% [34].

9.5 FUTURE PROSPECT IN H_2 PRODUCTION TECHNOLOGY

It is expected that the refineries may process crude oil with heavier and sourer characteristics, and the environmental regulations may become stricter in the future. In addition, different types of feedstock alternative to natural gas are becoming considered in hydrogen plants. These would increase the demand for hydrogen. The hydrotreatment processes would require more hydrogen in order to remove the impurities from the feedstocks prior to the subsequent catalytic processes that convert the feed into valuable products. The catalytic processes that consume hydrogen would require a higher amount of hydrogen as well to convert heavier hydrocarbons into liquid fuels in order to meet sales specifications. In addition, in some processes, the produced liquid fuel is further upgraded by using hydrogen. This indicates that hydrogen production may remain its importance in the following years.

The development of a catalyst with high activity and resistance to coke formation and sulfur poisoning is still a challenge [4]. Currently, nickel-based catalysts seem to be prevailing for steam reforming process. However, other types of active metals selected from transition or rare earth metals may gain more interest due to their higher resistance against coke and sulfur, so that the catalyst manufacturers may lead their focus more on these types of catalysts. Since the most important concern in this aspect is to increase resistance to coke formation according to the investigated patents, this subject may keep its significance in the following years.

Steam reforming of hydrocarbon is the most acceptable and efficient pathway to produce hydrogen. However, in the investigated patents, there are some studies that include CO_2 utilization in the process to overcome coking problems that may gain more interest in the following years.

Even though the existing steam reformer catalysts seem to have very high activity in terms of approaching the equilibrium conversion, the catalyst activity will remain as the major focus point for balancing the heat transfer and catalytic reaction. A catalyst with higher catalytic activity can convert the same amount of methane at a lower temperature into products at a given heat flux. In other words, for higher catalyst activity, the reactor temperatures, i.e., reactor tube wall temperatures, would be lower, which in turn would increase the operational life of the reactor tubes [4].

Annual distributions of the investigated patents between 2008 and 2014 in terms of the major topic classification are given in Figure 9.18. The main study area for hydrogen production catalysts was stability enhancement of the catalyst and was increasing throughout the years except for 2011. As emphasized earlier, the catalyst stability is important in order to obtain high catalytic activity for a longer period of time, which means longer catalyst life. Product yield and selectivity also gained interest and became one of the main focus points within the studies. Since the demand for hydrogen is increasing, researchers turn their attention to developing catalysts that would produce hydrogen in higher amounts and purity. Accordingly, the researchers have investigated different types of promoters in order to improve the stability of the catalysts in terms of resistance to coke formation and sulfur poisoning. Several studies have investigated different types of catalyst preparation methods and catalyst structures.

To conclude, the investigated studies show that steam reforming catalysts for hydrogen production have been developed in order to compensate for the demand for catalysts with higher activity and selectivity. To achieve that, novel catalyst designs and compositions have gained much interest that would maintain activity and stability with higher selectivity being insusceptible to severe operation conditions for a longer time. Therefore, development of steam reforming catalysts that could comply with different types of feedstock may remain as a focus of interest.

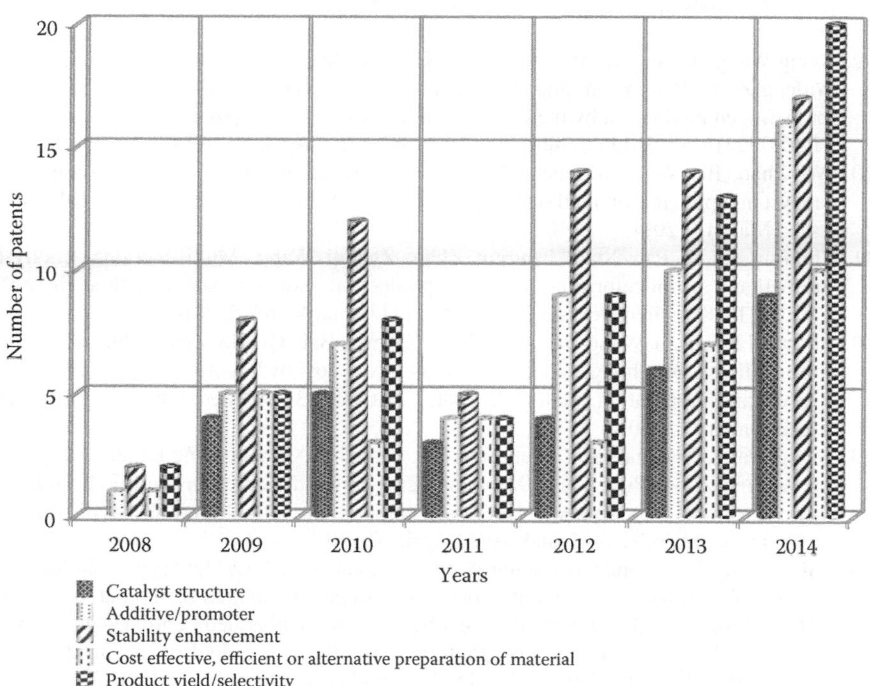

FIGURE 9.18 Annual distributions of the investigated patents in terms of the topic classification.

REFERENCES

1. James G. Speight, Baki Özüm, *Petroleum Refining Process*, Marcel Dekker Inc., New York, 2002, pp. 639–655.
2. James G. Speight, *The Desulfurization of Heavy Oils and Residua*, Marcel Dekker Inc., New York, 1999, 2nd Edition, pp. 387–396.
3. Jean-Louis Navarre, Hydrogen in the refinery an overview, 11th European Technical Seminar on Hydrogen Plants, 2010, Istanbul, Turkey.
4. Jens R. Rostrup-Nielsen, Thomas Rostrup-Nielsen, *Large-Scale Hydrogen Production*, Topsoe Technologies, 6th World Congress of Chemical Engineering, 2001, Melbourne, Australia.
5. Roland van Uffelen, The hydrogen plant flowsheet, 11th European Technical Seminar on Hydrogen Plants.
6. Peter Broadhurst, *The Fundamentals of Steam Reforming—Chemistry and Catalysis*, Johnson Matthey, 11th European Technical Seminar on Hydrogen Plants, 2010, Istanbul, Turkey.
7. Jin Soon Choi, Yun Jo Lee, Kyong Tae Kim, Ji Eun Min, Hae Goo Park, Ki Won Jun, Catalyst for the steam reforming of methane, method for preparing thereof and method for the steam reforming of methane in by-product gas containing sulphur using the catalyst, KR20140087239 A, filed November 28, 2012, and issued July 9, 2014.
8. Jin Soon Choi, Yun Jo Lee, Kyong Tae Kim, Ji Eun Min, Hae Goo Park, Ki Won Jun, Catalyst for the steam reforming of methane, method for preparing thereof and method for the steam reforming of methane included in by-product gas of iron manufacture using the catalyst, KR101386418 B1, filed December 28, 2012, and issued April 21, 2014.
9. Xiang Wang, Youhe Ma, Wenming Liu, Xianglan Xu, Xiuzhong Fang, Changqing Li, Wufeng Zhou, Ping Yuan, Xiaohong Chen, Anti-carbon-deposition Ni-based catalyst for hydrogen production by methane steam reforming and preparation method thereof, CN103752319 A, filed December 31, 2013, and issued April 30, 2014.
10. Wei Zhao, Bin Wu, Dongsheng Wang, Low-temperature methane steam reforming catalyst and preparation method thereof, CN103611541 A, filed December 6, 2013, and issued March 3, 2014.
11. Zhiming Zhou, Pan Xu, Changjun Zhao, Zhanqi Wang, Multifunctional catalyst for methane steam reforming hydrogen production and preparation method thereof, CN103611538 A, filed November 29, 2013, and issued March 5, 2014.
12. Jianbo Jiang, Hao Wang, Hongxia Xue, Zhimin Bai, Qinglu Zhao, Huandong Qi, Xiuling Jiang, Weizhong Liang, Hantao Yu, Low-cost hydrocarbon steam-reforming catalyst and preparation method thereof, CN103506134 A, filed June 27, 2012, and issued January 15, 2014.
13. Xiang Wang, Xiaojuan You, Wenming Liu, Xianglan Xu, Xiuzhong Fang, Changqing Li, Wufeng Zhou, Ping Yuan, Xiaohong Chen, Mixed rare-earth modified methane steam reformed nickel-based catalyst and preparation method thereof, CN103752320 A, filed December 31, 2013, and issued April 30, 2014.
14. Zhibin Yang, Weizhong Ding, Fang Wang, Application of NiO/MgO solid solution catalyst in coke oven coal gas steam reforming hydrogen production method at low water carbon ratio, CN102935998 A, filed October 11, 2012, and issued February 20, 2013.
15. Johannis Alouisius Zacharias Pieterse, Catalyst for hydrogen production, US20130302241 A1 filed November 15, 2011, and issued November 14, 2013.
16. Takami Susumu, Otsuka Hirofumi, Hirano Taketoku, Steam reforming catalyst for hydrocarbon compound and method for manufacturing the same, JP2013136047 A filed October 18, 2012, and issued July 11, 2013.

17. Ji Ho Yoo, Su Hyun Kim, Ho Kyung Choi, Dong Hyuk Chun, Young Joon Rhim, Jeong Hwan Lim, Si Hyun Lee, Sang Do Kim, A catalyst for steam reforming of hydrocarbons and method of steam reforming of hydrocarbons using the same, KR101306196 B1, filed April 10, 2013, and issued September 9, 2013.
18. Xing Zhu, Yane Zheng, Hua Wang, Yonggang Wei, Kongzhai Li, Yuhao Wang, Method for preparation of three-dimensional ordered macroporous Ce-Zr material by using PMMA as template, CN103409653 A, filed August 13, 2013, and issued November 27, 2013.
19. Hideki Koyanaka, Masahiko Tsujimoto, Catalysts reforming methane gases into hydrogen and methods for synthesizing the same, and methods for reforming methane gases using said catalysts, US20130195750 A1, filed June 24, 2011, and issued August 1, 2013.
20. Anatolij Vasil'evich Obysov, Aleksej Viktorovich Dulnev, Svjatoslav Mikhajlovich Sokolov, Natalja Anatolevna Levtrinskaja, Svetlana Gennadevna Dormidontova, Maksim Anatolevich Obysov, Catalyst for steam reforming hydrocarbons and method of producing said catalyst, RU2446879 C1, filed October 7, 2010, and issued April 10, 2012.
21. Viktor Ivanovich Jagodkin, Ljubov Nikolaevna Shmakova, Galina Vladimirovna Fokina, Nikolaj Ivanovich Murashov, Aleksandr Jakovlevich Vejnbender, Natalja Anatolevna Levtrinskaja, Pavel Viktorovich Jagodkin, Catalyst for steam reforming of C1–C4 methane hydrocarbons and method of preparing said catalyst, RU2462306 C1, filed June 1, 2011, and issued September 27, 2012.
22. Diwakar Garg, Frederick Carl Wilhelm, Supported catalyst and use thereof for reforming of steam and hydrocarbons, US8877673 B2, filed November 8, 2011, and issued November 4, 2014.
23. Yi Jiang, Xiaoxia Zhang, Junhe Chen, Juanyun Wang, Preparation method of novel high temperature-resistant carrier and catalyst, CN102755911 A, filed April 25, 2012, and issued December 31, 2012.
24. Katutoshi Nagaoka, Yuusaku Takita, Toshiya Wakatsuki, Catalyst for hydrocarbon reforming, its production method, and method for producing synthetic gas using the same, EP2198960 A1, filed October 8, 2008, and issued June 23, 2010.
25. Hideki Koyanaka, Masahiko Tsujimoto, Catalyst for reforming of methane gas into hydrogen, method for synthesis of the catalyst, and methane gas reforming method using the catalyst, WO2012002283 (A1)—2012-01-05, filed in June 24, 2011, issued in May 1, 2012.
26. Jin Young Kim, Myung Jin Kha, Won Chul Chang, Catalyst for steam reformation and hydrogen production apparatus using thereof, KR20110011276 A, filed July 28, 2009, and issued February 8, 2011.
27. Yasuhito Ogawa, Reforming catalyst for use in producing hydrogen, method of producing the same, and method of producing hydrogen using the same, JP2011083685 A, filed October 14, 2009, and issued April 28, 2011.
28. David James Irdsall, Mileta Babovic, Mikael Per Uno Carlsson, Samuel Arthur French, Michiel Nijemeisland, William Maurice Sengelow, Edmund Hugh Stitt, Shaped heterogeneous catalysts, US8563460 B2, filed August 24, 2009, and issued October 22, 2013.
29. Yong Wang, David P. Vanderwiel, Anna Lee Y. Tonkovich, A method and catalyst structure for steam reforming of a hydrocarbon, CA2380869 C, filed August 15, 2000, and issued October 9, 2012.
30. Katutoshi Nagaoka, Yuusaku Takita, Toshiya Wakatsuki, Hydrocarbon-reforming catalyst and process for producing synthesis gas using the same, US8178003 B2 filed October 8, 2008, and issued May 15, 2012.
31. Toshiya Wakatsuki, Composite oxide for hydrocarbon reforming catalyst, process for producing the same, and process for producing syngas using the same, US8475684 B2, filed July 17, 2009, and issued July 2, 2013.

32. Cheng Xianming, Hu Jianhang, Li Kongzhai, Wang Hua, Wei Yonggang, Yan Dongxia, Zhu Xing, Oxygen carrier for preparing hydrogen and synthesizing gas by reforming steam through two-step method, CN101786605 B, filed December 25, 2009, and issued July 4, 2012.
33. Dong Ju Moon, Sang Deuk Lee, Byung Gwon Lee, Dae Hyun Kim, Yun Joo Lee, Ki Poong Na, Myung Jun Kim, Jae Suk Choi, Rhodium metal modified nickel based steam reforming catalyst using hydrotalcite-like precursor and the method for producing thereof and the method for producing hydrogen gas using thereof, KR101207181 B1, filed February 6, 2009, and issued November 30, 2012.
34. Wang Lulu, Wu Sufang, Composite catalyst, preparation and application thereof, CN101829577 A, filed April 23, 2010, and issued September 15, 2010.
35. Yi Jiang, Denghua He, Jiyuan Cheng, Junhe Chen, Juanyun Wang, Wei Jiang, Xiaoxia Zhang, Zuolong Yu, Method for preparing hydrocarbon steam conversion catalyst of high anticaking carbon, CN101371993 A, filed August 12, 2008, and issued February 25, 2009.
36. Franklin D. Lomax, Jr., John Lettow, Aaron L. Wagner, Jon P. Wagner, Duane Myers, Catalyst for hydrogen generation through steam reforming of hydrocarbons, US7923409 B2, filed October 18, 2007, and issued April 12, 2011.
37. Chandra Ratnasamy, Yeping Cai, William M. Faris, Jürgen R. Ladebeck, Nickel on strontium-doped calcium aluminate catalyst for reforming, US7771586 B2, filed June 11, 2007, and issued August 10, 2010.
38. Shizhong Zhao, Yeping Cai, Xiao D. Hu, Jon P. Wagner, Jürgen Ladebeck, R. Steve Spivey, Promoted calcium-aluminate supported catalysts for synthesis gas generation, US7767619 B2, filed July 9, 2004, and issued August 3, 2010.
39. In Kyu Song, Seung Ju Han, Yong Ju Bang, Hyun Joo Lee, Mesoporous nickel–alumina–zirconia xerogel catalyst and production method of hydrogen by steam reforming of ethanol using said catalyst, KR20140030799 A, filed September 3, 2012, and issued March 12, 2014.
40. Jianli Su, A catalyst used for hydrogen production by ethanol steam reforming and a manufacturing technology of the catalyst, CN103638932 A, filed November 7, 2013, and issued March 19, 2014.
41. Jinlong Gong, Chengxi Zhang, Di Li, Xiao Wang, Tuo Wang, Nickel-loaded SBA-15 catalyst modified by cerium dioxide and preparation method and application of nickel-loaded SBA-15 catalyst modified by cerium dioxide, CN104001538 A, filed January 12, 2014, and issued August 27, 2014.
42. Wenju Wang, Peng Zhang, Bing Xing, Preparation method of nickel-modified calcium-based dual-functional particles, CN103752256 A, filed January 15, 2014, and issued April 30, 2014.
43. Yongdan Li, Guangming Zeng, Jingjing Shao, Preparation and application of shell-core-type perovskite-wrapping hydrotalcite-like-based oxide reforming hydrogen production catalyst, CN103111302 A, filed January 12, 2013, and issued May 22, 2013.
44. Yuan Liu, Zijun Wang, Lanthanum calcium iron cobalt calcium titanium ore type catalyst for oxidizing and reforming ethanol and method for preparing catalyst, CN102941099 A, filed November 22, 2012, and issued February 27, 2013.
45. Yuan Liu, Lijuan Zhang, Muzhaozi Yuan, Tingyu Xu, NiO/Cel-xPrxO_2 catalyst for ethanol steam reforming reaction and preparation method thereof, CN102500385 A, filed October 26, 2011, and issued June 20, 2012.
46. Ning Li, Lingyun Mo, Preparation method of catalysts for hydrogen production from ethanol steam reforming, CN102500384 A, filed October 18, 2011, and issued June 20, 2012.

47. Wen-Feng Wang, Jia-Lin Bi, Chen-Bin Wang, Chuin-Tih Yeh, Wenfeng Wang, Jialin Bi, Chenbin Wang, Chuintih Yeh, Process for producing hydrogen from ethanol under low temperature, TWI391321 B, filed July 1, 2010, and issued April 1, 2013.
48. Chen Shunquan, Liu Yuan, Perovskite catalyst used for steam reforming of oxygenated ethanol and preparation method thereof, CN101822989 B, filed May 19, 2010, and issued May 23, 2012.
49. Mori Nobuhiko, Nakamura Toshiyuki, Catalyst for reforming ethanol to produce hydrogen and method for producing hydrogen, JP2009000667 A, filed June 25, 2007, and issued January 8, 2009.
50. Mori Nobuhiko, Nakamura Toshiyuki, Method for producing hydrogen and ethanol reforming catalyst, JP2009102209 A, filed October 25, 2007, and issued May 14, 2009.
51. In Kyu Song, Min Hye Youn, Jeong Gil Seo, Nickel–zirconium–titanium complex metal oxides catalyst, preparation method thereof, and method for producing hydrogen by autothermal reforming of ethanol using the same, KR100836510 B1, filed March 8, 2007, and issued June 9, 2008.
52. Nobuhiko Mori, Toshiyuki Nakamura, Catalyst for reforming ethanol to produce hydrogen and method for producing hydrogen, JP5242086 B2, filed June 25, 2006, and issued July 24, 2013.
53. Jimmy A Faria, Jon M Nelson, Uchenna P Paul, Danielle K Smith, Renewable hydrogen production catalysts from oxygenated feedstocks, WO2014078119 A1, filed November 5, 2013, and issued May 22, 2014.
54. Dong Ju Moon, A Rong Kim, Hye Yong, Lee, Jong Wook Bae, Supported perovskite type catalysts for combined steam and carbon dioxide reforming with methane, KR101447681 B1, filed November 28, 2012, and issued October 7, 2014.
55. Dong Ju Moon, Tae Gyu Kim, Dea Il Park, A high-durability metal foam supported catalyst for competitive reforming steam–carbon dioxide and manufacture method thereof, KR20140065254 A, filed November 21, 2012, and issued May 29, 2014.
56. Chunliang An, Yongshan Liu, Preparation method for catalyst for hydrogen production by hydrocarbon reforming, CN102836721 B, filed June 20, 2011, and issued December 31, 2014.
57. Stephan Schunk, Andrian Milanov, Andreas Strasser, Guido Wasserschaff, Thomas Roussiere, Hexaaluminate-comprising catalyst for the reforming of hydrocarbons and a reforming process, US2014191449 A1, filed December 23, 2013, and issued July 10, 2014.
58. Feifei Cheng, Rongrong Hu, Changfeng Yan, Integral catalyst applied to biological oil reforming hydrogen production, preparation and application thereof, CN101757919 B, filed December 28, 2009, and issued December 26, 2012.
59. Joo-Hyoung Park, Heung-Soo Park, Hee-Dong Chun, Shee-Hyun Park, Nickel-based reforming catalyst for producing reduction gas for iron ore reduction and method for manufacturing same, reforming catalyst reaction and equipment for maximizing energy efficiency, and method for manufacturing reduction gas using same, WO2014104756 A1, filed December 26, 2013, and issued July 3, 2014.
60. Chen Hanping, Li Xuehui, Wang Junfeng, Wu Zhengshun, Zhang Kang, Nickel-base biomass tar reforming catalyst and preparation method thereof, CN101797507 B, filed April 2, 2010, and issued March 27, 2013.
61. Dong Ju Moon, Young Chul Kim, Hye Jeong Ok, Catalyst for reforming hydrocarbons, US20140148332 A1, filed June 4, 2013, and issued May 29, 2014.
62. Saito Shinji, Takahashi Ayumi, Tomishige Keiichi, Reforming catalyst, and method for manufacturing synthetic gas using the same, JP2010221160 A, filed March 24, 2009, and issued October 7, 2010.

63. Hideto Sato, Reforming catalyst, method for preparing same, and process for manufacturing synthesis gas, WO2013187436 A1, filed January 12, 2013, and issued December 19, 2013.
64. Dou Binlin, Catalyst for steam reformation hydrogen production with by-product glycerol of biological diesel oil and preparation method thereof, CN101342488 A, filed June 26, 2008, and issued January 14, 2009.
65. Yoshinori Saito, Hideto Sato, Catalyst for reforming hydrocarbon gas, method of manufacturing the same, and method of manufacturing synthesized gas, US20120161078 A1, filed March 2, 2012, and issued June 28, 2012.
66. Shigeru Hashimoto, Kenji Asami, Tar reforming catalyst, manufacturing method thereof, and steam reforming method for tar using the catalyst, JP4719194 B2, filed August 22, 2007, and issued July 6, 2011.
67. Young Kwon Park, Hyun Ju Park, Jong In Dong, Catalytic for steam reforming of tar produced from biomass gasification process and method of steam reforming of tar using the same, KR100986241 B1, filed October 23, 2007, and issued October 7, 2010.
68. Yoshinori Saito, Hideto Sato, Hydrocarbon gas reforming catalyst, method for producing same, and method for producing synthetic gas, EP2474355 A1, filed August 30, 2010, and issued July 11, 2012.
69. Hu Rongrong, Yan Changfeng, Luo Weimin, Catalyst for hydrogen production by bio-oil steam reforming and preparation method thereof, CN101444740 A, filed December 16, 2008, and issued June 3, 2009.
70. Yufu Li, Jianyue Gao, Qieyi Wu, Haitao Li, Pengwan Tao, Liangcong Yao, Xuliang Han, A low-grade alkalization hydrocarbons water vapour conversion catalyst and its preparing method, CN100463716 C, filed November 13, 2007, and issued September 25, 2009.
71. Quanxin Li, Yuan Lixia, Tongqi Ye, Yaqiong Chen, Yoshiaki Torimoto, Mitsuo Yamamoto, Mixed oxide catalyst and use thereof in steam reforming bio-oil hydrogen making, CN101306370 B, filed June 3, 2008, and issued March 16, 2011.
72. Hailong Xu, Jianhua Yu, Shaoping Xu, Shuqin Liu, Xiaoqin Yang, Biomass gasification tar vapor translation hydrogen making catalyst and preparation method thereof, CN101332428 B, filed July 17, 2008, and issued September 8, 2010.
73. Hyun Joo Lee, In Kyu Song, Yong Ju Bang, Seung Ju Han, Mesoporous nickel–alumina xerogel catalyst prepared using nano-carbon particle as a mould, preparation method thereof and production method of hydrogen by steam reforming of liquefied natural gas (LNG) using said catalyst, KR101405306 B1, filed January 12, 2012, and issued June 11, 2014.
74. In Kyu Song, Jeong Gil Seo, Min Hye Youn, Dong Ryul Park, A mesoporous nickel–alumina composite catalyst, preparation method thereof and production method of hydrogen gas by steam reforming of liquefied natural gas using said catalyst, KR20110005358 A, filed July 10, 2009, and issued January 18, 2011.
75. Kwan Young Lee, In Kyu Song, Jeong Gil Seo, Min Hye Youn, Jong Heop Yi, Preparation of mesoporous nickel-alumina catalyst by a sequential precipitation method, and method for hydrogen production by steam reforming of liquefied natural gas using said catalyst, KR20110002340 A, filed July 1, 2009, and issued January 7, 2011.
76. Changfeng Yan, Juan Li, Rongrong Hu, Weimin Luo, Method for fast preparing dimethyl ether steam reforming hydrogen production catalyst by virtue of alternate microwave, CN103551149 A, filed October 30, 2013, and issued in February 5, 2014.
77. Kwan Young Lee, In Kyu Song, Jeong Gil Seo, Min Hye Youn, Mesoporous Ni-alumina composite catalyst prepared using anionic surfactant, and method for hydrogen production by steam reforming of liquefied natural gas using said catalyst, KR20100044468 A, filed October 22, 2008, and issued April 30, 2010.

78. In Kyu Song, Jeong Gil Seo, Min Hye Youn, Ji Chul Jung, Hee Soo Kim, Dong Ryul Park, Nickel catalysts supported on mesoporous alumina prepared by using block copolymer as a structure-directing agent, preparation method thereof and production method of hydrogen gas by steam reforming of LNG using said catalysts, KR20100044319 A, filed October 22, 2008, and issued April 30, 2010.
79. In Kyu Song, Jeong Gil Seo, Min Hye Youn, Ji Chul Jung, A mesoporous nickel–alumina co-precipitated catalyst, preparation method thereof and production method of hydrogen gas by steam reforming of liquefied natural gas using said catalyst, KR100980591 B1, filed March 16, 2009, and issued September 6, 2010.
80. Kwan Young Lee, In Kyu Song, Jeong Gil Seo, Min Hye Youn, Pil Kim, Kyung Min Cho, Sun Young Park, Nickel catalysts supported on alumina-zirconia oxide complex for steam reforming of liquefied natural gas and preparation methods thereof, KR100891903 B1, filed October 12, 2007, and issued April 3, 2009.
81. In Kyu Song, Jeong Gil Seo, Min Hye Youn, Ho Won Lee, Joo Hyung Lee, Sang Hee Lee, Nickel catalyst supported of mesoporous alumina prepared by hydrothermal method using cationic surfactant as a templating agent, preparation method thereof and hydrogen production method by steam reforming of liquefied natural gas using said catalyst, KR20090100473 A, filed March 20, 2008, and issued September 24, 2009.
82. In Kyu Song, Jeong Gil Seo, Min Hye Youn, Metal catalysts supported on alumina xerogel support, preparation method thereof and hydrogen producing method by steam reforming of LNG using said catalyst, KR100833790 B1, filed February 28, 2007, and issued May 29, 2008.
83. Kwan Young Lee, In Kyu Song, Jeong Gil Seo, Min Hye Youn, Pil Kim, Supported catalyst for producing hydrogen gas by steam reforming reaction of liquefied natural gas, method for preparing the supported catalyst and method for producing hydrogen gas using the supported catalyst, KR20080043161 A, filed November 13, 2006, and issued May 16, 2008.
84. Kwan Young Lee, In Kyu Song, Jeong Gil Seo, Min Hye Youn, Jong Hyop Yi, Nickel–alumina xerogel catalysts for steam reforming of liquefied natural gas and methods of producing the same, KR20080113658 A, filed June 25, 2007, and issued December 31, 2008.
85. Shudong Wang, Liwei Pan, Lei Zhang, Changjun Ni, Methanol steam catalytic reforming hydrogen production catalyst and preparation method thereof, CN103566941 A, filed November 12, 2013, and issued February 12, 2014.
86. Chuin-Tih Yeh, Ming-Chun Tsai, Cu catalysts containing rare earth elements and method of producing hydrogen using the same, TW201223632 A, filed December 1, 2010, and issued June 16, 2012.
87. Yasuyuki Matsumura, Method for producing fine particle aggregate, steam-reforming catalyst, method for producing steam-reforming catalyst, and method for producing hydrogen, JP2012161787 A, filed January 11, 2012, and issued August 30, 2012.
88. Kwan Young Lee, Jae Jeong Kim, Oh Joong Kwon, Sang Hyun Ahn, Catalyst mixture for steam reforming reaction, KR20110059296 A, filed November 27, 2009, and issued June 2, 2011.
89. Hans Joerg Woelk, Alfred Hagemeyer, Frank Grossmann, Oliver Wegner, Nanocrystalline Copper Oxide and Method for the Production thereof, US8722009 B2, filed June 21, 2013, and issued May 13, 2014.
90. Shulian Li, Mei Yang, Fengjun Jiao, Guangwen Chen, Catalyst for producing hydrogen by methanol steam reforming and preparation and application, CN102145281 B, filed February 10, 2010, and issued January 30, 2013.
91. Guangwen Chen, Shulian Li, Fengjun Jiao, Quan Yuan, Composite oxide catalyst for producing hydrogen by reforming methanol steam, preparation and application thereof, CN101612563 B, filed June 25, 2008, and issued December 30, 2009.

92. Wei Shixin, Chen Zhangxin, Zhang Jie, Wu Xueqi, Tian Xianguo, Cai Chengwei, Qi Dexiang, Sun Yuanlong, Dong Tianlei, Natural gas steam reforming catalyst and preparation method thereof, CN101450312 A, filed November 29, 2007, and issued June 10, 2009.
93. Diwakar Garg, Shankar Nataraj, Kevin Boyle Fogash, James Richard O'leary, William Robert Licht, Sanjay Mehta, Eugene S. Genkin, Catalyst, process and apparatus for the adiabatic pre-reforming of natural gas, EP1616838 B1, filed July 5, 2005, and issued March 27, 2013.
94. Motoshige Yagyu, Kazuya Yamada, Hideki Nakamura, Shinichi Makino, Toshie Aizawa, Kimichika Fukushima, Dimethyl ether-reforming catalyst and its manufacturing method, JP2010012402 A, filed July 3, 2008, and issued January 21, 2010.
95. Tatsuki Ishihara, Osamu Makino, Masataka Kajiwara, Tatsuhiko Hashimoto, Seiichi Fujikawa, Takayuki Yoshikawa, Manufacturing method for water vapor reforming catalyst for LPG, and water vapor reforming catalyst for LPG, JP5119019 B2, filed March 19, 2008, and issued January 16, 2013.
96. Ekkehard Schwab, Andrian Milanov, Stephan Schunk, Thomas Roussiere, Guido Wasserschaff, Andreas Strasser, Process for producing a reforming catalyst and the reforming of methane, US20130116116 A1, filed November 7, 2012, and issued May 9, 2013.
97. Dong Ju Moon, Yun Ju Lee, Jae Sun Jung, Jin Hee Lee, Seung Hwan Lee, Bang Hee Kim, Hyun Jin Kim, Eun Hyeok Yang, Alkaline earth metal co-precipitated nickel-based catalyst for steam carbon dioxide reforming of natural gas, US2014339475 A1, filed August 20, 2014, and issued November 20, 2014.
98. Won Jun Cho, Jong Tae Chung, Yong Gi Mo, Hye Jin Yu, Catalysts for preparing syn-gas by steam-carbon dioxide reforming reaction and process for preparing syn-gas using same, KR20140075996 A, filed December 12, 2012, and issued June 20, 2014.
99. Dong Ju Moon, Eun Hyeok Yang, Jin Hee Lee, Hyun Jin Kim, Byoung Sung Ahn, Sang Woo Kim, Jae Sun Jung, Iron modified Ni-based perovskite type catalyst, preparing method thereof, and producing method of synthesis gas from combined steam CO_2 reforming of methane using the same, US20140145116 A1, filed June 4, 2013, and issued May 29, 2014.
100. Wonjun Cho, Jongtae Chung, Yonggi Mo, Hyejin Yu, Catalyst containing lanthanum for manufacturing synthetic gas through steam–carbon dioxide reforming, and method for manufacturing synthetic gas by using same, WO2014092482 A1, filed December 12, 2012, and issued June 19, 2014.
101. Dong Ju Moon, Jae Sun Jung, Eun Hyeok Yang, Sang Woo Kim, Jae Suk Lee, Bang Hee Kim, Jong Tae Jung, Hyun Jin Kim, Ga Ram Choi, Byoung Sung Ahn, Process for preparing nickel based catalysts for SCR of natural gas, US20140349836 A1, filed April 30, 2014, and issued November 27, 2014.
102. Hyun Seog Roh, Wang Lai Yoon, Yong Seog Seo, Dong Joo Seo, Un Ho Jung, Sang Ho Park, Young Jae Hwang, Kee Young Koo, High surface area spinel structured nano crystalline sized ymgo (1-y) Al_2O_3-supported nano-sized nickel reforming catalysts and their use for producing synthesis gas from combined steam and carbon dioxide reforming of natural gas, KR100892033 B1, filed October 8, 2007, and issued April 7, 2009.
103. Hongtao Jiang, Huiquan Li, Process for preparation of catalysts for making synthesis gas through tri-reforming of methane, CN100522366 C, filed July 28, 2006, and issued August 5, 2009.
104. Chunlin Li, Hengyong Xu, Shoufu Hou, Monolithic catalyst and use of the monolithic catalyst in natural gas reforming for synthesis gas preparation, CN102614903 A, filed January 28, 2011, and issued August 1, 2012.

105. Dong Ju Moon, Jung Shik Kang, Won Seok Nho, Dae-Hyun Kim, Sang Deuk Lee, Byung Gwon Lee, Ni-based catalyst for tri-reforming of methane and its catalysis application for the production of syngas, US20080260628 A1, filed April 17, 2007, and issued October 23, 2008.
106. Gabriel Wolfgang, Hanke Ingo, Catalyst design and preparation process for steam-reforming catalysts, US8349758 B2, filed September 25, 2008, and issued January 8, 2013.
107. Mark Robert Feaviour, Catalysts for use in steam reforming processes, US2014005042 A1, filed September 10, 2013, and issued January 2, 2014.
108. Norbert Ringer, Gerhard Selig, Hans-Joachim Müller, Catalyst support, US7799730 B2, filed April 27, 2006, and issued September 21, 2010.
109. Frederic Camy-Peyret, Fabrice Mathey, Optimized catalyst shape for steam methane reforming processes, WO2014053553 A1, filed October 2, 2013, and issued April 10, 2014.
110. Yan Feng, Liu Lihua, Shi Yixin, Tangjia Peng, Pang Governance, Zhang Ye Yin, Hydrocarbon steam reforming catalyst, CN201618570 U, filed February 10, 2010, and issued November 3, 2010.

Section III

Biorefineries—An Outlook

10 Recent Developments in Biorefinery Catalysis

Elif Ispir Gurbuz and Nazife Işık Semerci

CONTENTS

10.1 INTRODUCTION

Lignocellulosic biomass, as an abundant, cheap, and naturally occurring raw material, is an attractive starting point that offers promising platforms for the production of biofuels and biochemicals. Though many challenges exist for biochemical and thermochemical conversion routes due to the heterogeneity and recalcitrance of lignocellulosic feedstocks, catalysis brings opportunities for the selective processing of biomass. Accordingly, catalytic strategies used for the valorization of biomass-derived molecules targeting the production of higher alkanes and platform molecules to replace diesel/gasoline range fuels and crude oil–based intermediates, respectively, have gone through significant advancement in the last decade. In this chapter, we particularly highlight the recent advances in catalytic systems for biomass conversion to fuels and chemicals, explored for improving reactant conversions and product selectivities, while relying on the performance of typical catalytic reactions, e.g., dehydration, hydrogenation (HYD), aldol condensation, and hydrogenolysis. In addition, the significance of solvent selection in catalytic systems, leading to higher selectivity and allowing better separation and purification of carbohydrate derived platform chemicals, is explored. In terms of biofuel production, the recent catalytic systems developed for tar elimination in biomass gasification, high purity hydrogen production from syngas, SNG production, pyrolysis bio-oil upgrading, and aqueous phase conversion through platform molecules and proceeding C–C coupling reactions are discussed in this chapter. In addition, production of chemicals through "drop-in" and "emerging" strategies are both discussed as viable options, including specific examples for the formation of some important chemicals, such as adipic acid, 1,3-propanediol, 5-hydroxymethylfurfural (5-HMF), 2,5-furandicarboxylic acid (FDCA), and furfural.

The use of renewable resources for the generation of energy, fuels, and chemicals has attracted significant attention in recent years, because depletion of the finite fossil resources seems unavoidable in the long term as these resources are expended faster than can be regenerated. Furthermore, environmental concerns are growing with accelerated global warming rates with excessive consumption of fossil fuels, and political and economical instabilities are created with the limited availability of these resources. Renewable resources, such as wind, hydrothermal, solar, geothermal, and biomass, have been used for electricity and heat generation for centuries, but recent efforts have focused on increasing the efficiency of using these resources in the form of new technologies for the twenty-first century (Vennestrom et al. 2011). Among these renewable resources, biomass is unique as the energy it contains is stored as chemical bonds, and indeed biomass is the only renewable source of carbon atoms. This creates the potential for converting biomass to liquid fuels and chemicals, in a similar manner that petroleum is utilized today. A recent perspective on the efficient and sustainable use of bio-based carbon lies in the application of the "biorefining" concept (Cherubini and Stromman 2011). Similar to current petroleum refineries, a biorefinery aims to integrate various biomass conversion technologies to separate biomass resources into its building blocks and process these separate fractions to obtain a variety of products (i.e., fuels and chemicals) with maximum efficiency. For the production of energy and fuels from lignocellulose, it should be remembered that biofuels will only be a part of renewable energy solutions, as the direct conversion and storage of energy from other renewable resources, such as wind and solar, will also play important roles. However, the production of liquid fuels from lignocellulose will still be needed for transportation, especially for aviation, until longer-term technological developments are achieved, eliminating the need for liquid fuels. On the other hand, renewable chemical production will always rely on lignocellulose, the only renewable carbon source. Accordingly, two approaches are taken currently for the production of chemicals from lignocellulose: either converting lignocellulose into existing chemicals or generating new building blocks and end products for the chemical industry.

Establishing an efficient biorefinery will require many innovations in the life cycle of growing crops up to their final chemical conversion to fuels and chemicals, such as crop optimization through bioengineering, pretreatment of biomass, and selective and energy-efficient conversion of biomass-derived platform molecules to fuels and chemicals. In addition, biorefineries are expected to become energy- and cost-competitive with petroleum refineries, when lignocellulosic biomass, which does not compete with food resources, can be utilized in a manner that makes the most value of its three different constituents: cellulose, hemicellulose, and lignin. Two approaches have been exploited for this purpose. The three portions are processed together in the first approach through strategies like gasification and pyrolysis, offering simplicity of operation, and lower operating costs. The alternative approach is to separate the three fractions of lignocellulose from each other by adopting various pretreatment strategies and process them separately. This approach offers flexibility and allows for tailoring of each strategy to the different chemical and physical properties of the fraction, making it especially valuable for the production of different chemicals. The provided flexibility also entails the combination of biological and chemical approaches, as well as pulp and paper applications.

While many branches of science will be involved in establishing efficient biorefineries, catalysis, like in petroleum refining, will be at the heart of developing new technologies for fuel and chemical production. Catalytic conversion of lignocellulose and lignocellulose-derived intermediates, however, presents new challenges in comparison to catalytic conversion of petroleum derivatives due to the highly complex, oxygenated, and polar nature of biomass in comparison to the petrochemical apolar hydrocarbons. Therefore, selective conversion of lignocellulose components will require their controlled defunctionalization and deoxygenation (DDO). In addition, highly polar and functionalized components of lignocellulose mostly require their conversion in aqueous phase at high temperatures and pressures that put strains on catalyst performance and especially stability. Accordingly, this chapter explores recent advancements in catalytic conversion of lignocellulose to fuels and chemicals that aim to overcome these challenges, starting with a brief description of the three fractions of lignocellulose: cellulose, hemicellulose, and lignin, and the initial pretreatment strategies for converting these fractions to their building blocks to be used as feedstocks in these novel conversion strategies. More detailed descriptions of current and developing lignocellulose pretreatment strategies have been discussed previously (Mosier et al. 2005; Yang and Wyman 2008; Blanch et al. 2011).

10.1.1 Lignocellulose Pretreatment

Three major portions of lignocellulose include lignin, cellulose, and hemicellulose. Among these three portions, cellulose and hemicellulose are polymers of sugars (mainly glucose and xylose), which are viable feedstocks for the production of fuels and chemicals. Lignin is embedded in the cellulose–hemicellulose matrix and acts as a glue for holding the whole structure in rigidity. Besides, it serves as a vital barrier to biological and physical attacks (Brandt et al. 2013; Ragauskas et al. 2014). If desired, lignin can first be depolymerized by a pretreatment step so that the cellulose and hemicellulose portions can be accessed more easily for further upgrading (Mosier et al. 2005; Huber et al. 2006a). However, it is also common to employ pretreatment strategies, such as acid hydrolysis, in untreated biomass before the removal of lignin to depolymerize cellulose and hemicellulose (Mosier et al. 2005; Huber et al. 2006a), while leaving the lignin portion intact for further use.

10.1.1.1 Hemicellulose and Cellulose Deconstruction

Cellulose is a linear polymer composed of glucose units connected by β-(1,4) glycosidic bonds, providing the structure with rigid crystallinity that is recalcitrant to hydrolysis. These β-glycosidic bonds are cleaved upon the addition of water, forming glucose molecules (Lohmann 1990). The fibrous cellulose is covered by hemicellulose, an amorphous, branched polymer generally composed of five different sugar monomers: D-xylose (the most abundant), L-arabinose, D-galactose, D-glucose, and D-mannose (Lynd et al. 1991). These differences in structures of cellulose and hemicellulose lead to differences in their physical properties and reactivities. For instance, compared to hydrolysis of crystalline cellulose in acidic aqueous solutions, hemicellulose hydrolysis is an easier process and allows for high yields of sugars. In addition to acid hydrolysis processes, other pretreatment strategies have been developed such as ammonia fiber explosion (Bals et al. 2010), steam explosion (Tao et al. 2011), organosolv (Muurinen 2000; Zhao et al. 2009), hot water (Xing et al. 2011), and ionic liquid (IL) (Ståhlberg et al. 2011) pretreatment. These pretreatment techniques vary in the extent of hemicellulose removal and in the form in which hemicellulose is recovered (e.g., monomeric form or as oligomers) (Lau et al. 2009). Among these methods, acidic hydrolysis strategies are common due to their low costs and efficiencies. For example, high amounts of xylose, up to 95% yield, was obtained based on the hemicellulose fraction of corn stover using a 1 wt.% solution of sulfuric acid at 463 K (Tucker et al. 2003). Phosphoric (Robinson et al. 2004), maleic (Lee and Jeffries 2011), and oxalic acids (Lee and Jeffries 2011) have also been used successfully for hemicellulose deconstruction. The acid-catalyzed hydrolysis of cellulose to form glucose or equimolar amounts of levulinic and formic acids also typically involves a strong mineral acid catalyst but at concentrations greater than or temperatures higher (423–513 K) than those used for hemicellulose deconstruction (Kuster 1990; Girisuta et al. 2006; Binder and Raines 2010). Acid hydrolysis of lignocellulose is well established; however, its application to produce platform chemicals is still not ideal due to polymerization reactions that occur and moderate product yields (e.g., cellulose to levulinic acid yields are typically 55–60%) (Serrano-Ruiz et al. 2010).

10.1.1.2 Lignin Deconstruction

Lignin is the most abundant polymer possessing aromatic functionality on earth (Ragauskas et al. 2014). It is a three-dimensional and water-insoluble macromolecule, which is originated from polymerization of three monolignols (phenylpropane units), p-coumaryl, coniferyl, and sinapyl alcohols (Doherty et al. 2011; Brandt et al. 2013), incorporated into the structure through several linkages. (Ragauskas, Technical Review). β-O-4 ether linkage is the most common linkage, which constitutes almost 50% of the intersubunit bonds in lignin (Zakzeski et al. 2010; Brandt et al. 2013). Lignin, comprising 15–40 wt.% of the dry biomass, has a pivotal role in maintaining the structural integrity of the plant.

Delignification of wood has been the subject since "kraft pulping" was introduced into the literature in the late nineteenth century for papermaking. This patented process, in which wood is subjected to pretreatment by sodium hydroxide and sodium sulfide at temperatures 423–473 K, is the current commercial source of lignin (Chakar and Ragauskas 2004). Numerous efforts have been given for removal of lignin from biomass to enhance the accessibility of fermentable sugars to hydrolysis for bioethanol production. Alkaline reagents were found to be dramatically effective for lignin removal as high as 100% (Silverstein et al. 2007; Binod et al. 2012). In addition to alkaline reagents, chemical pretreatments such as ammonia recycle percolation (ARP), organosolv process, lime, and IL pretreatments were reported as leading technologies for lignin removal (Yang and Wyman 2008; Doherty et al. 2011). Lignin depolymerization through the superior solvation capacity of ILs, in combination with catalysis, is a current research direction that allows transformation of lignin to aromatic platform chemicals (Chatel and Rogers 2014). Based on its native aromatic network, lignin has great potential to be appreciated in place of a wide range of petroleum-derived products. However, due to its chemically stable polyphenolic structure, utilization of lignin is limited. Lignin, therefore, has been considered to be an ideal candidate for low-value applications. It was reported that 98% of lignin is either combusted or converted to kraft lignin and lignosulfonates (Chatel and Rogers 2014). The existing markets for lignosulfonates are dispersant, binder, and adhesive applications (Task 42 Biorefining 2014). On the other hand, vanillin production from lignosulfonates is an exception. Borregaard, a Norwegian firm, is the only company in the world producing wood-based vanillin and has been carrying on this process for over 50 years (Borregaard, Vanillin). Among several products, EuroVanillin Supreme is the one closest to natural product with minimum CO_2 footprint in vanillin (Smart Vanillin, EuroVanillin).

10.2 BIOFUELS FROM LIGNOCELLULOSIC BIOMASS

The research on valorization of lignocellulosic biomass to biofuels and value-added platform chemicals lies at the heart of biorefining. Processes developed for petroleum refining, however, are not relevant for biorefining, due to the unique composition of lignocellulosic biomass, containing overfunctionalized aromatics and long-chain carbohydrates. Due to the excess oxygen present, lignocellulosic feedstocks basically require DDO through catalytic removal of functional groups as well as controlled C–C coupling reactions to generate higher-range alkanes for gasoline and diesel fuel applications. On the other hand, the major strategy in petroleum refining processes has been adding functionality through oxidation, amination, and hydration (Alonso et al. 2010a). Catalysis research, therefore, has a vital role in developing processes for the efficient generation of biofuels, with enhanced quality and yield of products from a variety of feedstocks, to be able to rely on lignocellulosic biomass for meeting energy needs. Corma et al. reported three major routes to convert lignocellulosic biomass into value-added biofuels: gasification, pyrolysis, and hydrolysis. Gasification takes place in an oxidizing medium under elevated temperatures (>1100 K) for partial oxidation of carbonaceous material to produce useful gaseous products: hydrogen and syngas. Pyrolysis is the thermal decomposition of organic material in the absence of an oxidizing medium, yielding pyrolysis bio-oil in addition to solid and gaseous products. Hydrolysis leads to the production of monomeric sugars from biomass components, cellulose, and hemicellulose at lower temperatures in comparison to the former two processes (Huber et al. 2006a; Zhou et al. 2011).

Gasification, pyrolysis, and hydrolysis can be split into subprocesses that mainly lead to the production of bioethanol, hydrogen, gasoline, and diesel fuel range hydrocarbons. Although bioethanol production, which requires hydrolyzing lignocellulose, has attracted a lot of attention over the last decade, this approach presents several challenges. First, slightly corrosive nature and oxygenated structure of ethanol (35 wt.%) limit its compatibility with existing technologies in current vehicles. Second, biomass fractionation prior to hydrolysis is an essential step that makes the entire process more difficult compared to gasification and pyrolysis of biomass (Huber et al. 2006a). Third, the microorganisms that carry out fermentation during bioethanol production are extremely sensitive to the acid concentrations used in hydrolysis to optimize sugar yields (Larsson et al. 1999). Accordingly, the less robust nature of these biological routes attracts attention to catalytic processing strategies for biomass conversion. In addition, aqueous phase processing of biomass through catalytic routes introduces other significant advantages, such as ease of product separation, since catalytic transformations of carbohydrate-derived molecules yield either gaseous or hydrophobic liquid products. Therefore, this chapter will focus on catalysis-based strategies for conversion of lignocellulosic biomass to biofuels.

Figure 10.1 illustrates the major catalytic routes and corresponding products discussed in this section. First, biomass gasification is discussed in terms of catalytic

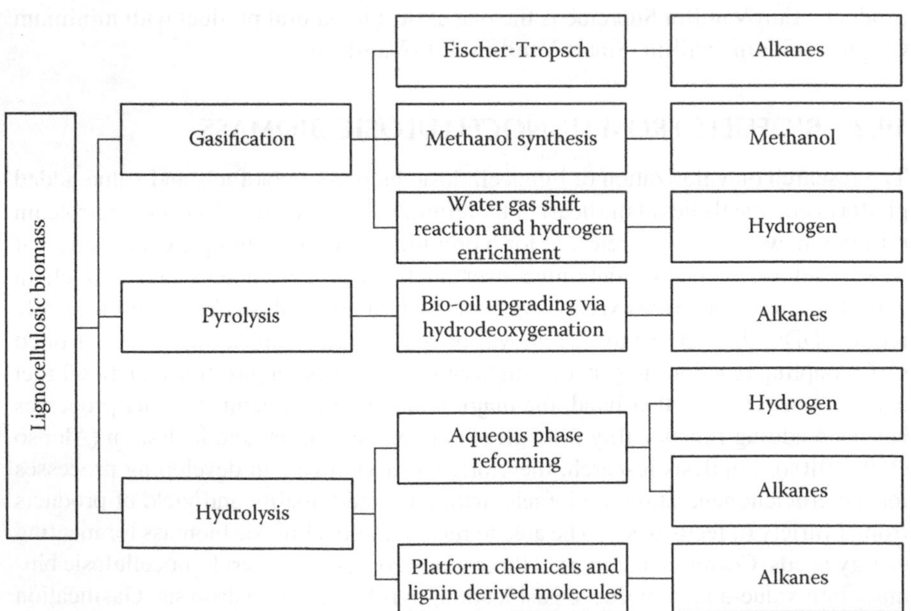

FIGURE 10.1 Major routes discussed, involving catalytic conversion of lignocellulosic biomass to biofuels.

systems favored for tar elimination, hydrogen enrichment, and synthetic natural gas (SNG) production. Second, the recent catalytic routes for upgrading pyrolysis bio-oil, which is generated directly from biomass or biomass-derived components, are addressed. Third, catalytic routes for conversion of biomass oxygenates (polyols and sugars) to hydrogen via aqueous phase reforming and gasoline/diesel range alkanes through partial reforming followed by C–C coupling reactions are discussed. Finally, the catalytic upgrading of various biomass derivatives (platform chemicals) and lignin-derived molecules to higher-range alkanes is considered.

10.2.1 Catalysis in Biomass Gasification

Gasification is a process that generates synthesis gas, composed of mainly CO and H_2, by partial oxidation of carbon-rich resources such as coal, petroleum, and biomass at temperatures above 1100 K in the presence of an oxidation medium such as air, oxygen, or steam (Serrano-Ruiz et al. 2010). Utilization of biomass as a renewable resource in gasification is an established technology that advanced significantly since World War II. Compared to coal, biomass is much easier to react due to its lower ignition temperatures (Huber et al. 2006a). Any lignocellulosic feedstock is suitable for biomass gasification; however, excessive moisture in the biomass creates problems associated with the thermal efficiency of the process (Alonso et al. 2010a). Transportation of low-energy density biomass to the gasification site and significant alkali content of the feedstocks, causing slagging and fouling in the reactor, are further drawbacks that raise economical and operational problems (Huber et al. 2006a; Serrano-Ruiz et al. 2010). The major challenge posed during biomass gasification is the formation of high molecular weight condensates, tars. Being primarily aromatics, tars are known to lower the quality of product gases by decreasing their energy content (Dayton 2002). A number of approaches have been proposed to reduce tar formation: designing suitable gasifiers, catalytic cracking (CRA), and modification of operation parameters. Among several types of gasifiers (fluidized bed, downdraft, and updraft gasifiers), downdraft gasifiers, where the oxidizer and biomass are both introduced from the top of the reactor, were reported to be highly favorable for tar elimination, owing to its effective combustion performance (Navarro et al. 2009). Noncatalytic tar removal through steam reforming can take place at temperatures exceeding 1173 K because of the endothermic nature of the reactions and high activation energies (Xu et al. 2010).

Catalytic tar CRA that takes place at lower process temperatures is the most effective route for removal of these organic impurities (Xu et al. 2010; Wang L. et al. 2014). Primarily, a tar CRA catalyst must be cheap, abundant, and easily disposable to comply with toxicity regulations (Nordgreen et al. 2012). Inexpensive dolomite ($CaMg(CO_3)_2$) and olivine ($(Mg,Fe)_2(SiO)_4$) and, especially, supported metal catalysts, such as Co and Ni, have been shown to exhibit greater performance for catalytic tar removal at reduced temperatures (Xu et al. 2010; Li et al. 2015). Calcined dolomites, CaO–MgO, showed better catalytic activity for tar CRA due to their higher surface area and oxide contents in comparison to their naturally occurring form (Xu et al. 2010). Hu et al. (2006) compared calcined dolomite and olivine as secondary catalysts with respect to their activity for decomposition of tar present in the product gas obtained upon steam gasification of apricot stones. Though calcined dolomite gave about twofold higher hydrogen yield compared to calcined olivine, the dolomite catalyst presented structural disadvantages for biomass gasification. The friable nature of the material is a drawback that limits its use in fluidized beds due to the possibility of pore blockage inside the reactor. Besides, partial pressure of CO_2 is a critical parameter for catalytic activity of calcined dolomite, because the catalyst may lose its activity at CO_2 pressures higher than the equilibrium calcination pressure of dolomite (Dayton 2002). However, olivine with a general formula is a more suitable option for fluidized-bed reactors owing to its high mechanical strength. Olivine has also been recognized as an appropriate support for Ni, which may be subjected to attrition and deactivation by coke formation and sintering (Mazumder and de Lasa 2014; Zhao et al. 2015). Almost 80% tar reduction was obtained when Ni–Ce modified olivine was used for gasification (in the pyrolysis mode) of white oak at 1073 K compared to unmodified olivine. Furthermore, the H_2/CO ratio obtained between 0.8 and 0.9 at both temperatures 923 and 1123 K indicated the possibility of conducting the process at a lower temperature range, 923–973 K, in the presence of Ni–Ce modified olivine (Cheah et al. 2013).

Ni catalysts supported on Al_2O_3 have also been shown to be promising regarding tar elimination and thus enhancement of syngas quality. Mazumder and de Lasa (2014) examined the performance of La_2O_3 promoted Ni/γ-Al_2O_3 (Sasol) catalyst for the conversion of biomass-derived model compounds to syngas and achieved roughly 100% and 81% conversions for glucose and 2-methoxy-4-methylphenol, respectively, at 923 K. These results were linked to Al_2O_3 modification by 5 wt.% La_2O_3 that provided higher Ni dispersion, surface area, and basicity for the support, as well as reduced coke deposition (0.0023 g coke/g catalyst) on the catalyst. Moreover, Ni supported on α-Al_2O_3 and ilmenite (iron titanium oxide) were compared as secondary catalysts for their potential in tar removal using a chemical looping reactor (CLR), where both biomass upgrading and catalyst regeneration took place (Lind et al. 2013). While 60% tar reduction was obtained with ilmenite at 1123 K, 95% tar reduction was achieved with Ni/α-Al_2O_3 at 1153 K. Furthermore, Jiang et al. (2012) reported outstanding performance for steam reforming of naphthalene as a model tar compound over Ni_5TiO_7 nanocrystals as a result of their high specific surface area. These nanocrystals were prepared by impregnation of Ni and Cu salts on a porous TiO_2 layer manufactured by plasma-electrolytic oxidation. The needle-shaped nanocrystals of Ni_5TiO_7 resulted in 100% naphthalene conversion at 1023 K, which is threefold and ninefold higher than the conversion obtained on a commercial Ni catalyst and olivine, respectively. In addition to Ni, Co also appeared as a promising tar decomposition catalyst. Co/Al/Mg catalysts with different atomic metal percentages were evaluated with regards to their catalytic performance for steam gasification of cedar wood (Wang L. et al. 2014). The catalysts were synthesized from hydrotalcite-like precursors that enabled improved interaction between the metals and the oxide supports. The highest catalytic activity was obtained by Co/Al/Mg (10/40/50 atomic metal%), which was the most effective one giving nearly complete tar decomposition at 823 K.

Similar to conversion of coal-derived syngas to value-added products, four major routes can be employed for valorization of biomass-derived syngas to liquid and gaseous fuels. The two common routes include generation of liquid alkanes suitable for diesel and gasoline applications through Fischer–Tropsch synthesis (FTS) and production of alcohols such as methanol. By eliminating the liquefaction step, direct utilization of syngas will provide electricity production for gas turbine engines and solid oxide fuel cells. Alternatively, syngas can be catalytically purified to hydrogen by subjecting the gaseous mixture to water-gas shift (WGS) reaction. Latter route also provides an opportunity for substitute natural gas (SNG) production (Shaddix 2011). The recent advances related to each catalytic route are described next.

A proper FTS catalyst should show high activity toward HYD of CO for generating higher-range hydrocarbons. The most efficient ones for this purpose are Fe, Co, Ni, and Ru catalysts. Despite having notably high catalytic activity toward formation of hydrocarbons, Ni and Ru are not preferable to be widely used at the industrial scale (Luque et al. 2012). Fe is the most economically attractive metal favoring the selectivity toward short-chain alkanes such as olefins, whereas Co has shown high selectivity to longer-chain hydrocarbons (paraffin) because of its lower susceptibility to deactivation. While Fe is able to operate under both low (473–513 K) and high (573–623 K) temperatures, FTS over Co is conducted only at low-temperature mode due to excess methane formation at higher temperatures (Luque et al. 2012; Jahangiri et al. 2014).

The identity and geometry of the support are critical parameters that enhance catalytic performance through providing better dispersion for the metal catalyst. Chen et al. (2008) reported confinement effects of iron oxide catalysts in carbon nanotubes (CNTs) and observed a remarkable change in FTS activity with respect to catalyst dispersion on CNTs. FTS was conducted at 543 K and 5.1 MPa with a syngas feed, having a volumetric H_2/CO ratio of 2. The yield of C_{5+} hydrocarbons were increased approximately twofold via encapsulating Fe_2O_3 inside CNTs compared to the performance of Fe_2O_3 dispersed on the outer walls of CNTs. The catalyst encapsulated inside the support enabled longer periods of contact between syngas and the active sites and accordingly resulted in longer-chain alkanes. Besides, a higher amount of iron carbide phase observed inside the CNTs played a promoting role in FTS activity.

Addition of structural or alkali promoters and metals to monometallic catalysts offers several benefits for adding value to biomass-derived syngas. Structural promotion of Fe-based FTS catalysts with Si, Al, or Zn alleviated the adverse effects of high CO_2 concentrations derived from gasification of southern pinewood chips (Sharma et al. 2014). CO_2 HYD (reverse reaction of WGS) competing with CO HYD can reduce the production of hydrocarbons and CO_2. Zn and Al promotion for Fe/Cu/K catalysts not only improved CO HYD but also provided higher catalyst stability against oxidation of the iron carbide phase at high levels of CO_2. FTS carried out at 553 K and 2.8 MPa in the presence of Fe/Cu/K/Zn resulted in the highest $CO+CO_2$ conversion (34%) and higher selectivity toward C_{12}–C_{18} and C_{19+} hydrocarbons: 13.9% and 4.6%, respectively.

Hydrogen enrichment of the gasification product gas, comprising carbon oxides, methane, hydrocarbons, and tar, is necessary for utilization of high purity hydrogen in gas turbines and solid oxide fuel cell applications (Shaddix 2011). Following the cleaning of biomass-derived syngas, steam reforming and WGS reactions usually take place to improve the final hydrogen concentrations, and thereafter, CO_2 removal leads to the production of highly pure hydrogen. The hydrogen concentration obtained upon WGS reaction together with steam reforming of methane and other hydrocarbons can be improved by absorption-enhanced reforming, which involves calcium oxide (CaO) as a CO_2 absorbent to shift the equilibrium toward hydrogen production (Navarro et al. 2009). CaO-based chemical looping gasification of pinewood sawdust conducted at 923 K not only increased hydrogen yield up to 78 vol% but also significantly reduced catalytic tar formation. CO_2 concentration was reduced to 4.98 vol% at the outlet stream (Udomsirichakorn et al. 2014). Similarly, the bifunctional Ni–Mg–Al–CaO catalyst enabled highly stable CO_2 absorption even after 20 consecutive runs of pyrolysis gasification of wood. Hydrogen production was ~20.2 mmol/g biomass, which corresponded to 63.7 mol% selectivity (Nahil et al. 2013). Furthermore, in situ CO_2 capture by calcined dolomite ($CaMg(CO_3)_2$) during gasification of chestnut wood dust in the presence of a Pd/Co–Ni catalyst resulted in the production of high purity hydrogen (~99.9 vol%) with ~84 mol% selectivity. A promising finding was that Pd promotion for a Co–Ni catalyst eliminated prereduction step, taking place between subsequent processes, required for regeneration of Co and Ni oxides. Regeneration of metal oxides through quick hydrogen production driven by Pt was right after followed by sorption-enhanced catalytic steam gasification (Fermoso et al. 2012).

Catalytic gasification in supercritical water is a recent hydrothermal process for the transformation of biomass to SNG, with high yields of methane and hydrogen. Minimization of energy loss that is encountered during vaporization of water in typical thermal processes, enhancement of mass transfer rates, and ease of product recovery are some processing advantages observed in supercritical fluid technologies developed for biomass conversion (Peterson et al. 2008). The catalysts, Ni and Ru supported on alumina or silica, were reported to be favorable for SNG production through HYD of CO or CO_2 during supercritical gasification of biomass (Onwudili and Williams 2013b). Gasification of woody biomass in supercritical water near 673 K and 30 MPa in the presence of hydrothermally stable Ru/C yielded 50 vol% CH_4 in a gaseous mixture containing H_2 and CO_2 as well, and resulted in minimized tar and char formation (Vogel et al. 2007). Besides, use of alkali reagents for biomass processing prior to gasification in supercritical water had a significant positive influence on hydrogen and methane yields since the deconstructed biomass, via alkali liquors, was more prone to hydrothermal gasification than the untreated biomass (Onwudili and Williams 2013a). Preprocessing of the pinewood sawdust in subcritical water with an alkaline compound, Na_2CO_3, prior to supercritical water gasification conducted at around 773 K and 30 MPa with an Ru/Al_2O_3 catalyst resulted in higher methane and hydrogen yields compared to the process including only gasification. Preprocessing of the biomass with Na_2CO_3 enhanced H_2 and CH_4 yields by 51% and 61%, respectively.

In addition to metals on supports, nanometal oxides were also recommended for transformation of biomass into useful fuels in hydrothermal media. Improved CH_4 and H_2 yields were observed in hydrothermal gasification of cellulose at 873 K via nano ZnO and SnO_2 in comparison to their bulk counterparts. This was linked to the increased relative surface area of the nano catalysts (Sinag et al. 2011).

10.2.2 Catalytic Upgrading of Bio-Oil Derived from Pyrolysis

Catalytic pyrolysis of biomass comprises a set of reactions that take place at around 773 K in the absence of air and transforms biomass primarily into H-rich volatile vapors, gaseous products, and carbon-rich solid product, char (Sanna and Andresen 2012). Following the rapid cooling and condensation (<1 s), a dark viscous liquid product referred to as bio-oil is obtained, which can be produced up to 70–80 wt.% of dry feed (Huber et al. 2006a; Ruddy et al. 2014). Rapid heating rates and finely ground biomass (<3 mm) are the essential points to facilitate heat transfer and thus maximize the oil yield in biomass pyrolysis (Huber et al. 2006a; Bridgwater 2012).

Pyrolysis bio-oil is a dark brown, viscous liquid substantially different from petroleum regarding its much higher oxygen and water content, 30–50% and 20–30%, respectively (Ruddy et al. 2014). Bio-oil is a mixture of oxygenated hydrocarbons as reported by Milne et al. (1997). Owing to its high oxygen content, low heating value, and its organic acid content, the crude bio-oil cannot directly be blended with gasoline (Ren et al. 2014). Therefore, bio-oil needs to be upgraded through several reaction routes to obtain gasoline- and diesel-grade compounds. Two major catalytic routes, hydrodeoxygenation (HDO) and zeolite CRA, have been described for bio-oil upgrading (Huber et al. 2006a). Catalytic upgrading of bio-oil through HDO involves a complex network of transformations, decarbonylation and decarboxylation (DCO), direct DDO, dealkoxylation (DAO), CRA, hydrocracking (HCR), HYD, demethylation (DME), and methyl transfer (MT) as reported by Ruddy et al. (2014). Upgrading through HDO results in the formation of water and saturated C–C bonds (Huber et al. 2006a). Principally, HDO of bio-oil takes place under elevated H_2 pressures, 7.5–30 MPa, and temperatures, 573–873 K (Huber et al. 2006a; Mortensen et al. 2011). In order to reduce the process costs brought by high H_2 consumption, advancements in catalyst development are imperative.

Bimetallic catalysts, Co–MoS_2 and Ni–MoS_2, have been found to be suitable for upgrading high sulfur containing bio-oils, however, under substantially high hydrogen pressures (5–20 MPa) (Ruddy et al. 2014). They further require H_2S co-feed to regenerate the sulfide state of the metals that get oxidized during reaction (Mortensen et al. 2011). Moreover, deactivation due to coking can be observed in transition metal sulfide catalysts via deposition of multi-oxygenated aromatics on the catalytic surface under severe conditions stated above. Despite their high costs, noble metal catalysts (supported Ru, Rh, Pd, and Pt) have the ability to operate under lower hydrogen pressures and temperatures for upgrading of bio-oil (Wang et al. 2013). Noble metal catalysts were also reported to have higher resistance to pyrolysis vapors containing a considerable amount of acid and water (Ruddy et al. 2014). Wang et al. (2013) tested the catalytic performance of Pt catalysts impregnated on various zeolite supports for HDO of the lignin model compounds at 473 K and 4 MPa at liquid space velocities changing between 1 and 6 h^{-1}. Pt on ZSM-5 (microporous zeolite) and MZSM-5 (mesoporous zeolite) exhibited much higher HDO activity for the lignin model compounds, cresol and guaiacol, compared to Pt/Al_2O_3. Especially, Pt/MZSM-5 resulted in 93% and 98% conversion upon HDO of cresol and guaiacol, respectively. This finding was strongly related to the efficient H_2 dissociation on Pt metal coupled with acidic nature of the mesoporous support, which allowed the diffusion and adsorption of large aromatic compounds. Moreover, Pt promoted with Mo on multiwalled carbon nanotube (MWCNT) support was used for vapor phase catalytic HDO of pyrolysis products from microcrystalline cellulose and poplar (Venkatakrishnan et al. 2015). According to the strategy proposed by Venkatakrishnan et al., condensation was conducted right after HDO of vapor phase pyrolysis products. However, condensation has been recognized to occur right before upgrading of pyrolysis bio-oil in the conventional process. The proposed strategy provided individual control of the fast hydropyrolysis and catalytic HDO regarding operation parameters. This strategy was a promising one due to the possibility of lower hydrogen consumption and generation of longer-chain hydrocarbons from pyrolysis bio-oil. In this regard, 73% of the carbon was recovered in the form of C_1–C_{8+} from cellulose, while 54% carbon recovery was obtained from poplar in the form of C_1–C_{8+}.

As alternatives to supported noble metal catalysts, naturally occurring minerals, serpentine, and olivine were explored for upgrading bio-oil derived from pyrolysis of spent grains (Sanna and Andresen 2012). Both the serpentine and olivine were activated via acid leaching to modify their surface structure. Hydrogen consumption was eliminated, and DCO was reported to be the major mechanism carried out by both catalysts for bio-oil upgrading. The highest oxygen removal, 50%, was achieved with activated olivine at 703–733 K. H-NMR analysis of upgraded bio-oil revealed that activated olivine favored the production of aliphatics with the formation of much less amount of aromatic and phenolic structures. Moreover, Ren et al. (2014) used bio-char as an inexpensive and non-hydrogen consuming catalyst for upgrading bio-oil derived from pyrolysis of the Douglas fir. The carboxylic functional group on the catalyst surface promoted the transformation of anhydrous glucose to aromatics via oligomerization, DCO, decarbonylation, and lignin conversion to aromatics by CRA and oligomerization.

Loop-oxide catalysis (LOC) is a promising approach enabling DDO of bio-oils on a low valence metal oxide that acts as a reducing agent. Hargus et al. (2014) reviewed this process and discussed its advantages over conventional HDO. This approach is basically composed of a two-step thermochemical cycle in which solar thermal energy indirectly promotes bio-oil upgrading. The metal oxide that is reduced by solar thermal energy shows high affinity toward oxygenation and thus drives DDO of bio-oil at temperatures around 500 K. The metal oxides (Fe_3O_4, ZnO, CdO, SnO, CeO_2) used in LOC are regenerated in each single repeat of the cycle. The authors also examined DDO of acetic acid on the ZnO catalyst through LOC. The selectivity toward acetaldehyde and acetone was favored with LOC, while the selectivity toward ethyl acetate and ethanol was minimized due to the absence of hydrogen.

10.2.3 Catalysis for Aqueous-Phase Reforming of Biomass-Derived Oxygenates

Aqueous-phase reforming (APR) is an attractive alternative to conventional steam reforming for high purity hydrogen production. APR was first introduced by Cortright et al. (2002) as a less energy-intensive process, which is carried out at temperatures around 500 K in a single reactor to convert renewable biomass-derived oxygenates, sugars, alcohols, and polyols (glycerol, ethylene glycol, sorbitol) to hydrogen and alkanes in the presence of supported metal catalysts. Compared to gas-phase steam reforming, APR conducted at low temperatures was reported to reduce the formation of undesirable products that lead to coke deposition and minimize CO production due to the promoted WGS reaction (Xiong et al. 2015). The major reaction pathways for hydrogen production by APR of oxygenated hydrocarbons over a metal catalyst were reported by Cortright et al. (2002). WGS reaction following the dehydrogenation of compounds and C–C bond cleavage favors the selectivity of hydrogen production. To obtain high H_2 yields, a proper APR catalyst must have high activity toward both C–C bond cleavage and WGS reactions and show low activity toward C–O bond cleavage and HYD of CO and CO_2 that lead to alkane formation (Huber et al. 2006b). Pt group metals were reported to be favorable for APR reactions (Ciftci et al. 2014), and Pt and Pd, which meet the aforementioned requirements, were shown as selective APR catalysts for hydrogen production from ethylene glycol (Davda et al. 2003; Shabaker et al. 2003).

Selection of metal promoters to increase product selectivity and a stable support to withstand hydrothermal conditions are the most critical parameters examined for APR of biomass oxygenates. The contribution of metal promoters on overall APR performance is significant since they can improve WGS activity, modify the selectivity to hydrogen production, and increase the catalyst stability. Huber et al. (2003) studied the effect of Sn promotion on the Raney-Ni catalyst for APR of sorbitol, glycerol, and ethylene glycerol. Sn promotion was effective for increasing hydrogen selectivity for each compound due to the presence of Sn on Ni-defect sites on which CO dissociation and methane formation occurred. Hydrogen selectivity was higher than 90% for APR of ethylene glycol over the Raney Ni–Sn catalyst at temperatures around 500 K. Moreover, WGS activity in APR of glycerol via carbon supported Pt and Rh nanoparticles was shown to increase with Re addition. This was due to the water dissociation promoted by Re in the bimetallic catalysts through stronger binding of OH on Re (Ciftci et al. 2014).

Carbon-based materials were shown as ideal candidates for supporting monometallic and bimetallic catalysts due to being more inert and hydrothermally stable compared to conventional metal oxide supports such as Al_2O_3 and SiO_2 (Kim et al. 2011). PdZn catalysts supported on carbon black (CB) and CNTs yielded CO free hydrogen production during APR of ethanol at 523 K and 6.5 MPa (Xiong et al. 2015). Superiority of CNT to CB regarding ethanol conversion and hydrogen production (2 moles hydrogen/mole ethanol) was linked to pore confinement effects, which provided proper location of PdZn in nanotubes. Besides, the CNT supported Ru catalyst, which exhibited great stability under acidic conditions, resulted in 97% conversion of 25 wt.% acetic acid solution in supercritical water at 673 K and 25 MPa (de Vlieger et al. 2014). Pt supported on ordered mesoporous carbon (OMC) performed better than Pt on activated carbon (AC) and alumina-based supports for hydrogen production in APR of 10 wt.% ethylene glycol (Kim et al. 2011). OMC retained its 3-D structure during the reaction and provided good dispersion for Pt clusters. Pt (7 wt.%) on OMC gave approximately twofold higher ethylene glycol conversion and hydrogen production compared to Pt/AC operated at 523 K and 4.6 MPa. Liu et al. (2010) reported outstanding performance of Fe_3O_4 as a support for Pd nanoparticles

(<3 nm) during APR of ethylene glycol in comparison to various supports, Fe_2O_3, NiO, Cr_2O_3, Al_2O_3, and ZrO_2. Pd/Fe_3O_4 did not lose its catalytic activity significantly during APR of ethylene glycol in 130 h. Hydrogen selectivity of the catalyst was at least 92% and alkane selectivity was below 0.1%.

Biomass-derived oxygenates can also serve as raw materials for the production of alkanes through successive catalytic steps. Combining aqueous phase processing of sugars and polyols with further reactions including aldol condensation and HYD/HDO can generate gasoline and diesel range alkanes. In 2008, Kunkes et al. (2008) proposed a process involving catalytic reactions that took place in sequential reactors for the production of higher-range alkanes from aqueous solutions of sorbitol and glucose. In the first step, conversion of glucose and sorbitol over a bifunctional catalyst, Pt–Re/C, at 503 K resulted in the formation of a variety of monofunctional hydrocarbons, such as alcohols, ketones, carboxylic acids, and heterocyclic compounds. During this first step, reactants (glucose and sorbitol) underwent both C–C cleavage to provide H_2 for the HYD reactions and C–O cleavage to form monofunctional hydrocarbons that separated out into an organic layer. This organic layer was primarily composed of ketones and heterocyclic compounds. Two routes were reported for the production of gasoline-grade components from the sorbitol-derived organic layer. The organic layer, which was hydrogenated over Ru/C at 433 K and 5.5 MPa H_2 and deoxygenated over H-ZSM at atmospheric pressure, yielded aromatic compounds with 38% conversion based on the carbon content of sorbitol. Besides, the secondary alcohols present in the organic phase were dehydrated over Nb to generate branched C_4–C_6 olefins and further oligomerized, and cracked over HZSM-5 to yield gasoline-grade components with 50% conversion. Alternatively, the organic layer was subjected to aldol condensation over $CuMg_{10}Al_7O_x$ at 573 K and 0.5 MPa H_2 to yield C_8–C_{12} species with 45% conversion. These components were almost completely deoxygenated and considered highly suitable for conversion to diesel-range components.

In another study in 2009, West et al. (2009) used a bifunctional catalyst, Pt supported on a niobium-based solid acid ($NbOPO_4$), for aqueous phase dehydration and HYD (APDH) of sorbitol at 530 K and 5.4 MPa H_2. While the acidic functionality in Pt/$NbOPO_4$ generated unsaturated C=C and C=O bonds through dehydration, the metal component resulted in HYD of these unsaturated bonds. At low space velocities (0.2–0.4 h^{-1}), higher than 50% conversion of sorbitol to C_1–C_6 alkanes was achieved on Pt/$NbOPO_4$ (West et al. 2009).

10.2.4 Catalytic Upgrading of Biomass-Derived Molecules for the Production of Higher-Range Hydrocarbons

HDO over noble metals can bring out advantages for valorization of cellulose and hemicellulose to higher alkanes that are suitable for jet and diesel fuel range. Xing et al. (2010) described a four-step process for the conversion of hemicellulose-based sugars extracted from northeastern hardwood to tridecane (C_{13}). Dehydration of xylose to furfural was followed by aldol condensation of furfural with acetone to form furfural–acetone–furfural trimers (F–Ac–F). Aldol condensation is a C–C coupling reaction and a critical step for initiating higher-range alkane production from HMF or furfural (Climent et al. 2014). Following the aldol condensation, F–Ac–F was hydrogenated to H–FAF at 383–403 K over an Ru/C catalyst. Lastly, the hydrogenated dimer, H–FAF, was subjected to HDO at 533 K and ~6.2 MPa H_2 over a bifunctional Pt/SiO_2–Al_2O_3 catalyst. H–FAF underwent dehydration on the acidic SiO_2–Al_2O_3 support forming C=C bonds, which were then saturated on Pt clusters. C_{12} and C_{13} were the most abundant alkanes produced from H–FAF with 91% yield (C%). Based on the theoretical amount of C_{13} that can be produced from xylose, 76% yield was obtained, which corresponded to 0.46 kg jet fuel/kg xylose.

In 2012, Corma et al. (2012) reported the Sylvan process, which provided high-quality diesel production from 2-methyl furan, Sylvan, which is obtained from HYD of furfural. Different approaches were proposed for converting Sylvan to trimerized precursors, which eventually could undergo HDO reaction on Pt clusters impregnated on different supports, resulting in the synthesis of higher-range alkanes compatible with conventional diesel. Trimerization of Sylvan was performed through hydroxyalkylation/alkylation reactions. In this regard, Sylvan was reacted with butanal at 323 K for 8 h on various zeolites for the synthesis of a diesel precursor, 1,1-bissylvylbutane. Delaminated zeolite, ITQ-2, resulted in the highest product yield, 90% with at least 93% product purity. Besides, three Sylvan molecules were reacted in the presence of sulfuric acid at 333 K to form another diesel precursor, 5,5-bissylvyl-2-pentanone. In 16 h of trimerization reaction, 94% product yield was achieved. HDO of 1,1-bissylvylbutane over Pt/C at 623 K and 5 MPa H_2 led to the formation of a mixture of n-C_9H_{20}, $C_{12}H_{26}$, $C_{14}H_{30}$, and $C_{14}H_{28}$ with 95% total selectivity. This mixture was referred to as Diesel 14. HDO of 5,5-bissylvyl-2-pentanone was carried out on Pt/C-TiO_2 at 623 K and 5 MPa. The reaction resulted in ~93% selectivity (Diesel 15). By taking the product yield obtained upon trimerization (94%) into account, Sylvan was efficiently converted to Diesel 15 with an 87% overall yield and a cetane number greater than 70.

In 2013, Sutton et al. (2013) reported one-pot synthesis of linear C_9–C_{15} alkanes from furan-derived precursors synthesized upon aldol condensation of HMF or furfural with acetone. Furan-derived precursors were subjected to HYD and ring opening on a Pd-based catalyst in the presence of aqueous acetic acid (10 vol%) for the synthesis of polyketones. These polyketones were then hydrodeoxygenated to linear alkanes using La(OTf)$_3$. These consecutive reactions carried out in a single reactor gave considerably high isolated product yields; 87%, 76%, and 65% were obtained for n-nonane, n-dodecane, and n-pentadecane, respectively. The process described is energy-efficient due to the elimination of intermediate isolation and mild reaction conditions used.

Gamma-valerolactone (GVL) is a renewable heterocyclic ester obtained from catalytic HYD of levulinic acid, an attractive platform chemical obtained from cellulose. Transformation of GVL-derived alkenes to C_{18}–C_{27} alkanes was reported by Alonso et al. (2010b). As shown in Scheme 10.1, the ring opening of GVL on Pd/Nb_2O_5 forming pentanoic acid was followed by ketonization of pentanoic acid on ceria–zirconia, which yielded 5-nonanone. 5-Nonanone was completely hydrogenated over Ru/C to 5-nonanol. The catalytic route including dehydration and oligomerization of 5-nonanol over Amberlyst-70 to long carbon-chain alkenes, followed by HYD on Ru/C, resulted in the production of alkanes suitable for use in diesel and jet fuel. This

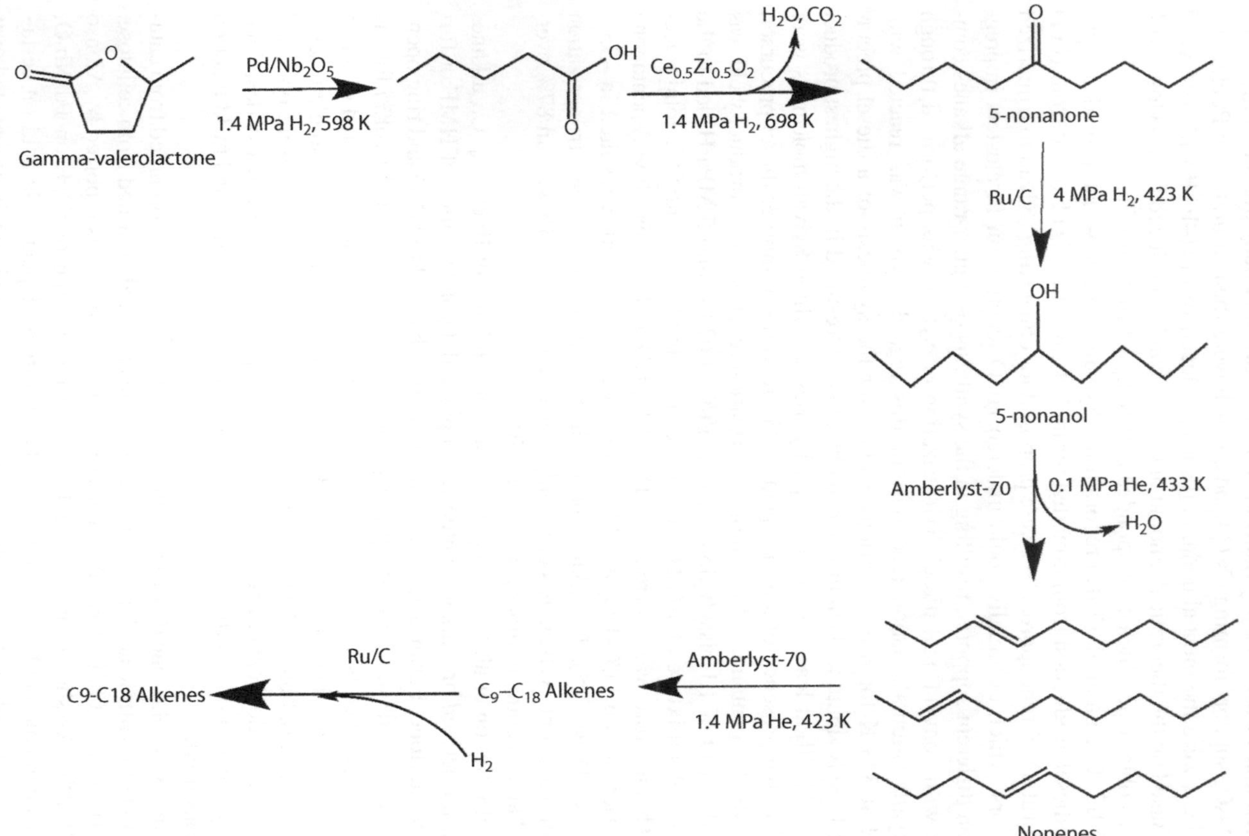

SCHEME 10.1 Reaction pathway for conversion of GVL to C_9–C_{18} alkanes through successive catalytic steps.

route is highly promising since 90% of nonene to alkane conversion was achieved. Alkanes were primarily composed of C_{18} with 71% by mass. Consequently, it was possible to produce 50 kg liquid hydrocarbons from 100 kg GVL by preserving 90% energy content of GVL.

Dumesic et al. proposed another attractive route for the transformation of GVL into alkenes with molecular weights suitable for transportation fuels. An integrated catalytic system consisting of two flow reactors, carrying out DCO and oligomerization as the major reactions, was used without any requirement for an external hydrogen source (Bond et al. 2010). As shown in Scheme 10.2, GVL was converted into an isomeric mixture of pentenoic acids through ring opening on SiO_2/Al_2O_3 and then to an equimolar mixture of butenes and CO_2 through DCO. Ninety-eight percent butene yield was achieved at 99% GVL conversion in the first reactor, which operated at 648 K and 3.6 MPa. Oligomerization of a mixture of butenes and a stoichiometric amount of CO_2 into alkenes was carried out on acid catalysts, HSZM-5 and Amberlyst-70, at 3.6 MPa and temperatures 498 and 443 K, respectively. Amberlyst-70 was shown to be a more effective catalyst than HZSM-5 considering its performance on the yield of C_8–C_{16} and C_{8+} alkenes and for minimizing unwanted CRA products. It was possible to reach more than 75% overall yield of C_{8+} alkenes with 12 g Amberlyst-70 catalyst, whereas it was only possible to obtain ~30% overall yield of C_{8+} alkenes using 14 g HZSM-5.

Reductive lignin depolymerization by HYD coupled with CRA, hydrolysis, or oxidation is an essential route for catalytic valorization of lignin into platform chemicals that can be converted to fuel-grade compounds. Phenol, unfunctionalized aromatics, such as benzene, toluene, xylene (B, T, X), and cycloalkanes are the major value-added compounds generated upon catalytic reduction of lignin (Zakzeski et al. 2010). Lignin depolymerization can be initiated through aforementioned catalytic routes; however, its upgrading into fuel-range aromatics involves primarily HDO. Noble metals on a variety of supports have been found as active candidates for converting lignin to high carbon number cycloalkanes (Ohta et al. 2011; Zhao and Lercher 2012; Yoon et al. 2013; Saidi et al. 2014; Zhang et al. 2014; Chen et al. 2015). Pt impregnated on various carbon supports (AC, mesoporous carbon, MWCNT) were screened for HDO of 4-propylphenol in water at 753 K and 4 MPa H_2 (Ohta et al. 2011). Pt on AC (Norit, AC(N)) resulted in the highest yield (97%) toward the deoxygenated product, propylcyclohexane (C_9). The same catalyst was also investigated for HDO of different guaiacols and syringols (characteristic products of lignin pyrolysis) at the same temperature and pressure conditions. Propylcyclohexane yield ranged between 65% and 78%, and the remaining products, propylcyclopentane (C_8) and methanol, were obtained with yields ranging 5–12% and 2–3%, respectively. In another study, a combination of Pd/C and HZSM-5 was shown to convert an aqueous mixture (14.2 wt.%) of nine different phenolic monomers into C_6–C_9 naphthenes (gasoline range) with 87% conversion (C%) at 473 K and 5 MPa H_2 in 4 h (Zhao and Lercher 2012). The same catalyst combination was used with the goal of converting phenolic dimers to cycloalkanes. C–O cleavage at the indicated conditions resulted in cyclohexane (C_6) and ethylcyclohexane (C_8) production with 46% and 54% yields, respectively. Besides, more than 95% conversion to C_{12}, C_{14}, and C_{16} bicycloalkanes was achieved by the removal of hydroxyl and ketone groups. Yoon et al. (2013)

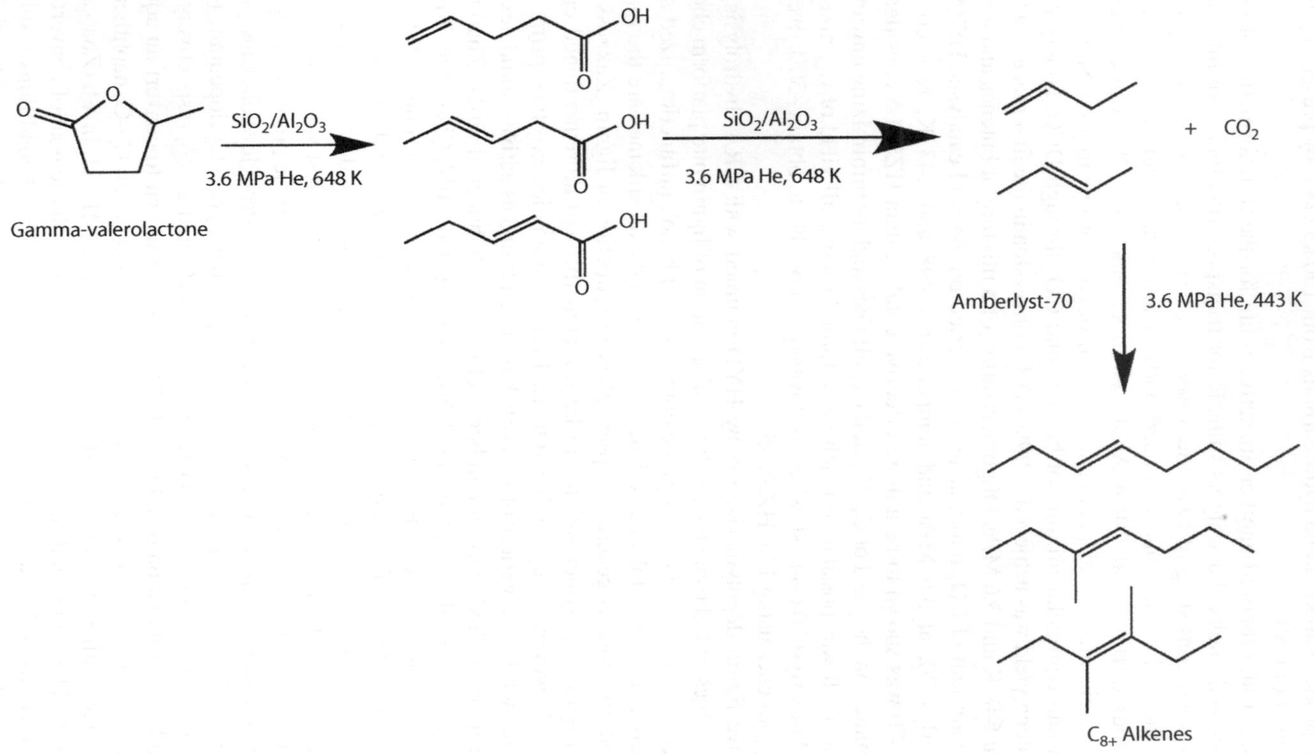

Gamma-valerolactone

SiO_2/Al_2O_3
3.6 MPa He, 648 K

OH
O

SiO_2/Al_2O_3
3.6 MPa He, 648 K

+ CO_2

Amberlyst-70
3.6 MPa He, 443 K

C_{8+} Alkenes

SCHEME 10.2 Reaction pathway for conversion of GVL to C_{8+} alkenes through successive catalytic steps.

suggested a two-step mechanism for the production of high carbon number dicyclohexyl hydrocarbons from a dimeric phenol, benzyl phenyl ether (BPE). In the first step, BPE was completely isomerized into phenol and benzyl phenols through ether cleavage on a silica–alumina aerogel at 373 K and 0.5 MPa H_2. GC/MS analysis of the product mixture showed that the dimers benzyltoluene and diphenylethane and the trimer dibenzylphenol were obtained. Products of this step were then hydrodeoxygenated using Ru impregnated on a commercial silica–alumina support at 523 K and 4 MPa H_2. Dicyclohexylmethane (C_{13}) was obtained as the most abundant product with 52.4% selectivity upon two-step HDO of BPE. Instead, direct HDO (one-step reaction) of BPE carried out at 523 K and 4 MPa H_2 yielded C_6 (methylcyclopentane and cyclohexane) and C_7 (methylcyclohexane and cycloheptane) with total selectivity of 96.3% and a minor amount of C_{13} (1.3%). In addition, Ru impregnated on HZSM-5 (Si/Al ratio of 25) was used for HDO of phenolic monomers and dimers at 473 K and 5 MPa H_2 (Zhang et al. 2014). While Brønsted acid sites in the pores of HZSM-5 catalyzed dehydration of phenolic compounds, Ru catalyzed HYD. HDO of phenolic monomers in the range of C_7–C_9 resulted in product selectivities higher than 89%. Phenolic dimers containing 4-O-5, α-O-4, β-5 phenylcoumaran ether linkages, and 5-5' aryl-aryl linkage were transformed into cycloalkanes with conversions higher than 99% at 493 K. Those having 4-O-5, α-O-4, and β-5 phenylcoumaran ether linkages yielded cycloalkanes with selectivities higher than 96%. Furthermore, Ru impregnated on different supports were explored in a biphasic H_2O/*n*-dodecane system for HDO of eugenol at 493 K and 5 MPa H_2 for 3 h (Chen et al. 2015). Among Ru-based catalysts, Ru on carbon nanotubes (Ru/CNT) exhibited superior performance for HDO of eugenol in comparison to Ru/AC, Ru/ZrO_2, and Ru/CeO_2 due to the unique morphology of CNT enabling better interaction for the metal catalyst and the substrate. Propylcyclohexane yields obtained upon HDO of eugenol using Ru/ZrO_2, Ru/AC, and Ru/CeO_2 were 46%, 22%, and 13%, respectively. Intermediate chemicals 4-propyl-cyclohexanol, 2-methoxy-4-propyl-cyclohexanol, and 2-methoxy-4-propylphenol were derived from eugenol in the presence of Ru/ZrO_2, Ru/AC, and Ru/CeO_2. Propylcyclohexane and propylcyclopentane were the abundant products obtained upon HDO of eugenol using Ru/CNT with 94% and 4% yields, respectively. In the same study, significant advantages of using a biphasic system for HDO of phenols were established regarding the selectivity and purification of cycloalkanes. Utilization of biphasic system for HDO of eugenol using Ru/CNT increased propylcyclohexane yield from ~57% (aqueous system) to 94%. When the same reaction was carried out in pure *n*-dodecane, propylcyclohexane yield was only 4% and oxygenated intermediate, 4-propylcyclohexanol was formed as the main product.

Above discussions mostly comprise catalytic valorization of lignin-derived molecules in aqueous phase catalytic systems (Ohta et al. 2011; Zhao and Lercher 2012; Yoon et al. 2013; Zhang et al. 2014; Chen et al. 2015). Alternatively, Yan et al. (2010) proposed a bifunctional catalytic system including Brønsted acidic ILs immobilized in nonfunctionalized ILs (used as solvents) and metallic nanoparticles (Rh, Ru, and Pt) operated at 403 K and 4 MPa H_2 for 4 h for the conversion of different phenolic compounds to cyclohexane. Eighty percent cyclohexane yield was achieved at 99% phenol conversion using the Rh catalyst in the acidic IL, 1-methyl-3-(ethyl-3-sulfonic acid)imidazolium triflate, which is immobilized in 1-butyl-3-methylimidazolium bis(trifluoromethylsulfonyl)imide (BMIMTF$_2$N). In addition, Bi et al. (2015) proposed a three-step reaction mechanism for the production of jet fuel-range alkanes from lignin, in which the second step, alkylation, was carried out in the IL, BMIMCl-AlCl$_3$. Lignin from wheat straw was subjected to depolymerization through pyrolysis using the HZSM-5 catalyst at 773 K. The product stream, consisting of low carbon number aromatics, in the range of C_6–C_8, was alkylated with light olefins (C_2H_4, C_3H_6, and C_4H_8) in the presence of BMIMCl-AlCl$_3$ at 333 K to produce higher carbon number aromatics. The resulting product was found to contain C_8–C_{15} aromatics with ~79 wt.% selectivity. HYD of these aromatics using Pd/C catalysts at 453 K yielded C_8–C_{15} cycloparaffins with ~80 wt.% selectivity. This final product, cyclic alkane biofuel (CABF), demonstrated notable similarities with the commercial/military jet fuels. While the heat of combustion for commercial/military jet fuels was reported to range between 40 and 43.3 MJ/kg, the heat of combustion was measured as ~43 MJ/kg for C_8–C_{15} cycloparaffins. Even the intermediate product, C_8–C_{15} aromatics, referred to as aromatic biofuel (ABF), met this requirement with a heat of combustion of ~46 MJ/kg. Both the CABF and the ABF were found to possess notably low oxygen and sulfur contents, and their carbon content (~85–90 wt.%) was comparable with that of commercial/military jet fuels (86.2–86.4 wt.%).

10.3 CHEMICALS FROM LIGNOCELLULOSIC BIOMASS

The majority of the chemicals produced in the chemical industry originate from fossil resources including petroleum and natural gas, and only about 5% of these fossil resources are sufficient for the production of petrochemicals utilized in the world for food, chemical, clothing, and manufacturing industries (Cherubini and Stromman 2011). As discussed in the previous section for energy production, fossil sources have also been most suitable for the production of chemicals, because they have been so far the least expensive and most readily available resources (Behr et al. 2008). In addition, many years of research and development in the conversion of fossil resources allowed the optimization of the production of certain building blocks (i.e., ethylene, propylene) of our current chemical industry. However, these fossil resources as mentioned earlier are losing their appeal as a result of continued increase in their prices, questions about their availability, environmental concerns, and a general demand for renewable or "green" fuel and chemical products from the consumers. The previous section showed that the utilization of biorenewable resources for the production of energy and fuels received a lot of attention from the scientific community, while less attention has been paid to the production of biochemicals. In addition, the majority of biomass used for energy production is burned to generate electricity. This approach has not been very sensible, because while the production of renewable energy can be realized from a range of resources such as solar, wind, hydrothermal, nuclear, and geothermal, production of chemicals requires a renewable carbon resource, which can only be obtained from biomass (Cherubini 2010). Even though only about 5% of petroleum is converted into chemicals, these generate roughly 50% of the profit in current petrochemical refineries, indicating the large economic potential in producing chemicals from biomass (Rinaldi and Schueth 2009). Accordingly, this section focuses on the developments of technologies on the conversion of various parts of biomass to chemicals.

10.3.1 Comparison of Fossil and Biomass Resources for the Production of Chemicals

Crude oil is distilled into its many fractions (i.e., gasoline, diesel fuel, naphtha, kerosene, lubricating oils, and asphalts) in an oil refinery. Among these fractions, the naphtha fraction is processed further to obtain a few platform chemicals, such as ethylene, propylene, butadiene, and aromatics, upon which the whole petrochemical industry is built (Wells 1999). The composition of biomass, however, is significantly different from the naphtha feedstock. In general, biomass feedstocks have an excess of oxygen and lack of hydrogen compared to naphtha (Wells 1999). Accordingly, olefins, key platform chemicals of the petrochemical industry, containing no oxygen, would require the complete DDO of the biomass feed, making biomass a poor starting point (Vennestrom et al. 2011). This also indicates that directly replacing the olefin market today with bio-based olefins may not be ideal, and biomass should be utilized through selective DDO reactions to form chemicals obtained in downstream of the petrochemical industry, therefore indirectly replacing olefins. Some examples of oxygen-rich chemicals in the petrochemical industry include ethylene glycol, acetic acid, adipic acid, and acrylic acid that could potentially be produced more efficiently from biomass resources compared to fossil resources. This strategy of converting biomass to already existing petroleum-derived intermediates can be referred to as the "drop-in" strategy (Figure 10.2) following Vennestrom et al. (2011). This strategy entails several advantages, as it requires minimal disruption to the downstream processing of these intermediates, such as in polymerization reactions for the formation of various plastics. However, the developed biomass-based process has to be cost-competitive to the already existing fossil resource-based technologies. As

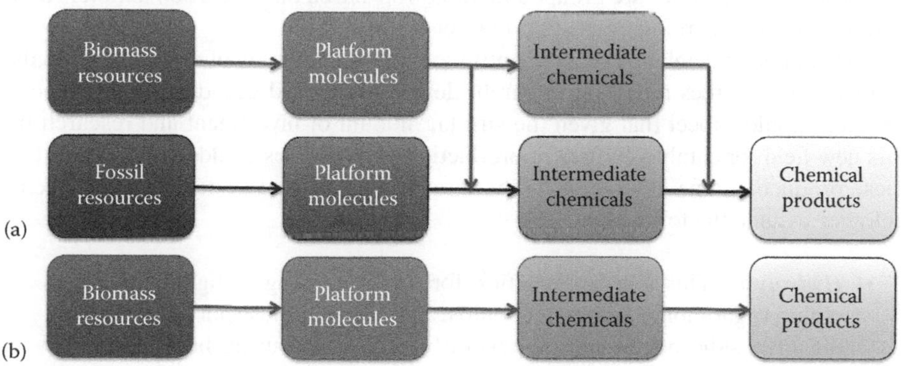

FIGURE 10.2 (a) Drop-in strategy for chemical products with existing markets and (b) emerging strategy for chemical products with emerging markets starting from biomass resources.

an alternative to a head-to-head substitution of petrochemicals in the "drop-in" strategy, biomass resources can be utilized to produce new platform chemicals that are more easily synthesized and that are closer to the initial biomass composition in what is referred to as the "emerging" strategies (Figure 10.2) (Vennestrom et al. 2011). While it is expected to be more difficult to implement these new intermediates in the existing chemical industry, the final products do not have to directly compete with existing products, and the inherent oxygen functionalities of the biomass feedstocks can provide flexibility on the range of possible platform chemicals. This approach would indicate a radical shift from the petrochemical industry and would require the development of not only new value chains but also novel production routes. However, at the same time, the final products must contain the same functional properties expected by the existing consumers (Marquardt et al. 2010). Therefore, the development of novel platform chemicals (and reaction routes) should be driven by the functional requirements of the existing applications to address the demand from the society. After the functional requirements are determined, the molecular structures of the potential chemicals that can satisfy these requirements need to be identified. This should be followed by the determination of appropriate reaction pathways, which would also identify the most suitable feedstock for the chemical conversions.

These considerations mainly apply to the carbohydrate fraction of lignocellulosic biomass, which are cellulose and hemicellulose fractions. Lignin portion of lignocellulose, not possessing a well-defined structure, poses some difficulties for performing selective chemistry (Vennestrom et al. 2011). Even though lignin has a potential to lead to a wide spectrum of oxygenated aromatics, carrying out selective bond cleavage reactions in order to maintain the monomeric lignin block structures is highly challenging (Holladay et al. 2007). Consequently, the best approach for getting value out of lignin currently is its conversion to synthesis gas through gasification, which is to be integrated within the biorefinery operations rather than a stand-alone application. The cellulose and hemicellulose portions of lignocellulose, on the other hand, can be converted to a wide spectrum of platform chemicals, and the most promising ones are grouped in Table 10.1 based on the carbon numbers and with the size of emerging or existing markets.

It should be remembered that the optimization of production of current chemicals from fossil resources relied on research, development, and decades of experience, and one should expect that given the similar amount of investment and research in this new field for catalysis, efficient production technologies could be developed for these products. Some key reactions that are being exploited to develop these technologies include the following:

- *Hydrolysis.* This is a key reaction for the processing of lignocellulose to obtain sugar monomers from cellulose and hemicellulose portions by cleaving glycosidic bonds, and is generally carried out using homogenous or solid acid catalysts.
- *Dehydration.* This reaction is catalyzed by liquid or solid acids and used to remove oxygen by generating water and forming C=C bonds.

TABLE 10.1
Examples of Chemicals (Based on Carbon Number) That Can Be Obtained from Biomass Resources through Existing or Emerging Technologies

Carbon Number	Chemical	Market Size (Mt year^{-1})	References
2	Ethylene	110	(Vennestrom et al. 2011)
	Ethanol	60	(Vennestrom et al. 2011)
	Acetic acid	9.0	(Vennestrom et al. 2011)
3	Lactic acid	*0.3–0.5*	(Vennestrom et al. 2011)
	3-Hydroxypropionic acid	*~0.6*	(Vennestrom et al. 2011)
	1,3-Propanediol	*0.1–0.6*	(Vennestrom et al. 2011)
4	Diacids (e.g., succinic and fumaric acids)	*0.1–0.6*	(Vennestrom et al. 2011)
5	Furfural	*0.3*	(Karinen et al. 2011)
	Levulinic acid	*~0.5*	(Vennestrom et al. 2011)
6	Gluconic acid	–	–
	5-Hydroxymethylfurfural	–	–
	2,5-Furan dicarboxylic acid	–	–
	Adipic acid	2.7	(Weissermel and Arpe 2003)
	Hexamethylenediamine	~1.4	Rennovia Markets

Note: Market sizes of emerging technologies are shown in italic.

- *HYD/reduction.* HYD reactions are catalyzed on metals to saturate C=C and C=O bonds. Acid groups can also be reduced on metals to form alcohol functionalities.
- *Hydrogenolysis/hydrodeoxydation.* Cleavage of C–O bonds is achieved either on metal (hydrogenolysis) or metal–acid bifunctional systems.
- *Oxidation.* Oxidation of the carbon atoms on metal catalysts is generally carried out for the synthesis of carboxylic acids.

10.3.2 Emerging Catalytic Processes for the Production of Existing or Alternative Chemicals from Biomass

Following the discussions on different approaches of incorporating catalytic strategies for the conversion of lignocellulose to chemicals, this section explores in detail examples of emerging catalytic processes for the production of existing or alternative chemicals that will ensure the formation of final products that contain the same functional properties expected by the existing consumers.

10.3.2.1 1,3-Propanediol (Existing Market)

1,3-Propandediol (1,3-PD) is a valuable chemical intermediate used in the production of resins, engine coolants, water-based inks, and most importantly polypropylene terephthalate (PPT), a polyester synthesized from 1,3-PD and teraphthalic acid. The current production of 1,3-PD relies on petroleum derivatives, such as ethylene oxide (through subsequent hydroformylation and HYD) or acrolein (through hydration followed by HYD) (Kraus 2008). Glycerol, being the smallest polyol readily available from biomass, has recently attracted significant attention as a renewable starting feedstock for the production of 1,3-PD. Glycerol is the backbone of triglycerides and thus released as a by-product from biodiesel production in significant amounts (1 kg of glycerol is obtained for every 100 kg of biodiesel) (Behr et al. 2008; Bozell and Petersen 2010). While catalytic conversion of glycerol to various products through different pathways is being investigated, production of 1,3-PD is commercially most interesting due to the existing market (ten Dam and Hanefeld 2011). DuPont recently reported the direct conversion of glycerol to 1,3-PD through fermentation (Emptage et al. 2013); however, this biological process suffers from low metabolic efficiency and poor compatibility with existing infrastructure of the chemical industry (ten Dam and Hanefeld 2011). Many research efforts have therefore been made for the selective production of 1,3-PD from glycerol using heterogeneous catalysts, but many reported attempts resulted in the selective production of 1,2-propandiol (1,2-PD) instead of 1,3-PD (Scheme 10.3), signifying that the transformation of glycerol to 1,3-PD is more challenging. Reaction conditions, as well as metal, acid, and base components of the solid catalysts, have been varied to reach a better understanding of the factors governing the selective reduction of glycerol to 1,3-PD. It has been suggested in many of these studies that the formation of 1,3-PD goes through the intermediate formation of 3-hydroxypropanal. This species is obtained by the endothermic elimination of the secondary alcohol group of glycerol, followed by a tautomerization reaction. This aldehyde species is then hydrogenated in an exothermic step to form the 1,3-PD species (ten Dam and Hanefeld 2011; Ruppert et al. 2012; Besson et al. 2014).

One of the first promising studies toward reaching a higher selectivity to 1,3-PD compared to 1,2-PD was carried out by Chaminand et al. (2004). Thirty-two percent

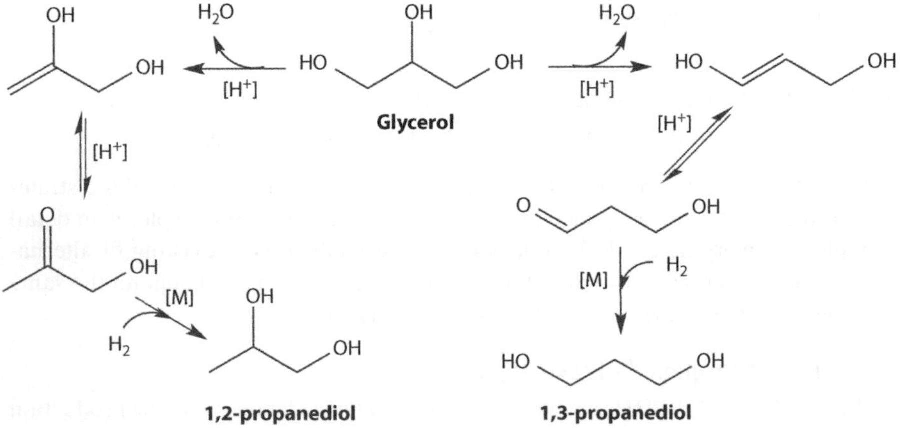

SCHEME 10.3 Conversion of glycerol to 1,2-propanediol and 1,3 propanediol.

glycerol conversion with 12% selectivity to 1,3-PD was obtained (with a 1,3-PD to 1,2-PD molar ratio of 2) using Rh/C in a slurry with tungstic acid and sulfolane. Following this work, many studies emerged for impregnation of several acidic supports with tungsten oxide and various transition metals. Kurosaka et al. (2008) reached notably higher glycerol conversion (86%) with 28% 1,3-PD selectivity using Pt and WO_x supported on ZrO_2 and 1,3-dimethyl-2-imidazolidinone (DMI) as solvent, which is more stable, polar, and aprotic compared to sulfolane. This study also explored different transition state metals, and selectivity to 1,3-PD seemed to increase in the order of Rh < Ru < Ir < Pd < Pt. Using $Pt/WO_3/ZrO_2$ as the catalyst, another study by Gong et al. (2009) explored the effect of the solvents. Specifically, polar, aprotic solvents, such as sulfolane and DMI, were compared to polar protic solvents, such as ethanol and water. This study claimed that aprotic solvents, in fact, resulted in the production of more 1,2-PD than 1,3-PD. It was suggested that a protic solvent, such as ethanol or water, improved 1,3-PD selectivity by facilitating the proton transfer from the heterogeneous acid site to the secondary alcohol functionality of glycerol, which is then eliminated. Glycerol conversion in water was measured to be lower than in ethanol due to equilibrium limitations of the DDO reaction, which can, however, be overcome by the rapid reduction of the enol intermediate (Gong et al. 2009). Qin et al. (2010) reached high glycerol conversion (70%), however still limited 1,3-PD selectivity (46%) in water on $Pt/WO_x/ZrO_2$ catalyst at 403 K using a fixed-bed reactor. Replacing ZrO_2 support with TiO_2 also resulted in similar selectivity toward 1,3-PD (44%), however, activity was measured to be low due to the low surface area of TiO_2, which could be overcome by impregnating TiO_2, WO_3, and Pt on SiO_2 (51% selectivity to 1,3-PD was achieved at 15% glycerol conversion) (Gong et al. 2010). Various other bimetallic catalysts supported on SiO_2, such as Rh-ReO_x, Rh-MoO_x, Rh-WO_x, and Ir-ReO_x, were studied to carry out selective DDO of aqueous phase glycerol, but the 1,3-PD yields were still quite limited (~20%), with 1,2-PD often produced in greater quantitites (Shimao et al. 2009; Nakagawa et al. 2010; Shinmi et al. 2010). In a more recent study by Oh et al. (2011), Pt deposited on sulfated zirconia was used to catalyze the conversion of glycerol to 1,3-PD. Fifty-six percent yield to 1,3-PD was obtained at 443 K in DMI solvent with 73 bar H_2 (Oh et al. 2011). The authors suggested that the deposition of Pt on sulfated zirconia increased the number of Brønsted acid sites through Pt catalyzed hydrogen spillover, and these acid sites were responsible for preferentially protonating the secondary alcohol group in glycerol. Following this work, Mizugaki et al. (2012) converted glycerol without any additives at 453 K under 30 bar H_2 on a Pt–AlO_x/WO_3 catalyst with a 40% yield of 1,3-PD at 90% glycerol conversion. The catalyst was reused four times without significant deactivation, and the authors proposed that AlO_x preferentially bonded the primary OH groups of glycerol to form Al-alkoxides, allowing the subsequent hydrogenolysis of the unbound secondary alcohol group by Pt particles. The research group furthered their study by depositing Pt and WO_x on boehmite (monohydroxy aluminum oxide, AlOOH), which has a higher number of Al–OH groups to promote Al–alkoxide formation through the primary alcohol groups of glycerol (Arundhathi et al. 2013). The authors could reach 66% yield of 1,3-PD at complete conversion of glycerol in aqueous phase at 453 K and 50 bar H_2 in a batch system. As all these studies showcase, the formation of 1,3-PD is not as straightforward as it seems; however, significant progress has been made over the last couple of years, and substantial research efforts are made to further improve the selective conversion of glycerol.

10.3.2.2 Adipic Acid (Existing Market)

Adipic acid is a versatile building block in the chemical, pharmaceutical, and food industries, and one of the most common commodity chemicals worldwide, with a projected market size of more than 6 million lbs. by 2017 (Weissermel and Arpe 2003). Adipic acid has use in the production of polyester, polyurethane, polyvinyl chloride, cosmetics, adhesives, lubricants, and insecticides, but its primary use lies in the production of the polyamide Nylon-6,6 (Casanova et al. 2009). Currently, adipic acid is produced from the catalytic oxidation of a mixture of cyclohexanol and cyclohexanone obtained from petroleum via an inefficient catalytic process (Castellan et al. 1991; Chaminand et al. 2004; Welch et al. 2005; Hermans et al. 2006; Casanova et al. 2009; Cherubini 2010; Gursel et al. 2012). As a result, adipic acid has attracted a significant amount of attention from the academic and industrial catalysis community as a potential "drop-in" chemical to be produced in a more efficient and sustainable manner from lignocellulose (Van de Vyver and Roman-Leshkov 2013). Many US-based start-up companies, such as Verdezyne, Rennovia, and BioAmber, have declared ongoing initiatives to enter the adipic acid market. Several processes have been developed starting from glucose involving different intermediates, such as glucaric acid, 5-HMF, and GVL. Rennovia, a US start-up company that specializes in the development of sustainable production routes of chemicals from biomass, disclosed a patent for the production of adipic acid from glucose through the intermediate production of glucaric acid (Scheme 10.4a) (Cherubini and Stromman 2011). Oxidation of glucose to glucaric acid in aqueous phase, in this process, is carried out on Pt-based catalysts at 363 K and 5 bar of O_2, in absence of any base, reaching high yields (~70%). Following its production, glucaric acid undergoes a HDO reaction on bimetallic Pt–Rh based catalysts in the presence of HBr and acetic acid at 413 K and 49 bar of H_2 to form adipic acid with yields up to 90%.

Production of HMF from the cellulose portion of lignocellulosic biomass and C_6 sugars (i.e., glucose and fructose) obtained from hydrolysis of cellulose has attracted significant attention in the catalysis research community over the past few years as will be discussed in the next section (Climent et al. 2011; Karinen et al. 2011; van Putten et al. 2013). Accordingly, several academic and industrial research groups developed pathways for the conversion of HMF to adipic acid. For example, Boussie et al. (2010) at Rennovia Inc. in 2010 patented a two-step process for the production of adipic acid from HMF through the formation of 2,5-FDCA. FDCA could be formed from HMF on Au/CeO_2 and Au/TiO_2 catalysts in aqueous phase through the intermediate formations of 5-hydroxymethyl-2-furancarboxylic acid and 5-formyl-2-furancarboxylic acid. Following its formation, FDCA was first hydrogenated to tetrahydrofuran-2,5-dicarboxylic acid (THFDCA) on Pd/SiO_2 under 52 bar H_2 at 413 K with a yield of 88%. Finally, THFDCA was hydrogenated to adipic acid, using Pd/SiO_2 or Rh/SiO_2 in the presence of HBr or HI in acetic acid, under 49 bar H_2 at 433 K, yielding 99% adipic acid (Scheme 10.4b).

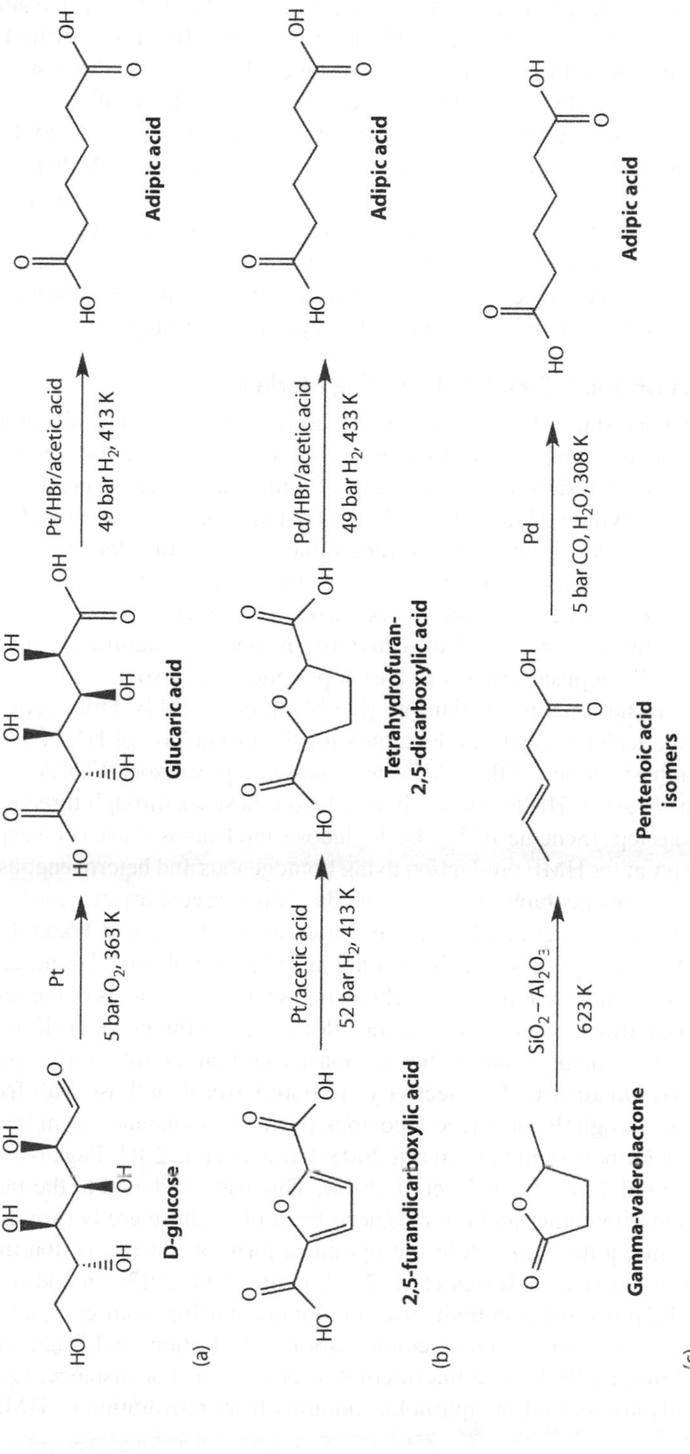

SCHEME 10.4 Alternative production pathways of adipic acid starting from (a) glucose, (b) 2,5-furandicarboxylic, and (c) gamma-valerolactone.

Another attractive route for the production of adipic acid is through the ring opening of GVL. GVL can be derived from the catalytic HYD of levulinic acid, an important platform chemical derived mainly from cellulose (Bond et al. 2010; Lange et al. 2010; Qi and Horvath 2012). Recently, Wong et al. proposed a three-step catalytic process for the production of adipic acid from GVL (Scheme 10.4c) (Wong et al. 2012). In the first step, a mixture of isomers of pentenoic acid is obtained through reactive distillation of GVL in the presence of a SiO_2–Al_2O_3 catalyst. Pentenoic acids then undergo carbonylation reaction on supported Pd catalysts to form adipic acid, which is then precipitated through crystallization, allowing the unreacted pentenoic acids to be recycled. Adipic acid yields of 22–48% were obtained in a single pass after collection of crude adipic acid. Adipic acid was the only product detected with ^{13}C NMR and gas chromatography prior to the crystallization step.

10.3.2.3 5-HMF and 2,5-FDCA (Emerging Markets)

The dehydration product of hexoses, 5-HMF is a heterocyclic furanic molecule that has both aldehyde and alcohol functionalities. HMF can be either oxidized to a dicarboxylic acid or reduced to a diol, both of which are excellent precursors for the synthesis of polymers (Wang T. et al. 2014). Therefore, this unique chemical structure of HMF makes it an attractive renewable starting point for the formation of functional or direct replacement alternatives to the current commodity chemicals. For example, HMF can be converted to six-carbon monomers, such as 2,5-FDCA, 5-hydroxymethylfuranoic acid, 2,5-diformyl furan, and 2,5-diaminomethylfuran, that can potentially replace crude oil–based polymer precursors, such as adipic acid, alkydiols, or hexamethylenediamine (Dashtban et al. 2014). This section will first focus on the technological developments for the production of HMF, and then explore the synthesis of one of the HMF-based polymer precursors, FDCA.

As mentioned earlier, HMF can be obtained from hexoses through three successive dehydration steps (Scheme 10.5a). Both glucose and fructose have been explored as the starting point for HMF production using homogenous and heterogeneous catalysts, in aqueous, organic, biphasic, and IL media. Many recent review articles from Hu et al. (2012), Karinen et al. (2011), van Putten et al. (2013), and Wang T. et al. (2014) focused on the synthesis of this promising platform chemical summarizing recent advances in understanding the mechanistic aspects, and effects of the reaction conditions and catalyst effects, as well as the pilot and precommercial HMF production status. Many of the previous studies demonstrated that the rate of glucose conversion and corresponding HMF selectivity are both lower than those with fructose conversion, even though the preferred feedstock due to its abundance from lignocellulose is glucose (Ribeiro and Schuchardt 2003; Climent et al. 2011; Pagan-Torres et al. 2012; Gallo et al. 2013; Wang T. et al. 2014). This was attributed to the fact that the cyclic furanose tautomer and the pyranose form of fructose are both present in aqueous phase in equilibrium, while the pyranose form of glucose is dominant at equilibrium in water (Pagan-Torres et al. 2012; Gallo et al. 2013). In addition, the selectivity of HMF is limited in both cases (and more starting from glucose) due to various other side reactions, such as condensation, rehydration, and fragmentation reactions involving HMF, hexose intermediates, and water. For instance, levulinic and formic acids are formed in equimolar amounts from rehydration of HMF, and

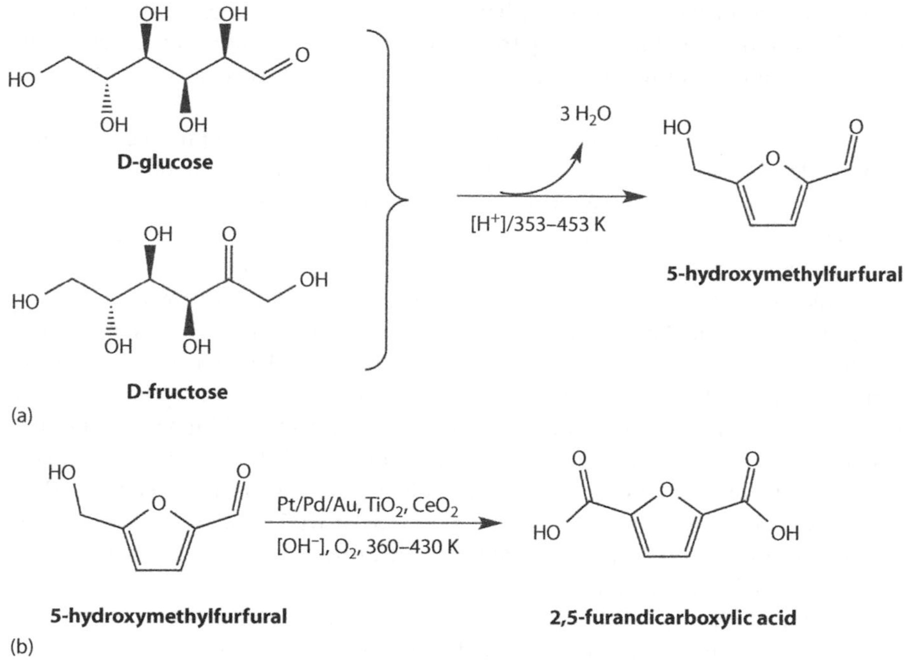

SCHEME 10.5 Conversion of (a) hexoses to 5-HMF and (b) HMF to 2,5-FDCA.

condensation and polymerization reactions among HMF and hexose intermediates lead to formation of insoluble humins (Karinen et al. 2011). As a result, to develop an efficient formation strategy for HMF, research has focused on identifying catalysts and reaction/operation conditions to limit the unwanted side reactions, as well as integrating glucose isomerization step to fructose with fructose dehydration step.

It was observed in many studies that HMF yields were poor regardless of the acid catalyst used if water was used as the solvent. However, the limited solubility of hexoses in most organic solvents led researchers to work with high boiling point aprotic solvents, such as DMSO, which could solubilize these sugars (Climent et al. 2011; Wang T. et al. 2014). Ion-exchange resins were among the first solid acid catalysts explored for the production of HMF. Nakamura and Morikawa (1980) used a strongly acidic ion-exchange resin, Diaion PK-216, very early on and achieved 90% HMF yield from fructose at 353 K in dimethyl sulfoxide (DMSO). While it results in excellent selectivity, the use of DMSO as the solvent is problematic due to its high boiling point and difficult separation from HMF. Accordingly, since then, research has focused on identifying systems that lead to high HMF selectivity as well as a facilitated HMF separation. Simeonov et al. (2012) used crystallizable salts as a solvent to facilitate HMF separation. Yields higher than 92% were obtained from fructose in tetraethylammonium bromide with Amberlyst-15 at 373 K (Simeonov et al. 2012). Various zeolites, such as beta (Rivalier et al. 1995), Y (Rivalier et al. 1995), faujasite (Rivalier et al. 1995), and mordenite (Moreau et al. 1996), have also been evaluated for the conversion of hexoses to HMF. The highest yield (69%) has been obtained on mordenite starting from fructose in a continuous flow reactor with countercurrent flows of water and methyl isobutyl ketone (MIBK) at 438 K (Rivalier et al. 1995).

A similar approach for achieving high yields of HMF in the presence of water is to employ biphasic rector systems. In this way, HMF is continuously removed from the aqueous acidic phase during reaction, avoiding its acid catalyzed consecutive reactions, including reactions with water. In the presence of an organic phase, homogeneous mineral acids can be used to catalyze the reaction in the aqueous phase, and they can be recovered and recycled by separating the aqueous and organic phases after reaction. An important factor in choosing extractive solvents in biphasic systems is the partitioning of HMF between two phases. This partitioning can be altered by using different extractive solvents or saturating the aqueous layer with an inorganic salt, which is referred to as the salting-out effect, and higher partitioning of HMF to the organic layer generally leads to higher HMF selectivities (Roman-Leshkov and Dumesic 2009). For example, Chheda et al. (2007) achieved 85% HMF yield at 95% fructose conversion starting from a 10 wt.% fructose solution in an aqueous layer of 1:1 (w/w) water:DMSO with HCl (pH = 1), and an extracting layer of 7:3 (w/w) MIBK:2-butanol.

As mentioned earlier, HMF yields obtained from direct glucose conversion using homogeneous or heterogeneous Brønsted acids are considerably lower (less than 25%) than those from fructose in aqueous as well as organic solvents. However, since glucose is the more abundant and cheaper feedstock, researchers explored other options, such as coupling the isomerization reaction of glucose to fructose with fructose dehydration reaction to form HMF in a single-pot system. In 2011, a tin exchanged beta zeolite (Sn-BEA) was found to be active and selective for isomerization of glucose to fructose (Nikolla et al. 2011). Nikolla et al. (2011) combined Sn-BEA (Lewis acid for isomerization) and HCl (Brønsted acid for dehydration) in a water–tetrahydrofuran (THF) biphasic system and achieved 72% HMF selectivity at 79% glucose conversion at 453 K. In another study, Fan et al. (2011) used a heteropolyacid salt ($Ag_3PW_{12}O_{40}$) that has both Lewis and Brønsted acidity in a water–MIBK biphasic system and achieved 76% HMF yield at 90% glucose conversion at 403 K. Lewis acidic metal salts (e.g., $AlCl_3$, $SnCl_4$, $GaCl_3$, $YbCl_3$) have been combined with HCl in a study by Pagan-Torres et al. (2012) to catalyze glucose dehydration through the intermediate fructose formation in water-sec-butylphenol biphasic solutions, and the highest HMF yield was ~60% when $AlCl_3$ was used as the Lewis acid. In a more recent work, Gallo et al. (2013) combined the Lewis acid Sn-BEA and Brønsted acid Amberlyst-70 for the conversion of glucose in a monophasic system using GVL, gamma-hexalactone, or THF (each solvent contained 10 wt.% water), and reached 64%, 59%, and 70% HMF selectivity, respectively, at ~90% glucose conversion at 443 K, eliminating the need for biphasic systems or corrosive mineral acids.

Despite many studies on changing the solvent systems and composition, only a few studies aimed to develop new catalysts to improve HMF yields from fructose as well as glucose. One example is from Alamillo et al. (2013), who showed that HMF yields could be improved through controlling the fructose tautomer equilibrium by creating a microenvironment where poly(vinylpyrrolidone) is cross-linked in the mesopores of acid-functionalized silica.

In addition to papers and patents on laboratory-scale production of HMF, some work on pilot plant-scale production of HMF has been reported and patented. In a recent example, Evonik Degussa GmbH carried out a kilogram-scale production of high purity HMF (>99%) from fructose by first formation and purification of a derivative of HMF, 5-acetoxymethylfurfural (AMF), followed by its conversion to HMF and final purification. An acidic ion-exchange resin (Dowex 50WX8-200) was used for the dehydration of fructose at 383 K forming HMF, which was then reacted further in the presence of acetic acid and 4-(N,N-dimethylamino)pyridine to form AMF. AMF could be separated more energy efficiently, compared to HMF, through consecutive distillation steps. After the purification of AMF, HMF was obtained by the hydrolysis of AMF in methanol using a strongly basic ion-exchange resin (Amberlyst-A26 OH) at room temperature. About 7 kg of high purity HMF could be obtained from 39 kg of fructose (Reichert et al. 2008).

FDCA is one of the most promising chemicals that can be obtained from HMF, as it has been demonstrated that polyethylene furanoate (PEF), produced from the coupling of ethylene glycol and FDCA, has comparable chemical, physical, and mechanical properties and therefore has the capability to replace polyethylene terephthalate (PET) (Eerhart et al. 2012). The oxidation of HMF to FDCA (see Scheme 10.5b) with air or oxygen has been studied on several different catalysts. Different catalytic systems using Pt-, Pd-, Ru-, or Au-based heterogeneous catalysts have been reported for the selective oxidation of HMF to FDCA in aqueous phase. Gorbanev et al. (2009) achieved 71% FDCA yield at full conversion using Au/TiO_2 in the presence of homogenous base at ambient temperature. Casanova et al. (2009) could achieve 99% FDCA yield on Au/CeO_2 in aqueous phase in the presence of base. Pasini et al. (2011) used bimetallic Au–Cu particles supported on TiO_2 to achieve 99% yield of FDCA at 368 K. Davis et al. (2011) also achieved high FDCA yields (~80%) with molecular oxygen on various supported precious metals (Pt/C, Pd/C, and Au/TiO_2) in aqueous phase at high pH conditions. All these systems needed the addition of excess amounts of base, which was shown to improve both activity and selectivity. The presence of base was also found to be beneficial in terms of decreasing HMF degradation rates; however, like mineral acids, homogeneous bases increase cost of operation and product separation. Therefore, research efforts have been made to carry out base-free oxidation of HMF. Accordingly, Saha et al. (2013) reported 79% FDCA yield from HMF at complete conversion using a Fe^{III}–porphyrin complex supported on porous carbon, free of base at 373 K and 10 bar air. Gupta et al. (2011) replaced the homogeneous base with a basic support, hydrotalcite, and could reach >90% FDCA yield from HMF in aqueous phase on hydrotalcite-supported gold nanoparticles at 368 K under atmospheric oxygen pressure. Gorbanev et al. (2011a,b) studied the oxidation of HMF using ruthenium deposited on several different supports and obtained best results using a basic support, MgO, in the absence of base. However, these solid base catalysts were found to suffer from the metals leaching and possibly forming homogeneous bases in the aqueous solution (Gorbanev et al. 2011a,b). Direct synthesis of FDCA from fructose was explored in literature as well. Only 25% FDCA yield from fructose could be obtained using Pt–Bi/C in combination with a solid acid in water–MIBK solution, while high yields (71%) could be achieved using a Co(acac)–SiO_2 bifunctional catalyst at 433 K and 20 bar air pressure (Ribeiro and Schuchardt 2003).

10.3.2.4 Furfural (Emerging/Existing Market)

Hemicellulose, comprising a substantial amount of biomass (20–35%), needs to be efficiently utilized to establish a cost-competitive biorefining strategy; however, compared to cellulose, fewer studies address the conversion of hemicellulose to chemicals. Among the products that can be obtained from C_5 sugars, furfural can be used to replace crude oil–based organics for the production of resins, lubricants, adhesives, and plastics. Furfural can also be converted to other valuable chemicals, such as furfuryl alcohol, tetrahydrofurfuryl alcohol, furanoic acid, and THF (Climent et al. 2011; Lange et al. 2012; Dashtban et al. 2014). The industrial production of furfural is exclusively from lignocellulose (pentoses), with no other available synthetic routes from petroleum. However, the current production methods do not comply well with the principles of green and sustainable biorefining. The first commercial process for furfural production was achieved in 1921 by Quaker Oats (Dashtban et al. 2014) by using aqueous sulfuric acid in a batch mode at 443–458 K to achieve limited (40–50%) furfural yields. Following the Quaker Oats process, other production processes including Westpro-modified Huaxia Technology, Supra yield, and Vedernikov's single-step furfural production were developed in batch or continuous operation (Karinen et al. 2011). However, the recovery of furfural and the mineral acid is problematic in these commercial processes with single-phase operation. In addition to the dehydration reaction, various other side reactions, such as fragmentation, condensation, and resinification, take place, decreasing furfural yields, similar to HMF, as shown in Scheme 10.6. Acid catalyzed fragmentation of furfural results in the formation of smaller molecules, such as formic acid, formaldehyde, acetaldehyde, and lactic acid (Karinen et al. 2011). Resinification reaction can take place among furfural molecules. In condensation reactions, furfural reacts with pentose molecules or intermediates. Products of both resinification and condensation reactions lead to the formation of high molecular weight solid humin products (Karinen et al. 2011). Therefore, the reaction conditions need to be optimized to minimize humin production, which would in turn maximize furfural yields.

Furfural production has been studied using solid acid catalysts as well as biphasic systems as improvement on the current furfural production processes. High furfural yields (~90%) have been achieved in several previous studies using mineral acids and salts in biphasic systems with organic solvents, such as MIBK, 2-butanol, THF, and sec-butylphenol (Chheda et al. 2007; Xing et al. 2010; Guerbuez et al. 2012). However, with biphasic systems, the use of mineral acids poses environmental and cost-related issues, and mixing and separating the two phases in industrial scale may become very challenging. Therefore, this section focuses on the advances made in furfural production using solid acid catalysts in monophasic aqueous or organic solvents.

Even though biomass-derived sugars are best dissolved in aqueous solutions, many solid acid catalysts, such as zeolites, are not active in liquid water, and leaching

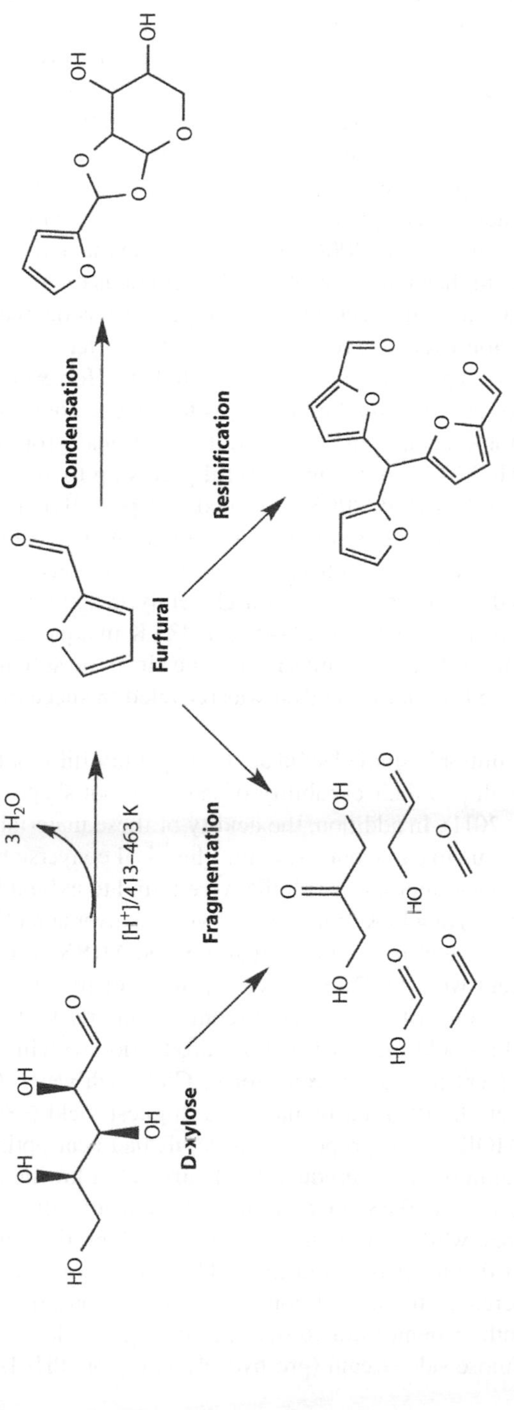

SCHEME 10.6 Dehydration of xylose to furfural and side reactions of furfural and pentose intermediates.

of acid functionalities due to hydrolysis at high reaction temperatures in aqueous environments is a significant challenge. In addition, water has been found to accelerate side reactions and decrease furfural selectivity (Dunlop 1948). Therefore, different organic compounds such as toluene, acetone, DMSO, and MIBK have been studied in addition to water as solvents for the formation of furfural over solid acid catalysts. Ion-exchange resins have been studied commonly for sugar dehydration reactions as discussed earlier for HMF formation. Operation temperature limits of most of these resins (loss of stability due to degradation and leaching has been observed for most at temperatures above 423 K), however, constraint their application for furfural production, as higher temperatures were found to increase furfural selectivity (Roman-Leshkov et al. 2006). This is because the activation enthalpy of xylose dehydration is higher than that of parallel condensation and resinification reactions, causing low furfural selectivities as temperature is decreased (O'Neill et al. 2009). Among the ion-exchange resins, Amberlyst-70 offers the highest thermal stability with an operating temperature limit at 463 K. However, limited furfural yields (~30%) were obtained on Amberlyst-70 for xylose conversion in aqueous phase at 448 K. Sulfonic acids were exploited as functional groups on other supports, for example, MCM-41, producing furfural yields up to 76% at 413 K using DMSO as the solvent (Dias et al. 2005). Sulfated and persulfated zirconia, mesoporous sulfated zirconia, as well as sulfated and persulfated zirconia supported on ordered mesoporous silica have also been explored for the conversion of xylose to furfural. The modified zirconias resulted in high catalytic activities; however, poor selectivities result in furfural yields of 22–46% at 433 K in aqueous phase. In addition, the stability of the sulfated zirconias is an issue since the sulfate concentration on the catalyst decreased when the catalyst was recycled in successive experiments (Dias et al. 2007).

One of the most commonly studied solid acid catalyst families for the dehydration of xylose are zeolites due to their capability of carrying out shape-selective chemistries (Karinen et al. 2011). In addition, the acidity of these materials can be varied to find the optimum value to carry out a specific chemical conversion (Moreau et al. 1998). Faujasite (FAU) and mordenite (MOR) were found to exhibit high selectivity toward furfural (82–96%) at 443 K at moderate xylose conversion (43% and 34% on FAU and MOR, respectively) in mixtures of water and MIBK or toluene (Moreau et al. 1998). In a recent work by Dumesic et al., many of the aforementioned and similar catalysts (i.e., MOR, BEA, sulfated zirconia, Amberlyst-70, Nafion SAC-13, and sulfonated SBA-15) could carry out furfural production with high yields (>70%) using GVL as the solvent instead of water (or in GVL-rich water–GVL mixtures) at 448 K (Guerbuez et al. 2013). Particularly, the highest yield (~80%) of furfural was achieved using MOR. It was proposed that MOR had near-optimal dimensions (0.65 nm channel opening) for the production of furfural, allowing the diffusion of xylose (molecular diameter of 0.68 nm) and furfural (molecular diameter of 0.57 nm) in and out of the pores, while inhibiting undesired condensation and resinification reactions in the pores due to spatial hindrance. This study also determined that the presence of water increased the rate of condensation reactions, decreasing furfural yields. Finally, the authors demonstrated that this strategy could be implemented to process the hemicellulose side stream (pre-hydrolysis liquor [PHL]) in a dissolving pulp facility. Hemicellulose contained in poplar wood chips was dissolved by hot water extraction to form the PHL (containing mostly xylose, xylose oligomers, and glucose). Following that, approximately 90% of the water in the PHL stream was evaporated, and GVL was added to achieve 10 wt.% H_2O in the GVL solvent. The yield of furfural from xylose and oligomers of xylose in the PHL-derived feed was 75%, only slightly lower than the value of 80% obtained from pure xylose.

10.4 CONCLUDING REMARKS

Establishing a sustainable platform for the production of energy, fuels, and chemicals from biomass, the only renewable resource of carbon, has become of increasing importance, as our dependence on fossil fuels, which are expended faster than they are produced, raises environmental, political, and economic concerns. Catalysis will play a vital role in developing biomass-based technologies, as it did with the development of petroleum-based technologies; however, it will have to overcome a new set of challenges to achieve the desired scientific and technological advancements to create efficient biorefining strategies. The two major challenges for the conversion biomass to liquid fuels include reducing the high oxygen content of the starting material to increase the energy density and forming C–C bonds, such that hydrocarbons of suitable molecular weight and volatility can be obtained. Two distinct processing components are generally utilized to tackle these challenges. First, biomass is deconstructed forming gas phase or liquid phase platform molecules, through gasification (synthesis gas), pyrolysis (bio-oil), or aqueous phase conversion pathways (sugars, acids, alcohols, furanic species, etc.). These platform molecules are then upgraded by C–C coupling reactions (e.g., FTS, oligomerization, and aldol condensation) to form liquid hydrocarbons. As for the production of chemicals, the overfunctionalized nature of biomass also brings out chemical challenges. Selective oxygen removal reactions (e.g., dehydration, hydrogenolysis, etc.) are required to obtain platform molecules, such as alcohols, ketones, acids, and furanic species, which can be upgraded to a wide range of chemical products. In addition, the presence of an aqueous reaction medium in most of the biomass conversion reactions challenges the use of conventional catalysts that are not necessarily hydrothermally stable.

As displayed in this chapter, establishing energy-efficient and cost-competitive biorefineries requires getting the most value out of the three portions of lignocellulosic feeds: cellulose, hemicellulose, and lignin. Although the conversion of cellulose and hemicellulose portions to fuels and chemicals (e.g., adipic acid, 5-HMF, 2,5-FDCA, and furfural) are being explored extensively, the recent literature on lignin valorization for chemical production is not fulfilling and is still in its infancy. This is due to the complex structure of lignin and depolymerization chemistry required to form aromatic platform chemicals. In the near future, we expect that the conversion of lignin to value-added chemicals will undergo significant advancements. In addition, catalysis research on biomass processing will focus on exploration of novel systems with improved catalyst stability and recycling, as well as facilitated product separation and recovery. We believe that these prospects will bring numerous environmental and economic benefits.

REFERENCES

Alamillo, R., A. J. Crisci, J. M. R. Gallo, S. L. Scott, and J. A. Dumesic. 2013. "A Tailored Microenvironment for Catalytic Biomass Conversion in Inorganic-Organic Nanoreactors." *Angewandte Chemie—International Edition* 52 (39):10349–10351. doi: 10.1002/anie.201304693.

Alonso, D. M., J. Q. Bond, and J. A. Dumesic. 2010a. "Catalytic Conversion of Biomass to Biofuels." *Green Chemistry* 12 (9):1493–1513. doi: 10.1039/c004654j.

Alonso, D. M., J. Q. Bond, J. C. Serrano-Ruiz, and J. A. Dumesic. 2010b. "Production of Liquid Hydrocarbon Transportation Fuels by Oligomerization of Biomass-Derived C-9 Alkenes." *Green Chemistry* 12 (6):992–999. doi: 10.1039/c001899f.

Arundhathi, R., T. Mizugaki, T. Mitsudome, K. Jitsukawa, and K. Kaneda. 2013. "Highly Selective Hydrogenolysis of Glycerol to 1,3-Propanediol over a Boehmite-Supported Platinum/Tungsten Catalyst." *Chemsuschem* 6 (8):1345–1347. doi: 10.1002/cssc.201300196.

Bals, B., C. Rogers, M. J. Jin, V. Balan, and B. Dale. 2010. "Evaluation of Ammonia Fibre Expansion (AFEX) Pretreatment for Enzymatic Hydrolysis of Switchgrass Harvested in Different Seasons and Locations." *Biotechnology for Biofuels* 3. doi: 10.1186/1754-6834-3-1.

Behr, A., J. Eilting, K. Irawadi, J. Leschinski, and F. Lindner. 2008. "Improved Utilisation of Renewable Resources: New Important Derivatives of Glycerol." *Green Chemistry* 10 (1):13–30. doi: 10.1039/b710561d.

Besson, M., P. Gallezot, and C. Pinel. 2014. "Conversion of Biomass into Chemicals over Metal Catalysts." *Chemical Reviews* 114 (3):1827–1870. doi: 10.1021/cr4002269.

Bi, P., J. Wang, Y. Zhang, P. Jiang, X. Wu, J. Liu, H. Xue, T. Wang, and Q. Li. 2015. "From Lignin to Cycloparaffins and Aromatics: Directional Synthesis of Jet and Diesel Fuel Range Biofuels Using Biomass." *Bioresource Technology* 183:10–17. doi: 10.1016/j.biortech.2015.02.023.

Binder, J. B., and R. T. Raines. 2010. "Fermentable Sugars by Chemical Hydrolysis of Biomass." *Proceedings of the National Academy of Sciences of the United States of America* 107 (10):4516–4521. doi: 10.1073/pnas.0912073107.

Binod, P., M. Kuttiraja, M. Archana, K. U. Janu, R. Sindhu, R. K. Sukumaran, and A. Pandey. 2012. "High Temperature Pretreatment and Hydrolysis of Cotton Stalk for Producing Sugars for Bioethanol Production." *Fuel* 92 (1):340–345. doi: 10.1016/j.fuel.2011.07.044.

Blanch, H. W., B. A. Simmons, and D. Klein-Marcuschamer. 2011. "Biomass Deconstruction to Sugars." *Biotechnology Journal* 6 (9):1086–1102. doi: 10.1002/biot.201000180.

Bond, J. Q., D. M. Alonso, D. Wang, R. M. West, and J. A. Dumesic. 2010. "Integrated Catalytic Conversion of gamma-Valerolactone to Liquid Alkenes for Transportation Fuels." *Science* 327 (5969):1110–1114. doi: 10.1126/science.1184362.

Borregaard. The flavor that carries—Vanillin for 50 years, http://www.borregaard.com/News/The-flavor-that-carries-Vanillin-for-50-years (accessed March 8, 2015).

Boussie, T. R., E. L. Dias, Z. M. Fresco, V. J. Murphy, J. Shoemaker, R. Archer, and H. Jiang. 2010. Production of adipic acid and derivatives from carbohydrate-containing materials, US Patent 0317823, filed June 11, 2010, and published December 16, 2010.

Bozell, J. J., and G. R. Petersen. 2010. "Technology Development for the Production of Biobased Products from Biorefinery Carbohydrates—The US Department of Energy's 'Top 10' Revisited." *Green Chemistry* 12 (4):539–554. doi: 10.1039/b922014c.

Brandt, A., J. Grasvik, J. P. Hallett, and T. Welton. 2013. "Deconstruction of Lignocellulosic Biomass with Ionic Liquids." *Green Chemistry* 15 (3):550–583. doi: 10.1039/c2gc36364j.

Bridgwater, A. V. 2012. "Review of Fast Pyrolysis of Biomass and Product Upgrading." *Biomass & Bioenergy* 38:68–94. doi: 10.1016/j.biombioe.2011.01.048.

Casanova, O., S. Iborra, and A. Corma. 2009. "Biomass into Chemicals: Aerobic Oxidation of 5-Hydroxymethyl-2-Furfural into 2,5-Furandicarboxylic Acid with Gold Nanoparticle Catalysts." *Chemsuschem* 2 (12):1138–1144. doi: 10.1002/cssc.200900137.

Castellan, A., J. C. J. Bart, and S. Cavallaro. 1991. "Industrial-Production and Use of Adipic Acid." *Catalysis Today* 9 (3):237–254. doi: 10.1016/0920-5861(91)80049-f.

Chaminand, J., L. Djakovitch, P. Gallezot, P. Marion, C. Pinel, and C. Rosier. 2004. "Glycerol Hydrogenolysis on Heterogeneous Catalysts." *Green Chemistry* 6 (8):359–361. doi: 10.1039/b407378a.

Chakar, F. S., and A. J. Ragauskas. 2004. "Review of Current and Future Softwood Kraft Lignin Process Chemistry." *Industrial Crops and Products* 20 (2):131–141. doi: 10.1016/j.indcrop.2004.04.016.

Chatel, G., and R. D. Rogers. 2014. "Review: Oxidation of Lignin Using Ionic Liquids-An Innovative Strategy to Produce Renewable Chemicals." *Acs Sustainable Chemistry & Engineering* 2 (3):322–339. doi: 10.1021/sc4004086.

Cheah, S., K. R. Gaston, Y. O. Parent, M. W. Jarvis, T. B. Vinzant, K. M. Smith, N. E. Thornburg, M. R. Nimlos, and K. A. Magrini-Bair. 2013. "Nickel Cerium Olivine Catalyst for Catalytic Gasification of Biomass." *Applied Catalysis B—Environmental* 134:34–45. doi: 10.1016/j.apcatb.2012.12.022.

Chen, M.-Y., Y.-B. Huang, H. Pang, X.-X. Liu, and Y. Fu. 2015. "Hydrodeoxygenation of Lignin-Derived Phenols into Alkanes over Carbon Nanotube Supported Ru Catalysts in Biphasic Systems." *Green Chemistry* 17 (3):1710–1717. doi: 10.1039/c4gc01992j.

Chen, W., Z. Fan, X. Pan, and X. Bao. 2008. "Effect of Confinement in Carbon Nanotubes on the Activity of Fischer–Tropsch Iron Catalyst." *Journal of the American Chemical Society* 130 (29):9414–9419. doi: 10.1021/ja8008192.

Cherubini, F. 2010. "The Biorefinery Concept: Using Biomass instead of Oil for Producing Energy and Chemicals." *Energy Conversion and Management* 51 (7):1412–1421. doi: 10.1016/j.enconman.2010.01.015.

Cherubini, F., and A. H. Stromman. 2010. "Production of Biofuels and Biochemicals from Lignocellulosic Biomass: Estimation of Maximum Theoretical Yields and Efficiencies Using Matrix Algebra." *Energy & Fuels* 24:2657–2666. doi: 10.1021/ef901379s.

Cherubini, F., and A. H. Stromman. 2011. "Chemicals from Lignocellulosic Biomass: Opportunities, Perspectives, and Potential of Biorefinery Systems." *Biofuels Bioproducts & Biorefining* 5 (5):548–561. doi: 10.1002/bbb.297.

Chheda, J. N., Y. Roman-Leshkov, and J. A. Dumesic. 2007. "Production of 5-Hydroxymethylfurfural and Furfural by Dehydration of Biomass-Derived Mono- and Poly-saccharides." *Green Chemistry* 9 (4):342–350. doi: 10.1039/b611568c.

Ciftci, A., D. A. J. Michel Ligthart, and E. J. M. Hensen. 2014. "Aqueous Phase Reforming of Glycerol over Re-Promoted Pt and Rh Catalysts." *Green Chemistry* 16 (2):853–863. doi: 10.1039/c3gc42046a.

Climent, M. J., A. Corma, and S. Iborra. 2011. "Converting Carbohydrates to Bulk Chemicals and Fine Chemicals over Heterogeneous Catalysts." *Green Chemistry* 13 (3):520–540. doi: 10.1039/c0gc00639d.

Climent, M. J., A. Corma, and S. Iborra. 2014. "Conversion of Biomass Platform Molecules into Fuel Additives and Liquid Hydrocarbon Fuels." *Green Chemistry* 16 (2):516–547. doi: 10.1039/c3gc41492b.

Corma, A., O. de la Torre, and M. Renz. 2012. "Production of High Quality Diesel from Cellulose and Hemicellulose by the Sylvan Process: Catalysts and Process Variables." *Energy & Environmental Science* 5 (4):6328–6344. doi: 10.1039/c2ee02778j.

Cortright, R. D., R. R. Davda, and J. A. Dumesic. 2002. "Hydrogen from Catalytic Reforming of Biomass-Derived Hydrocarbons in Liquid Water." *Nature* 418 (6901):964–967. doi: 10.1038/nature01009.

Dashtban, M., A. Gilbert, and P. Fatehi. 2014. "Recent Advancements in the Production of Hydroxymethylfurfural." *Rsc Advances* 4 (4):2037–2050. doi: 10.1039/c3ra45396k.

Davda, R. R., J. W. Shabaker, G. W. Huber, R. D. Cortright, and J. A. Dumesic. 2003. "Aqueous-Phase Reforming of Ethylene Glycol on Silica-Supported Metal Catalysts." *Applied Catalysis B—Environmental* 43 (1):13–26. doi: 10.1016/s0926-3373(02)00277-1.

Davis, S. E., L. R. Houk, E. C. Tamargo, A. K. Datye, and R. J. Davis. 2011. "Oxidation of 5-Hydroxymethylfurfural over Supported Pt, Pd and Au Catalysts." *Catalysis Today* 160 (1):55–60. doi: 10.1016/j.cattod.2010.06.004.

Dayton, D. 2002. "A Review of the Literature on Catalytic Biomass Tar Destruction." Research Paper, National Renewable Energy Laboratory, http://www.nrel.gov/doc/fy03osti/32815.pdf (accessed February 13, 2015).

de Vlieger, D. J. M., L. Lefferts, and K. Seshan. 2014. "Ru Decorated Carbon Nanotubes—A Promising Catalyst for Reforming Bio-Based Acetic Acid in the Aqueous Phase." *Green Chemistry* 16 (2):864–874. doi: 10.1039/c3gc41922c.

Dias, A. S., M. Pillinger, and A. A. Valente. 2005. "Dehydration of Xylose into Furfural over Micro-mesoporous Sulfonic Acid Catalysts." *Journal of Catalysis* 229 (2):414–423. doi: 10.1016/j.jcat.2004.11.016.

Dias, A. S., S. Lima, M. Pillinger, and A. A. Valente. 2007. "Modified Versions of Sulfated Zirconia as Catalysts for the Conversion of Xylose to Furfural." *Catalysis Letters* 114 (3–4):151–160. doi: 10.1007/s10562-007-9052-6.

Doherty, W. O. S., P. Mousavioun, and C. M. Fellows. 2011. "Value-Adding to Cellulosic Ethanol: Lignin Polymers." *Industrial Crops and Products* 33 (2):259–276. doi: 10.1016/j.indcrop.2010.10.022.

Dunlop, A. P. 1948. "Furfural Formation and Behavior." *Industrial and Engineering Chemistry* 40 (2):204–209. doi: 10.1021/ie50458a006.

Eerhart, A. J. J. E., A. P. C. Faaij, and M. K. Patel. 2012. "Replacing Fossil Based PET with Biobased PEF; Process Analysis, Energy and GHG Balance." *Energy & Environmental Science* 5 (4):6407–6422. doi: 10.1039/c2ee02480b.

Emptage, M., S. L. Haynie, L. A. Laffend, J. P. Pucci, and G. Whited. 2003. Process for the biological production of 1,3-propanediol with high titer, US Patent 6514733, filed August 18, 2000, and published February 4, 2003.

Fan, C., H. Guan, H. Zhang, J. Wang, S. Wang, and X. Wang. 2011. "Conversion of Fructose and Glucose into 5-Hydroxymethylfurfural Catalyzed by a Solid Heteropolyacid Salt." *Biomass & Bioenergy* 35 (7):2659–2665. doi: 10.1016/j.biombioe.2011.03.004.

Fermoso, J., F. Rubiera, and D. Chen. 2012. "Sorption Enhanced Catalytic Steam Gasification Process: A Direct Route from Lignocellulosic Biomass to High Purity Hydrogen." *Energy & Environmental Science* 5 (4):6358–6367. doi: 10.1039/c2ee02593k.

Gallo, J. M. R., D. M. Alonso, M. A. Mellmer, and J. A. Dumesic. 2013. "Production and Upgrading of 5-Hydroxymethylfurfural Using Heterogeneous Catalysts and Biomass-Derived Solvents." *Green Chemistry* 15 (1):85–90. doi: 10.1039/c2gc36536g.

Girisuta, B., L. Janssen, and H. J. Heeres. 2006. "A Kinetic Study on the Conversion of Glucose to Levulinic Acid." *Chemical Engineering Research & Design* 84 (A5):339–349. doi: 10.1205/cherd05038.

Gong, L., Y. Lue, Y. Ding, R. Lin, J. Li, W. Dong, T. Wang, and W. Chen. 2009. "Solvent Effect on Selective Dehydroxylation of Glycerol to 1,3-Propanediol over a Pt/WO_3/ZrO_2 Catalyst." *Chinese Journal of Catalysis* 30 (12):1189–1191.

Gong, L., Y. Lu, Y. Ding, R. Lin, J. Li, W. Dong, T. Wang, and W. Chen. 2010. "Selective Hydrogenolysis of Glycerol to 1,3-Propanediol over a Pt/WO_3/TiO_2/SiO_2 Catalyst in Aqueous Media." *Applied Catalysis A—General* 390 (1–2):119–126. doi: 10.1016/j.apcata.2010.10.002.

Gorbanev, Y. Y., S. K. Klitgaard, J. M. Woodley, C. H. Christensen, and A. Riisager. 2009. "Gold-Catalyzed Aerobic Oxidation of 5-Hydroxymethylfurfural in Water at Ambient Temperature." *Chemsuschem* 2 (7):672–675. doi: 10.1002/cssc.200900059.

Gorbanev, Y. Y., S. Kegaens, and A. Riisager. 2011a. "Effect of Support in Heterogeneous Ruthenium Catalysts Used for the Selective Aerobic Oxidation of HMF in Water." *Topics in Catalysis* 54 (16–18):1318–1324. doi: 10.1007/s11244-011-9754-2.

Gorbanev, Y. Y., S. Kegnaes, and A. Riisager. 2011b. "Selective Aerobic Oxidation of 5-Hydroxymethylfurfural in Water Over Solid Ruthenium Hydroxide Catalysts with Magnesium-Based Supports." *Catalysis Letters* 141 (12):1752–1760. doi: 10.1007/s10562-011-0707-y.

Guerbuez, E. I., S. G. Wettstein, and J. A. Dumesic. 2012. "Conversion of Hemicellulose to Furfural and Levulinic Acid Using Biphasic Reactors with Alkylphenol Solvents." *Chemsuschem* 5 (2):383–387. doi: 10.1002/cssc.201100608.

Guerbuez, E. I., J. M. R. Gallo, D. M. Alonso, S. G. Wettstein, W. Y. Lim, and J. A. Dumesic. 2013. "Conversion of Hemicellulose into Furfural Using Solid Acid Catalysts in gamma-Valerolactone." *Angewandte Chemie—International Edition* 52 (4):1270–1274. doi: 10.1002/anie.201207334.

Gupta, N. K., S. Nishimura, A. Takagaki, and K. Ebitani. 2011. "Hydrotalcite-Supported Gold-Nanoparticle-Catalyzed Highly Efficient Base-Free Aqueous Oxidation of 5-Hydroxymethylfurfural into 2,5-Furandicarboxylic Acid under Atmospheric Oxygen Pressure." *Green Chemistry* 13 (4):824–827. doi: 10.1039/c0gc00911c.

Gursel, I. V., Q. Wang, T. Noel, and V. Hessel. 2012. "Process-Design Intensification—Direct Synthesis of Adipic Acid in Flow." *Pres 2012: 15th International Conference on Process Integration, Modelling and Optimisation for Energy Saving and Pollution Reduction* 29:565–571. doi: 10.3303/cet1229095.

Hargus, C., R. Michalsky, and A. A. Peterson. 2014. "Looped-Oxide Catalysis: A Solar Thermal Approach to Bio-oil Deoxygenation." *Energy & Environmental Science* 7 (10):3122–3134. doi: 10.1039/c4ee01684j.

Hermans, I., P. A. Jacobs, and J. Peeters. 2006. "To the Core of Autocatalysis in Cyclohexane Autoxidation." *Chemistry—A European Journal* 12 (16):4229–4240. doi: 10.1002/chem.200600189.

Holladay, J. E., J.J. Bozell, J.F. White, and D. Johnson. 2007. Top value-added chemicals from biomass volume II—Results of screening for potential candidates from biorefinery lignin, http://www1.eere.energy.gov/bioenergy/pdfs/pnnl-16983.pdf (accessed March 3, 2015).

Hu, G., S. P. Xu, S. G. Li, C. R. Xiao, and S. Q. Liu. 2006. "Steam Gasification of Apricot Stones with Olivine and Dolomite as Downstream Catalysts." *Fuel Processing Technology* 87 (5):375–382. doi: 10.1016/j.fuproc.2005.07.008.

Hu, L., G. Zhao, W. Hao, X. Tang, Y. Sun, L. Lin, and S. Liu. 2012. "Catalytic Conversion of Biomass-Derived Carbohydrates into Fuels and Chemicals via Furanic Aldehydes." *Rsc Advances* 2 (30):11184–11206. doi: 10.1039/c2ra21811a.

Huber, G. W., J. W. Shabaker, and J. A. Dumesic. 2003. "Raney Ni–Sn Catalyst for H-2 Production from Biomass-Derived Hydrocarbons." *Science* 300 (5628):2075–2077. doi: 10.1126/science.1085597.

Huber, G. W., S. Iborra, and A. Corma. 2006a. "Synthesis of Transportation Fuels from Biomass: Chemistry, Catalysts, and Engineering." *Chemical Reviews* 106 (9):4044–4098. doi: 10.1021/cr068360d.

Huber, G. W., J. W. Shabaker, S. T. Evans, and J. A. Dumesic. 2006b. "Aqueous-Phase Reforming of Ethylene Glycol over Supported Pt and Pd Bimetallic Catalysts." *Applied Catalysis B—Environmental* 62 (3–4):226–235. doi: 10.1016/j.apcatb.2005.07.010.

Jahangiri, H., J. Bennett, P. Mahjoubi, K. Wilson, and S. Gu. 2014. "A Review of Advanced Catalyst Development for Fischer–Tropsch Synthesis of Hydrocarbons from Biomass Derived Syn-gas." *Catalysis Science & Technology* 4 (8):2210–2229. doi: 10.1039/c4cy00327f.

Jiang, X., L. Zhang, S. Wybornov, T. Staedler, D. Hein, F. Wiedenmann, W. Krumm, V. Rudnev, and I. Lukiyanchuk. 2012. "Highly Efficient Nanoarchitectured Ni_5TiO_7 Catalyst for Biomass Gasification." *Acs Applied Materials & Interfaces* 4 (8):4062–4066. doi: 10.1021/am3008449.

Karinen, R., K. Vilonen, and M. Niemela. 2011. "Biorefining: Heterogeneously Catalyzed Reactions of Carbohydrates for the Production of Furfural and Hydroxymethylfurfural." *Chemsuschem* 4 (8):1002–1016. doi: 10.1002/cssc.201000375.

Kim, T.-W., H.-D. Kim, K.-E. Jeong, H.-J. Chae, S.-Y. Jeong, C.-H. Lee, and C.-U. Kim. 2011. "Catalytic Production of Hydrogen through Aqueous-Phase Reforming over Platinum/Ordered Mesoporous Carbon Catalysts." *Green Chemistry* 13 (7):1718–1728. doi: 10.1039/c1gc15235a.

Kraus, G. A. 2008. "Synthetic Methods for the Preparation of 1,3-Propanediol." *Clean-Soil Air Water* 36 (8):648–651. doi: 10.1002/clen.200800084.

Kunkes, E. L., D. A. Simonetti, R. M. West, J. C. Serrano-Ruiz, C. A. Gartner, and J. A. Dumesic. 2008. "Catalytic Conversion of Biomass to Monofunctional Hydrocarbons and Targeted Liquid-Fuel Classes." *Science* 322 (5900):417–421. doi: 10.1126/science.1159210.

Kurosaka, T., H. Maruyama, I. Naribayashi, and Y. Sasaki. 2008. "Production of 1,3-Propanediol by Hydrogenolysis of Glycerol Catalyzed by Pt/WO(3)/ZrO(2)." *Catalysis Communications* 9 (6):1360–1363. doi: 10.1016/j.catcom.2007.11.034.

Kuster, B. F. M. 1990. "5-Hydroxymethylfurfural (HMF)—A Review Focusing on Its Manufacture." *Starch-Starke* 42 (8):314–321. doi: 10.1002/star.19900420808.

Lange, J.-P., R. Price, P. M. Ayoub, J. Louis, L. Petrus, L. Clarke, and H. Gosselink. 2010. "Valeric Biofuels: A Platform of Cellulosic Transportation Fuels." *Angewandte Chemie—International Edition* 49 (26):4479–4483. doi: 10.1002/anie.201000655.

Lange, J.-P., E. van der Heide, J. van Buijtenen, and R. Price. 2012. "FurfuraluA Promising Platform for Lignocellulosic Biofuels." *Chemsuschem* 5 (1):150–166. doi: 10.1002/cssc.201100648.

Larsson, S., E. Palmqvist, B. Hahn-Hagerdal, C. Tengborg, K. Stenberg, G. Zacchi, and N. O. Nilvebrant. 1999. "The Generation of Fermentation Inhibitors during Dilute Acid Hydrolysis of Softwood." *Enzyme and Microbial Technology* 24 (3–4):151–159. doi: 10.1016/s0141-0229(98)00101-x.

Lau, M. W., C. Gunawan, and B. E. Dale. 2009. "The Impacts of Pretreatment on the Fermentability of Pretreated Lignocellulosic Biomass: A Comparative Evaluation Between Ammonia Fiber Expansion and Dilute Acid Pretreatment." *Biotechnology for Biofuels* 2. doi: 10.1186/1754-6834-2-30.

Lee, J. W., and T. W. Jeffries. 2011. "Efficiencies of Acid Catalysts in the Hydrolysis of Lignocellulosic Biomass Over a Range of Combined Severity Factors." *Bioresource Technology* 102 (10):5884–5890. doi: 10.1016/j.biortech.2011.02.048.

Li, D., M. Tamura, Y. Nakagawa, and K. Tomishige. 2015. "Metal Catalysts for Steam Reforming of Tar Derived from the Gasification of Lignocellulosic Biomass." *Bioresource Technology* 178:53–64. doi: 10.1016/j.biortech.2014.10.010.

Lind, F., N. Berguerand, M. Seemann, and H. Thunman. 2013. "Ilmenite and Nickel as Catalysts for Upgrading of Raw Gas Derived from Biomass Gasification." *Energy & Fuels* 27 (2):997–1007. doi: 10.1021/ef302091w.

Liu, J., B. Sun, J. Hu, Y. Pei, H. Li, and M. Qiao. 2010. "Aqueous-Phase Reforming of Ethylene Glycol to Hydrogen on Pd/Fe_3O_4 Catalyst Prepared by Co-precipitation: Metal–Support Interaction and Excellent Intrinsic Activity." *Journal of Catalysis* 274 (2):287–295. doi: 10.1016/j.jcat.2010.07.014.

Lohmann, D. 1990. "Structural Diversity and Functional Versatility of Polysaccharides." *Novel Biodegradable Microbial Polymers* 186:333–348. doi: 10.1007/978-94-009-2129-0_27.

Luque, R., A. R. de la Osa, J. M. Campelo, A. A. Romero, J. L. Valverde, and P. Sanchez. 2012. "Design and Development of Catalysts for Biomass-to-Liquid-Fischer–Tropsch (BTL-FT) Processes for Biofuels Production." *Energy & Environmental Science* 5 (1):5186–5202. doi: 10.1039/c1ee02238e.

Lynd, L. R., J. H. Cushman, R. J. Nichols, and C. E. Wyman. 1991. "Fuel Ethanol from Cellulosic Biomass." *Science* 251 (4999):1318–1323. doi: 10.1126/science.251.4999.1318.

Marquardt, W., A. Harwardt, M. Hechinger, K. Kraemer, J. Viell, and A. Voll. 2010. "The Biorenewables Opportunity—Toward Next Generation Process and Product Systems." *Aiche Journal* 56 (9):2228–2235. doi: 10.1002/aic.12380.

Mazumder, J., and H. I. de Lasa. 2014. "Ni Catalysts for Steam Gasification of Biomass: Effect of La_2O_3 Loading." *Catalysis Today* 237:100–110. doi: 10.1016/j.cattod.2014.02.015.

Milne, T., F. Agblevor, M. Davis, S. Deutch, and D. Johnson. 1997. "A Review of the Chemical Composition of Fast-Pyrolysis Oils from Biomass." *Developments in Thermochemical Biomass Conversion* 409–424. Springer, the Netherlands.

Mizugaki, T., T. Yamakawa, R. Arundhathi, T. Mitsudome, K. Jitsukawa, and K. Kanede. 2012. "Selective Hydrogenolysis of Glycerol to 1,3-Propanediol Catalyzed by Pt Nanoparticles AlO_x/WO_3." *Chemistry Letters* 41 (12):1720–1722. doi: 10.1246/cl.2012.1720.

Moreau, C., R. Durand, S. Razigade, J. Duhamet, P. Faugeras, P. Rivalier, P. Ros, and G. Avignon. 1996. "Dehydration of Fructose to 5-Hydroxymethylfurfural over H-Mordenites." *Applied Catalysis A—General* 145 (1–2):211–224. doi: 10.1016/0926-860x(96)00136-6.

Moreau, C., R. Durand, D. Peyron, J. Duhamet, and P. Rivalier. 1998. "Selective Preparation of Furfural from Xylose over Microporous Solid Acid Catalysts." *Industrial Crops and Products* 7 (2–3):95–99. doi: 10.1016/s0926-6690(97)00037-x.

Mortensen, P. M., J. D. Grunwaldt, P. A. Jensen, K. G. Knudsen, and A. D. Jensen. 2011. "A Review of Catalytic Upgrading of Bio-oil to Engine Fuels." *Applied Catalysis A—General* 407 (1–2):1–19. doi: 10.1016/j.apcata.2011.08.046.

Mosier, N., C. Wyman, B. Dale, R. Elander, Y. Y. Lee, M. Holtzapple, and M. Ladisch. 2005. "Features of Promising Technologies for Pretreatment of Lignocellulosic Biomass." *Bioresource Technology* 96 (6):673–686. doi: 10.1016/j.biortech.2004.06.025.

Muurinen, E. 2000. "A review and distillation study related to peroxyacid pulping." Academic Dissertation presented in the University of Oulu, http://herkules.oulu.fi/isbn9514256611/isbn9514256611.pdf (accessed February 20, 2015).

Nahil, M. A., X. Wang, C. Wu, H. Yang, H. Chen, and P. T. Williams. 2013. "Novel Bi-functional Ni–Mg–Al–CaO Catalyst for Catalytic Gasification of Biomass for Hydrogen Production with in situ CO_2 Adsorption." *Rsc Advances* 3 (16):5583–5590. doi: 10.1039/c3ra40576a.

Nakagawa, Y., Y. Shinmi, S. Koso, and K. Tomishige. 2010. "Direct Hydrogenolysis of Glycerol into 1,3-Propanediol over Rhenium-Modified Iridium Catalyst." *Journal of Catalysis* 272 (2):191–194. doi: 10.1016/j.jcat.2010.04.009.

Nakamura, Y., and S. Morikawa. 1980. "The Dehydration of D-Fructose to 5-Hydroxymethyl-2-Furaldehyde." *Bulletin of the Chemical Society of Japan* 53 (12):3705–3706. doi: 10.1246/bcsj.53.3705.

Navarro, R. M., M. C. Sanchez-Sanchez, M. C. Alvarez-Galvan, F. del Valle, and J. L. G. Fierro. 2009. "Hydrogen Production from Renewable Sources: Biomass and Photocatalytic Opportunities." *Energy & Environmental Science* 2 (1):35–54. doi: 10.1039/b808138g.

Nikolla, E., Y. Roman-Leshkov, M. Moliner, and M. E. Davis. 2011. "'One-Pot' Synthesis of 5-(Hydroxymethyl)furfural from Carbohydrates Using Tin-Beta Zeolite." *Acs Catalysis* 1 (4):408–410. doi: 10.1021/cs2000544.

Nordgreen, T., V. Nemanova, K. Engvall, and K. Sjostrom. 2012. "Iron-Based Materials as Tar Depletion Catalysts in Biomass Gasification: Dependency on Oxygen Potential." *Fuel* 95 (1):71–78. doi: 10.1016/j.fuel.2011.06.002.

Oh, J., S. Dash, and H. Lee. 2011. “Selective Conversion of Glycerol to 1,3-Propanediol Using Pt-Sulfated Zirconia.” *Green Chemistry* 13 (8):2004–2007. doi: 10.1039/c1gc15263g.

Ohta, H., H. Kobayashi, K. Hara, and A. Fukuoka. 2011. “Hydrodeoxygenation of Phenols as Lignin Models under Acid-Free Conditions with Carbon-Supported Platinum Catalysts.” *Chemical Communications* 47 (44):12209–12211. doi: 10.1039/c1cc14859a.

O'Neill, R., M. Najeeb Ahmad, L. Vanoye, and F. Aiouache. 2009. “Kinetics of Aqueous Phase Dehydration of Xylose into Furfural Catalyzed by ZSM-5 Zeolite.” *Industrial & Engineering Chemistry Research* 48 (9):4300–4306. doi: 10.1021/ie801599k.

Onwudili, J. A., and P. T. Williams. 2013a. “Enhanced Methane and Hydrogen Yields from Catalytic Supercritical Water Gasification of Pine Wood Sawdust via Pre-processing in Subcritical Water.” *Rsc Advances* 3 (30):12432–12442. doi: 10.1039/c3ra41362d.

Onwudili, J. A., and P. T. Williams. 2013b. “Hydrogen and Methane Selectivity during Alkaline Supercritical Water Gasification of Biomass with Ruthenium–Alumina Catalyst.” *Applied Catalysis B—Environmental* 132:70–79. doi: 10.1016/j.apcatb.2012.11.033.

Pagan-Torres, Y. J., T. Wang, J. M. R. Gallo, B. H. Shanks, and J. A. Dumesic. 2012. “Production of 5-Hydroxymethylfurfural from Glucose Using a Combination of Lewis and Bronsted Acid Catalysts in Water in a Biphasic Reactor with an Alkylphenol Solvent.” *Acs Catalysis* 2 (6):930–934. doi: 10.1021/cs300192z.

Pasini, T., M. Piccinini, M. Blosi, R. Bonelli, S. Albonetti, N. Dimitratos, J. A. Lopez-Sanchez, M. Sankar, Q. He, C. J. Kiely, G. J. Hutchings, and F. Cavani. 2011. “Selective Oxidation of 5-Hydroxymethyl-2-Furfural Using Supported Gold–Copper Nanoparticles.” *Green Chemistry* 13 (8):2091–2099. doi: 10.1039/c1gc15355b.

Peterson, A. A., F. Vogel, R. P. Lachance, M. Froeling, M. J. Antal, Jr., and J. W. Tester. 2008. “Thermochemical Biofuel Production in Hydrothermal Media: A Review of Sub- and Supercritical Water Technologies.” *Energy & Environmental Science* 1 (1):32–65. doi: 10.1039/b810100k.

Qi, L., and I. T. Horvath. 2012. “Catalytic Conversion of Fructose to gamma-Valerolactone in gamma-Valerolactone.” *Acs Catalysis* 2 (11):2247–2249. doi: 10.1021/cs300428f.

Qin, L.-Z., M.-J. Song, and C.-L. Chen. 2010. “Aqueous-Phase Deoxygenation of Glycerol to 1,3-Propanediol over $Pt/WO_3/ZrO_2$ Catalysts in a Fixed-Bed Reactor.” *Green Chemistry* 12 (8):1466–1472. doi: 10.1039/c0gc00005a.

Ragauskas, A. “Lignin Overview”. Technical Review, The Institute of Paper Science and Technology, Georgia Institute of Technology, http://www.ipst.gatech.edu/faculty/ragauskas_art/technical_reviews/lignin%20overview.pdf (accessed March 8, 2015).

Ragauskas, A. J., G. T. Beckham, M. J. Biddy, R. Chandra, F. Chen, M. F. Davis, B. H. Davison, R. A. Dixon, P. Gilna, M. Keller, P. Langan, A. K. Naskar, J. N. Saddler, T. J. Tschaplinski, G. A. Tuskan, and C. E. Wyman. 2014. “Lignin Valorization: Improving Lignin Processing in the Biorefinery.” *Science* 344 (6185):709–719. doi: 10.1126/science.1246843.

Reichert, D., M. Sarich, and F. Merz. 2008. Method for producing enantiomer 5-hydroxymethylfurfural with 5-acyloxymethylfurfural as intermediate, EP1958944, filed February 7, 2008, and published August 20, 2008.

Ren, S., H. Lei, L. Wang, Q. Bu, S. Chen, and J. Wu. 2014. “Hydrocarbon and Hydrogen-Rich Syngas Production by Biomass Catalytic Pyrolysis and Bio-oil Upgrading over Biochar Catalysts.” *Rsc Advances* 4 (21):10731–10737. doi: 10.1039/c4ra00122b.

Rennovia Markets, http://www.rennovia.com/markets/ (accessed February 8, 2015).

Ribeiro, M. L., and U. Schuchardt. 2003. “Cooperative Effect of Cobalt Acetylacetonate and Silica in the Catalytic Cyclization and Oxidation of Fructose to 2,5-Furandicarboxylic Acid.” *Catalysis Communications* 4 (2):83–86. doi: 10.1016/s1566-7367(02)00261-3.

Rinaldi, R., and F. Schueth. 2009. “Design of Solid Catalysts for the Conversion of Biomass.” *Energy & Environmental Science* 2 (6):610–626. doi: 10.1039/b902668a.

Rivalier, P., J. Duhamet, C. Moreau, and R. Durand. 1995. "Development of a Continuous Catalytic Heterogeneous Column Reactor with Simultaneous Extraction of an Intermediate Product by an Organic-Solvent Circulating in Countercurrent Manner with the Aqueous-Phase." *Catalysis Today* 24 (1–2):165–171. doi: 10.1016/0920-5861(95)00026-c.

Robinson, J. M., C. E. Burgess, M. A. Bently, C. D. Brasher, B. O. Horne, D. M. Lillard, J. M. Macias, H. D. Mandal, S. C. Mills, K. D. O'Hara, J. T. Pon, A. F. Raigoza, E. H. Sanchez, and J. S. Villarreal. 2004. "The Use of Catalytic Hydrogenation to Intercept Carbohydrates in a Dilute Acid Hydrolysis of Biomass to Effect a Clean Separation from Lignin." *Biomass and Bioenergy* 26:473–483.

Roman-Leshkov, Y., and J. A. Dumesic. 2009. "Solvent Effects on Fructose Dehydration to 5-Hydroxymethylfurfural in Biphasic Systems Saturated with Inorganic Salts." *Topics in Catalysis* 52 (3):297–303. doi: 10.1007/s11244-008-9166-0.

Roman-Leshkov, Y., J. N. Chheda, and J. A. Dumesic. 2006. "Phase Modifiers Promote Efficient Production of Hydroxymethylfurfural from Fructose." *Science* 312 (5782):1933–1937. doi: 10.1126/science.1126337.

Ruddy, D. A., J. A. Schaidle, J. R. Ferrell III, J. Wang, L. Moens, and J. E. Hensley. 2014. "Recent Advances in Heterogeneous Catalysts for Bio-oil Upgrading via "Ex situ Catalytic Fast Pyrolysis": Catalyst Development through the Study of Model Compounds." *Green Chemistry* 16 (2):454–490. doi: 10.1039/c3gc41354c.

Ruppert, A. M., K. Weinberg, and R. Palkovits. 2012. "Hydrogenolysis Goes Bio: From Carbohydrates and Sugar Alcohols to Platform Chemicals." *Angewandte Chemie—International Edition* 51 (11):2564–2601. doi: 10.1002/anie.201105125.

Saha, B., D. Gupta, M. M. Abu-Omar, A. Modak, and A. Bhaumik. 2013. "Porphyrin-Based Porous Organic Polymer-Supported Iron(III) Catalyst for Efficient Aerobic Oxidation of 5-Hydroxymethyl-Furfural into 2,5-Furandicarboxylic Acid." *Journal of Catalysis* 299:316–320. doi: 10.1016/j.jcat.2012.12.024.

Saidi, M., F. Samimi, D. Karimipourfard, T. Nimmanwudipong, B. C. Gates, and M. R. Rahimpour. 2014. "Upgrading of Lignin-Derived Bio-oils by Catalytic Hydrodeoxygenation." *Energy & Environmental Science* 7 (1):103–129. doi: 10.1039/c3ee43081b.

Sanna, A., and J. M. Andresen. 2012. "Bio-oil Deoxygenation by Catalytic Pyrolysis: New Catalysts for the Conversion of Biomass into Densified and Deoxygenated Bio-oil." *Chemsuschem* 5 (10):1944–1957. doi: 10.1002/cssc.201200245.

Serrano-Ruiz, J. C., R. M. West, and J. A. Dumesic. 2010. "Catalytic Conversion of Renewable Biomass Resources to Fuels and Chemicals." *Annual Review of Chemical and Bio molecular Engineering* 1:79–100. doi: 10.1146/annurev-chembioeng-073009-100935.

Shabaker, J. W., G. W. Huber, R. R. Davda, R. D. Cortright, and J. A. Dumesic. 2003. "Aqueous-Phase Reforming of Ethylene Glycol over Supported Platinum Catalysts." *Catalysis Letters* 88 (1–2):1–8. doi: 10.1023/a:1023538917186.

Shaddix, C. R. 2011. "Chapter 7 Advances in Gasification for Biofuel Production." In *Chemical and Biochemical Catalysis for Next Generation Biofuels*, 136–155. The Royal Society of Chemistry, UK.

Sharma, P., T. Elder, L. H. Groom, and J. J. Spivey. 2014. "Effect of Structural Promoters on Fe-Based Fischer–Tropsch Synthesis of Biomass Derived Syngas." *Topics in Catalysis* 57 (6–9):526–537. doi: 10.1007/s11244-013-0209-9.

Shimao, A., S. Koso, N. Ueda, Y. Shinmi, I. Furikado, and K. Tomishige. 2009. "Promoting Effect of Re Addition to Rh/SiO_2 on Glycerol Hydrogenolysis." *Chemistry Letters* 38 (6):540–541. doi: 10.1246/cl.2009.540.

Shinmi, Y., S. Koso, T. Kubota, Y. Nakagawa, and K. Tomishige. 2010. "Modification of Rh/SiO_2 Catalyst for the Hydrogenolysis of Glycerol in Water." *Applied Catalysis B—Environmental* 94 (3–4):318–326. doi: 10.1016/j.apcatb.2009.11.021.

Silverstein, R. A., Y. Chen, R. R. Sharma-Shivappa, M. D. Boyette, and J. Osborne. 2007. "A Comparison of Chemical Pretreatment Methods for Improving Saccharification of Cotton Stalks." *Bioresource Technology* 98 (16):3000–3011. doi: 10.1016/j.biortech.2006.10.022.

Simeonov, S. P., J. A. S. Coelho, and C. A. M. Afonso. 2012. "An Integrated Approach for the Production and Isolation of 5-Hydroxymethylfurfural from Carbohydrates." *Chemsuschem* 5 (8):1388–1391. doi: 10.1002/cssc.201200236.

Sinag, A., T. Yumak, V. Balci, and A. Kruse. 2011. "Catalytic Hydrothermal Conversion of Cellulose over SnO_2 and ZnO Nanoparticle Catalysts." *Journal of Supercritical Fluids* 56 (2):179–185. doi: 10.1016/j.supflu.2011.01.002.

Smart Vanillin, http://www.smartvanillin.com/euro-vanillin-pure.html (accessed March 8, 2015).

Stahlberg, T., W. J. Fu, J. M. Woodley, and A. Riisager. 2011. "Synthesis of 5-(Hydroxymethyl) furfural in Ionic Liquids: Paving the Way to Renewable Chemicals." *Chemsuschem* 4 (4):451–458. doi: 10.1002/cssc.201000374.

Sutton, A. D., F. D. Waldie, R. Wu, M. Schlaf, L. A. 'Pete' Silks, III, and J. C. Gordon. 2013. "The Hydrodeoxygenation of Bioderived Furans into Alkanes." *Nature Chemistry* 5 (5):428–432. doi: 10.1038/nchem.1609.

Tao, L., A. Aden, R. T. Elander, V. R. Pallapolu, Y. Y. Lee, R. J. Garlock, V. Balan, B. E. Dale, Y. Kim, N. S. Mosier, M. R. Ladisch, M. Falls, M. T. Holtzapple, R. Sierra, J. Shi, M. A. Ebrik, T. Redmond, B. Yang, C. E. Wyman, B. Hames, S. Thomas, and R. E. Warner. 2011. "Process and Technoeconomic Analysis of Leading Pretreatment Technologies for Lignocellulosic Ethanol Production Using Switchgrass." *Bioresource Technology* 102 (24):11105–11114. doi: 10.1016/j.biortech.2011.07.051.

Task 42 Biorefinery. "Bio-based Chemicals". International Energy Agency, http://www.ieabioenergy.com/wp-content/uploads/2013/10/Task-42-Biobased-Chemicals-value-added-products-from-biorefineries.pdf (accessed March 8, 2015).

ten Dam, J., and U. Hanefeld. 2011. "Renewable Chemicals: Dehydroxylation of Glycerol and Polyols." *Chemsuschem* 4 (8):1017–1034. doi: 10.1002/cssc.201100162.

Tucker, M. P., K. H. Kim, M. M. Newman, and Q. A. Nguyen. 2003. "Effects of Temperature and Moisture on Dilute-Acid Steam Explosion Pretreatment of Corn Stover and Cellulase Enzyme Digestibility." *Applied Biochemistry and Biotechnology* 105:165–177. doi: 10.1385/abab:105:1-3:165.

Udomsirichakorn, J., P. Basu, P. A. Salam, and B. Acharya. 2014. "CaO-Based Chemical Looping Gasification of Biomass for Hydrogen-Enriched Gas Production with In situ CO_2 Capture and Tar Reduction." *Fuel Processing Technology* 127:7–12. doi: 10.1016/j.fuproc.2014.06.007.

Van de Vyver, S., and Y. Roman-Leshkov. 2013. "Emerging Catalytic Processes for the Production of Adipic Acid." *Catalysis Science & Technology* 3 (6):1465–1479. doi: 10.1039/c3cy20728e.

van Putten, R.-J., J. C. van der Waal, E. de Jong, C. B. Rasrendra, H. J. Heeres, and J. G. de Vries. 2013. "Hydroxymethylfurfural, A Versatile Platform Chemical Made from Renewable Resources." *Chemical Reviews* 113 (3):1499–1597. doi: 10.1021/cr300182k.

Venkatakrishnan, V. K., W. Nicholas Delgass, F. H. Ribeiro, and R. Agrawal. 2015. "Oxygen Removal from Intact Biomass to Produce Liquid Fuel Range Hydrocarbons via Fast-Hydropyrolysis and Vapor-Phase Catalytic Hydrodeoxygenation." *Green Chemistry* 17 (1):178–183. doi: 10.1039/c4gc01746c.

Vennestrom, P. N. R., C. M. Osmundsen, C. H. Christensen, and E. Taarning. 2011. "Beyond Petrochemicals: The Renewable Chemicals Industry." *Angewandte Chemie—International Edition* 50 (45):10502–10509. doi: 10.1002/anie.201102117.

Vogel, F., M. H. Waldner, A. A. Rouff, and S. Rabe. 2007. "Synthetic Natural Gas from Biomass by Catalytic Conversion in Supercritical Water." *Green Chemistry* 9 (6):616–619. doi: 10.1039/b614601e.

Wang, Y., J. Wu, and S. Wang. 2013. "Hydrodeoxygenation of Bio-oil over Pt-Based Supported Catalysts: Importance of Mesopores and Acidity of the Support to Compounds with Different Oxygen Contents." *Rsc Advances* 3 (31):12635–12640. doi: 10.1039/c3ra41405a.

Wang, L., D. Li, H. Watanabe, M. Tamura, Y. Nakagawa, and K. Tomishige. 2014. "Catalytic Performance and Characterization of Co/Mg/Al Catalysts Prepared from Hydrotalcite-Like Precursors for the Steam Gasification of Biomass." *Applied Catalysis B—Environmental* 150:82–92. doi: 10.1016/j.apcatb.2013.12.002.

Wang, T., M. W. Nolte, and B. H. Shanks. 2014. "Catalytic Dehydration of C-6 Carbohydrates for the Production of Hydroxymethylfurfural (HMF) as a Versatile Platform Chemical." *Green Chemistry* 16 (2):548–572. doi: 10.1039/c3gc41365a.

Weissermel, K., and H.-J. Arpe. 2003. "Frontmatter." In *Industrial Organic Chemistry*, I–XIX. Wiley-VCH Verlag GmbH, Weinheim, Germany.

Welch, A., N. R. Shiju, I. D. Watts, G. Sankar, S. Nikitenko, and W. Bras. 2005. "Epoxidation of Cyclohexene over Crystalline and Amorphous Titanosilicate Catalysts." *Catalysis Letters* 105 (3–4):179–182. doi: 10.1007/s10562-005-8688-3.

Wells, G. M. 1999. *Handbook of Petrochemicals and Processes*. Ashgate, UK.

West, R. M., M. H. Tucker, D. J. Braden, and J. A. Dumesic. 2009. "Production of Alkanes from Biomass Derived Carbohydrates on Bi-functional Catalysts Employing Niobium-Based Supports." *Catalysis Communications* 10 (13):1743–1746. doi: 10.1016/j.catcom.2009.05.021.

Wong, K., C. Li, L. Stubbs, M. Van Meurs, D. G. Anak Kumbang, S. C. Y. Lim, and E. Drent. 2012. Synthesis of Diacids, WO/2012/134397, file March 28, 2012, Published February 5, 2014.

Xing, R., W. Qi, and G. W. Huber. 2011. "Production of Furfural and Carboxylic Acids from Waste Aqueous Hemicellulose Solutions from the Pulp and Paper and Cellulosic Ethanol Industries." *Energy & Environmental Science* 4 (6):2193–2205. doi: 10.1039/c1ee01022k.

Xing, R., A. V. Subrahmanyam, H. Olcay, W. Qi, G. P. van Walsum, H. Pendse, and G. W. Huber. 2010. "Production of Jet and Diesel Fuel Range Alkanes from Waste Hemicellulose-Derived Aqueous Solutions." *Green Chemistry* 12 (11):1933–1946. doi: 10.1039/c0gc00263a.

Xiong, H., A. DeLaRiva, Y. Wang, and A. K. Datye. 2015. "Low-Temperature Aqueous-Phase Reforming of Ethanol on Bimetallic PdZn Catalysts." *Catalysis Science & Technology* 5 (1):254–263. doi: 10.1039/c4cy00914b.

Xu, C., J. Donald, E. Byambajav, and Y. Ohtsuka. 2010. "Recent Advances in Catalysts for Hot-Gas Removal of Tar and NH_3 from Biomass Gasification." *Fuel* 89 (8):1784–1795. doi: 10.1016/j.fuel.2010.02.014.

Yan, N., Y. Yuan, R. Dykeman, Y. Kou, and P. J. Dyson. 2010. "Hydrodeoxygenation of Lignin-Derived Phenols into Alkanes by Using Nanoparticle Catalysts Combined with Bronsted Acidic Ionic Liquids." *Angewandte Chemie—International Edition* 49 (32):5549–5553. doi: 10.1002/anie.201001531.

Yang, B., and C. E. Wyman. 2008. "Pretreatment: The Key to Unlocking Low Cost Cellulosic Ethanol." *Biofuels Bioproducts & Biorefining-Biofpr* 2 (1):26–40. doi: 10.1002/bbb.49.

Yoon, J. S., Y. Lee, J. Ryu, Y.-A. Kim, E. D. Park, J.-W. Choi, J.-M. Ha, D. J. Suh, and H. Lee. 2013. "Production of High Carbon Number Hydrocarbon Fuels from a Lignin-Derived alpha-O-4 Phenolic Dimer, Benzyl Phenyl Ether, via Isomerization of Ether to Alcohols on High-Surface-Area Silica–Alumina Aerogel Catalysts." *Applied Catalysis B—Environmental* 142:668–676. doi: 10.1016/j.apcatb.2013.05.039.

Zakzeski, J., P. C. A. Bruijnincx, A. L. Jongerius, and B. M. Weckhuysen. 2010. "The Catalytic Valorization of Lignin for the Production of Renewable Chemicals." *Chemical Reviews* 110 (6):3552–3599. doi: 10.1021/cr900354u.

Zhang, W., J. Chen, R. Liu, S. Wang, L. Chen, and K. Li. 2014. "Hydrodeoxygenation of Lignin-Derived Phenolic Monomers and Dimers to Alkane Fuels over Bifunctional Zeolite-Supported Metal Catalysts." *Acs Sustainable Chemistry & Engineering* 2 (4):683–691. doi: 10.1021/sc400401n.

Zhao, C., and J. A. Lercher. 2012. "Selective Hydrodeoxygenation of Lignin-Derived Phenolic Monomers and Dimers to Cycloalkanes on Pd/C and HZSM-5 Catalysts." *Chemcatchem* 4 (1):64–68. doi: 10.1002/cctc.201100273.

Zhao, X. B., K. K. Cheng, and D. H. Liu. 2009. "Organosolv Pretreatment of Lignocellulosic Biomass for Enzymatic Hydrolysis." *Applied Microbiology and Biotechnology* 82 (5):815–827. doi: 10.1007/s00253-009-1883-1.

Zhao, Z., N. Lakshminarayanan, S. L. Swartz, G. B. Arkenberg, L. G. Felix, R. B. Slimane, C. C. Choi, and U. S. Ozkan. 2015. "Characterization of Olivine-Supported Nickel Silicate as Potential Catalysts for Tar Removal from Biomass Gasification." *Applied Catalysis a-General* 489:42–50. doi: 10.1016/j.apcata.2014.10.011.

Zhou, C. H., X. Xia, C. X. Lin, D. S. Tong, and J. Beltramini. 2011. "Catalytic Conversion of Lignocellulosic Biomass to Fine Chemicals and Fuels." *Chemical Society Reviews* 40 (11):5588–617. doi: 10.1039/c1cs15124j.

11 Catalyst Needs and Perspective for Integrating Biorefineries within the Refinery Value Chain

Paola Lanzafame, Siglinda Perathoner, and Gabriele Centi

CONTENTS

11.1 INTRODUCTION

There are many key drivers motivating a transition from fossil- to bio-based refineries, between which the more relevant are the following [1–17]:

i. Feedstock change—trend to renewable raw materials: (a) securing supply (shortage of fossil feedstocks); (b) rising costs of fossil feedstocks; and (c) autonomy of supply
ii. Sustainability—(a) neutral/positive CO_2 balance (reduce greenhouse gas [GHG] emissions); (b) energy, water, resources, and environmental burden; and (c) lower impact on environment of mobility
iii. (Socio)-political will—(a) to establish a bio-based economy; (b) to keep the country in a leading position in White Biotech; and (c) to support agriculture and preserve rural development
iv. Economics—(a) new/innovative products and (b) cheaper production processes

About a decade ago, it was thus expected that these drivers would induce a fast transition to a new bio-based economy, with biorefineries at the core of this new ecosystem. Today, however, the relevance and role of a bio-based economy in the future economic scenario are under revision due to different reasons, from the revision of forecasts about the shortage of fossil fuels (and/or high market cost) to the reconsideration of the expected benefits. Biofuel production and consumption have soared over the last decade in response to market forces, and especially government policies, such as subsidies and mandated use of biofuels, based often on an overestimation of benefits related to the environment, energy security, and rural development. It becomes increasingly evident that except in some specific cases, the cost-effectiveness of achieving these goals under current subsidy schemes is often low.

While a decade ago the alternatives were quite limited, this is not more true currently. The reduction of the emissions of GHGs was a main driver to push the use of biofuels through various political actions such as mandated fuel mix, subsidies, renewable and carbon targets. However, the use of bioenergy allows, on average, to save about 50% of CO_2 emissions with respect to fossil fuels, although there are many parameters affecting the life cycle analysis (LCA) of the use of bioenergy. Solar fuels, which relevance is continuously increasing [11–16], offer today a better cost-effectiveness perspective, although they are still not possible to produce on a large scale [13–16]. Methanol produced from the reuse of CO_2 and renewable energy has a potentially larger impact in terms of saving CO_2 emissions as well as resource and energy efficiency with respect to biofuel, with a Gton potential contribution in addressing GHG emissions, well comparable with the impact of biofuels but with better cost-effectiveness [18]. In addition, there is actually an increasing understanding about the impact of bio-based production on the environment [19,20]:

i. Long payback time, e.g., the time delay before the emissions from bioenergy systems will have reached a break-even point compared with the fossil fuel systems. It ranges from 20 to over 100 years for bio-energy, while it can be estimated for renewable methanol to be lower than 10 years. Therefore, alternatives to biofuels, such as renewable methanol or equivalent fuels from CO_2, are potentially faster and more effective in addressing the GHG emission challenge.
ii. High impact on soil quality and water ecosystem due to the intensive agro-productions to sustain biofuel manufacture. The impact in terms of wastewater is over one order of magnitude larger in comparison to the production of fossil (or solar) fuels.

There are also new elements pushing a revision of the role of biorefineries in the future (sustainable) energy scenario [11,12,21]:

1. New possibilities exist today to produce energy vectors not based on fossil-fuel raw materials.
2. The production of renewable energy (not based on biomass) increased far beyond expectations.
3. New routes are opened for non-energy use of biomass, higher-added-value production, and rural valorization.

4. Biomass transport and intensive monoculture agriculture necessary for biofuel production are bottlenecks for large biorefinery plants.
5. Growing and processing aquatic biomass, which shows better efficiencies and lower impact on the environment with respect to intensive terrestrial production of biomass (on equivalent production of fuels), is becoming a feasible possibility to produce biofuels.
6. Cost of fossil fuels has been maintained to low values to make costs of second-generation (2G) biofuels from lignocellulosic unacceptable.
7. The need to produce large amounts of H_2 to synthetize drop-in biofuels (hydrocarbons) from oxygen-rich biomass substrates (sugars, etc.) has been underestimated in terms of impact.
8. New models of symbiosis and value chain integration require reconsidering the model of biofuel production.

The predicted scenario of biofuel expansion and production of a decade ago has largely failed. For example, the International Energy Agency (IEA) in recent reports remarked how the world biofuel production showed the end of rapid expansion and the growth rate is currently significantly lower than expected [22]. The target of 8% biofuel share in transport energy for the year 2025 is currently revised to be within the 4–5% range, for example.

It is thus necessary to reconsider the model of future biorefineries and the perspective for integrating biorefineries within the refinery value chain, with consequent change in terms of catalyst needs to realize these new models. Although both bio- and chemocatalysis are utilized in biorefineries, especially the first, discussion here is limited prevailing to chemocatalysis and especially using solid (heterogeneous) catalysts. Some recent reviews on biocatalysis for biorefinery have been made by Patel et al. [23], Dutta and Wu [24], and Teter et al. [25], among others. Several reviews have been also published on the use of catalysis in biorefinery; a selection of them is given in Refs. [26–36]. However, the model of biorefineries, and as a consequence the priorities also in terms of catalysis needs, is evolving, as commented above and discussed in more detail by Lanzafame et al. [11–16]. There is thus the need to comment again on these aspects, from the perspective of directions and evolving scenario for catalysis in biorefineries and their integration with current refineries, in order to evaluate possibilities and anticipate needs for research. In fact, the transition from the current fossil-fuel centered energy system to the future sustainable energy will require a relatively long time (at least three decades), and it is thus of pivotal relevance to preserve as much as possible current very large investments in the energy infrastructure. The key aspects to make effective this transition (aspects often not considered enough in many of the current reviews of the topic) [21] are as follows:

i. Biofuels are not the final but the transitional solution to move to a better sustainable energy future.
ii. *Drop-in* fuels, i.e., fuels that integrate within the current energy infrastructure minimizing as much as possible the changes, are the only solution to reduce the costs of change.

This has clear implication in terms of technologies, type of fuels, and characteristics of biorefinery. We must mention, however, that there are contrasting ideas for the future and dynamics of change. This is somewhat obvious for such a complex topic as that of biorefineries, with large interest and investments. On the other hand, history teaches that a transition to a new economy is typically characterized from a change in companies, because it is often difficult to predict the game-changer elements [37]. We thus will not discuss strengths and weaknesses of these different thoughts, also because multiple models may coexist simultaneously, depending on the different relevance given to drivers such as GHG emissions, energy security, agriculture production, etc., in different world regions. We thus limit the discussion here on our vision to the future of biorefineries [11–16], although in the general context of dynamics of evolution of this sector [38–41]. The aim is to evidence new opportunities and remark area on which research can open new perspectives.

11.2 EVOLUTION OF THE CONCEPT OF BIOREFINERIES AND ITS INTEGRATION WITH THE CURRENT REFINERY VALUE CHAIN

There is not a unique definition of biorefinery. IEA Bioenergy Task 42 has developed the following definition for biorefinery: it is the sustainable processing of biomass into a spectrum of marketable products and energy. Biorefinery can be a facility, a process, a plant, or even a cluster of facilities. From a practical perspective, this concept translates to the different kinds of biorefineries [41]. In the classification system, IEA Bioenergy Task 42 has differentiated among mechanical pretreatments (extraction, fractionation, separation), thermochemical conversions, chemical conversions, enzymatic conversions, and microbial (fermentation, both aerobic and anaerobic) conversions. All kinds of biomass from forestry, agriculture, aquaculture, and residues from industry and households including wood, agricultural crops, organic residues (both plant and animal derived), forest residues, and aquatic biomass (algae and seaweeds) can be potentially used in a biorefinery.

According to the Joint European Biorefinery Vision for 2030 and European Biorefinery Joint Strategic Research Roadmap [42], a first classification of the different types of biorefineries can be made according to four main criteria:

- Integration of biorefining into existing industrial value chains (bottom-up approach) or development of new industrial value chains (top-down approach)
- Upstream or downstream integration
- Feedstock: biomass produced locally or imported biomass or intermediates
- Biorefinery scale

These criteria evidence how the specific model of biorefinery to be developed is based on specific sector requirements or geographical constraints: biomass type, quality and availability, infrastructure, local industries, etc. The choice of technology options (processes, feedstocks, location, and scale) for each biorefinery will be specific. Elements for the decisions are available industrial equipment, technological and industrial know-how or access to biomass, etc. Biorefinery development will be driven by the demands of leading players from sectors such as agroindustries, forest-based industry, the energy and biofuel sectors, and the chemical industry. The integration of biorefineries into industrial value chains will be driven by either upstream players (producers and transformers of biomass) or downstream customers (producers of intermediates and final products). In principle, each biorefinery is somewhat unique even if the input and output can be formally similar. This is also what occurs in oil refineries. Nevertheless, the different biorefinery concepts could be lumped into some main classes [42] based on the type of feedstock.

- *Starch and sugar biorefineries*: processing starch crops, such as cereals (e.g., wheat or maize) and potatoes, or sugar crops, such as sugar beet or sugar cane.
- *Oilseed biorefineries*: currently produce mainly food and feed ingredients, biodiesel and oleochemicals from oilseeds such as rape, sunflower, and soybean.
- *Green biorefineries*: processing wet biomass, such as grass, clover, lucerne, or alfalfa. They are pressed to obtain two separate product streams: fiber-rich press juice and nutrient-rich pressed cake.
- *Lignocellulosic biorefineries*: processing a range of lignocellulosic biomass via thermochemical, chemiocatalytic, or biochemical routes.
- *Aquatic (marine) biorefineries*: processing aquatic biomass—microalgae and seaweed.

The platform chemicals (e.g., C5/C6 sugars, syngas, biogas) are intermediates that are able to connect different biorefinery systems and their processes, although may be themselves the final product. The number of involved platforms is an indication of the system complexity. While a decade ago it was supposed that many (about 15–20) platform molecules will be central to biorefinery, this number has been drastically reduced currently. Two groups of products generate from biorefinery processes: energy vectors (e.g., bioethanol, biodiesel, synthetic biofuels) and chemicals (e.g., chemicals, materials). Food and feed may be also produced. The two main feedstock groups are "energy crops" from agriculture (e.g., starch crops, short rotation forestry) and "biomass residues" from agriculture, forestry, trade, and industry (e.g., straw, bark, wood chips from forest residues, used cooking oils, waste streams from biomass processing). Figure 11.1 reports an overview of the different main raw materials and outputs from biorefinery, as well as the main process operations. The latter can be differentiated among four main conversion processes: biochemical (e.g., fermentation, enzymatic conversion), thermochemical (e.g., gasification, pyrolysis), chemical (e.g., acid hydrolysis, synthesis, esterification), and mechanical processes (e.g., fractionation, pressing, size reduction). The biorefinery systems are classified by quoting the involved platforms, products, feedstocks, and the processes. Two examples of classifications are the following: (1) oil biorefinery using oilseed crops for biodiesel, glycerin, and feed, and (2) C6 sugar platform biorefinery for bioethanol and animal feed from starch crops. Although these schemes are general, they evidence the complexity of treatments necessary to convert different types of biomass substrates.

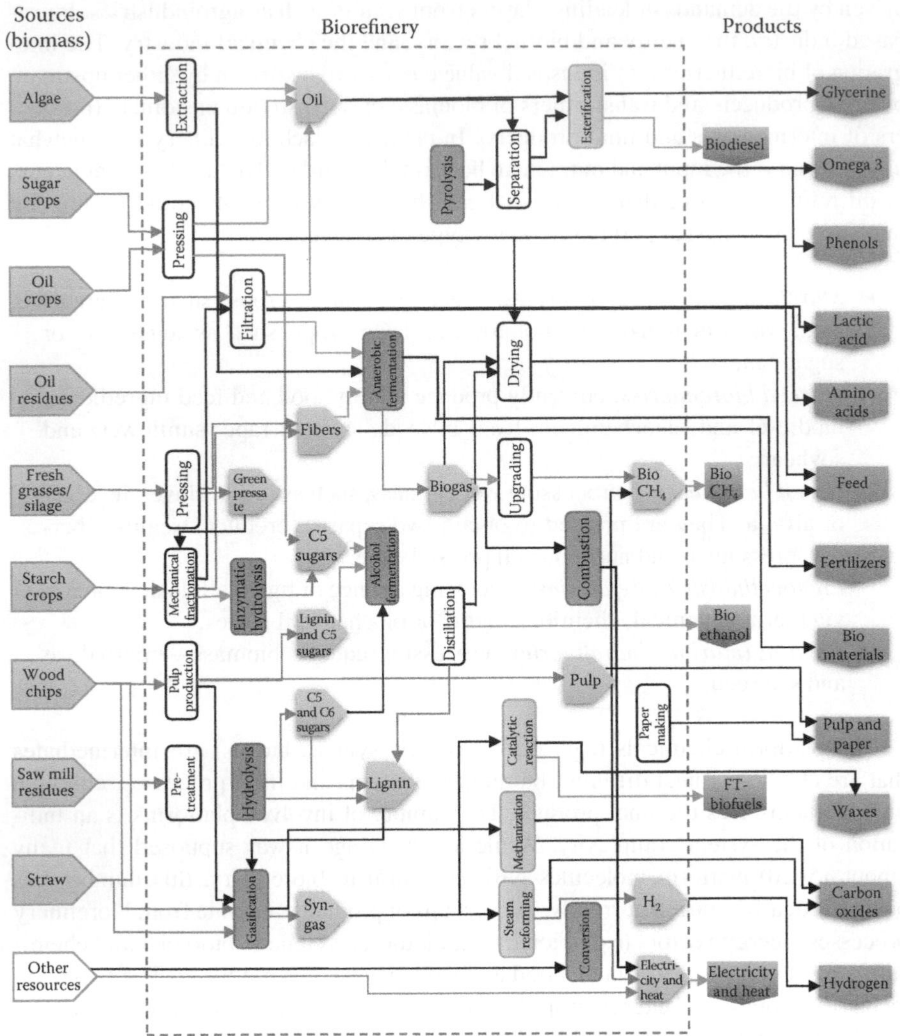

FIGURE 11.1 Simplified overall scheme of the different feeds, output products, and main biorefinery operations. (Reprinted from *Industrial Biorefineries and White Biotechnology*, Pandey A, Höfer R, Taherzadeh M, Nampoothiri M, Larroche C (Eds.), Chapter 1, p. 3. Figure 1.2, de Jong E, Jungmeier G, Copyright 2015, with permission from Elsevier.)

This complexity is much higher with respect to oil, which is composed of several thousands of chemicals, but which can be lumped into a limited number of classes of chemicals. In addition, each type of biomass substrate requires different procedures and technologies. As a consequence, many different models and concepts of biorefineries can be foreseen, with a large variety of technologies necessary to develop affecting in turn the fixed costs and profitability of the different specific routes. This is one of the main weaknesses of biorefinery, together with costs of producing, collecting, and transporting biomass. This is one of the reasons why a revised concept of biorefinery, as will be mentioned later, is necessary for the future. Sustainability is another critical element that will drive the effective implementation on a large scale of the different biorefinery concepts [43].

As remarked before, the commercial implementation of bioroutes of production appears less promising today than a few years ago for a number of reasons, ranging from economics to sustainability [11–16]. It is thus necessary to revise the concept of biorefineries to identify priority paths and technologies. Figure 11.2 illustrates this concept. With respect to the current situation, largely based on first-generation (1G) biorefineries (e.g., bioethanol, starch, pulp, and paper) and some petrochemical sites with individual biotech plants, typical data presented in literature and discussions in conferences show that the future trend is represented by the passage to 2G biorefineries (not based on raw materials in competition with food, like lignocellulosic raw materials) or even in a longer term to 3G biorefineries, and to fully integrated biotechnological and chemical production sites. While this already happened, even with a much slower rate with respect to what supposed a decade ago, we should remark that the future of bio-based production will be based also on the development of new advanced concepts, particularly (1) solar biorefineries, (2) biofactories (characterized by small-size, process intensification [PI] and chemical production at the core of plant design), and (3) symbiosis bioproduction, e.g., highly integrated with surrounding products in terms of raw materials and output.

The new integrative plant concept in moving from the current to the future biorefinery models will be based on the following key elements: (1) expansion of raw material basis, (2) improvement of sustainability, (3) better efficiency of use of resource and energy, and (4) expansion of product range. At the core is the development of new PI concepts, and the development of very efficient small-scale and flexible processes, adapted to operate with a range of input raw materials and type of products. Full biomass use and symbiosis at the regional level are further critical aspects of the future of biorefineries, with a high degree of regional specialization, in addition

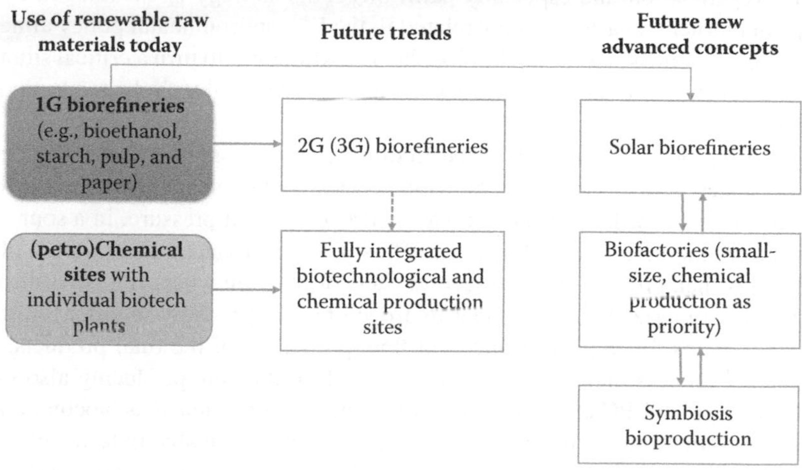

FIGURE 11.2 Simplified scheme of the status and future trends in integration of fossil-to-bio-based refineries.

to process technologies that are easy to be adapted at different process schemes. The development of new catalysts that fit these new requirements and process operations, as well optimally integrate with biocatalytic operations, is a key element to realize in this future vision.

New models of biorefineries are presented, for example, by Matrica, a joint venture between Versalis (eni group) and Novamont. The project in Porto Torres (Sardinia, Italy) plans to produce various chemicals starting from an upstream process of oxidative cleavage with ozone of vegetable oils. This plant has a planned production capacity of 35 ktons/year (thus small size), with the driving element of the initiative being to decommission an old industrial site (petrochemical plant) to develop a cleaner production with better market perspectives. The social factor is thus driving the creation of the new biorefinery model strongly linked to territory and as a recovery strategy for a local economy.

This concept is planned to be applied in other similar situations in Italy. An example is given, in particular, by the Green Refinery model developed by eni to "reinvent petroleum refineries" [44]. The background element is that the European refining industry is living a major economic crisis, characterized by (1) very low operating margins, (2) bear fuel market, and (3) refining installed overcapacity. These general structural crisis elements lead to the closure of the eni Venice Refinery due to its simple process scheme made by hydroskimming plus visbreaker/thermal cracker, without any catalytic conversion units.

The refinery was situated in an area already suffering from the closure of many plants, and thus further enhancing an already dramatic social situation from the employment perspective. Eni took thus the opportunity to invest in the innovative Green Refinery project to convert the existing Venice Refinery into a "biorefinery" (Figure 11.3), able to produce a new generation of very high-quality biofuels, mainly green diesel but also green jet, green naphtha, and green LPG, each one exploitable as biocomponents for transportation fuels, starting from biological feedstocks (mainly vegetable oil and especially palm oil). This strategy is encouraged by the European biofuel scenario, strongly related to the EU environmental policy aimed to reduce the CO_2 emissions. In the Venice site, the strategy is to turn a critical situation into the opportunity of entering the new business of renewable fuels in a leadership position.

The core process technology is the Ecofining™ process, which consists of two reaction stages: a first one where the triglycerides contained in the biological feedstock are completely deoxygenated under hydrogen partial pressure, in a sour environment, producing a mix of linear paraffins, CO_2, and water. The product of the first stage is then processed in the second stage of reaction where this mix of linear paraffins is isomerized, always under hydrogen partial pressure, in order to branch the linear chains for improving the cold flow properties of the final products. The Ecofining™ process maximizes the green diesel production, producing also green naphtha and green LPG and optionally green jet, each one valued as biocomponents for transportation fuels. The latter, however, is the more valuable in terms of added value. One of the challenges in catalysts development is thus to maximize the production of this energy product.

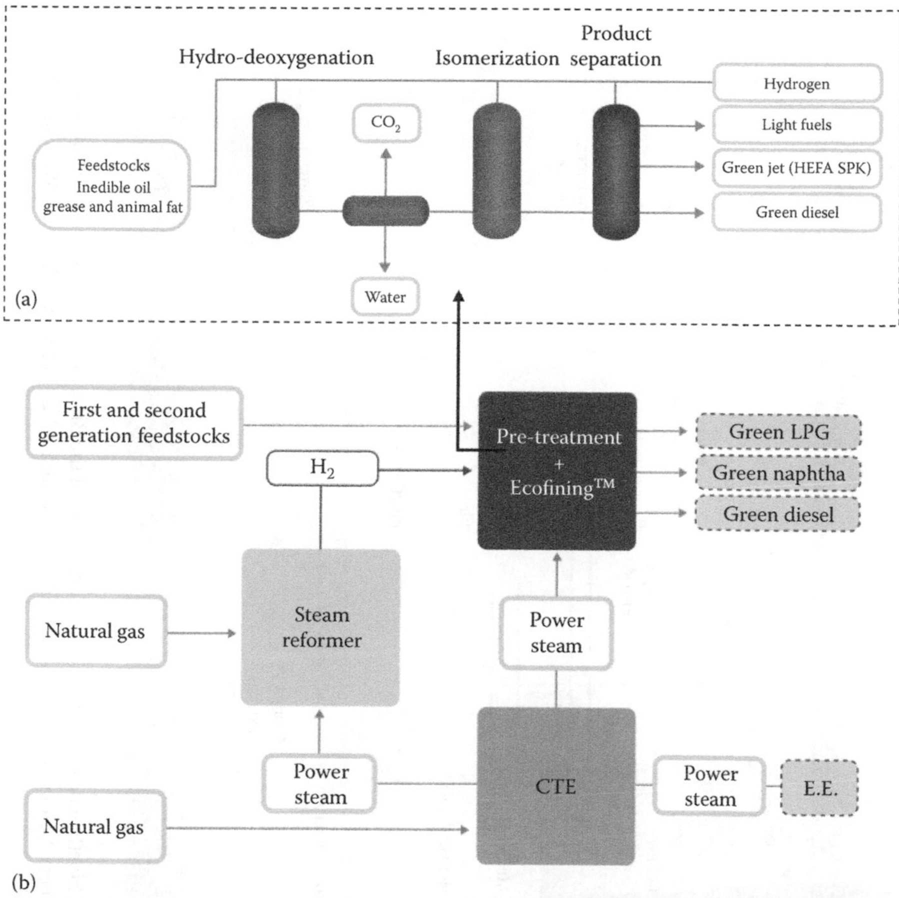

FIGURE 11.3 (a) Simplified Ecofining™ process flow scheme and (b) eni Green Refinery configuration. (Adapted from eni Brochure, "Green Refinery: Reinventing petroleum refineries," https://www.eni.com/en_IT/attachments/azienda/attivita-strategie/refining-marketing/eni_Green-Refinery_esecutivo.pdf.)

Another example of integration between fossil fuel- and bio-based refinery is the Biorefinery Leuna (Linde-KCA-Dresden GmbH) [17]. The eni Green Refinery strengths are as follows: (1) it uses an existing dismissing refinery to decrease the fixed costs for the biorefinery and (2) it provides an alternative strategy to reduce the social (employment) impact of the closure of an old conventional refinery plant. In the case of the biorefinery Leuna, the motivations are different and based especially on the advantage for bio-based processes upon integration: (1) use of existing site facilities (infrastructure, utilities, services), (2) integrated energy concept, and (3) support in approval processes. One of the vision elements for the "Biorefinery Leuna" design is the integration of bioethylene in the value chain. The plant concept [17] is presented in Figure 11.4. Bioethanol, produced via 2G (lignocellulosic

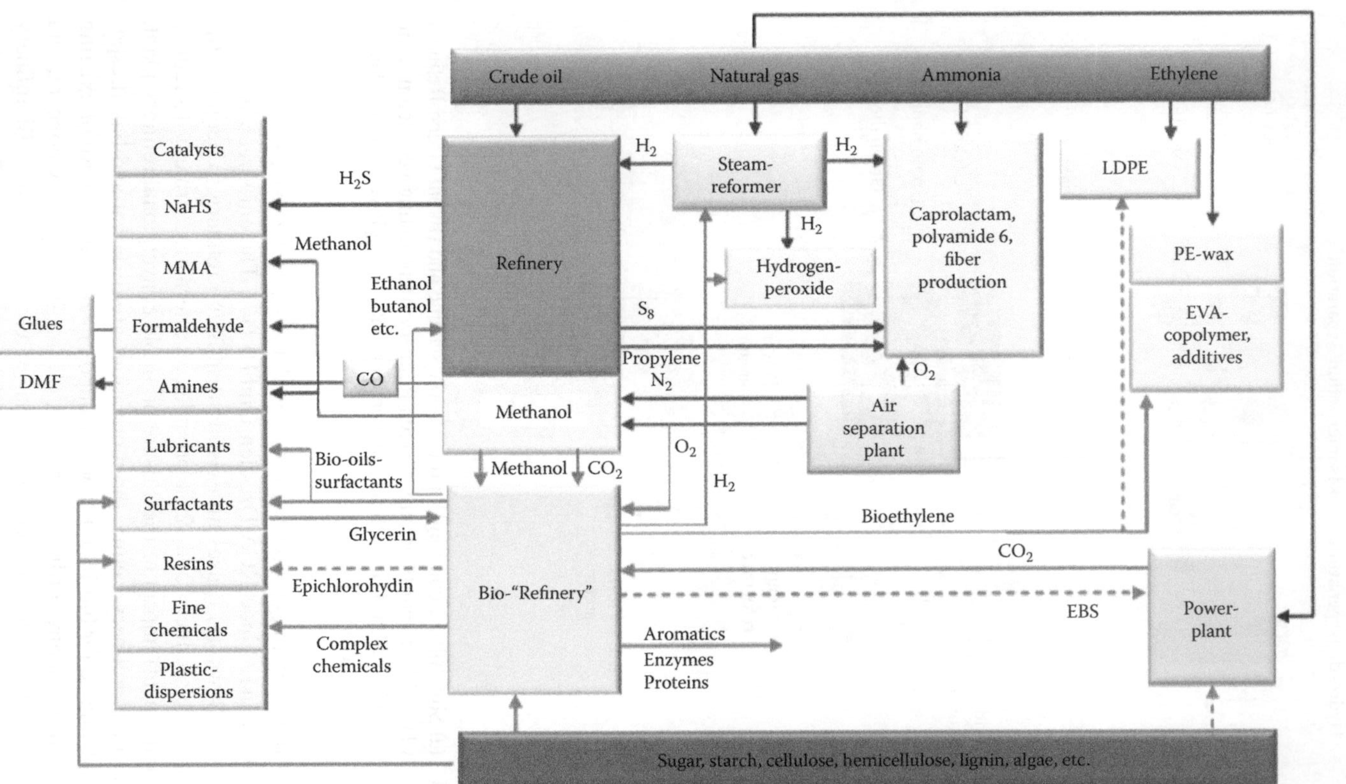

FIGURE 11.4 Block diagram scheme of the integration of bioethylene production in the value chain of the Biorefinery Leuna. (Adapted from Welteroth U, From fossil- to bio-based refineries. Case Study Biorefinery Leuna, presented at IEA-Bioenergy Task 42 Biorefinery—German Workshop Worms, September 15, 2009.)

biomass) conventional enzymatic treatment and fermentation or (in a longer term) via gasification and syngas fermentation, is at the core of biorefinery integration. Via catalytic dehydration, ethanol is then converted to (bio)-ethylene and then further converted using existing infrastructure (based on the integration between energy and chemical production from fossil fuels). Other products from biomass conversion are also produced (see Figure 11.4). The strength of this process scheme is the integration between energy and chemical production and between fossil fuel- and bio-based production in order to reduce the costs. The large-scale production of bioethylene via bioethanol in 2G processes is still not economical in Europe, but this model of biorefinery evidences one of the possible directions to proceed.

This example of the Biorefinery Leuna well evidences the idea that the future of biorefineries will be centered on four elements:

- Smaller size (to adapt better at the regional level and minimize biomass transport) and lower capital investments with short time amortization
- More flexible production, e.g., possibility to fast switch between chemical/fuel production to adapt better to market
- Production of higher-added-value products, but in an integrated full biomass use
- Value chain integration and symbiosis

Although the concept of biorefinery already includes the idea of biomass processing into a spectrum of products (food, feed, chemicals, fuels, and materials), it is evident that this is strongly linked with the oil refinery concept: (1) large scale, (2) relatively limited range of products with low added value, and (3) limited flexibility in terms of products. We thus prefer to indicate this new model of biorefinery as biofactory [11–16] to remark the aspect that chemical production is becoming the main target, with energy products (mainly higher-value additives rather than direct components for fuel blending) as a side element. This model focuses at integrating within the current oil-centered refinery and chemical production, rather than to substitute it, as hidden in many current models and discussions on biorefinery and bio-based economy. Oil will still be abundant in the next half century, and a successful transition to a more sustainable economy is possible only when the barrier to change (e.g., the necessary economic investments, but also other elements that delay the introduction of new market products) is reduced. This mandatorily requires that the new processes/technologies can be inserted readily (or with minor changes) within the current infrastructure (drop-in products and processes/technologies).

The process design should allow a flexible switch between chemical versus energy products in order to adapt better to the fast-changing market opportunities. The new model is thus different not only in terms of targets but also in size, plant design, and reference market. Keywords characterizing the new models of biorefinery will be (1) small scale, (2) flexible production, (3) high-added-value products, and (4) chemical production as the core element. The reasons are the need to integrate better at the regional level, the need to have a production more suited to follow a fast evolving scenario, low capital investment, more efficient intensification of the processes, and overall lower impact on the environment. This has clear consequences also in terms of catalysis needs to enable these new models of biorefinery.

One of the key elements in the new models of biorefineries is also the need for integrated CO_2 reuse and introduction of renewable energy in the biorefinery value chain. These are aspects not typically considered in current biorefinery models, but which are a key element to move to a sustainable, low-carbon economy [13–16]. Sharifzadeh et al. [45] showed the possibility to increase biomass to fuel conversion from 55% to 73% when CO_2 utilization is integrated in biorefinery. In biorefineries where chemicals are the main product (biofactories), the scope is instead different and the target is the optimal integration of CO_2 utilization within the value chain. Symbiosis with near-lying factories is another emerging element characterizing the new models of bioeconomy. There are different possibilities of efficient symbiosis, but an interesting option is the use of waste from other productions (wastewater and CO_2, for example, in advanced microalgae processes) to enhance the energy efficiency and reduce environmental impact and CO_2 emissions of a productive district.

There are two interesting new models for advanced biorefineries/biofactories, which are emerging as a new opportunity [11–16]:

i. Bioproduction of olefin and other base raw materials
ii. Development of flexible production of chemicals and fuels

The first is centered on the production of base raw materials for chemical production, while the second focuses on intermediate and high-added-value chemical products, including monomers for polymerization, but with flexible type of production for a rapid switch to produce eventually fuel additives, depending on the market opportunities. Two of the elements characterizing both models are the full use of the biomass and PI (for efficient small-scale production).

We focus the discussion here only on the first model because it provides good indications on different alternative paths that should be provided and the related catalysis needs, as well as how this may change the future of integration between oil- and bio-based refineries.

11.3 OLEFIN BIOREFINERY AND RELATED CATALYSIS NEEDS

Currently, light olefins are produced principally by steam cracking of naphtha or in minor amount of natural gas, i.e., using fossil fuels. This is a very energy-intensive process. The pyrolysis section of a naphtha steam cracker alone consumes approximately 65% of the total process energy and accounts for approximately 75% of the total exergy loss. The specific emission factor (CO_2 Mt/Mt light olefin) depends on the starting feedstock but ranges between 1.2 and 1.8. Light olefins can be produced from different sources (crude oil, natural gas, coal, biomass and biowaste such as recycled plastics, and CO_2). Olefins are also a side product of the fluid catalytic cracking (FCC) process in refinery but are usually utilized inside the refinery for alkylation or oligomerization processes. The alternative process is the dehydration of ethanol produced from biomass fermentation or the indirect route of production of syngas from biomass pyrolysis/gasification. Syngas can then be converted to methanol/dimethylether (DME) via conventional catalysts and further converted to olefins via the methanol-to-olefin (MTO) process (Scheme 11.1). A selective conversion to propylene (methanol-to-propylene [MTP]) is also possible. Olefins can be further interconverted via metathesis reaction.

New routes under development are based on the Fischer–Tropsch-to-olefin (FTO) reaction by using syngas or directly CO_2/H_2 mixtures [41] (Scheme 11.1). To make the process sustainable, H_2 should be produced using renewable energy sources (rH_2). An alternative path is based on a first step of reverse water gas shift (RWGS) from CO/rH_2 mixtures, followed by gas fermentation of the $CO/H_2/CO_2$ mixture to produce ethanol. LanzaTech has developed already some semicommercial units for the second step of ethanol production, although productivity is still limited and ethanol has to be recovered from the solution. Ethanol could then be dehydrated to ethylene. Renewable H_2 could derive also from processes using microorganisms, such as cyanobacteria, able to produce H_2 using sunlight.

Ethanol to ethylene is currently the main practical possibility [46]. The process is utilized by Braskem, Dow, and Solvay Indupa in Brazil. Industrial catalysts for this reaction are mainly based on zeolites, which should be doped to tune acidity. Selectivity is already quite high (over 98%), but an improvement of space yields is necessary. The transformation of bioethanol to C3 and C4 olefins/diolefins via the so-called Lebedev process (oxidative dimerization to butadiene) (Scheme 11.1) is more challenging, but the selectivities are still not enough, even if good [47,48]. Propylene may be instead produced from ethylene via metathesis [49]. Butadiene

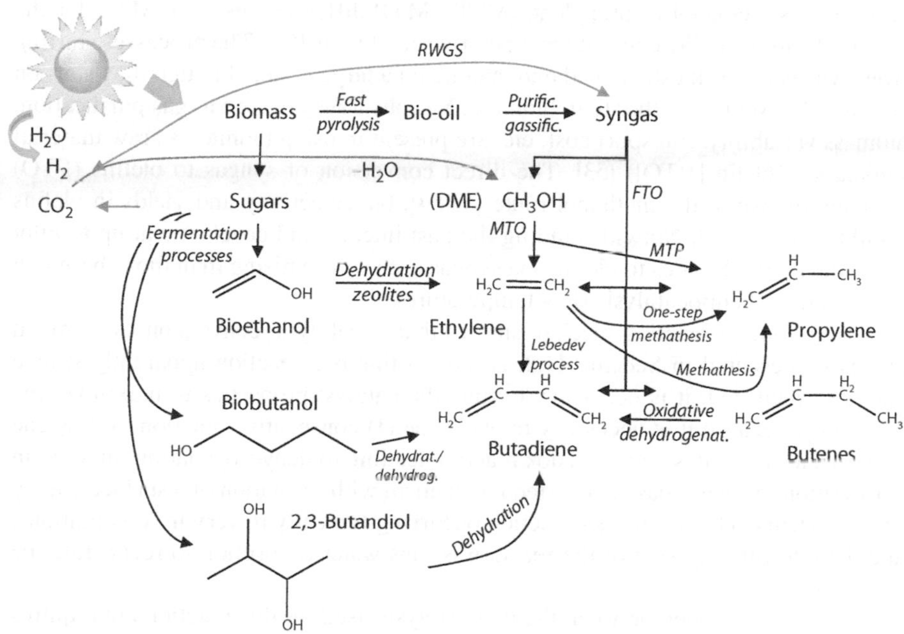

SCHEME 11.1 Different routes to synthetize light olefins from non-fossil fuels and renewable (solar) energy. (Adapted from Lanzafame P et al., *Catal. Today* 2014, 234: 2.)

can also be produced from biobutanol and 2,3-butanediol obtained by fermentation [48]. Propylene could be synthetized from 1- or 2-propanol, or from 1,2-propanediol [50], the latter being obtained by catalytic conversion of sugars or other platform molecules (glycerol, lactic acid). The chemiocatalytic route is preferable over the fermentation paths [51] because the production by fermentation of alcohols and glycols other than ethanol has still too low productivities, the enzyme being sensitive to inhibition by product concentrations and operating under stressed conditions.

There are various companies quite active in this area:

- Polyol Chemical Inc. (propylene and ethylene glycols from sorbitol/glucose).
- LanzaTech (2,3-butanediol by gas fermentation and in joint venture with Invista to then produce butadiene).
- Versalis in joint venture with Genomatica to produce butadiene via 2,3-butanediol obtained by fermentation. Genecor in joint venture with Goodyear is developing a bio-isoprene production process.
- Glycos Biotechnologies is developing a process to produce isoprene from crude glycerine.
- Global Bioenergies and Gevo/Lanxess are developing processes to produce isobutene from glucose or from isobutanol, respectively. The latter is produced by fermentation.

An alternative possibility is to produce syngas from biomass and then convert methanol to olefins via methanol-to-olefin (MTO) process or variations of this process such as methanol-to-propylene (MTP). MTP differentiates from MTO for the type of the used zeolite catalyst and operative conditions [52]. The processes starting from syngas are well established and operate on a large scale, although they start on coal or other sources rather than biomass. Various issues related to gas purification, biomass variability, transport cost, etc. are present in using biomass as raw material (biomass-to-olefin [BTO]) [53]. The direct conversion of syngas to olefins (FTO) is an alternative to the methanol route [53,54], but selectivity and yields in olefins should be improved. Notwithstanding the past interest and development up to pilot units, the thermal routes to olefins via syngas are less promising than those based on fermentation/chemiocatalysis (low temperature).

The simplified reaction mechanism of ethanol catalytic conversion over mixed oxides is presented in Scheme 11.2. Although this is a reaction apparently simple and acid-catalyzed, it is necessary to tune the catalyst properties to maximize the selectivity, because it is necessary to avoid the (1) consecutive reaction of ethylene with surface acid sites and (2) redox reaction leading to dehydrogenation rather than dehydration. An acid–base concerted mechanism with formation of a surface ethoxy species occurs. This is an easy reaction occurring typically in very mild conditions, and the rate-limiting step of the reaction is thus water desorption to regenerate the active site.

γ-Alumina has been between the first catalysts used for this reaction but requires a high reaction temperature (450°C) resulting in relatively low ethylene yield (about 80%). Doped alumina with KOH and/or ZnO or $MgO–Al_2O_3/SiO_2$ mixed oxides (Syndol catalyst) were used by companies such as Phillips Oil Co. and Halcon SD,

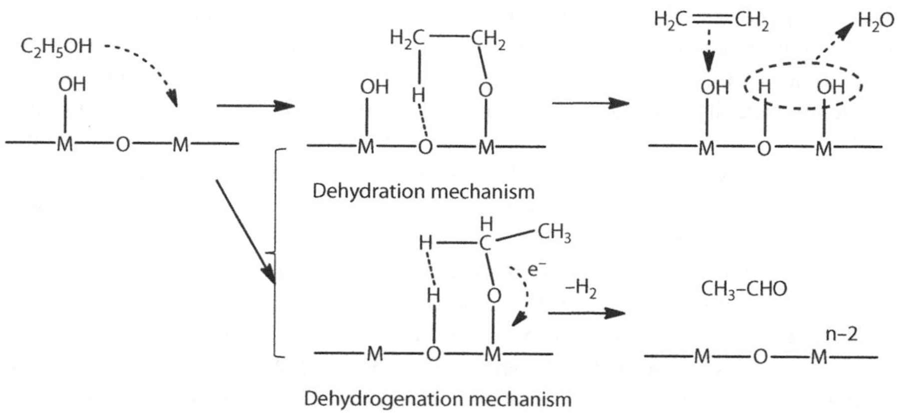

Dehydration mechanism

Dehydrogenation mechanism

SCHEME 11.2 Simplified reaction mechanism of ethanol conversion over mixed oxide catalysts. (Adapted from Lanzafame P et al., *Chem. Soc. Rev.* 2014, 43: 7562.)

respectively, to increase the selectivity [46]. Further modifications with other dopants resulted in selectivities of over 98–99%, although still high reaction temperatures were necessary.

Zeolites, particularly HZSM-5, were a second class of catalysts used for ethanol dehydration [55–59]. The main advantage is the activity at lower temperatures (200–300°C). At 300°C, HZSM-5 can reach an ethanol conversion level of 98% and 95% ethylene selectivity. The main disadvantage of HZSM-5 is its acidity, which reduces its stability and coking resistance. Modification with phosphorous to reduce acidity improves both selectivity (over 99%) and stability [60]. Modification with La (eventually as a codopant with P) leads also to interesting results. With almost 100% ethylene selectivity and ethanol conversion at low temperatures (about 240°C), 0.5% La-2% P-HZSM-5 is currently one of the best catalysts for industrial use. Due to diffusional limitations, the use of nanosized zeolites leads to better results. Also SAPO zeolites, such as the SAPO-34 that is between the best catalysts for MTO reaction together with HZSM-5, also show good performance in ethanol dehydration.

A third class of catalysts investigated was heteropolyacids. Particularly, $Ag_3PW_{12}O_{40}$ has demonstrated high catalytic ability, making it a promising catalyst for the dehydration of ethanol to ethylene, but its high acidity reduces its stability. This catalyst gives 99.2% selectivity at 100% ethanol conversion in rather mild conditions (220°C, GHSV = 6000 h^{-1}) [61], but long-term stability has to be demonstrated. As a comparison, an industrial catalyst such as the cited SynDol (Halcon) gives comparable performances (96.8% selectivity at 99% conversion) but requires higher temperatures (450°C, LHSV = 26–234 h^{-1}) [55].

Therefore, recent developments in catalysts [46,55], particularly nanosized HZSM-5, which has a 99% ethylene yield at 240°C and a lifespan of over 630 h before ethylene selectivity decreased to below 98%, and in heteropolyacids such as $Ag_3PW_{12}O_{40}$, achieving over 99% ethylene yield at temperatures as low as 220°C, have significantly improved performances over currently used catalysts in industrial plants for ethanol dehydration. It should be remarked that the profitability of the process depends essentially on the cost of production of bioethanol, which may vary considerably depending on the raw materials and technology of production. Energy integration of the process is also critical. Actual ethanol to ethylene plants have a production capacity of about one order of magnitude lower than that of typical steam cracking plants, but this aspect could be an advantage in terms of on-purpose plants. PI is one of the ways to make profitable also small- to medium-size plants.

A different route, evidenced in Scheme 11.1, to produce olefins in biorefineries is to use the CO_2 produced in fermentation processes and H_2 produced from renewable resources: from photoreforming or aqueous phase reforming of biowaste, water photo(electro)catalytic splitting, and/or biocatalytic processes [62–68]. There are different possible routes to produce light olefins from CO_2 and renewable H_2, the main being summarized in Scheme 11.1 [13–16]. RWGS reaction is typically promoted from the same catalysts of the consecutive steps (methanol synthesis or Fischer–Tropsch [FT] reaction), and thus, a single reactor/catalyst could be used. However, a direct route converting CO_2 without involving the RWGS step is preferable because reversibility of the latter limits the performances. There are two main paths to light olefins: (1) a direct route from syngas ($CO + H_2$) using modified FT catalysts and (2) an indirect two-step route via the intermediate formation of methanol [14,53]. In this indirect route, a conventional commercial methanol catalyst is used for the first step and small-pore zeolites (CHA or MFI framework-type) for the second MTO step. In the presence of an acid catalyst, two methanol molecules could be dehydrated to DME, which can be also converted to light olefins (it is an intermediate in the process) [69,70].

Being current methanol catalysts active both in RWGS reaction and in the methanol synthesis, it may be apparently the same to start from CO/H_2 or CO_2/H_2 mixtures. However, the productivities in the second case are typically one third of those using syngas (CO/H_2), even if the addition of small amounts of CO_2 (less than 3–4%) to syngas promotes the methanol synthesis rate. There are two main motivations. CO_2 is more oxidant than CO and thus in large amounts alters the surface active state of the catalysts ($Cu/ZnO/Al_2O_3$-based materials). The water formed in the RWGS reaction inhibits the reaction. It is thus convenient to use two reactors in series, with intermediate removal of water from the stream. The alternative is to use a reactor approach with in situ removal of water (catalytic distillation, membrane reactor). It is also possible to combine the catalysts for methanol to the zeolite for MTO to have in one step the direct formation of light olefins from CO_2 and H_2.

The direct FTO conversion of CO_2 and renewable H_2 to light olefins is the most attracting route [71–73]. The probability for the selective formation of lower olefins increases with temperature (in the 200–400°C temperature range, the typical one for FT reaction) and decreases at higher pressures and H_2:CO ratios in the feed. Olefins can also be incorporated into the growing chain involving a metallocyclobutane transition state followed by β-H transfer to form an α-olefin. It is thus necessary to prevent the readsorption of olefins, which increases the formation of longer-chain compounds. Shorter contact times are preferable, but also the choice of the reactor is important. Operations in liquid phase (slurry-type reactors) allow to limit olefin readsorption and surface overheating due to the exothermic reaction and thus to maximize the yields of lower olefins. In general, overall, yields up to over 55% in

C2–C4 olefins have been observed, but together with C2–C4 alkanes, methane, and C5+ products. These conditions are still not satisfactory and further improvement would be necessary, besides starting from CO_2 instead of CO would be preferable. There is thus a need of further R&D on catalysis to develop and exploit this route.

Olefins, when primarily formed, can be then converted by various routes to follow market demands, as shown in Scheme 11.1. Butadiene, for example, is a product with a current market demand higher than production, as a consequence of the change from naphtha cracking to cracking of ethane due to increased shale gas production and decrease in ethane costs. This has stimulated the interest in bio-based routes for the production of butadiene, particularly for on-purpose applications, even though the current perspective is of a return back to naphtha cracking, with a reduced interest in the development of alternative routes for C4+ olefins/diolefins. In addition, the catalytic dehydrogenation or oxidative dehydrogenation of *n*-butane is a strong competitive route.

There are some main possibilities to produce butadiene by alternative (non-fossil-fuel based) routes (Scheme 11.1). The process to convert ethanol into butadiene has been known for a long time, although it is still rather inefficient. Ethanol is converted into acetaldehyde after which an aldolization is performed followed by dehydration. A target yield of over 90%, at temperatures below about 450°C, is required for industrial exploitability as well as good stability of the catalyst. While, in general, yields were not satisfactory (below 70%), a catalyst based on magnesia–silica, doped with Na_2O, has been reported to have nearly 90% butadiene yield (although not reproduced by other researchers) [74]. There is thus the need to develop improved catalysts for this reaction, as well as to understand a number of fundamental aspects (reaction mechanism, structure–activity–selectivity relationship, etc.) [47]. Posada et al. [75] indicate the conversion of ethanol to ethylene or 1,3-butadiene as promising routes for an integrated biorefinery concept, in contrast to other possibilities (for example, ethanol conversion to acetic acid, *n*-butanol, isobutylene, hydrogen, and acetone).

Scheme 11.1 evidences also other possible routes to produce butadiene. It must be remarked that different companies are also exploring all these possibilities, which are market-conditioned from the availability at low cost of the raw material. Butanol dehydration to butenes and butadiene has been presented as a value route [76], but current manufacture cost of butanol is still high. In addition, further catalytic conversion to butadiene is also difficult. While ethanol to ethylene conversion is a dehydration reaction, butanol conversion to butadiene requires an oxidative dehydration/dehydrogenation mechanism. Data are quite limited on this reaction and related catalysts.

The third alternative is the conversion of butanediol to butadiene, which is investigated by companies such as LanzaTech and Genomatica/Versalis. 1,3-, 1,4-, or 2,3-butanediol could be dehydrated to butadiene over acid catalysts, but various by-products (unsaturated alcohols, ketones, etc.) form and the reaction is thus more challenging with respect to ethanol dehydration. Acetone–butanol–ethanol (ABE) fermentation with wild and genetically modified strains (from the *Clostridium* family) is known for a long time but has received renewed interest recently. However, there are still many aspects to improve in order to produce *n*-butanol at commercially attractive prices, such as (1) improving yields of butanol, (2) expanding substrate utilization, and (3) minimizing energy consumption during separation and purification. Cost of *n*-butanol is thus still high and therefore it is necessary to develop microorganisms able to give the selective fermentation to butanol to make the synthesis competitive.

Solid acids such as SiO_2–Al_2O_3, Al_2O_3, ZrO_2, and TiO_2 convert 1,3-butanediol (1,3-BDO) depending on their acid properties [77]. Strong acid catalysts (SiO_2–Al_2O_3) catalyze the dehydration of 1,3-butanediol at reaction temperatures below 250°C, while weak acid catalysts (ZrO_2 and TiO_2) require temperatures above 325°C. SiO_2–Al_2O_3 catalyzes the dehydration of 1,3-butanediol into unsaturated alcohols. The latter are then dehydrated into 1,3-butadiene. Alumina alone instead forms 4-methyl-1,3-dioxane, which is the acetal compound of 1,3-butanediol and formaldehyde. Several compounds were produced over TiO_2 and ZrO_2 owing to the side reactions such as dehydrogenation and hydrogenation. On strong solid acids, the butadiene selectivity is still unsatisfactory and thus more research is needed. On the other hand, the method could be interesting for the synthesis of unsaturated alcohols (raw materials for the synthesis of various fine and specialty chemicals for applications as medicines, perfumes, agricultural products). Over weak basic oxides such as CeO_2 at 325°C, 3-buten-2-ol and trans-2-buten-1-ol are produced with selectivities of about 58% and 36%, respectively [78].

11.4 CONCLUSIONS

The concept of biorefineries is evolving because the nexus between chemistry and energy is changing, driven from the demand for sustainable energy and chemistry production. We have discussed here some of the possibilities for new sustainable biorefineries, with focus on the production of olefins from biomass sources and some aspects of the possibility of exploitation and valorization of CO_2 within biorefineries and the opportunities to integrate renewable energy sources in the biorefinery production (solar biorefineries). There are various other aspects that were discussed elsewhere [11–16]. We should remark that the integration between oil- and bio-based refineries and of the new model of refineries, including the solar one, are topics with great interest and seeing rapid progress. We have thus given here only some glimpses to evidence some of the possibilities and the need to think "out of the box" to look at the new challenges, and as a consequence to the new needs for catalysis to enable these possibilities.

There is thus a changing panorama for bio-based production and its integration within current refinery schemes, with new opportunities offered by the integrated utilization of CO_2 and renewable energy to move to a sustainable and low-carbon bioeconomy. There are technologies to enable this future, and legislative regulation and incentives should better recognize the possibilities offered. However, this is a clear path necessary to follow for substituting fossil fuels and to develop a low carbon society.

ACKNOWLEDGMENTS

The authors acknowledge the PRIN10-11 projects "Mechanisms of activation of CO_2 for the design of new materials for energy and resource efficiency" and "Innovative processes for the conversion of algal biomass for the production of jet fuel and green diesel" for the financial support and the EU IAPP project no. 324292 BIOFUR (Biopolymers and Biofuels from Furan-Based Building Blocks), in the frame of which part of this work was realized.

REFERENCES

1. Zhang Y-H P, *Energy Sci. Eng.* 2013, 1: 27.
2. Zhang Y-H P, *Process Biochem.* 2011, 46: 2091.
3. Chen H-G, Zhang Y-H P, *Renew. Sustain. Energy Rev.* 2015, 47: 117.
4. Kamm B, *Pure Appl. Chem.* 2014, 86: 821.
5. Morais A R C, da Costa Lopes A M, Bogel-Lukasik R, *Chem. Rev.* 2015, 115: 3.
6. McCormick K, *Biofuels* 2014, 5: 191.
7. Ioanna V, Posten C, *Biotechn. J.* 2014, 9: 739.
8. Kazmi A, Kamm B, Henke S, Theuvsen L, Hoefer R, *RSC Green Chem. Ser.* 2012, 14: 1.
9. Waldron K W (ed.), *Advances in Biorefineries. Biomass and Waste Supply Chain Exploitation*, Woodhead Pub. Series in Energy, Cambridge, UK 2014.
10. Stuart P R, El-Halwagi M M (eds.), *Integrated Biorefineries: Design, Analysis, and Optimization*, CRC Press, Boca Raton, FL 2012.
11. Lanzafame P, Centi G, Perathoner S, *Catal. Today* 2014, 234: 2.
12. Centi G, Lanzafame P, Perathoner S, *Catal. Today* 2011, 167: 14.
13. Centi G, Perathoner S, *J. Chinese Chem. Soc.* 2014, 61: 712.
14. Lanzafame P, Centi G, Perathoner S, *Chem. Soc. Rev.* 2014, 43: 7562.
15. Abate S, Lanzafame P, Perathoner S, Centi G, *ChemSusChem* 2015, 8: 2854.
16. Perathoner S, Centi G, *ChemSusChem* 2014, 7: 1274.
17. Welteroth U, From fossil- to bio-based refineries. Case Study Biorefinery Leuna, presented at IEA-Bioenergy Task 42 Biorefinery—German Workshop Worms, September 15, 2009.
18. Barbato L, Centi G, Iaquaniello G, Mangiapane A, Perathoner S, *Energy Technol.* 2014, 2: 453.
19. Martinez-Hernandez E, Campbell G, Sadhukhan J, *Chem. Eng. Res. Des.* 2013, 91: 1418.
20. Demirbas A, In: *Biorefineries. For Biomass Upgrading Facilities*, Springer, London 2010, Chapter 10, p. 227.
21. Centi G, van Santen R A, In: *Catalysis for Renewables: From Feedstock to Energy Production*, Wiley-VCH, Weinheim, Germany 2008, p. 387.
22. International Energy Agency (IEA), Renewable Energy Medium-Term Market Report 2014, IEA, Paris 2014.
23. Patel B, Dechatiwongse P, Hellgardt K, *Pan Stanford Ser. Biocatal.* 2015, 1: 1065.
24. Dutta S, Wu K C-W, *Green Chem.* 2014, 16: 4615.
25. Teter S A, Xu F, Nedwin G E, Cherry J R, In: *Biorefineries—Industrial Processes and Products*, Kamm B, Gruber P R, Kamm M (eds.), Wiley-VCH, Weinheim, Germany 2006, Vol. 1, p. 357.
26. Luque R, *Current Green Chem.* 2015, 2: 90–95.
27. Ma R, Xu Y, Zhang X, *ChemSusChem* 2015, 8: 24.
28. Ciriminna R, Demma Cara P, Lopez-Sanchez J A, Pagliaro M, *ChemCatChem* 2014, 6: 3053.
29. Jacobs P A, Dusselier M, Sels B F, *Angew. Chemie, Int. Ed.* 2014, 53: 8621.
30. Sousa-Aguiar E F, Appel L G, Zonetti P C, Fraga A do C, Bicudo A A, Fonseca I, *Catal. Today* 2014, 234: 13.

31. Cejka J, Centi G, Perez-Pariente J, Roth W J, *Catal. Today* 2012, 179: 2.
32. Kobayashi H, Komanoya T, Guha S K, Hara K, Fukuoka A, *Appl. Catal., A: Gen.* 2011, 409–410: 13.
33. Tran N H, Bartlett J R, Kannangara G S K, Milev A S, Volk H, Wilson M A, *Fuel* 2010, 89: 265.
34. Stocker M, *Angew. Chemie, Int. Ed.* 2008, 47: 9200.
35. Huber G W, Corma A, *Angew. Chemie, Int. Ed.* 2007, 46: 7184.
36. Huber G W, Iborra S, Corma A, *Chem. Rev.* 2006, 106: 4044.
37. Cavani F, Centi G, Perathoner S, Trifiro F, *Sustainable Industrial Chemistry: Principles, Tools and Industrial Examples*, Wiley-VCH, Weinheim, Germany 2009.
38. Haro P, Ollero P, Villanueva Perales Al L, Vidal-Barrero F, *Biofuels Bioprod. Biorefin.* 2013, 7: 551.
39. Kamm B, Gruber P R, Kamm M, *Biorefineries—Industrial Processes and Products: Status Quo and Future Directions*, Wiley-VCH, Weinheim, Germany 2010.
40. Aresta M, Dibenedetto A, Dumeignil F, *Biorefinery: From Biomass to Chemicals and Fuels*, de Gruyter, Berlin, Germany 2012.
41. de Jong E, Jungmeier G, In: *Industrial Biorefineries and White Biotechnology*, Pandey A, Höfer R, Taherzadeh M, Nampoothiri M, Larroche C (eds.), Elsevier, Amsterdam, The Netherlands 2015, Chapter 1, p. 3.
42. Luguel C, Joint European Biorefinery Vision for 2030, and European Biorefinery Joint Strategic Research Roadmap/Star-Colibri (2011). http://www.star-colibri.eu/publications.
43. Zinoviev S, Müller-Langer F, Das P, Bertero N, Fornasiero P, Kaltschmitt M, Centi G, Miertus S, *ChemSusChem*, 2010, 3: 1106.
44. eni Brochure "Green Refinery: Reinventing petroleum refineries," https://www.eni.com/en_IT/attachments/azienda/attivita-strategie/refining-marketing/eni_Green-Refinery_esecutivo.pdf.
45. Sharifzadeh M, Wang L, Shah N, *Renew. Sustain. Energy Rev.* 2015, 47: 151.
46. Zhang M, Yu Y, *Ind. Eng. Chem. Res.* 2013, 52: 9505.
47. Angelici C, Weckhuysen B M, Bruijnincx P C A, *ChemSusChem* 2013, 6: 1595.
48. Makshina E V, Dusselier M, Janssens W, Degrève J, Jacobs P A, Sels B F, *Chem. Soc. Rev.* 2014, 43: 7917.
49. Behkish A, Wang S, Candela L, Ruszkay J, *Oil, Gas* 2010, 36: 29.
50. Rodriguez B A, Stowers C C, Pham V, Cox B M, *Green Chem.* 2014, 16: 1066.
51. Marinas A, Bruijnincx P, Ftouni J, Urbano F J, Pinel C, *Catal. Today* 2015, 239: 31.
52. Gupta R P, Tuli D K, Malhotra R K, *Chem. Industry Digest* 2011, 24: 86.
53. Torres Galvis H M, de Jong K P, *ACS Catal.* 2013, 3: 2130.
54. Centi G, Quadrelli E A, Perathoner S, *Energy Env. Sci.* 2013, 6: 1711.
55. Fan D, Dai D-J, Wu H-S, *Materials* 2013, 6: 101.
56. Phung T K, Busca G, *Chem. Eng. J.* 2015, 272: 92.
57. Phung T K, Proietti Hernandez L, Lagazzo A, Busca G, *Appl. Catal., A: Gen.* 2015, 493: 77.
58. Sheng Q, Guo S, Ling K, Zhao L, *J. Braz. Chem. Soc.* 2014, 25: 1365.
59. Xin H, Li X, Fang Y, Yi X, Hu W, Chu Y, Zhang F, Zheng A, Zhang H, Li X, *J. Catal.* 2014, 312: 204.
60. Huang Y, Dong X, Li M, Yu Y, *Catal. Sci. Technol.* 2015, 5: 1093.
61. Gurgul J, Zimowska M, Mucha D, Socha R P, Matachowski L, *J. Mol. Catal. A* 2011, 351: 1.
62. Ampelli C, Genovese C, Passalacqua R, Perathoner S, Centi G, *Appl. Thermal Eng.* 2014, 70: 1270.
63. Ampelli C, Passalacqua R, Genovese C, Perathoner S, Centi G, Montini T, Gombac V, Delgado Jaen J J, Fornasiero P, *RSC Adv.* 2013, 3: 21776.

64. El Doukkali M, Iriondo A, Cambra J F, Arias P L, *Topics Catal.* 2014, 57: 1066.
65. Kalamaras C M, Efstathiou A M, *Conference Papers in Science* 2013, 1.
66. de Poulpiquet A, Ranava D, Monsalve K, Giudici-Orticoni M-T, Lojou E, *ChemElectroChem* 2014, 1: 1724.
67. Trchounian K, Trchounian A, *Appl. Energy* 2015, 156: 174.
68. Chandrasekhar K, Lee Y-J, Lee D-W, *Int. J. Mol. Sci.* 2015, 16: 8266.
69. Ghavipour M, Behbahani R M, Rostami R B, Lemraski A S, *J. Nat. Gas Sci. Eng.* 2014, 21: 532.
70. Hirota Y, Murata K, Miyamoto M, Egashira Y, Nishiyama N, *Catal. Lett.* 2010, 140: 22.
71. Centi G, Iaquaniello G, Perathoner S, *ChemSusChem* 2011, 4: 1265.
72. Zhang J, Lu S, Su X, Fan S, Ma Q, Zhao T, *J. CO_2 Utilization* 2015, Ahead of Print. doi:10.1016/j.jcou.2015.05.004.
73. Satthawong R, Koizumi N, Song C, Prasassarakich P, *Catal. Today* 2015, 251: 34.
74. Ohnishi R, Akimoto T, Tanabe K, *Chem. Commun.* 1985, 1613.
75. Posada J A, Patel A D, Roes A, Blok K, Faaij A P C, Patel M K, *Bioresource Technol.* 2013, 135: 490.
76. Mascal M, *Biofuels Bioprod. Biorefin.* 2012, 6: 483.
77. Ichikawa N, Sato S, Takahashi R, Sodesawa T, *J. Molec. Catal. A: Chem.* 2006, 256: 106.
78. Sato S, Sato F, Gotoh H, Yamada Y, *ACS Catal.* 2013, 3: 721.

Index

Page numbers followed by f and t indicate figures and tables, respectively.

F

G

H

I

Q

R

V

W

X

Y

Z

Printed and bound by CPI Group (UK) Ltd, Croydon, CR0 4YY

01/11/2024

01782614-0013